Proceedings of the 34th International MATADOR Conference

Springer

London
Berlin
Heidelberg
New York
Hong Kong
Milan
Paris
Tokyo

Srichand Hinduja (Ed.)

Proceedings of the 34th International MATADOR Conference

Formerly The International Machine Tool Design and Research Conferences

With 339 Figures

 Springer

Srichand Hinduja, BE, MSc, PhD
UMIST, Department of Mechanical, Aerospace and Manufacturing Engineering,
Sackville Street, Manchester, M60 1QD, UK

British Library Cataloguing in Publication Data
International MATADOR Conference (34th : Manchester, England)
  Proceedings of the 34th International MATADOR Conference :
  formerly the International Machine Tool Design and Research Conferences
  1.Manufacturing processes - Congresses 2. Computer integrated
  manufacturing systems - Congresses 3.Manufacturing
  processes - Automation - Congresses 4.Machine-tools - Congresses
  I.Title II. Hinduja, Srichand
  621.9'02
ISBN-13: 978-1-4471-1169-6   e-ISBN-13: 978-1-4471-0647-0
DOI: 10.1007/978-1-4471-0647-0

Library of Congress Cataloging-in-Publication Data
A catalog record for this book is available from the Library of Congress

ISBN-13: 978-1-4471-1169-6 Springer London Berlin Heidelberg
Springer is a part of Springer Science+Business Media
springeronline.com

Typesetting: Camera ready by contributors

69/3830-543210 Printed on acid-free paper SPIN 11012504

Proceedings of the
Thirty-Fourth International

MATADOR

Conference

Organising Committee
Professor S Hinduja *(Chairman)*
Professor D R Hayhurst
Professor L. Li
Dr J Atkinson
Dr R G Hannam
Dr A W Labib
Dr P Mativenga
Dr S Mekid
Dr D J Petty
Dr M A Sheikh

Organising Secretary
Mrs C Collins

UMIST

Manufacturing Division
Department of Mechanical, Aerospace and Manufacturing
Engineering
University of Manchester Institute of Science and Technology

Foreword

by the

Conference Chairman

It is my pleasure to introduce the Proceedings for the 34th International MATADOR Conference. The Proceedings include 73 refereed papers submitted from several countries on different continents, thus making the Conference truly international.

The MATADOR Conference was established in 1959, which makes it one of the longest established conferences in the field of manufacturing in the world. The first conference dealt with metal working fundamentals in cutting, grinding and forming, and the design of the metal working machines for this purpose. Since those early days, research in manufacturing has switched its attention to advanced manufacturing technologies and modern management techniques, precision engineering, use of intelligent software in design, modelling and control of manufacturing processes, processing of materials using lasers, and rapid prototyping. This conference reflects this switch of emphasis. It is now broadly divided into two halves, one dealing with the original themes of this Conference, and the other with the above-mentioned advances in manufacturing technology and manufacturing systems.

I hope that you will find the papers interesting and stimulating.

Professor Srichand Hinduja
Manchester, 2004

Contents

BIO-ENGINEERING

Fabrication of Bio-mimetic Artificial Bone Scaffolds Using A Novel Rapid Prototyping-based Technique

Jianhong Chen[1] Lijun Xu[2] Joon Hock Yeo[3] Lijun Jiang[4] Jiemo Tian[5]

[1]Nanyang Technological University, Singapore, 639798, mjhchen@ntu.edu.sg
[2]Nanyang Technological University, Singapore, 639798, mljxu@ntu.edu.sg
3Nanyang Technological University, Singapore, 639798, mjhyeo@ntu.edu.sg
[4]Institute for Infocomm Research, Singapore, 119613, ljjiang@i2r.a-star.edu.sg
[5]Tsinghua University, Beijing, China, 100084, intjm@tsinghua.edu.cn

Abstract: It is important to fabricate bio-mimetic artificial bone (BMAB) with the similar structure and properties as human bones. An innovative concept for manufacturing the custom-tailored BMAB has been proposed, which is achieved by layer slurry gelation. The bone scaffolds with complex and delicate structure are produced using the computer-controlled three-dimensional (3D) rapid prototyping (RP) system. Sodium alginate and calcium chloride are adopted in a gelling procedure by trial-and-error. In order to obtain well-dispersed and non-agglomerated ceramic slurry, certain alumina is added as dispersant for the hydroxyapatite (HA) suspensions in order to improve HA rheological properties and maximize its solid phase content. With the proposed methods, the porous HA bio-mimetic artificial bone scaffolds with good mechanical properties can be fabricated.

1. Introduction

With the development of social civilization, the value of human life has been widely recognized. However, many diseases, accidents, and wars lead to a lot of injuries for the human body, bone injuries has made many people disabled and put heavy burden for both the family and society. Therefore, it is ideal to reconstruct the destroyed bone using the other part of the bone in the patient-self. However, this may cause new injury to the patient and increase the infection possibility. It is also restricted by limited source of bones. The method of using animal bone will cause prompt rejection reaction. Though using skeleton of other people that is similar to self-graft in shape may be accepted by the patient at the early phase, it is very difficult to avoid rejection in the long term. The general substitute materials for human bone, like ceramic, metal or high molecular material, can't satisfy the human bone's requirements for the density, structure and quality. Therefore, it is desirable to develop a new artificial material to restore human bone injury and rebuild the bone's functionality.

It is well known that the human bone possesses complex shape, whose inanimate matter basis is HA with a reticulated structure and a large numbers of mutual transfixions. Therefore, the artificial bio-alive bone made by HA can establish good bio-affinity for bone tissue, which can bond with natural bone by chemical link and with exchange of certain componental elements via natural metabolism after its implantation. In addition, the HA with a porous structure analogous to human bone, holds a dypass through which the bone tissue and other tissue are able to grow inside so that natural bone and the artificial bio-alive bone can trespass and interlace each other on their interface. This will cause the abduction of osteogenesis and gradually fuse the implanted artificial bio-alive bone and the natural bone [1]. Therefore, the prevailing substitute for bio-alive osseous tissue is HA based bio-ceramic and their composite material.

In early 1980s, rapid prototyping technology emerged in the hi-tech manufacture industry [2]. Since this technique can fabricate products with complex structure and individuation at a small batch, it can be realized in one design-manufacture process with high flexibility. The products with different shapes can be obtained by only modifying the computer-aided design (CAD) model, hence shorten the production cycle.

Researchers from Dayton University explored how to convert papers into HA/glass with laminated object manufacturing. First, HA/Glass is cranked out to slice using conventional tape casting method, Then, the substitute bone, which is expected to be replaced by new bone thru bio-degradability, is obtained based on the CAD file acquired by computer tomography (CT) scanning. In Michigan University, based on selective laser technology, RP has been implemented by blending the HA powder with UV solidify-able acrylate monomer. G. Allen Brady, Tien-Min Chu, and John W. Hallolan studied stereolithography methods by using ultraviolet curable suspension of ceramic powders based on Al_2O_3, SiO_2 and $Ca_{10}(PO_4)_6(OH)_2$ (Hydroxyapatite) [3]. In MIT, Michael J. Cima developed a microstructure planting part by three-dimensional printing technology [4]. Other researchers tried to directly produce replant-able bone using HA ceramic powder wrapped with resin adopting selective laser sintering technology [5]. For the time being, RP for ceramic auto-prototyping are mainly based on the commercial RP system with slight improvement. Therefore, the effect is not good.

2. Theory of 3D Gel-lamination for Fabricating Artificial Human Bone

The theory of gel-casting, a new prototyping method to fabricate high-quality ceramics with complex shape, is described as following: First, slurry of the mixture of the ceramic powers and the monomer-water liquor are injected to a mould. After the slurry agglomerates, grains of ceramic powders will solidify at original position and become ceramic body. Then, remove the die, dry and sinter it. It adopts organic additive, so there is no positive ion left after sintering. Gel-casting can make a series of ceramic powders with different compound, from alumina to silicon nitride. The shortcomings of the method are inconvenient for mould making and high cost for small volume of products. In view of these various existing methods of RP, a new RP method for ceramic material, named 3D gel-lamination, which combines with gel-casting technology, is developed.

3D gel-lamination is a branch of RP, which adopts the idea of the layering manufacturing. At first the human bone is scanned by CT, and then rebuild the human bone 3D space model using 3D shaping software, to obtain standard construction database of human bone. Customizing the bio-mimetic artificial bone for patients, we only need to scan characteristic parameters of the patient's bone with the CT, and compare it with the standard database, and then the 3D space model for the patient will be created automatically.

Utilizing special software to delaminate 3D model, we gain a series of 2D plane data. Then, shaping control software transforms 2D data into control signals for the movements of the 3D RP machine (Figure.1) and for the coordination of the spray head and each motorial axis. In order to form gelatin, free radical initiator has been sprayed to the particular surface of ceramics slurry, layer by layer, until whole porous HA body is completed. After pre-burning, the organism has been taken away from HA with higher temperature sintering, thus the porous HA is obtained.

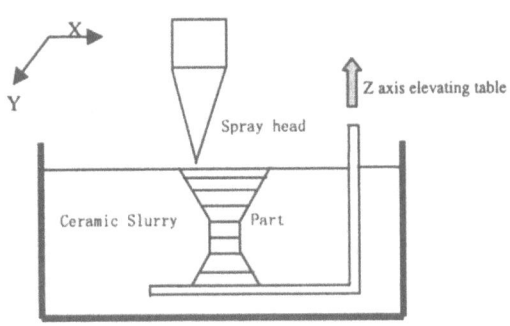

Figure.1 Illustrative diagram

3. Experiments and Results

From the viewpoint of shaping ceramic body, the basic technique of the 3D gel-lamination method is gel-casting, it has the advantage of higher body strength before desiccation and after desiccation, and a little catalyst used. For every liter slurry needs only 0.1-0.3g/L free radical initiator, and 3-4mL/L catalyst. This makes it possible to use a little amount of free radical initiator and catalyst for partial gelatin.

For the system ceramic slurry, the gelatin system must satisfy the following requirements: first each composition inside the system should not be poisonous; second the physics and chemistry performance of free radical initiator should not erode the spray head; third the gelatin reaction should be rapid without the catalyst; and fourth the process should be completed quickly at normal temperature; and finally the gelatin should have a certain strength.

After trial-and-error and studying on various gelatin system, we find that sodium alginate and calcium chloride system has features like, high speed and strength of gelation, no need of catalyst for fast gelatin reaction at normal temperature, avirulence, and easy to degrease. Hence, it is suitable for the solid free shaping.

The key technique of this method is to make the ceramic slurry with high solid phase content and lower viscosity. HA aqueous suspension as ceramic slurry is used in our experiments. However due to its specific surface characteristics, the HA's grains is not easily distributed in the aqueous solution. It can't meet the demand of net size shaping when the HA solid phase content reaches 25% in volume with bad liquidity.

In order to improve liquidity and enhance its solid phase content, we add the biology inertia composition alumina into HA slurry without destroying the biology

6

alive of HA. As the alumina slurry has the best liquidity, its viscosity is very low even within the scope of 65% in volume. Figure.2 shows the HA powder and alumina powder Zeta potential contrast curves. When the HA and alumina volume ratio ranges from 80/20 to 90/10, the slurry solid phase content will be larger than 40% in volume, while the viscosity also meet shaping requirement. Meanwhile, because single-phase HA ceramic has lower strength and poor tenacity, adding alumina will enhance the strength of artificial bone and its tenacity.

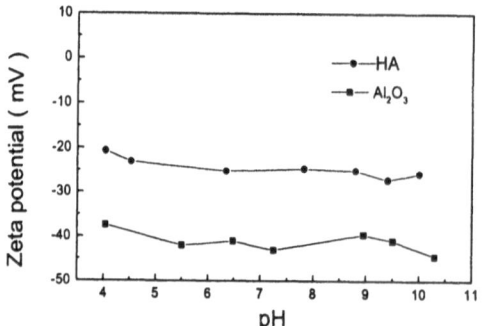

Figure. 2 The Zeta potential curves of HA and alumina

In our experiments, we found that the form of the powder has important impact upon slurry liquidity. Under the same conditions, we can obtain the high solid phase content slurry after many times of calcine. It is verified that ceramic powder contains a great deal of loosen holes by analyzing the grains without calcine using electron microscope. While preparing the slurry, it contains a great deal of solvent, thereby reducing the solid phase content. Comparing the HA powder of Zeta potential without calcine with those through 24 hours calcine at 900 Celsius, we found that the later one has increased Zeta potential absolute value, and its performance of diffuseness also improves obviously.

Figure 3 shows that the viscidity curves of two kinds of HA powder. The larger grains have important impact on the improvement of rheological properties of HA slurry.

The important craft parameters affecting ceramic body shaping quality for 3D gel-lamination method include: solid phase content of ceramic slurry, viscosity of ceramic slurry, concentration of free radical initiator and filling ratio, thickness of layer etc.

The density of shaping body part is determined by the ceramic powder solid phase content in slurry, which requires that the system needs higher content of solid phase

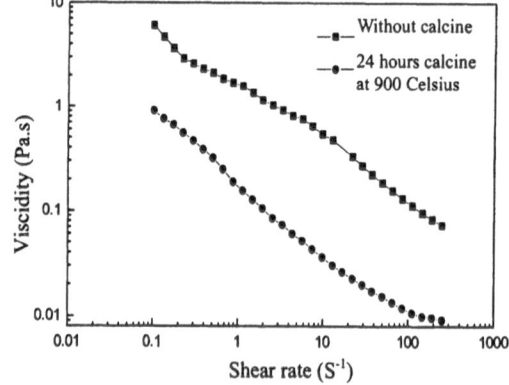

Figure.3 Viscidity curves of two kinds of HA powder

of 40-60% in volume, comparing to 25-50% in volume with slip casting and spray drying. Lower solid phase content can lead to lower density directly, even collapse of body while degreased and sintered.

To obtain accurate and well-proportioned layer of ceramic slurry, the correct value of ceramic slurry viscosity must be chosen. When the HA and alumina volume ratio is between 80/20 and 90/10, the content of slurry solid phase should be larger than 40% in volume to meet the system requirements.

Concentration of free radical initiator affects the strength of gelatin directly, the higher the concentration, the higher the gelatin intension. Free radical initiators, actually some small discrete dots, are sprayed to the plane through spray head. It will diffuse to all directions when a drop of it is sprayed upon ceramics slurry so that adjacent dots can join together. Therefore, high filling ratio will induce high quantity in unit area. The joint intension is also higher, but diffuseness of part outline become more serious and vice versa. So appropriate filling ratio must be selected to satisfy the joint strength with least boundary diffuseness. We could obtain least diffuseness and high intension of body from the experiments, when free radical initiator concentration is 1.5mol and filling ratio reaches 80%.

Obviously the value of layer thickness is relative to the length of ceramic body shaping time and accuracy of body. Experiment results indicate that layer thickness is reasonable between 0.1-0.3 mm.

By means of adding the effective composition alumina, the HA/alumina composite slurry with required liquidity is obtained. Meanwhile, by adjusting system craft parameters, we can fabricate the cervical vertebra petal and the finger bone successfully.

4. Conclusion

The 3D gel-lamination experiments demonstrate that this system is suitable for fabrication of ceramic green parts with complex geometric shape. A majority of porous in HA body has been automatically fabricated by the system, but internal tiny hole can be produced in sintering via adding pore forming material. With low cost and simple craft, this equipment can quickly customize bone restorations that meet the clinic requirements according to the patient individual difference. Moreover, this approach shows the potential for fabrication of other ceramic components, such as silicon carbide, silicon nitride, alumina, zirconia and etc. This method can also be used widely in other applications.

References

[1] Christel Klein, K.de Groot, et al. 1992, The 4th World Biomaterials Congress, Berlin Germany, April, P24-28.
[2] Ashely S.: 1991, Rapid Prototyping System, Mechanical Engineering, Vol.113, No.4, P34-43.
[3] G. Allen Brady, et al. 1996, Curing Behavior of Ceramic Resins for StereoLithography, Solid Freeform Fabrication Proceedings, P403-410.
[4] Yoo, J., Cima M., et al. Cerma. Eng. 1995,And Sci. Proc., Vol.16, No.5, P755-762.
[5] Curtis Griffin. 1996, Rapid Prototyping of Structural Ceramic Components Using Selective Laser Sintering. Materials Technology, V11n2, P48-49.

A Robotic System for Non-invasive Treatment of Urological Organs Accessible Through Abdominal Window

S.Chauhan, R.Mishra and J.R Li

Robotics Research Center, Div. of Mechatronic and Design,
School of Mechanical and Production Engineering,
Nanyang Technological University, Singapore 639798.
Mcsunita@ntu.edu.sg

Abstract: In this paper, design and development of a non-invasive surgical system of deep-seated cancers of the organs, accessible through abdominal acoustic window such as liver, kidneys, prostate etc., is described. The use of High Intensity Focused Ultrasound (HIFU) as a non-invasive surgical modality is demonstrated. The robot, called FUSBOT-US, works partially in a water tank and is devised to guide a specially designed 'end-effector' through a pre-determined trajectory. The end-effector comprises a custom designed assembly of multiple HIFU transducers operating in a water tank. A 5-axis PC based controller is used for controlling various sub-sections of the system within a pre-defined constrained work envelope. A user friendly Graphical User Interface is designed to enable image-guided supervisory control by end users (clinicians). The end-point accuracy of the designed system at the target is accomplished within ± 0.5mm.

1. Introduction

In the recent years, Minimally Invasive Surgery (MIS) and Image Guided Surgery (IGS) have greatly changed the traditional surgical protocols adopted in open surgery for many medical procedures [1-3]. The applications of robots and computer integration in medicine range from simplistic laboratory robots for tool positioning, to highly complex surgical robots that carry out surgical procedures under computer control. Most of the robotic systems for surgery aim at assisting in minimal invasive surgical procedures. One such surgical robot is the *da Vinci*® surgical system, which has successfully completed hundreds of cardiac, general and other types of procedures [2]. Other notable examples are *AESOP* and *ZEUS* robotic surgical systems [3].

HIFU (High Intensity Focused Ultrasound), alternatively called as FUS (Focal Ultrasound Surgery), uses Ultrasound wave as the therapeutic resource for thermal necrosis. It works on the principle of converting mechanical energy into heat energy. Focused high-intensity ultrasound energy generates a sharp impulse and produces a temperature in the range of 60-80 °C, which could lead to instantaneous coagulative necrosis in the target tissue within its focal zone. Experimental studies and preliminary clinical data are available from various laboratories worldwide and have shown promising result [4-6].

The lethal dosages used in Focal Ultrasound Surgery (FUS) can produce irreversible damage in the normal tissue, the localization and targeting the surgical tools (HIFU transducers, in this case) are critical to the success of treatment. An accurate and repeatable positioning, in a defined spatial configuration with respect to the patient (target organ site), is desirable. For a target tumor area, which may be larger in size than the composite focus, the probe assembly needs to scan over the entire volume of the lesion. For this purpose, robotic techniques can be employed for precise and accurate mechanical manipulation in given spatial configuration. A robotic system for neuro-surgical application has been devised and explained in [7].

The present paper deals with design of a system, which target cancers that can be accessed through trans-abdominal route. These include mainly, cancers of the liver, lung, kidney and trans-abdominal approach to the prostate. A multi-transducer [8] robotic HIFU system and control for the treatment will be described in the following sections. The use of both single and multiple probes has been considered.

2. System Overview

The robotic system developed for cancer ablation in this research is comprised of following sub-sections: 1) End-effector jig; 2) Base manipulator and 3) Surgical planning and operating interface.

2.1 End-effector jig

The generation of HIFU energy for tumor ablation in the present system is resulted by the use of a set of multiple transducers, operating at 2 MHz frequency and a focal depth of 64 mm/80 mm. The transducers are arranged in a jig fixture (as shown in Figure 1), which is affixed as an end-effector for the robotic system (as explained in next section). The orientation of HIFU probes with respect to each other in a defined spatial configuration can be pre-planned such that the superimposed region is of desired intensity and dimensions, and is targeted in the diagnosed lump/tumor [7,8]. Single probe can be directly affixed in a probe holder mounted on the central shaft of the manipulator. The multiple probe end-effector jig is designed both for planar and spatial configuration as shown in Figure 1. While in the former the probes can be mounted on a plate through precise angular holes, the latter has a central shaft and collars (arms) of equal diameter to secure the transducers. In both cases, the composite focal point is always achieved along the central axis.

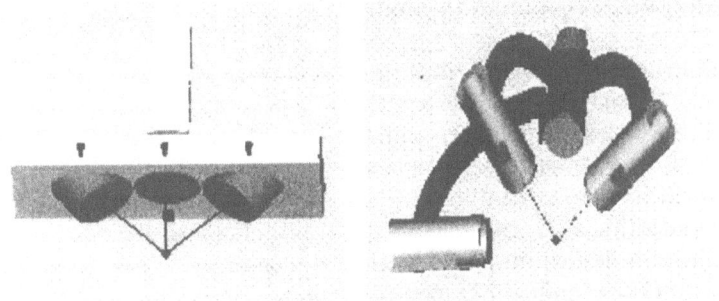

Figure 1: An end-effector jig with 3-probe holders in pre-defined, planar and spatial configurations.

2.2 Robotic manipulator

The robotic system can position HIFU probes accurately and repeatably in a given point in space. The dimensions and reach of this system is based upon human anthropometry data. The clinical set-up is proposed such that the patient lies in a prone position over a surgical bed, integrated with a water bolus/tank underneath the torso region (coupling media). The robotic manipulator is made to manipulate end-effector jig inside the tank targeted upwards at a desired angle into the patient. The schematic diagram of manipulator axes end-effector jig placement is shown in Figure-2, along with its work envelope dimensions. The robotic manipulator comprises of 5-degrees of freedom including positioning in x,y,z coordinates and an arrangement at the end-effector for angular positioning. HIFU module jig is mounted on the central shaft and can be deployed in a desired region within the tank.

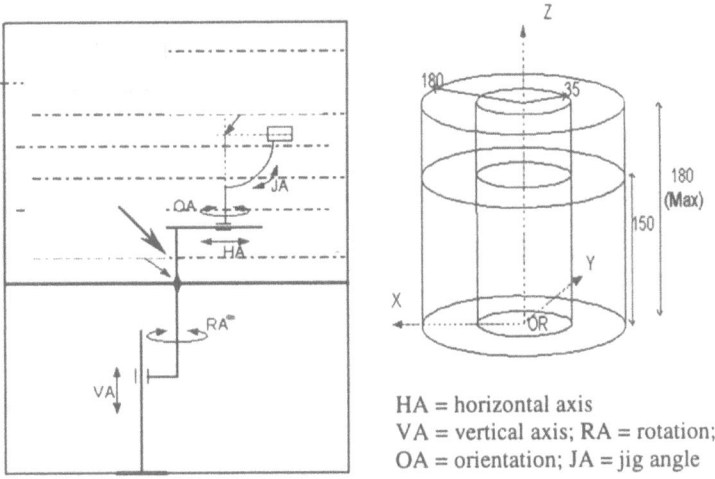

HA = horizontal axis
VA = vertical axis; RA = rotation;
OA = orientation; JA = jig angle

Figure 2: Schematic diagram of the robotic manipulator integrated in a water tank (left) and the robot's workspace (right)

2.3 Surgical planning and control

A diagnostic ultrasound scan is first taken to identify the position of the target and then suitably orient the robotic module, which positions and scans a single or, multiple HIFU beams to focus into the desired target region. A Galil 5-axis controller is used for trajectory programming and controlling individual axis of the robotic module and its registration within the tank. A custom designed graphical user interface (Figure 3) integrates surgical planning and supervisory control for robot axes in a desired manner. The surgeon demarcates a desired position using mouse cursor on an on-line 2-D ultrasound image. Robot kinematics and workspace are mapped into the 2D image on screen, which controls the end-effector so as to target the energy to the corresponding position in the tissue in conjunction with the HIFU control sub-system.

3. Results and Discussions

The integrated system is tested using a suitably designed calibration box for individual axis and joint end-point accuracy to be within 0.5mm, which is conceived to be sufficient with respect to the specified ellipsoid lesion size as measured for a given transducer configuration during experiments.

For *in vitro* experiments, fresh samples of pork adipose tissue, liver, kidneys were acquired and mounting onto the tissue holder positioned on the top surface of the water tank through an operating window. Experiments were conducted using single and three identical HIFU probes, mounted on the end-effector jig of the robot and with various protocols for dosage delivery. After HIFU exposure, the tissue was removed from the holder and excised along the exposed planes to check for induced lesion. One such representative result is shown in Figure 3. No off-focal hot-spots were seen in the overlying region in any of the samples and the lesions with same protocol were of consistent size and shape. Complete tumor ablation is possible by sweeping out the targeted tumor tissue region.

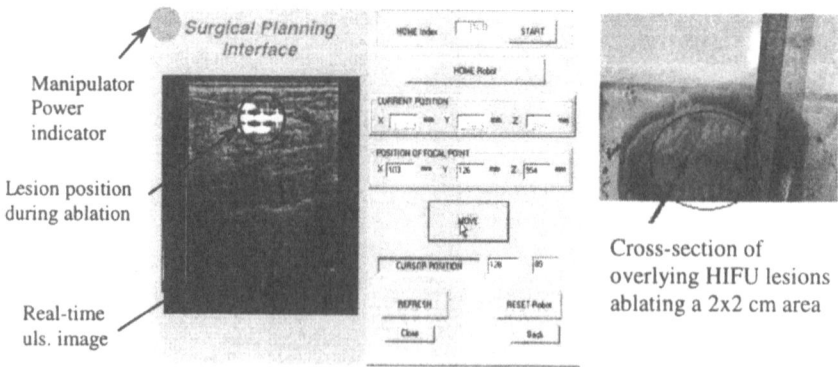

Figure 3. Interactive GUI

4. Conclusions

A robotic prototype system for HIFU induced non-invasive surgery is designed and developed. The end-point accuracy of mechanical manipulator has been measured to be within ± 0.5mm. The laboratory trials using animal tissue have shown that the robotic assisted HIFU approach with single or multiple transducers is able to accurately produce a refined and consistent lesion size as compared to conventional, manually controlled approach.

HIFU modality may provide an alternative non-invasive solution at an early stage when the size of the tumor is still small and the carcinoma has not spread beyond the original site. With this procedure, both risks and complication arising from open surgery can be avoided.

5. Acknowledgements

The authors would like to thankfully acknowledge the support from Ministry of Education, Singapore and Agency for Science, Technology and Research, Singapore for jointly funding the project. Many thanks also go to our clinical collaborators, at Department of Urology, University Clinic, Mannheim, University of Heidelberg, Germany.

References

[1]. Taylor, RH , Mittelstadt, BD, Paul, HA, Hanson, W, Kaza-nzides, P, Zuhars, JF, Williamson, B, Musits, BL, Glass-man, E and Bargar, WL. An Image-Directed Robotic System for Precise Orthopaedic Surgery. IEEE Trans. Robotics and Automation. 10(3):261-275, June, 1994.

[2]. Hoznek, A., Zaki, S., Samadi, D et al, Robotic assisted kidney transplantation: an initial experience. Journal of Urology, 167, pp.1604-1606, 2002.

[3]. Boyd, W.D., Kiaii, B., Kodera, K. et al. Early experience with robotically assisted internal thoracic artery harvest. Surgical Laparoscopy & Endoscopy, 12(1), pp.52-57, 2002.

[4]. G. ter Haar, I. Rivens, L.Chen and S. Riddler, 1991, High intensity focused ultrasound for the treatment of rat tumors, Physics in Medicine and Biology, pp. 1495-1501.

[5]. J.Y. Chapelon, A. Gelet, J. Margonari, F. Gorry, Y. Theillere and D. Cathignol, 1996, High intensity ultrasound ablation of canine prostate, INSERM U.28 and Urology Dept of. Herroit Hospital, Lyon, France.

[6]. Arefiev et al, 1999, Ultrasound-induced tissue ablation: Studies on isolated, perfused porcine liver, Ultrasound in Medicine and Biology, INSERM, Lyon, Fr, page 1033-1043.

[7]. B.L. Davies, S. Chauhan and M.J., Lowe, 1998, A robotic approach to HIFU based neurosurgery, Computer Science, No. 1496, UK, October, pp.386-396.

[8]. S. Chauhan, M.J.S. Lowe and B. L. Davies, A multiple focused probe approach for HIFU based surgery, Journal of Ultrasonics (Medicine and Biology), UK 2001, Vol.39, pp.33-44.

Improving Noninvasive Blood Glucose Measurement Accuracy by Applying Genetic Algorithm to Partial Least Square Regression Model

Lijun Xu[1], Jianhong Chen[2], Xiqin Zhang[3], Joon Hock Yeo[4], Lijun Jiang[5]

[1]Nanyang Technological University, Singapore, 639798, mljxu@ntu.edu.sg
[2]Nanyang Technological University, Singapore, 639798, mjhchen@ntu.edu.sg
[3]Nanyang Technological University, Singapore, 639798, mxqzhang@ntu.edu.sg
[4]Nanyang Technological University, Singapore, 639798, mjhyeo@ntu.edu.sg
[5]Institute for Infocomm Research, Singapore, 119613, ljjiang@i2r.a-star.edu.sg

Abstract: Near infrared (NIR) absorption spectroscopy is a promising technique to non-invasively quantify blood glucose level. In order to extract the glucose signal out of the noisy background, Partial Least Squares (PLS) was utilized to create calibration models that relate the absorption spectra to glucose concentrations. A research grade Fourier Transformed Infrared (FTIR) spectrometer configured with a NIR quartz beam-splitter was used in this investigation. Genetic Algorithm (GA) was implemented to search the most appropriate modeling parameters such as wavelengths within NIR range for PLS regression. Using GA method to optimize the wavelength selection by applying the PLS-based calibration model could greatly enhance the prediction capacity and improve the measurement accuracy.

1. Background

About 16 million people in the United States and 100 million people worldwide suffer from diabetes [1][2]. In the realm of diseases, it is the third biggest killer and can lead to blindness, kidney failure, cardiovascular disease, and serious infection. Therefore, it is essential for diabetics to maintain good health, which requires frequent monitoring of blood glucose and treatment with diet, medication, or insulin injection. Unfortunately, current "finger-stick" method is invasive, requiring the patient to draw blood from the finger for test. This process is both painful and expensive, not only inconvenient, messy, but also carries a risk of infection. One solution to this problem is the development of a painless and convenient noninvasive optical glucose monitoring that allows fast and frequent monitoring of blood glucose levels.

2. Methods

Near-infrared (NIR), the spectral region between 780nm (12,800 cm^{-1}) and 2,500nm (4,000 cm^{-1}), is characterized by broad and overlapping spectral peaks produced by the overtones and combinations of molecular fundamental vibration modes. Absorption of NIR by the body is glucose-dependent; therefore, NIR absorption can be used to quantify blood glucose. Optical absorption techniques for quantification of glucose are based on selective absorption of light by the molecule, which is described by the Beer-Lambert law:

$$I = I_0 e^{-\varepsilon CL}$$

where I_0 is the intensity of incident optical radiation, I is the transmitted intensity, ε is the molar extinction coefficient, C is the molar concentration and L is the path-length.

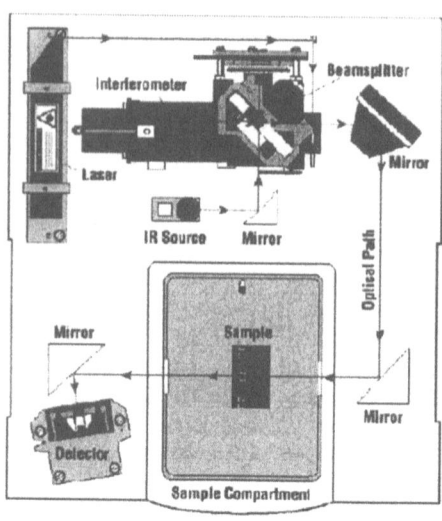

Fig. 1 Fourier transformed infrared spectrometer

Fig. 1 illustrates the principle of Fourier Transformed Infrared (FTIR) Spectrometer. After the sample absorbs some of the light, FTIR converts the sample's interferogram into a single-beam spectrum that shows the signal intensity for each frequency point in the range. After background data is removed, this spectrum can be viewed in transmittance or absorbance.

Noninvasive measurement of blood glucose levels in diabetics using NIR absorption spectroscopy is complex [3][4][6]. The measurement requires a thorough understanding of the sample and an application of methods from a broad field of disciplines including tissue optics, spectroscopy, chemometrics, signal processing, scattering theory and analytical chemistry. Besides the main interference posed by body temperature and absorption of water, protein and other contents in blood, there are some other obstacles. The glucose absorption spectrum in the NIR region is very weak; and the non-linear scattering of the near-infrared light complicates the extraction of the blood glucose signal as the near-infrared light propagates through the dynamically varying tissue. Some other measurement difficulties include light-tissue interaction, presence of interferents in high concentrations, inhomogeneous distribution of glucose in the body, and unknown path length of light in skin. Therefore the challenge lies in how to accurately extract the glucose specific information from the highly overlapping spectrum covering a wide spectral band. Complex problems like measuring accuracy, condition, as well as the physical interpretation are key factors that need further attention before a complete solution is achieved. Furthermore, advanced signal processing and multivariate analysis is necessary for the extraction of the spectral contribution due to glucose amid the complex and varying background signals.

Sophisticated chemometrics methods – particularly multivariate analysis – have been employed to extract the glucose signal out of the noise. Quantification of glucose within this complex mixture is done using multiple wavelengths and complex mathematical procedures. Multivariate techniques are used to create calibration models that relate the absorption spectra to glucose concentrations.

Because of serious overlapping of components spectrum within NIR range and collinearity problems between spectrum data, it is not desirable to setup calibration model and carry out prediction by using Univariate Linear Regression (ULR) and Multiple Linear Regression (MLR). Multivariate calibration models have come into wide use in implementing quantitative spectroscopic analyses due to their ability to overcome deviations from the Beer-Lambert law caused by effects such as overlapping spectral bands and interactions between components [9]. Currently one of the popular multivariate analysis methods applied for glucose measurement is Partial Least Squares (PLS) [4][5][7], which models both the X- and Y-matrices simultaneously to find the latent variables in X that will best predict the latent variables in Y. Fig. 2 depicts PLS procedure.

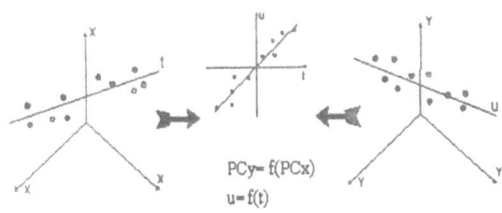

PCy= f(PCx)
u= f(t)

Fig. 2 PLS procedure

PLS method combines principle component analysis and regression analysis together, and it can overcome the collinearity problems between spectrum data. There is no limitation on the number of wavelength employed when setting up calibration model; all spectrum data can be used to set up calibration model. However, it is inevitable spectrum information contain noise, in order to improve prediction capacity it is beneficial to employ spectrum range of high Signal to Noise Ratio (SNR) instead of those of low SNR. Thus it is necessary to carry out wavelength selection to obtain suitable wavelengths for calibration model. Given the capability of computing models based on the use of multiple wavelengths, optimization issues arise regarding which specific wavelengths should be included [9]. Due to huge amount of spectra data and the complexity of the wavelength combination, normal approximation and systematic methods that solve a global optimization problem is not effective in this application.

Such kind of optimization problem could be dealt with Genetic Algorithm (GA) [8][9]. The optimized wavelengths can be used to set up calibration model based on PLS regression, thus improve the prediction capacity.

As with typical heuristic methods, GA's take advantage of not only heuristic gradient ascension (selection & crossover) but also semi-random exploration (mutations). Main advantages of GA's include that they are simple to understand and to implement, robust and early give a good near-solution.

GA's are inspired from biological processes (i.e. cells' division, DNA, crossover, mutation, ...) which are generalizations of natural genetic search mechanisms. The underlining idea is to generate successive sets of solutions (generations), making each new generation inheriting properties from the best solutions of the precedent. In order to perform such a step, we have to select the best solutions and mix them together (crossover). A GA typically including the following:
a) Generate a first generation with random parameters.
b) Evaluate the fitness value of all individuals of the generation.

18

c) Crossover the best individuals with high fitness value together to get the new generation (children).
d) Make random mutation across the new generation.
e) To determine if the termination conditions satisfied (if satisfied, exit; otherwise, go back to b).

3. Experiments and Results

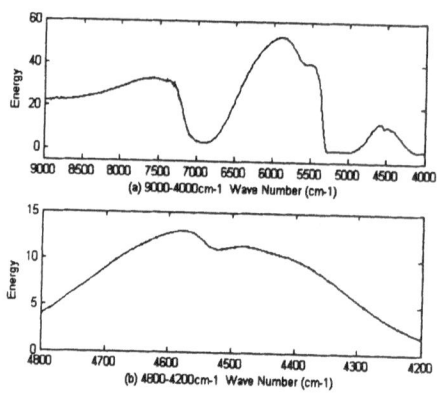

Fig. 3 Water-glucose solution spectrum
(a) 9000~4000cm^{-1} (b) 4800~4200cm^{-1}

A research grade FTIR spectrometer (Model: Perkin-Elmer GX2000, USA) was used in this investigation. The spectrum data under study was within 9000~4000cm^{-1} with the resolution of 0.1cm^{-1}. The known concentration water-glucose solution was pumped by a peristaltic pump (Model: Heiodolph PD5201, Germany), and conveyed to a demountable FTIR liquid sample cell with CaF$_2$ windows, in which a 1mm spacer was applied in this study to guarantee 1mm path-length. A liquid-nitrogen cooled InSb photo-detector was employed to sense the NIR light which passed through liquid sample cell in transmission mode, single beam spectrum were recorded through FTIR accompany software from Perkin-Elmer. Fourteen samples ranged from 50mg/dL to 400mg/dL were investigated in this study. Fig. 3 showed the water-glucose solution spectrum of two wavelength ranges (9000~4000 cm^{-1} and 4800~4200 cm^{-1}).

After acquiring the spectrum data, GA method was applied to carry out wavelength selection optimization. In this study, every wavelength was encoded as one gene and thus every combination of wavelengths formed one chromosome of an individual. Every gene was a binary value, which represented the state of existence of its corresponding wavelength. Fitness function to evaluate individuals was defined as:

$$F = g \text{ (RMSEP)}$$

where RMSEP (Root Mean Square Error of Prediction) is the objective value calculated through PLS regression, which is based on the combination of selected wavelengths; g transforms the objective value RMSEP to a non-negative number and F is the resulting relative fitness. This mapping is always necessary when the objective value is to be minimized as the lower objective values corresponds to fitter individuals. In this study cross validation method was applied in PLS regression to obtain RMSEP for selected wavelengths combination. The purpose of GA is to seek optimal combination of wavelengths so that the corresponding RMSEP could be minimized, thus the prediction capacity improved.

When implementing GA method, selection method and crossover method were configured as stochastic universal sampling and multi-point crossover respectively, with generation gap of 0.9 and crossover rate of 0.7; the probability of any one element of a chromosome being mutated was set to approximately 0.5. Population in one certain generation was assumed to contain 30 individuals.

A maximum number of generations (1000) and a minimum of relative variation (0.1%) within 100 recent generations were pre-specified as termination conditions of GA.

The above-mentioned two wavelength ranges were studied respectively to compare the RMSEP values obtained through PLS regression models with and without GA optimization. The positions of optimized wavelengths were shown in Fig. 4.

The comparison results were depicted in Table 1. Within the wavelength range 9000~4000 cm^{-1}, 66 wavelengths were selected from 501 wavelengths through GA optimization, RMSEP was reduced from 22.65 mg/dL to 1.87 mg/dL and

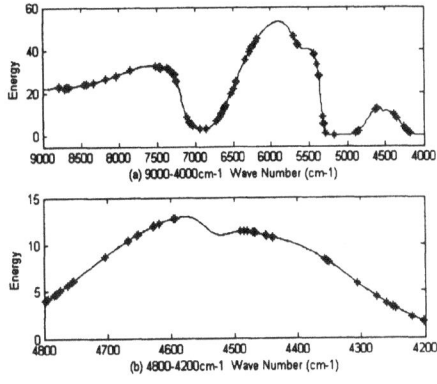

Fig. 4 Positions of optimized wavelengths (a) 9000~4000cm^{-1} (b) 4800~4200cm^{-1}

correlation of prediction was improved from 0.975 to 0.999. While in the wavelength range 4800~4200 cm^{-1}, by virtue of GA method, 39 wavelengths were retrieved out of 301 wavelengths; using such optimized wavelengths the PLS regression model ameliorated the correlation of prediction from 0.966 to 0.999 and decreased the RMSEP from 26.64 mg/dL to 2.97 mg/dL.

Table 1: Comparison results

Wavelength Range (cm^{-1})	Interval (cm^{-1})	Method	Number of Wavelength	RMSEP (mg/dL)	Correlation of Prediction
9000~4000	10	PLS	501	22.65	0.975
		GA + PLS	66	1.87	0.999
4800~4200	2	PLS	301	26.64	0.966
		GA + PLS	39	2.97	0.999

4. Conclusion

Using GA method to optimize the wavelength selection, applying to PLS-based multivariate calibration model could greatly enhance the prediction capacity and improve the measurement accuracy. Thus this method can be effectively applied in the non-invasive blood glucose monitoring using NIR absorption spectroscopy.

5. Acknowledgements

The present study was supported by research grant funding JT ARC2/01, from Agency for Science, Technology and Research / Ministry of Education, Singapore.

References

[1] Roger J. McNichols, Gerard L. Cote, 2000, *Optical glucose sensing in biological fluids: an overview*, Journal of Biomedical Optics, **5** (1), 5-16.

[2] R. W. Waynant, V. M.Chenault, 1998, *Overview of Non-Invasive Fluid Glucose Measurement Using Optical Techniques to Maintain Glucose Control in Diabetes Mellitus*, IEEE Lasers and Electro-Optics Society Newsletter, **12** (2), 3-6.

[3] Jason J. Burmeister, Mark A. Arnold, Gary W. Small, 1998, *Spectroscopic Considerations for Noninvasive Blood Glucose Measurements with Near Infrared Spectroscopy*, IEEE Lasers and Electro-Optics Society Newsletter, **12** (2), 6-9.

[4] Mark A. Arnold, Gary W. Small, 1998, *Data Handling Issues for Near-Infrared Glucose Measurements*, IEEE Lasers and Electro-Optics Society Newsletter, **12** (2), 16-18.

[5] F. M Ham., I. N. Kostanic., G. M. Cohen., et al, 1997, *Determination of glucose concentrations in an aqueous matrix from NIR spectra using optimal time-domain filtering and partial least-squares regression*, IEEE Transactions on Biomedical Engineering, **44** (6), 475 -485.

[6] H. M. Heise, 1996, *Technology for non-invasive monitoring of glucose*, 18th Annual International Conference of the IEEE Engineering in Medicine and Biology Society, Amsterdam, **5**, 2159 -2161.

[7] F. M Ham., G. M. Cohen.., K. Patel K., et al, 1994, *Multivariate determination of glucose using NIR spectra of human blood serum*, 16th Annual International Conference of the IEEE Engineering in Medicine and Biology Society, **2**, 818 -819.

[8] R. E. Shaffer, G. W. Small, M. A. Arnold, 1996, *Genetic Algorithm-Based Protocol for Coupling Digital Filtering and Partial Least-Squares Regression: Application to the Near-Infrared Analysis of Glucose in Biological Matrices*, Anal. Chem. **68**, 2663-2675.

[9] A. S. Bangalore, R. E. Shaffer, G. W. Small, et al, 1996, *Genetic Algorithm-Based Method for Selecting Wavelengths and Model Size for Use with Partial Least-Squares Regression: Application to Near-Infrared Spectroscopy*, Anal. Chem. **68**, 4200-4212.

Robotic System for Ablation of Deep-seated Skull Base Cancers - A Feasibility Study

Sunita Chauhan, Ming Yeong Teo and Wendy Teo

School of Mechanical & Production Engineering,
Nanyang Technological University, Singapore 639798.
MCSunita@ntu.edu.sg

Abstract: The fact that human brain has intricate composition and physiology, its surgery has always been considered as a medical challenge. This led to the need and subsequent development of various minimally invasive procedures for this medical specialty. This paper describes the feasibility of implementing a non-invasive modality, namely High Intensity Focused Ultrasound (HIFU), to treat cancers in the skull base region using robotic means. It proposes a method of effective energy transfer and design of a specialized end-effector along with an interfacing jig for integrating HIFU system to a custom designed robot, called Neurobot developed for drilling bone at the skull base. It is proposed to deploy focal ultrasound end-effector after the drilling application and on creating desired craniotomy with the precision Hexapod system (integrated in the Neurobot). The accuracy of Hexapod system is within ± 0.5 mm. The feasibility of using the base manipulator and HIFU energy transmission control strategies is studied separately and an integrated (using multiple probes) end-effector module is devised and tested for intended functionality. HIFU can be applied either as a stand-alone modality or, to treat the residual cancer regions after surgery as decided by the surgeon based upon the distinct need of a particular case.

1. Introduction

Minimally invasive and non-invasive modalities are being widely investigated as alternatives to traditional open surgery. Also, with the introduction of Computer Assisted Surgery (CAS) and minimally invasive surgical methods, the risks involved in open surgery have been greatly reduced. Moreover, the CAS procedures involve accurate planning, registration and navigation issues, both during planning and intervention stages. Brain cancer, though not as prevalent as cancers of the liver, breast, prostate etc., is one of the most lethal types of cancers. This is because brain, being the critical unit of the central nervous system, controls the functioning of all the other organs. Since brain tissue is not regenerative, it is vital that minimal damage is caused to normal brain tissue during surgical removal of the tumor volume.

Presently, the conventional modes of treatment of brain tumors are surgery, radiation therapy and chemotherapy, with open surgery being the basic form of treatment for removal of brain tumors. Minimally invasive neurosurgery, such as stereotactic surgery involves opening of the skull followed by tumor/cancer excision. Basically, stereotactic surgery is used for precise point-in-space applications such as biopsy extended its use for volumetric excision is still being perfected. Any such surgical intervention in the brain may cause loss of cerebral-spinal-fluid and severe anatomical changes (commonly known as *brain-shift*) resulting in errors in registration. Also, during volumetric removal of the tumors, it is highly desirable to avoid the functional areas in the overlying tissue, which may result into very complex trajectories of the surgical tools.

In the turn of the 20th century, development in volumetric stereotaxis and radiosurgery particularly, stereotactic and Gamma Knife radiosurgery sparked interest in the research and development towards minimally invasive and non-invasive surgery [1,2]. Radiation therapy makes use of highly powered rays to damage cancer cells and stop them from growing. It is used when either surgery is not possible or, when there is residual cancer after surgery. Stereotactic radiosurgery involves aiming high-energy rays from various angles with the use of a stereotactic reference frame that is clamped to the skull of the patient [3]. Gamma knife radiosurgery developed during 1950's by Lars Leksell, is one of non-invasive treatments being used in brain cancer treatment, which does not require a craniotomy. Although, the radiosurgical treatments have shown very successful and promising results, the use of ionizing radiation poses a potential health hazard. Hence, a non-radiation based non-invasive surgical method such as Focal Ultrasound Surgery (FUS) using high intensity ultrasound to coagulate cancerous cells withholds a high potential for treatment of brain cancers.

In this paper, we propose this non-invasive modality for surgery of the abnormalities in deep-seated targets of the brain through a precise craniotomy using a customized robotic approach. Since, in this process, no incision is required in the brain tissue, the concerns related to inaccuracy and imprecision due to loss in registration will be minimized. Following sections would include our methodology, design of the base robot and its specialized end-effector for HIFU deployment.

2. Methodology

While the developments in application of HIFU in other organs are relatively easy, application of completely non-invasive procedure of HIFU in neurosurgery is met with the fundamental barrier, the skull. The skull not only absorbs much of the ultrasound energy and causes overheating but also distorts the beams. In the case of brain tissue, which is encased in the bony envelope of the skull, direct application of HIFU is difficult as bone reflects a large amount of the energy. In order to reduce heating up the skull, several efforts are being carried out using phased array techniques that can largely reduce the heat absorbed by the skull [4]. As temperature elevation near bone structures may cause thermal damage to the bone, direct application of the transducer on the *dura mater* is adopted in this work. Therefore, it is necessary to create access craniotomies of optimized size to place the applicator through an appropriate couplant.

2.1 Multi-probe HIFU

In an earlier study by one of the authors [5], small multiple HIFU probes working in unison were shown to minimize off-focal hot spots that are prevalent in the use of conventional large spherical ultrasound transducers as well as improving the flexibility in the lesion parameters as compared to those achieved by the latter. It was also shown that the integration of robotic techniques with a specialized custom designed stereotactic frame, it is possible to localize the energy in the target accurately and precisely [6]. In the present work, we retain our methodology of the use of multiple probes and the use of robotic means both for creating desired craniotomies as well as precise HIFU energy deployment, however, without the use of streotactic frame. The design of this robotic system, named Neurobot is targeted to the surgical applications in the skull base area.

2.2 Deep-seated skull base targets

Skull base refers to the area of the skull that provides the base on which the brain rests, as shown in Figure1. It essentially contains the eye orbital and ear canals. As one of the most complex regions of the human anatomy, vital blood vessels and major cranial nerves pass through the skull base. Skull Base Surgery (SBS) refers to the surgical techniques required to access the brain through this part of the skull.

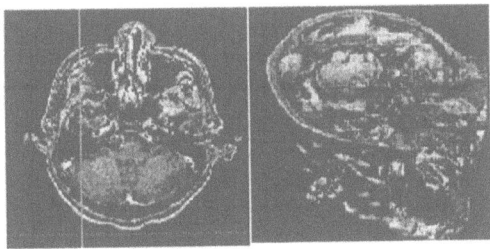

Figure 1: Skull base area of a patient's head

3. System Overview and Detailed Design

A robotic system, Neurobot is devised at MPE. It is designed to help neuro- and ENT surgeons to achieve quick (from 5 - 8 hours to about two hours) and precise drilling of bone at the skull-base. Skull base surgery is a time consuming and tiring procedure due to the complexity of this region. Hence much more effort is required to avoid complications. The surgical process is divided into two phases, the drilling phase followed by surgery. The cavity generated by the robot is used to access deep-seated brain area (inaccessible by other routes) for surgical intervention using manual means. However, means of deploying HIFU energy need to be integrated as a detachable end-effector for non-invasive surgery. Neurobot system and its various counterparts are explained in the following sections.

3.1 The positioning system

Neurobot, as shown in Figure 2a, is made up of a 4-Degree-of-Freedom (DOF) base manipulator which augments the position of another 6-DOF fine-positioning system [7]. The 6 DOF is provided by the micro-positioning parallel manipulator, Hexapod, where positioning of the tip of the drill is monitored by an optical tracker. A linear guide (7th DOF) displaces the drill forward into the skull. The base positioning system is used to position the Neurobot at an optimal distance near to the skull of the patient. The 7th DOF then advances the drill until it just touches the skull. The desired motion of drill to remove the skull bone is performed by the Hexapod. For a detailed description of the system and drilling trajectory, please refer to [8]. The kinematics of the Hexapod system are illustrated in Figure 2b.

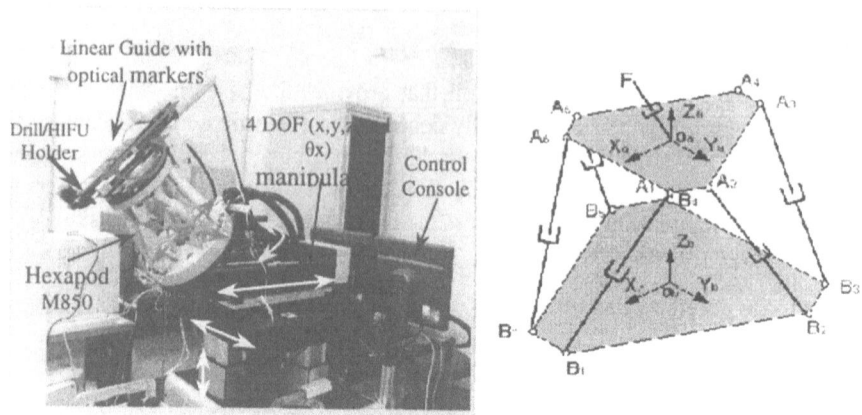

Figure 2: (a) Neurobot system, (b) Hexapod kinematics.

3.2 Pre-operative planning and image guidance

The surgical pre-operative planner utilizes MRI (Magnetic Resonance Imaging) and CT (Computed Tomography) images scanned from the patient for the motion planning. The Neurobot's image-guided system (combination of CT and MRI images) presents the information in 3-D to the surgeon. The surgeon pre-operatively defines the areas that are to be avoided and the system automatically proceeds to generate the required path using a proprietary path-generation algorithm [7]. After verification of the path, the robot moves the tool to the optimal position with respect to the patient and perform the drilling sequence in accordance to the pre-planned path. An optical tracking system, OPTOTRAK®, tracks the displacements of infrared markers placed both on the hexapod mobile platform and on the patient. The OPTOTRAK offers a resolution of 0.01mm at a 2.5m distance, which is suitable for the surgical requirement.

3.3 End-effector modules

The Neurobot end-effector is made up of detachable units, which can be mechanically locked in position on a base-plate sharing a common reference.

3.3.1 Drill-unit end-effector

In order to create the craniotomy, the drill unit end-effector is mounted on the 7th DOF. A Force/Torque sensor is installed between the actuator and end-effector to monitor the drilling forces. The Hexapod has a high accuracy of 0.5um in the vertical direction and 1.0um in the horizontal direction. The 7th DOF is designed with an accuracy of 5µm.

Presently, Neurobot does not have real-time registration for soft tissues within the skull. However, being designed specifically for skull removal during skull-based surgery, it satisfies most of the kinematics requirements for manipulation of the HIFU-effector for brain cancer treatment. Thus, the design of the HIFU effector and interfacing jig is based on the design restrictions of mounting tool on Neurobot. In the present work, a multiple HIFU probe end-effector replaces the surgical drill attachment while retaining the same registration information with respect to the patient (brain). It is designed for interfacing with the robot for remote ablation of brain cancers through access craniotomies of optimized size through an appropriate couplant (degassed water).

3.3.2 HIFU-effector

The HIFU-effector comprises of a base-plate on which the desired number of HIFU probes (with a chosen specification) are mounted. This base-plate is made to move as a piston (with a motorized axis), in and out, within an outer water-tight cylindrical envelope (as shown in Figure 3). In order to avoid energy reflection during exposure, the material chosen for base-plate as well as cylinder is perpex. An inter-probe angle in the range of 25° - 40° is chosen as deduced in the previous studies to investigate the efficiency of the multiple probes arrangement [5] using similar probes. The cylinder is filled with degassed water and a miniature pump to regulate water during procedure.

The front end of the cylindrical bolus is coupled with a flexible bellow, which ends with a compliant natural latex cover for coupling with the brain tissue through a controlled craniotomy. The bellow provides adjustment along the axis of the transducer of the order of 4cm. The schematic and design of this module is illustrated in Figure 3. At the rear surface of the cylinder, an attachment for easy interfacing on the drill-holder of the base robot is provided. This module is rigidly coupled at the drill-holder mounted on the Hexapod and is actuated with the 7th DOF of the base robot, thus maintaining the original accuracy and registration of the Neurobot at the remote point intersection of the multiple probes (targeted at remote site inside the brain).

The intersection of foci of individual probes and the effective lesion size is tested within ±0.5mm, both using mechanical calibration as well as ultrasound scanning in A-mode, prior to mounting the base-plate within the HIFU-effector. The end-effector, thus fabricated, is a relatively compact and lightweight design for application of multiple probes approach. Various tests have been conducted to verify the desired functionality in laboratory phantoms (not reported in this paper).

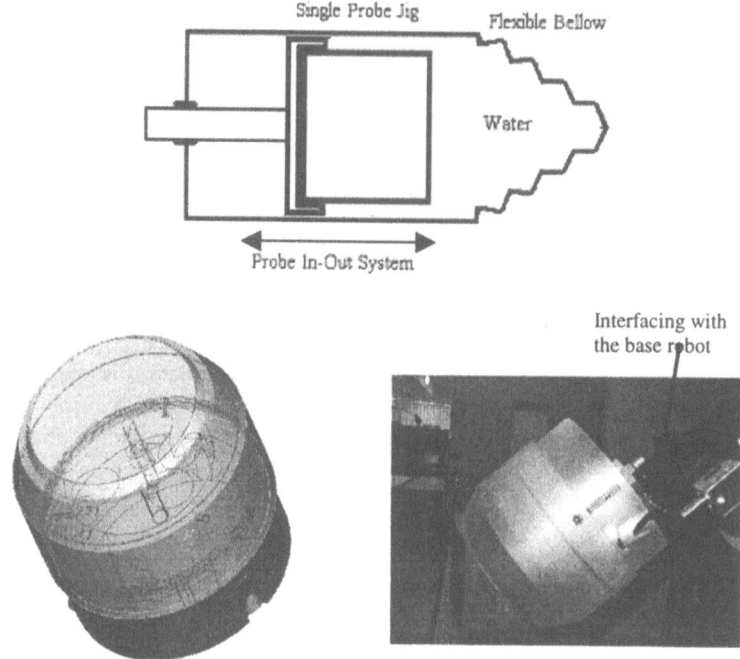

Figure 3: Schematic diagram of HIFU-effector and the finished design.

HIFU energy control and firing system is a separate module, which is de-coupled from robot control for safety reasons. This module governs HIFU parameters on-line such as the power in the beam, frequency, duration of exposure, triggering control etc. Both single probe and multiple probe jigs are adopted for experimental trials. For HIFU-effector, an on-line ultrasound image-guided system presents the information in 3-D to the surgeon along with real-time ultrasound imaging.

4. Conclusion

The present study is a feasibility effort for integrating a specialized end-effector for ablation of remote sites in the brain. Though extensive studies have been done separately both for Skull base surgery and HIFU in the past with promising results, the hybrid system is aimed to provide a non-invasive means to ablate localized remote/deep-seated targets in the brain. We are now testing this system in laboratory phantoms for efficacy of energy transmission.

5. Acknowledgements

The authors would like to thankfully acknowledge the support from Ministry of Education and Agency for Science, Technology and Research, Singapore for generous funding support.

References

[1]. A. Paul "Surgical Robot In Endoprosthetics. How CASPAR Assists On The Hip" MMW Fortschr Med. 1999 Aug 19; pp. 141

[2]. Taylor, R. H., G. Fichtinger, P. Jensen, and C. Riviere, "Medical Robotics and Computer-Integrated Surgery: Information-driven Systems for 21st Century Operating Rooms". *Japanese Journal of Computer-Assisted Surgery*, 2000.

[3]. Douglas Kondziolka, Atul Patel, Dade Lunsford et al., Stereotactic Radiosurgery Plus Whole Brain Radiotherapy versus Radiotherapy Alone for Patients with Multiple Brain Metastases, *Int J Radiation Oncology Biol Phys* 45(2):427-434, 1999.

[4]. Smith, N. B., Temkin, J. M., Shapiro, F. and Hynynen, K. Thermal effects of focused ultrasound energy on bone tissue. Ultrasound Med Biol 27(10):1427-1433, 2001.

[5]. S. Chauhan, 2001, Field modeling for multiple focused ultrasound transducers M2VIP, 8th IEEE International Conference on Mechatronics and Machine Vision in Practice, Hong Kong 27-29th Aug. 2001.

[6]. B. L. Davies, S. Chauhan and M. J., Lowe, A robotic approach to HIFU based neurosurgery, Computer Science, No. 1496, UK, October, pp.386-396, 1998.

[7]. Bai S. P., Teo M. Y. A Robotic Neuro-Surgery System And Its Calibration By Using A Motion Tracking System, 11th IEEE International Workshop On Robot And Human Interactive Communication 2002, Berlin, Germany, 25 – 27 September, 2002. Proceedings IEEE ROMAN, Pp 436 – 441, 2002.

[8]. Sim C., Teo M. Y., Ng W. S., Yeo T. T. Development Of A Robotic System For Skull Base Surgery, World Congress Of High-Tech Medicine, 15-20 Oct'00, Hanover, Germany, Pp 165 – 171, 2000.

COMPUTER-AIDED ENGINEERING AND PROCESS PLANNING

A Web-based Collaborative Feature Modeling System Framework

S-H Tang [1], Y-S Ma [1] [*] and Chen G [2]

[1] DRC, School of MPE, Nanyang Technological University, Singapore
[2] CAD/CAM Lab, School of MPE, Nanyang Technological University, Singapore
[*1] mysma@ntu.edu.sg

Abstract: In this paper, a web-based collaborative feature modeling system framework is proposed. To integrate CAx applications, a four-layer information model is proposed. STEP (STandard for the Exchange of Product model data) is extended to build the product model structure for information sharing. Mapping mechanisms are also investigated to convert the EXPRESS-defined information types into the database schemas. The generic feature representation and geometrical data representation in database are given. Mechanisms for feature validation are explained.

1. Introduction

In a collaborative engineering environment, engineering tasks are always carried out by a group of distributed engineers who may use different applications. Therefore, information sharing among product development team members becomes a bottleneck. Although much research work [2, 3, 4, 5, 6] and commercial products [7, 8] have been carried out in this area, problems still exist. They fall into the following two aspects, information loss and data conflict.

Although many proposed systems claim to be CAD-neutral based on STEP and CORBA, they lack the necessary interoperability so far. In the process of data exchange, useful information such as features is often lost. Therefore, these systems are not completely information sharing. On the other hand, CAD data is often stored in a file format, which means duplicated data and potential conflicts.

2. System Architecture

To enable information sharing among CAx applications, a web-based, database-driven, and feature-oriented system architecture is proposed as shown in Figure 1. The proposed system includes clients, application servers and a database server. The application servers include a web server, an application object server and a feature object server.

The web server contains a multiview data access interface (MDAI), a security manager and a session manager. MDAI provides shared access for multiple users. It can instantiate different views for different users according to the user's requirements. Security manager is used to prevent unauthorized access to the product data. The session manager is responsible for controlling concurrent access by multiple users of the same data. The application object server provides different application packages for different users on the basis of feature object server. The product model manager is responsible for organizing information for multiple applications according to the user's requirements. This information includes feature

32

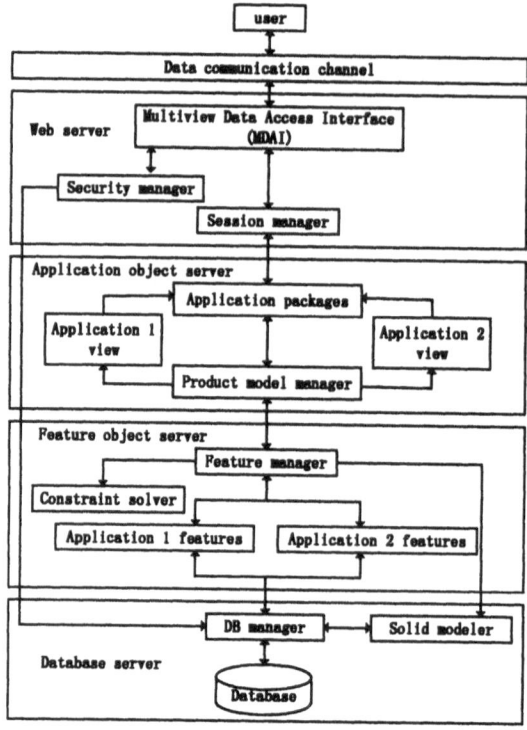

Figure 1. Overall system architecture

model and solid model (B-rep). To maintain the feature semantics during a modelling operation, such as adding, deleting and modifying features, the feature manager will call the constraint solver and the solid modeler to validate the feature. Details for feature validation are explained in section 5. The constraint solver can check the validation of all constraints, which are part of the feature definition. The geometric modeller can validate feature geometry. The database server provides physical storage for all kinds of data including product model data, security management data and so on. Within the database, geometrical data and features for different applications are stored as data elements across tables so that they can be reorganized with great flexibility. The solid modeller provides general functions such as geometry construction, modification, and computation to support higher-level feature modelling.

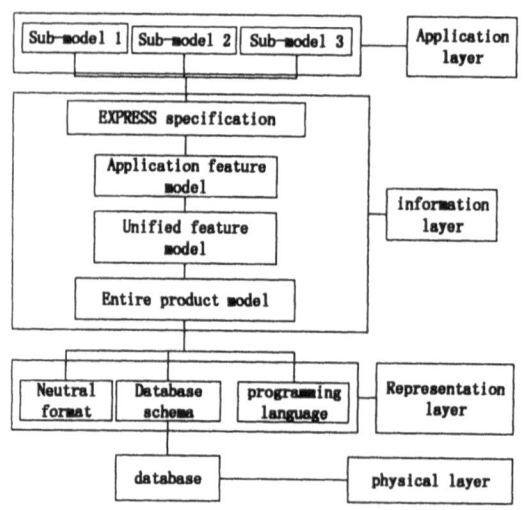

Figure 2. Four-layer information model

3. Information Model

In this research, the information model is built on the basis of an extended STEP framework. To achieve integration among CAx applications, the sharing of a common product model is crucial. The shared product model provides different views for various applications. Based on Zha's work [9], we propose the four-layer information model as showed in Figure 2. The four layers are application, information, representation and physical layers. The information layer contains four components, i.e. EXPRESS specification, application features, unified features and entire product model (EPM).

EPM describes information across applications, and contains the domain classification ontology and metadata; the detailed high level feature objects are organized by different sub-models in the application feature layer. Application features can provide specific view of the EPM. Next, a unified feature model [10] is used to specify the generic feature-modelling framework for common definitions of different application features.

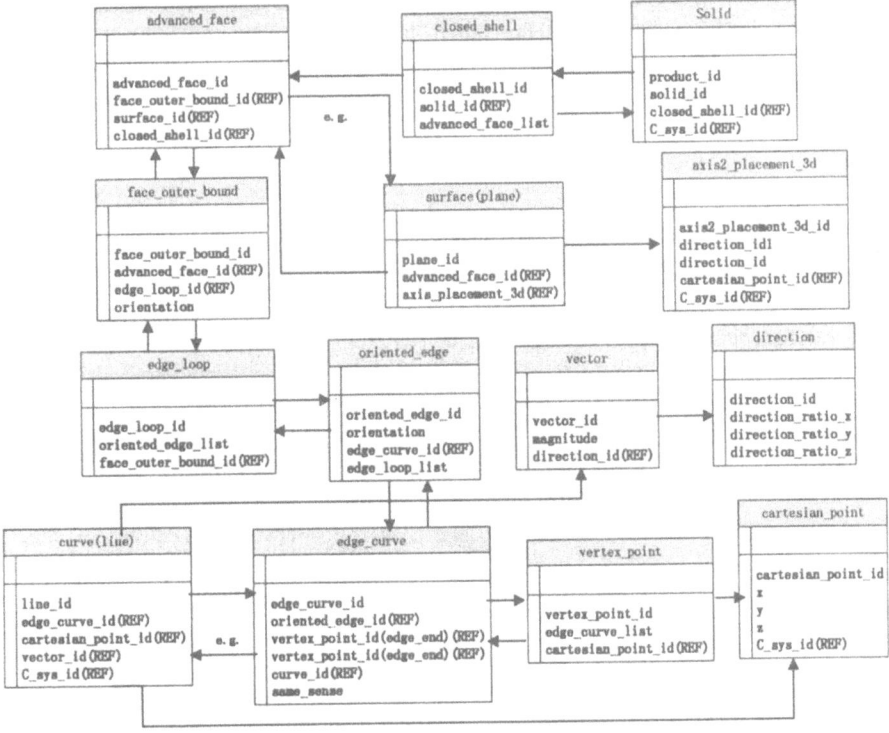

Figure3. Partial database schema for geometrical representation

Unified feature model allows different applications to define features with different configurations based on the predefined candidate types, such as geometrical representations, and the common processing methods. All the contents of EPM, application feature model and unified feature model are described in EXPRESS language.

4. Database Schemas

Under the four-layer information model structure, the entities at different levels shall be mapped to schema definitions for a potential comprehensive product database such that arbitrary feature object structure can be represented. Details of mapping mechanism are given in [11].

A partial geometrical database schema is created according to STEP 42 [12] as shown in Figure 3. All attributes with suffix *id* (but without *REF*) represent object identifier (OID). A built-in data type called a *REF* represents the reference to OID. An arrow here represents such *REF* relationship between object types.

The generic feature representation in database can be expressed as in Figure 4 under the framework of a unified feature model. A feature has *feature_id, product_id* and *domain* as its attribute. A feature also contains a list of referenced entities, a list of constraints and a list of parameters. Parameters can be uniquely identified by a *parameter_id* from the parameter table. *Referenced_entity* of feature includes entities (e.g. solid, faces, edges and vertices, etc.) or other features. *Entity_id* can uniquely identify the referenced entities stored in the entity table. A constraint of a feature can be uniquely identified by *constraint_id*. *Constrained_entity_list* identifies constrained entity from the entity table by *entity_id* and *entity_type*.

5. Maintenance of Feature Validity

Feature validity must be checked during feature modelling operations in order to maintain the feature semantics. A feature is valid as long as the feature satisfies all the relevant constraints and the feature geometry is valid. After each feature modelling operation, the solid modeller will be called to validate feature geometry. Then the feature manager will call constraints solver to check all the relevant constraints to determine if all features are valid. The constraint manager maintains all constraints in a constraint graph for EPM, which contains sub-graphs for specific application views. The constraint manager solves the constraints by calling the corresponding solvers according to

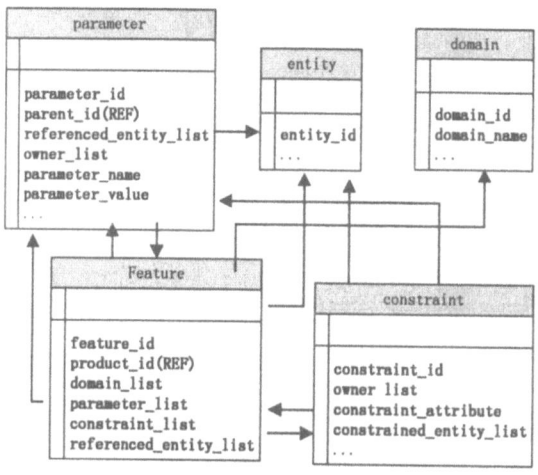

Figure 4. Generic feature representation in database

different constraint types. For example, the SkyBlue algorithm [13] can be used to solve local algebraic constraints in the design domain. If a conflict of intra-application constraints occurs, the local constraints solver can determine automatically which constraint should be satisfied first according to the value of *constraint_strength*, which is an attribute of constraint. It is an enumeration data type, which may include several levels, such as *required, strong, medium or weak*. Inter-application constraints can also be solved under the control of constraint manager according to the value of *domain_strength*. Its value, which regulates priority sequence of different domains, is predefined. Any conflict of inter-application constraints will be detected by the constraint manager, which can trigger the relevant applications and the constraint solver to re-evaluate the product model according to the *domain_strength*. Only when all the constraints are checked and the feature geometry is validated, does feature validation finish.

6. Case Study

A case study is carried out to testify whether product and process information can be well managed using the proposed database schema. The geometrical entities (e.g. shell, advanced_face, etc.) of an example part (block with through_slot feature) are explained in Figure 5. Adopting the proposed feature representation schema, the example part can be expressed in a database as shown in Figure 6.

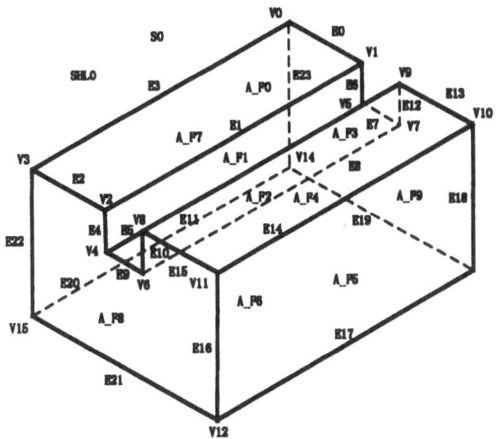

Figure 5 Geometrical representation of example part

Functions for managing product information, such as *save*, *restore*, and *validate*, have to be developed. These functions are used to organize information for different application views according to users' requirements. Here, we only briefly explain the *save* and *restore* algorithms. Part information, which includes geometrical data, features and others, is represented as ENTITIES. ENTITY is a virtual class; it represents common data and functionality that is mandatory in all classes that represent permanent objects. *Save* algorithm can be expressed in step as follows: (a) Create an empty entity list and add the part to be saved to the list; (b) Get all entities such as solid, shell and so on from the part and add them to a graph map so that object pointers can be fixed as unique database object IDs; (c) Use such object pointers to call *save*

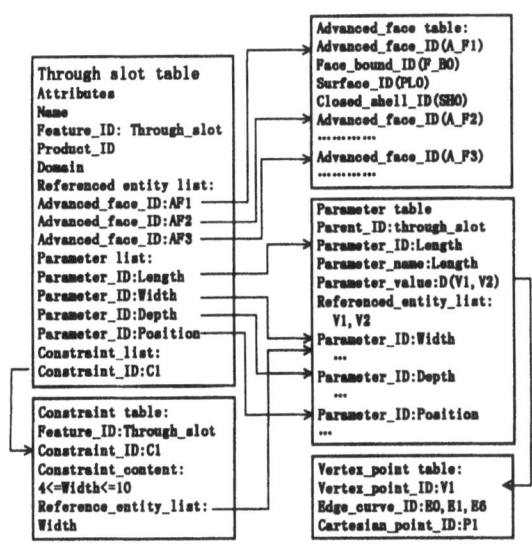

Figure 6 Through_slot feature in database

functions of the specific class (e.g. *point.save* (), *vertex.save* ()) to save part data to the database.

Restore algorithm has the following steps: (a) all the entity of a part are retrieved from the database by searching their linked Object IDs; (b) reconstruct new objects; (c) add all the entities to a newly generated object graph map; (d) convert these IDs to genuine pointers; (e) create an entity list and add all the entities to the list to form the part.

7. Conclusion

In this paper, a framework is proposed to enable information sharing among CAx applications. The proposed four-layer information model can integrate different applications with EPM, and allow the manipulation of application-specific information within sub-models. STEP information model has been extended; product and process information can be organized for multiple applications with great flexibility. A generic feature representation schema and a geometrical data representation schema in databases are given. Mechanism to maintain feature validation is described.

References

[1] *Industrial Automation Systems and Integration — Product Data Representation and Exchange* — Part 1: Overview and Fundamental Principles, ISO 10303-1:1994 (E), ISO, Geneva, 1994.
[2] H J Helpenstein, *CAD Geometry Data Exchange Using STEP*, ECSC-EEC-EAEC, Brussels-Luxembourg, 1993.
[3] Y P Zhang, *An Internet based STEP Data Exchange Framework for Virtual Enterprises*, Computers in Industry, **41**, 51-63, 2000.
[4] T Dereli and H Filiz, *A note on the use of STEP for interfacing design to process planning*, Computer-Aided Design, **34**, 1075-1085, 2002.
[5] R Bidarra, W F Bronsvoort, *Semantic feature modeling*, Computer-Aided Design, **32**, 201–225, 2000.
[6] J Kim and S Han, *Encapsulation of geometric functions for ship structural CAD using a STEP database as native storage*, Computer-Aided Design, **35**, 1161–1170, 2003.
[7] *ACS software*. http://www.acssoftware.com/.
[8] J Emmel, *OneSpace-Integrating Collaboration Technology and Enterprise PDM*, Technical Whitepaper of CoCreate Software GambH, 2000.
[9] X F Zha, H Du, *A PDES/STEP-based Model and System for Concurrent Integrated Design and Assembly Planning*, Computer-Aided Design, **34**, 1087-1110, 2002.
[10] G Chen, *Unified Feature Model for the Integration of CAD and CAX*, First-year report, School of MPE, 2003.
[11] S-H Tang., Y–S Ma and G Chen, *A Feature-oriented Database Framework for Web-based CAx Applications*, accepted by conference of CAD04, 2004.
[12] *Industrial Automation Systems and Integration — Product Data Representation and Exchange* — Part 42: Integrated Generic Resources: Geometric and Topological Representation, ISO 10303-42:1994 (E), ISO, Geneva, 1994.
[13] M Sannella. *The SkyBlue constraint solver*, Technical Report 92-07-02, Department of Computer Science and Engineering, University of Washington, 1993.

Extendible Detection, Classification and Process Planning of Machining Features

O. Owodunni and S. Hinduja

Department of Mechanical, Aerospace and Manufacturing Engineering,
University of Manchester Institute of Science and Technology (UMIST),
Manchester, United Kingdom. s.hinduja@umist.ac.uk

Abstract: This paper presents a computer system for the detection, classification and process planning of machining features within an extendible framework. Feature face sets are extracted from a 3D CAD model using a feature detection module which has been developed by systematic synthesis and combination of several detection methods. Feature completion and volume formation are then carried out are without prejudging machining process planning decisions. In the third step, the machining volumes are given names linked to machining process options and resources within a framework that allows extendibility without additional programming effort. Results of one of the several components used to test the system are presented.

1. Introduction

Automatic Feature Recognition (AFR) and Computer-Aided Process Planning (CAPP) have been identified in the manufacturing community as two key blocks in the integration of Computer-Aided Design (CAD) and Computer-Aided Manufacture (CAM). For this reason, research in the areas of feature recognition and CAPP has been carried out for over two decades.

In feature recognition systems, features are extracted from conventional CAD data thus making feature recognition an integrating link between CAD and CAPP. From such initial contributions of researchers like Kyprianou in 1980 [1], several feature recognition approaches which include graph-based, hint-based, volume-decomposition and cell-based methods have been developed. Other works [2, 3] provide more detail review of these approaches.

CAPP research which predates that of AFR involves the use of computer technology to obtain a plan which consist of such information as machining processes/operations and their sequences, set-ups, workholding, machine tools, cutting tools, cutting conditions, toolpath and CNC codes required for machining a component. Approaches have included variant, generative and their hybrids. Detailed reviews of CAPP and several systems that have been developed especially in the academic community are presented by other researchers [4, 5].

A major requirement for AFR and process planning is to have them extendible without requiring additional programming effort on the part of the users. This paper attempts to address this need by adopting a systematic approach in developing a system which facilitates extendibility in AFR and CAPP. The modules of the system consisting of feature detection, completion of face sets and building of feature volumes, classification of feature volumes and extendible process planning are presented in sections 2 to 5 of the paper. Section 6 presents some results and sections 7 draws some conclusions on the research as well as suggestions for future research.

2. Detection of Feature Face Sets

For the purpose of exploring the best feature detection procedure, all the possible face sets are systematically generated and algorithms for detecting them are designed and implemented, basing those algorithms on basic geometric reasoning procedures. These face sets and their algorithms, which vary from the simplest to the most complex, are the minimum and maximum concavity methods, single-face and multiple-face inner loop methods, the visible face set and cavity methods.

In the visible face set method, only faces that are mutually visible to one another constitute a feature set. The algorithm for realising this concept starts with a list of feature face sets, which is initialised to zero. As the list of faces on the component is traversed, faces, which are mutually visible to the current face set, are added to the list.

The minimal concave face set consist of faces that are linked by concave edges only. The procedure for detecting minimal face sets starts with a face that has not been visited in a breadth-first search and has a concave edge. Other faces, which have not been visited and are concavely related to the current list of faces to be visited are added to the feature face list. Faces that are convexly related are forbidden from participating and if they are already in a feature list are removed.

The maximum concave face set includes all faces that can be reached through at least one concave edge. The algorithm for the maximum concave face set is similar to that for the minimal face set, except that forbidding or removing convexly related faces does not apply.

The next face set in complexity is the face set that arises from a convex inner loop that is contained wholly in a single face. This type of face set has been prominent from the earliest feature recognition research [1]. The face set is determined by collecting all faces, starting from the face adjacent but not the parent face of the starting loop, into a set that are connected to the starting face. The procedure for a set stops when no more face is encountered or when a convex inner loop is encountered.

The concept of a face set from an inner loop on a single face can be extended conceptually to a loop of edges straddling several faces. This face set is referred to as multiple-face inner loop set. The algorithm for realising this set of faces is based on marking faces adjacent to a list of non-convex hull edges and collecting connected marked faces into a list.

The cavity face set forms the next type of face set in order of increasing complexity. This face set consists of all faces that are not part of the 3D convex hull of the component. These face sets are determined by marking faces that are not enveloping faces (3D convex hull faces) on the component and collecting connected ones into a list.

Further details of these algorithms, the comparative analysis carried out on their performance and the systematic methods of obtaining an optimal composite method from them are presented in another work [6].

3. Completion of Feature Volumes

The feature faces detected are usually incomplete due to interacting features and also need to be formed into volumes. Though several researchers have addressed the problem there is still the need to determine which of the several possible volumes need to be formed without making premature assumption of the application. In this research, a volume building algorithm developed in another UMIST research [7] project has been modified to allow an investigation of the mentioned issue. The result shows that the best approach which minimises the amount of features formed while not making any application assumptions is obtained when features are overlapping volumes which can be combined by simple Boolean operations to obtain any application volume.

4. Classification of Feature Volumes

Feature classification usually involves the existence a feature taxonomy, library of feature definitions and a procedure of matching features in a component with the library features. Despite the efforts of researchers [8-10], issues such as completeness of feature taxonomy, automatic extendibility of a feature library and formalisation of the classification procedure still need to be addressed.

4.1 Complete feature taxonomy

A systematic method of enumerating all possible basic features and analysing the data set is required in order to establish a complete taxonomy of basic features. One such method is to use a property such as the closure of the feature shape in the local x, y and z directions of the feature. By giving numerical values to indicate the degree of closure of the feature (or its enclosing box) in each direction, a triad xyz can be formed. A value of 2 is assigned if the feature is closed in both the positive and negative directions of the axis, 1 if closed in one and 0 if open in both. For example, the through slot in Figure 1 is assigned a code of 210. The xyz triad can be used to exhaustively enumerate all geometrically feasible primitive features and since it can be interpreted as a $r\theta z$ triad, rotational features can also be catered for. The number (N) of basic features combinatorially possible is given by equation 1.

$$N = n_x.n_y.n_z \qquad (1)$$

where n_x, n_y and n_z are the number of possible closures in the x, y and z directions respectively. Since $n_x = n_y = n_z = 3$, N = 27. Of the 27 combinatorially possible cases, 24 are combinations of 7 triads while three are unique. Thus there are 10 feasible cases. These 10 types of prismatic and rotational features are represented in Figure 1 for subtractive features. The number of access directions n_a can be derived from the triad by using equation 2.

$$n_a = 6-(x+y+z) \qquad (2)$$

Name of Feature	Prismatic		Rotational	
Free Space (000), Surface (100)	None 		None 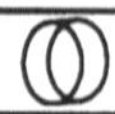	
Step (110), Notch (111)				 Not feasible
Bottomless Slot (200), Through Slot (210)				
Blind Slot (211), Hole (220)			Not feasible 	
Pocket (221), Enclosure (222)				

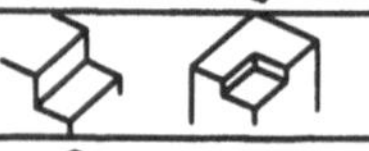

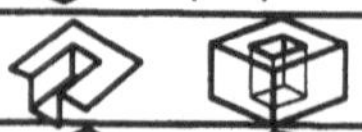

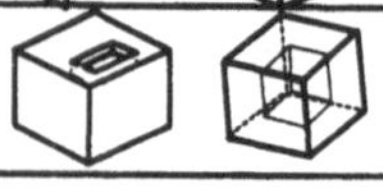

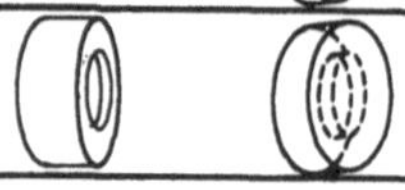

Figure 1: Taxonomy of prismatic and rotational machining features

4.2 Extendible library of feature definitions

A feature is defined as a list of attributes. The attributes include feature name, feature type, lists of feature invariants, feature faces, relationships between feature faces, feature parameters, possible machining operations that can manufacture the feature. Some of the feature invariants are: number of faces in the feature; number of faces of a particular type; and the number of face pairs that are parallel, perpendicular, concentric, oblique, tangent, co-axial etc. to each other. Geometric reasoning routines have been developed to automatically extract these attributes from a solid model representation of a feature. This provides a means to determine the properties required to define a feature in a systematic manner rather than the usual heuristic method and allows the feature library to be extendible without the need for additional programming. Also conflicts can be avoided in the database of feature templates through automatic checks when a new feature is defined.

4.3 Extendible and formalised feature matching procedure

In classifying a feature volume, the definition for its volume is determined and this feature definition is matched with the templates in the feature library. The classification of features is carried out at different levels. At the first level, a feature is classified using the taxonomy presented in section 4.1. The next level classifies the feature using the feature invariants presented in section 4.2. At the third level, an exact matching algorithm is used to determine a feature in the library whose shape exactly matches that of the feature. This third level has been approached by other researchers using heuristics. Initially the authors also adopted a heuristic approach so that they can study the matching process and hence formalise it.

By studying the matching process, a graph-theoretical property which enables the canonical labelling of features has been observed. This property is that once the first two faces of a feature graph are established, the arrangement of the whole graph is canonically fixed. The canonical ordering of the faces is used to generate a unique code for the feature and matching can then be more efficiently accomplished by simply comparing these codes.

5. Extendible Machining Process Planning Module

The extendible CAPP consists of machining process, machining operations, cutting tool and machine tool sub-modules, each which exists on three levels. The first level is the meta-class which allows user-defined classes constituting the second level to be achieved and the third level are the instances.

Every feature definition is associated with a machining process in a way that does not imply any process planning decision. For example, if a hole feature is associated with a hole making machining process the decision as to which hole making processes is utilised is taken at the process selection stage.

Each machining process consists of a number of sets of operations, the volume generated in realising the process, process tolerances and availability of the manufacturing resources. In the hole example, one of the sets may contain a centre drilling operation, a hole starting operation, a hole enlargement operation and a reaming operation. Another set may omit the starting centre drilling operation and use a boring operation instead of a reaming operation. Machining processes are defined from a library of existing machining operations. If a machining operation is required which does not exist, it can be created without programming.

Machining operations are defined in this work in terms of the specific tool and machine tool types used for achieving them. When machining operations are created, resources from the library of tools and machine tools are used. A cutting tool or machine tool that does not exist in the library can also be created without requiring programming.

Tools and machine tools are defined as meta-classes and classes. The meta-class allows new types of tools and machine tools to be added to the library. The while the classes allow specific cases of these tools to be created.

6. Results

The system described in this paper has been implemented in C++ using Spatial technology's ACIS geometry modeling on a SUN Ultra 1 workstation. Several components, mostly from the national design repository [11] have been tested. The results of one of these are presented in figures 2 and 3.

In figure 2 the face sets extracted by the various detection methods are presented. It would be noticed that some of the face sets such as F1-F3, F20 are not acceptable without further processing. This shows the need for a composite method of the basic detection methods and an appropriate sequence to obtain the best face sets. The feature volumes from the face sets are shown in figure 3. As can be observed from features: F6 and F10; F9 and F14 and F11-F13, the volumes are overlapping and can be composed by simple boolean operations to obtain the specific application volumes depending on criteria such as machining sequence.

When the classification stage starts with an empty feature library, the first volume F3 was classified at the top level, as a general pocket and the user was offered the option of adding the feature to the library. The user's acceptance resulted in the automatic determination of the feature description from its volume. In addition to the feature name (flat-bottom hole, in this case), the user can define the parameters of the feature which, in the case of F3, are the diameter and depth. After defining the parameters, the consistency mechanism checks that a feature with the same shape does not exist in the feature library and the extendible CAPP module enables a machining process to be attached to the feature definition.

When feature volumes F1, F4, F7, F8, F15, F16, F17 and F19 were processed, the system was able to identify them as flat-bottom holes, having learnt the definition of this feature type from F3 and their parameters were automatically determined. A similar procedure was followed to automatically define and then classify features: F6 and F9 as through holes; F11, F12 and F13 as round end slots; F10 and F14 as filleted blind slots and F0 as a filleted rectangular pocket. F19 is classified at the top level as a general open pocket. After classification of features, machining process options linked with each feature instance are automatically obtained.

7. Conclusions

This research has investigated research issues in AFR and CAPP. The investigation has resulted in the proposal of a methodology for the systematic development of AFR and CAPP systems, which result in improved performance and are extendible without requiring additional programming effort. Areas of improvement, some of which are the subject of ongoing research, include having a more robust feature completion and volume formation procedure and greater details in the extendible CAPP module.

References

[1]   Kyprianou, L.K., Shape classification in computer aided design, Ph.D. Dissertation, Kings College, University of Cambridge, U.K., 1980

[2]   Han, J-H, Pratt, M. and Regli, W. C., 2000,   Manufacturing feature recognition from solid models: A status report, IEEE Trans. Robotics and Automation, Vol 16, No 6, 782-796.

[3]   Owodunni O. and S.Hinduja, Evaluation of existing and new feature recognition algorithms: Parts 1: theory and implementation, Proc Instn Mech Engrs Vol 216, Part B, J Engineering Manufacture. August, 2002, 839-851

[4]   Tan W., 2000, Integration of Process Planning and Scheduling-a review, Journal of Intelligent Manufacturing, Vol. 11, 51-63

[5]   Marri H. B., Gunasekaran A. and Grieve R. J., 1998, Computer-Aided Process Planning: A State of Art, int J Adv Manuf Tech, Vol 12, 261-268

[6]   Owodunni, O.O., 2001, A systematic evaluation and   improvement of feature recognition methods in 3D components, PhD thesis, UMIST, UK.

[7]   Sandiford, D. and Hinduja, S., 2001, Construction of   feature volumes using intersection of adjacent surfaces, CAD, Vol 33, No 6, 455-473.

[8]   Gaines, D. M. and Hayes, C. C., 1999, C-C: a customisable feature recognisor, CAD, Vol 31, No 2, 85-100.

[9]   Gindy, N. N. Z., 1989, A hierarchical structure for form features, International Journal of Production research, Vol. 27, No. 12, 2089-2103.

[10]Sakurai, H. and Gossard, D.C., 1990, Recognising shape features in solid models, IEEE Computer Graphics and Applications, Vol. 10, No 9, 22-32.

[11]Ftp site of the National Design Repository, ftp://elib.cme.nist.gov/pub/subject/pptb/repository.

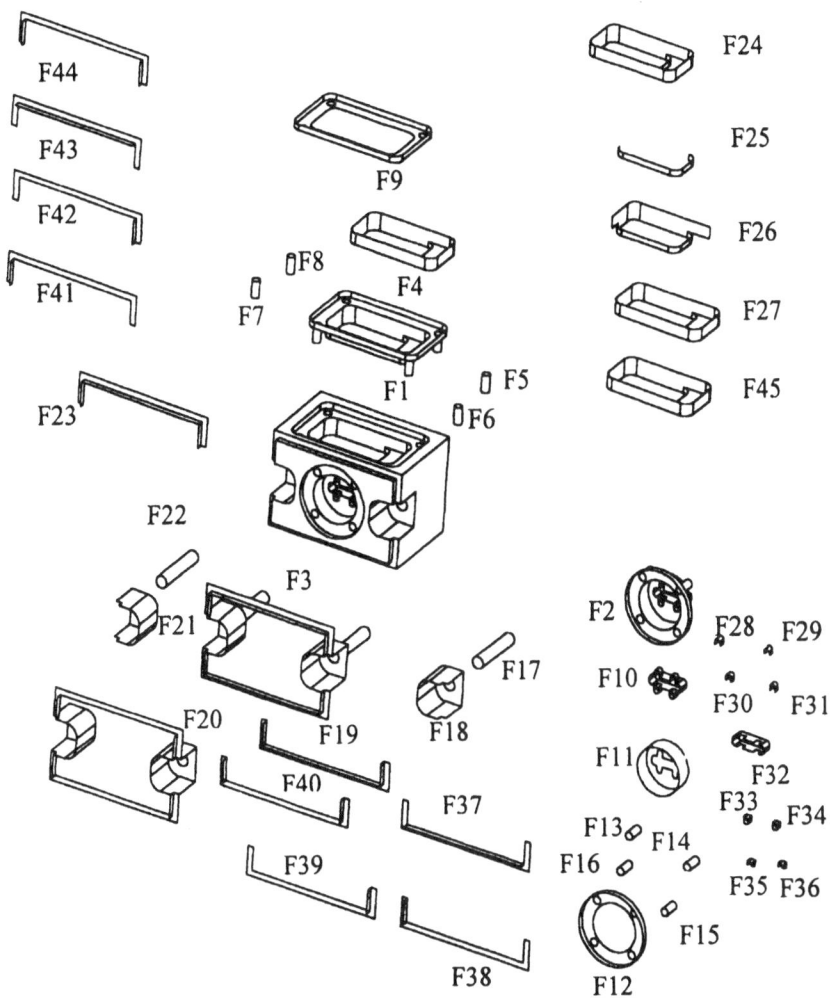

Figure 2: Face sets detected in test component by various methods

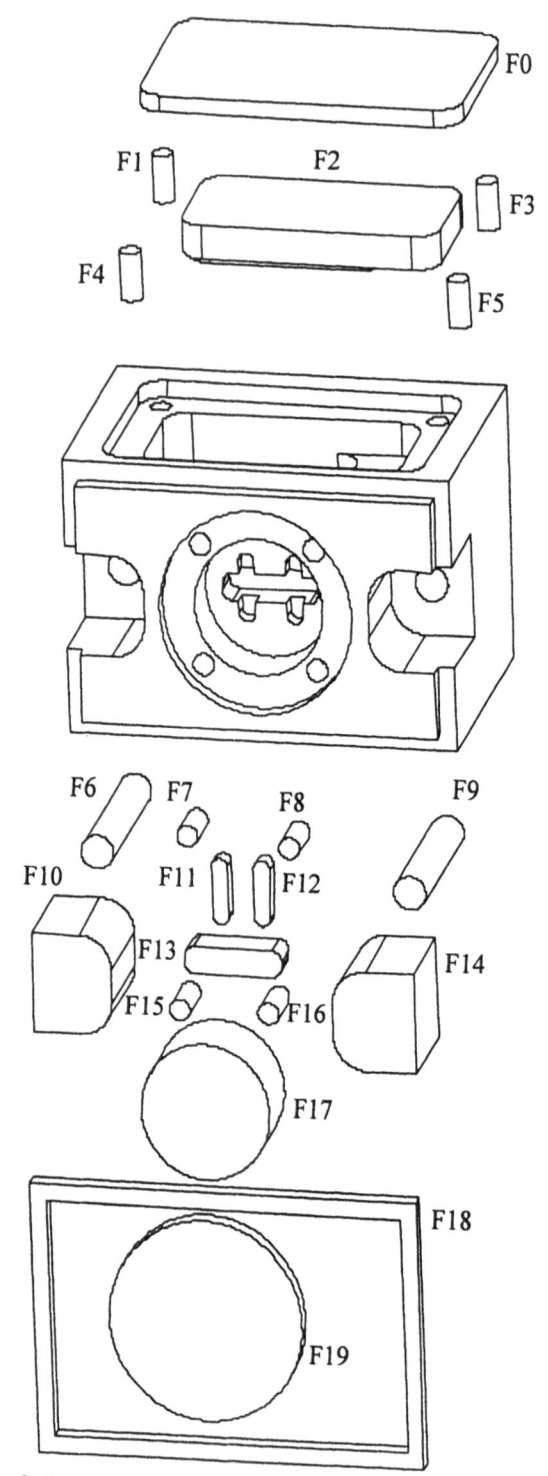

Figure 3. Feature volumes for test component

A STEP-Compliant "Adaptor" for Linking CAPP with CNC

H Wang and X Xu

Department of Mechanical Engineering, School of Engineering
The University of Auckland
Private Bag 92019, Auckland, New Zealand

Abstract: Combination of a series of STEP APs and the new ISO 14649 STEP-NC forms a unified data format for the entire product's lifecycle. The research reported in this paper is a framework for adapting STEP-compliant manufacturing information for a CNC machine. This information is represented in STEP-NC format containing entities such as "Workplan", "Workingstep" and "Machining Features", as stipulated in ISO 14649. This type of information is assumed to be machine-independent, hence it is the information about "what-to-do" rather than "how-to-do". This system can generate the "how-to-do" information when a machine tool is chosen for the job. There are three parts in this system, a native CNC System knowledge database, a Translator and a Human-Machine interface.

1. Introduction

Increasingly, manufacturing industry is being led by a group of smaller, more agile corporations that spring up for specific purposes, exist while the market sustains the new product and then gracefully disband as the market changes. To meet the requirements of modern manufacturing systems, the standard of ISO 10303 STEP offers manufacturers a convenient method for exchanging product data between heterogeneous CAD, CAPP and CAM systems. In 2003, STEP was extended to NC programming through a new standard - ISO 14649 [1] often known as STEP-NC. The introduction of STEP-NC is set to change the manufacturing industry dramatically, as solid geometrical models enriched with manufacturing information will be made available for the first time to machine controllers and operators.

It is understood that STEP together with STEP-NC can now provide a global data exchange format that covers the whole product lifecycle [2]. This is done by constructing a data exchange model via STEP-related APs (Application Protocols) throughout the manufacturing activities from design to process plan and manufacturing. In the design phase, STEP AP-203 [3] and AP-214 [4] are used as a neutral data format to exchange design data between different CAD systems. In the process planning phase, the CAPP system is connected to a generic database which includes information about machine tools and cutting tools of different types. The outputs from this type of CAPP system can take the format of STEP-NC. However, when the generic STEP-NC program is implemented on the native CNC machine system, the native manufacturing knowledge has to be considered. This paper introduces a native STEP-NC mapping system (Native STEP-NC program Adaptor) to fulfil this function. The adaptor is built in three parts: Native CNC knowledge database, Translator and Human-Machine Interface.

2. Literature Review

The global research in the areas of STEP-NC has remained highly visible with a number of major projects co-ordinated and conducted across different countries in the same region as well as on a truly international scale. There are three types of projects, those carried out (a) on the international scale, (b) across a few countries in the same region and (c) within a country. On the international scale, the IMS (Intelligent Manufacturing System) STEP-NC project [5], endorsed in November 2001, entails a truly international package of actions with research partners from four different regions: European Union, Korea, Switzerland and USA. They covered the manufacturers of all systems related to the data interface (CAM systems, controls, and machine tools), the users and academic institutions. Because of the Workingstep based feature in STEP-NC, many projects have incorporated prototype/commercial CAPP systems or CAM systems with process planning functionalities, i.e. STEPturn [6] from WZL RWTH, Germany, ST-Plan [7] from STEP Tools Inc. USA, Shop-Floor Programming (SFP) system [8] from NRL-SNT (National Research Laboratory for STEP-NC Technology), Korea, and the AB-CAM system [9] from Loughborough University, UK.

3. Proposed Research

The system being developed in the Manufacturing Systems Group at the University of Auckland has three objectives: to make product data interchangeable, product information flow seamless and the system itself independent of any CAD/CAM systems. STEP-NC is adopted as the "common language" for NC manufacturing as is STEP for the design processes. Figure 1 depicts the information flow in the system which is accompanied by different "derivatives" of STEP standards at different stages, hence both product data interchangeability and seamless product information flow can be achieved.

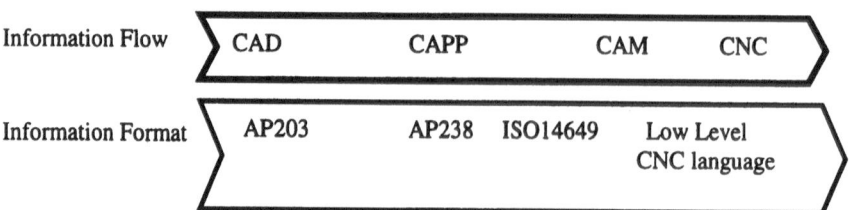

Figure 1 Integrating design with manufacturing via STEP/STEP-NC

The first move towards a "common language" based product development is to work with design models in STEP (AP 203) format. A product model defined in terms of pure STEP geometry is re-interpreted in terms of the manufacturing features stipulated by STEP AP224. Then, macro process planning is carried out based on these manufacturing features. The outcome of this macro process planning is a STEP-NC (AP238) [10] file, i.e. an AIM (Application Interpreted Model) file. This file includes information about the design model itself, the stock, its manufacturing features, tool/fixture requirements, the manufacturing process sequence, etc. It is self-documenting, machine-independent and allows complete safety checking. This type of information can be described as "what-to-do" information, meaning only manufacturing tasks are described, but not the way(s) of completing them. Therefore, information in AIM can also be called "Generic STEP-NC information" that is hardware-independent. This type of "what-to-do" information is then used for micro-process planning to generate so-called "how-to-do" information, which may be documented in an ISO 14649 file, i.e. an ARM (Application Reference Model) file. This type of information then becomes machine-specific. Hence, information in ARM can also be called "Native STEP-NC information". It is this information that will be used as input to a STEP-NC enabled CNC controller that can understand information such as Workplan(s) and Workingstep(s) and issue low level NC commands to drive the machine tool. Corresponding to the four phases illustrated in Figure 1, the system consists of four main modules (Figure 2), (A1) Generation of Generic STEP-NC Programs; (A2) Generation of Native STEP-NC Programs, (A3) Generation of Native CNC Commands and (A4) Execution of the Process Plan on a CNC Machine.

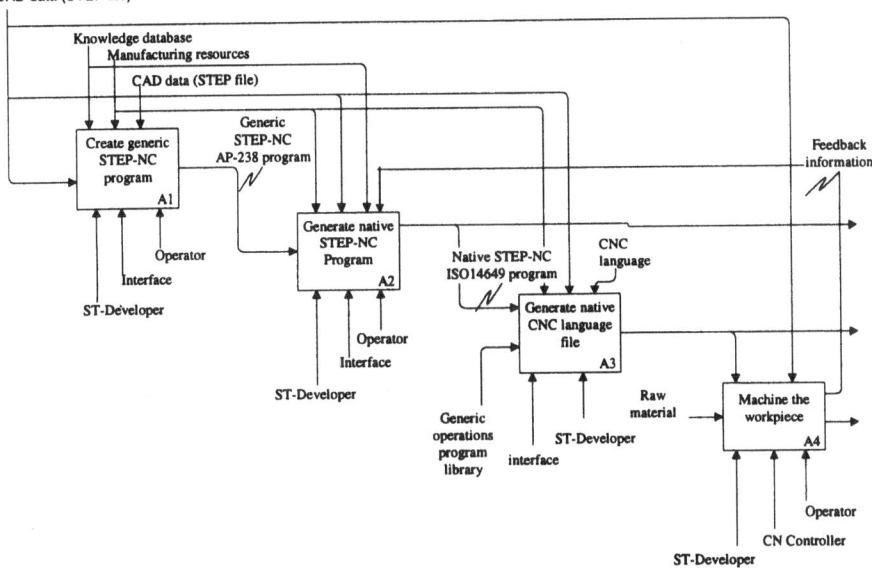

Figure 2 IDEF0 diagram of the STEP-compliant CAD/CAPP/CAM/CNC system

4. STEP-Compliant Adaptor

This paper focuses on the development of the second module (A2), trying to "map" the generic STEP-NC program to a native one, hence the name "STEP-Compliant Adaptor". In other words, the main task is to convert "what-to-do" information in a generic STEP-NC program, to the "how-to-do" information for a specific CNC machine. Different CNC machines have varying machining capabilities and native parameters. Until a native STEP-NC program that takes into account all the information specific to the CNC system is generated, it is not possible to execute the Workplan(s) specified in a STEP-NC program. The mapping process is also a "checking" process to evaluate the manufacturability of the job stipulated by a generic STEP-NC program.

4.1 Basic structure of the STEP-compliant adaptor

There are three functional units in the STEP-Compliant Adaptor, a native CNC knowledge database, a Translator and a Human-Machine interface. The CNC database stores detailed information about the machine tools, cutting tools, setup, etc. whereas the Translator maps the generic information (stored in a generic STEP-NC program) to the specific information (stored in a native CNC system resource Database) so as to generate a new NC program specific to, or customized for, the targeted NC machine tool. The HM interface displays the details of the "what-to-do" information stored in the program.

4.2 Modelling native manufacturing facilities

CNC machining centres are vendor-specific and vary in their hardware configuration and control software. Hundreds of specific parameters are preloaded onto a machine tool during production, catering for different machine functions. In order for the Adaptor to work with different machine tools, it is essential to develop a database structure that is capable of modelling the native information of different machine tools. Once this database structure is available, the translator can be developed with ease. Whenever a new machine tool arrives, the only task is to store the information of the new machine tool into the native database. The translator, which only depends on the structure of the database, does not need alterations.

In search for an effective means of constructing such a database, we found that the canonical characteristics model developed by the National Institute of Standards and Technology (NIST) in their Enhanced Machine Controller (EMC) project [11] is a viable concept. It is a high-level conceptual data model built for a testbed to evaluate open-architecture machine controllers. The canonical machining functions provide a single set of interpreter source code for the EMC controller to control 3 to 6-axis machining centres. In the research reported herein, the native manufacturing database was developed based on a sub-set of these canonical machining functions. There are two categories of information about a machining centre that need to be modelled in the database, (a) mechanical components of a machining centre, i.e. the physical configuration of a machining centre and (b) the activities of the machining centre that may be controlled, and the data that is used in control, i.e. control and data components of a machining centre.

The Database has an explicit and simple data structure, making it easier for a machine operator to use. It is built using Microsoft Access. There are four sub-databases in the system, the Machine Resource database, Cutting Tool database,

Set-up Resource database and Material (workpiece) Resource database. ISO/DIS 14649 Part 12/1 [12] has been followed to define the cutter parameters for the lathe cutting tool database (Figure 3). The tool database provides the information for tool set-up in an NC program. Therefore, information such as tool type, tool geometry and tool expected life is also included.

4.3 Translator

The Translator is an important part of the Adaptor, which relates the process plan in STEP-NC format to the potential manufacturing facilities stored in the database. In other words, the Translator tries to "interpret" the STEP-NC entities in terms of the native machining functions present in the database. Whether or not a complete interpretation is possible will also indicate if the given process plan is achievable or not.

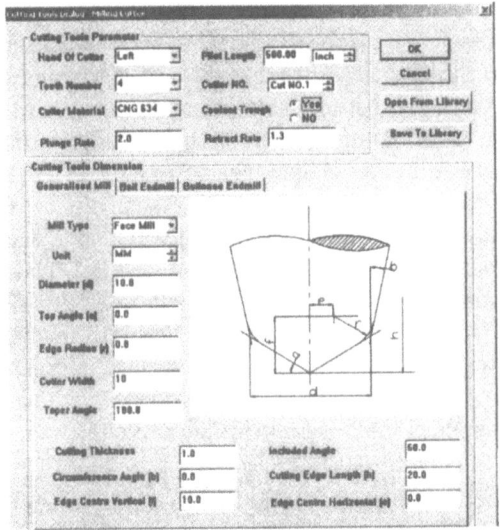

Figure 3 Interface for cutting tool resource database

The Interpreter takes STEP AP 238 physical file as an input. The output, which mainly consists of Workingsteps, is used directly as the input for the Translator. Alternatively, the Workingsteps can be modified manually by an operator through an interface to match the capabilities of a native machine tool in the database. In the translator, the decomposed generic STEP-NC program is mapped to the information stored in the native knowledge database. The outcome is a collection of Workingsteps organized under one or more than one Workplan, for one or more than one machine tool present in the database.

4.4 Human-Machine interfaces

The Human-Machine interfaces act as a console that the operator of a CNC system can use to (a) load a STEP-NC part program, (b) set up the native knowledge database and (c) generate a native STEP-NC program. It is Web-enabled so that various databases can be accessed through Internet, e.g. STEP-NC part program database, native knowledge database and STEP repositories. The interface can also provide the functions to edit both generic and native STEP-NC programs. When the tool-path simulation reveals an error, the operator can easily locate and fix it.

5. Conclusions

The proposed STEP-Compliant "Adaptor" provides a robust link between CAD/CAM system and CNC machines. The functional architecture is derived from the analysis of information contents of ISO 14649 and generic information included in common machine centres. The native manufacturing databases are built based on the canonical characteristics of machining centres. In doing so, the process of mapping "what-to-do" information (in an AIM model) to "how-to-do" information (in an ARM model) can be made easier. Unlike other systems, this system has preserved the unique characteristics of STEP-NC philosophy, i.e. hardware-independence of its code. The system also has the ability to interface with a process planner and/or machine operator for tasks such as amending a process plan and updating the databases. The current focus of research is on incorporating information documented by the draft STEP AP 240 standard [13]. This is because STEP AP 240 includes not only the required information for supporting NC programming, but also the Shop floor information such as machine setup and part loading instructions. Work on connecting this system with a Web-enabled, STEP-compliant CAPP system [14] is also underway.

References

[1] ISO 14649-1: 2003, *Data model for Computerized Numerical Controllers: Part 1 Overview and fundamental principles.*

[2] Xu X.W. and He Q., 2004, Striving for a total integration of CAD, CAPP, CAM and CNC. *Robotics and Computer Integrated Manufacturing*, Vol 20, no 2, pp 101-109

[3] ISO 10303-203: 1994, *Industrial automation systems and integration – Product data representation and exchange – Part 203: Application protocol: Configuration controlled 3D designs of mechanical parts and assemblies*

[4] ISO 10303-214: 1994, *Industrial automation systems and integration – Product data representation and exchange – Part 214: Application protocol: Core data for automotive mechanical design processes.*

[5] IMS STEP-NC consortium, 2003, *Technical Report 3 of IMS Project (97006) STEP-Compliant Data Interface for Numerical Controls (STEP-NC)*

[6] Weck M. and Wolf J., 2002, ISO 14649 Provides Information for Sophisticated and Flexible Numerically Controlled Production

[7] STEP Tools. Inc, http://www.steptools.com/stix/

[8] Suh S.H., Lee B.E., Chung D.H. et al, 2003, Architecture and implementation of a shop-floor programming system for STEP-compliant CNC. *Computer-Aided Design*, 35, 1069-1083

[9] Newman S.T., Allen R.D. and Rosso Jr. R.S.U. 2002, CAD/CAM solutions for STEP Compliant CNC Manufacture, *Proc of the 1st CIRP (UK) Seminar on Digital Enterprise Technology*. School of Engineering, University of Durham, UK

[10] ISO 10303-238: 2003, *Industrial automation systems and integration – Product data representation and exchange – Part 238: Application Protocols: Application interpreted model for computerized numerical controllers.*

[11] Kramer T.R., Proctor, F.M. and Messina E., 2000, *The NIST RS274/NGC Interpreter - Version 3*

[12] ISO/DIS 14649-12/1: 2003, *Data model for Computerised Numerical Controllers, Part 121: Tools For Turning Machines*

[13] ISO/DIS 10303-240: 2003, *Industrial automation systems and integration – Product data representation and exchange – Part 240: Application protocol: process plan for machined products.*

[14] Xu X. and Mao J., 2004, A STEP-Compliant Collaborative Product Development System, *Proc of the 33rd International Conference on Computers and Industrial Engineering*, March 25-27, 2004, Ramada Plaza Oriental Hotel, Jeju, Korea. CIE598

A Method of Clamping Optimisation for Machining Fixture

X Chen, Y Wang, Q Liu, N Gindy

School of Mechanical, Materials, Manufacturing Engineering and Management,
The University of Nottingham, UK NG7 2RD. xun.chen@nottingham.ac.uk

Abstract: Clamping plays a vital role in fixturing processes. A good clamping should satisfy the requirement of immobility and stability with minimum clamping force. The optimisation of clamping includes (1) finding feasible clamping position in terms of the immobility requirement; (2) finding the best clamp among the feasible clamping positions under constraint of the stability. As an example, the potential optimal fixture layouts for an aerofoil of turbine blade are presented. The method considers the optimisation of fixture layouts at a global level and is suitable for the workpieces with complex shapes.

1. Introduction

A machining fixture has two basic functions: a) to locate the component to the right position so the machining process can be finished; b) to hold the component tightly so that it will not move during the machining. Therefore, the fixture layout should be designed in such a way that the workpiece holding would possess repeatability and immobility. Besides, the workpiece should always contact fixture locators during clamping and machining, otherwise the contact status between workpiece and fixture locators would change, resulting in an unstable system. The stability of fixturing has been studied in previous researches [1-4].

Asada and By [5] addressed the conditions of the deterministic positioning and proved that the position repeatability of a workpiece depended only on the fixture locators rather than on clamps. Fixture layout is a critical influential factor to the fixture performance. R. A. Martin et al [6-8] proposed a 3-2-1 locator scheme to find the clamping region and minimal clamping forces. A near optimal locator scheme can be obtained from the screw theory [9]. Subsequently, the extended screw theory [10] was utilised to find the admissible position of clamps so that form closure was achieved. The maximum required clamping force was minimized by expressing the contact wrench, placed inside a convex polygon, as a positive combination of basic contact wrench that located at the vertices. With consideration of the fixture-workpiece compliance, De Meter [10] addressed an approach to compute the minimum required pre-load necessary to prevent workpiece slip from fixture elements throughout the machining process. The most current works [1-2, 6-8, 11-16] are based on the situations that (1) the fixturing surfaces are flat or cylindrical; (2) the fixturing positions are predefined. Thus the results are normally local enhancement rather than global optimisation. There is little attention on the relation between the magnitude of clamping forces and the number of clamps in a fixture.

This paper presents a method for the fixture optimisation in considering the fixturing location accuracy, repeatability, immobility and stability. The locator scheme is decided from location accuracy and repeatability. The feasible clamping positions are identified in terms of immobility. The optimal clamping position is where the minimum clamping forces and deformations are obtained whilst the stability is satisfied. The relationship between clamping positions and force intensity was investigated to determine the number of clamps required. Fixture layouts for a turbine blade are given as examples to demonstrate optimal fixturing in 2D space. All of the results and algorithms in this paper can be extended to a three dimensional space.

2. Optimisation of Fixture Layout

Fixture layout optimisation has two aspects: locator optimisation and clamp optimisation. Locating schemes may be enumerated to present a group of local optimal locator schemes according to the repeatability and accuracy of workpiece positioning. For each locator scheme, the enumeration of clamps can be undertaken in order to find the best clamping set. Searches were carried out using a Matlab program. Two assumptions were used: 1) no friction exists at the contact points between fixture element and workpiece; 2) cutting forces are static loads.

During the clamping, the forces and moments exerted on the workpiece should be in equilibrium. This can be expressed as:

$$\mathbf{GF_L} + \mathbf{CF_C} = 0 \tag{1}$$

Where $\mathbf{G}$ is the so called fixturing matrix, $\mathbf{G} = [\mathbf{N_L^1}, ..., \mathbf{N_L^i}, ..., \mathbf{N_L^n}]$, $\mathbf{N_L^i} = [\mathbf{n_L^i}, \mathbf{r_L^i} \times \mathbf{n_L^i}]'$, $\mathbf{F_L} = [f_L^1, ..., f_L^i, ..., f_L^n]'$. $\mathbf{C}$ is the clamping matrix and $\mathbf{C} = [\mathbf{N_C^1}, ..., \mathbf{N_C^j}, ..., \mathbf{N_C^m}]$, $\mathbf{N_C^j} = [\mathbf{n_C^j}, \mathbf{r_C^j} \times \mathbf{n_C^j}]'$, and $\mathbf{F_C} = [f_C^1, ..., f_C^j, ..., f_C^m]'$. Where n is the number of locators required for positioning a part ($n = 6$ for 3D space and $n = 3$ for 2D situation); m is the number of clamps; $\mathbf{n_L^i}$ and $\mathbf{n_C^j}$ are the unit normal vectors of the ith locator and j^{th} clamp towards the workpiece; $\mathbf{r_L^i}$ and $\mathbf{r_C^j}$ are the position vectors of the ith locator and j^{th} clamp; f_L^i and f_C^j are the magnitudes of the reacting forces of locators and clamping forces respectively. The relationship of force-deformation between fixture elements and workpiece can be represented by visual springs with certain stiffness [11, 12]. Assume the stiffness of the spring of the ith locator is k_i, deformation at the locator is X_i, then reaction force magnitude $f_L^i = k_i X_i$. By defining the contact stiffness matrix $\mathbf{K} = \text{diag}[k_1, ..., k_i, ..., k_n]$, and deformations matrix $\mathbf{X} = [X_1, ..., X_i, ..., X_n]'$, Equation (1) can be rewritten as:

$$\mathbf{GKX} + \mathbf{CF_C} = 0 \tag{2}$$

The necessary condition of the repeatability of workpiece location is that deformation $\mathbf{X}$ should be deterministic under action of arbitrary clamping forces. In other words, there is an unique solution for Equation (2). Since $\mathbf{K}$ is always positive, the determinant of $\mathbf{G}$ should be non-zero:

$$\|\mathbf{G}\| \neq 0 \tag{3}$$

In order to get high accuracy of location, X should be as small as possible. This means the norm of X should be minimum under arbitrary clamping forces. It has been proved by A. Atkinson [17] that for arbitrary clamping force, minimise $\|X\|^2$ is equivalent to minimise $\|K^{-1}G^{-1}\|^2 = (\|K^{-1}\|)^2 \, (\|G^{-1}\|)^2$. So, the location accuracy requirement indicates that the determinant of G should have maximum value:

$$Max \, (\|G\|^2) \tag{4}$$

With assumption of frictionless and point contact, the immobility is purely relevant to the workpiece geometry. Lakshminaryana [18] has proved that the sufficient and necessary condition of form closure is the existence of positive solution of Equation (1). In other words, if there is least one clamp that pushes the workpiece to contact all the locators, the workpiece is immobilised.

The optimal clamp position is selected to ensure the stability of fixturing. This means that clamping forces should make the workpiece contact all the fixture elements (locators and clamp) during the entire machining process. Therefore, clamping optimisation is to minimise total clamping forces on the workpiece. The optimisation of clumping function may be written as

$$\text{Minimize} \quad \sum_{j=1}^{m} (f_C^j)$$

$$\text{Subject to} \quad GF_L + CF_C + MF_M = 0 \tag{5}$$

$$\text{Bounds} \quad 0 < F_L, \, 0 < F_C < F_e, \, 0 < F_M$$

Where M is the machining matrix, and $M = [M^n, M^t]$ and $M^n = [n_M^n, r_M \times n_M^n]'$, $M^t = [n_M^t, r_M \times n_M^t]'$. n_M^n, n_M^t and r_M are the unit normal vector, unit tangential vector and positional vector of machining force respectively; $F_M = [F_M^n, F_M^t]'$, F_M^n and F_M^t are the magnitudes of the normal machining force and tangential force respectively; F_e is the magnitude of the force that will cause outset of plastic deformation in the contact zone.

3. Implementation of Fixture Optimisation

Assume a discrete point set P_set consists of n_p points distributed evenly on the boundary surface of a workpiece. Three locators are needed to ensure $\|G\| \neq 0$ in a 2D space. We assume three locators are arranged in sequence L_1, L_2 and L_3 within the P_set array. While L_1 is moving along the boundary surface of the workpiece, an optimal arrangement of L_2^* and L_3^* is determined with respect to L_1 position from the maximum determinant of all possible fixturing matrixes G. Therefore a locator set LOC is generated in accordance with the position of locator L_1 in the P_set array.

$$LOC = [Loc^1, ..., Loc^i, ..., Loc^{n_p}] \tag{6}$$

Where $Loc^i = [L_1^i, L_2^i{}^*, L_3^i{}^*]'$, L_1 is put on i^{th} point of P_set, and $L_2^i{}^*$ and $L_3^i{}^*$ is the optimal positions of L_2^i and L_3^i to get $\| G_{max}^i \|^2$.

As mentioned before, the requirement of the clamping position is that clamping force should push the workpiece to contact all the locators. While satisfying Equation 1, the immobility condition should be:

$$\mathbf{F_L} = [f_{L1}^i, f_{L2}^i*, f_{L3}^i*] > 0 \text{ and } F_C = [f_C^j] > 0 \tag{7}$$

Where $f_{L1}^i, f_{L2}^i*, f_{L3}^i*$ are the magnitudes of reaction forces of locators L_1^i, L_2^i*, and L_3^i* in Loc^i respectively. f_C^j is the magnitude of clamping force acting on the jth point of **P_set**.

The loca of the machining surface is discretised into a set of n_m points. Matrix **Mach** is composed of the machining matrix of the points on the loca of the machining surface:

$$\mathbf{Mach} = [\mathbf{M}_1, ... \mathbf{M}_k ... \mathbf{M}_{n_m}] \tag{8}$$

Where M_k is the machining matrix of the kth point on the machining path. For each locator set Loc^i, the minimal clamping forces are calculated for a stability check.

To reduce the stress concentration and plastic deformation in the clamping contact region, multiple clamps may be required. The clamping optimisation takes into account the general situation when one or more clamps are required.

4. Example

The fixture layouts with three locators and one clamp (3L/1C), and three locators and two clamps (3L/2C) are investigated for the cross-section of a turbine blade aerofoil. As an example, the normal grinding force is 1500N, and the tangential grinding force is around one 10th of the normal grinding force. The clamping force is presented as a ratio to the normal machining force. In accordance with every locator set Loc^i, the determinant and the minimal clamping forces of both 3L/1C and 3L/2C are shown in Figure 1. The results show that large deduction of clamping force may be achieved in some locations if the 3L/2C layout is used.

By considering feasible clamping structures, three best fixture layouts with high determinants and low clamping forces were identified from Figure 1. They are Loc^4, Loc^{34} and Loc^{37}. Figure 2 shows these layout positions, and their clamping deformations under required minimum clamp forces. It can be seen that the deformation of Loc^4 is larger than the others though its clamping force is the lowest. This is because the clamp force at Loc^4 is not directly pointed to the locators. This situation indicates that optimal clamping should be a comprehensive consideration of both clamping forces and deformations. If the weights on the location accuracy, clamping force and clamping deformation were equal, then the best clamp layout would be Loc^{34}.

4. Conclusions

A method of global optimisation can be used for fixture layout design to satisfy the constraints of repeatability, immobility and stability. The method is feasible for the workpiece with complex holding surfaces. The clamping optimisation should aim at minimising locating error, clamping force and deformation whilst satisfying the stability requirement. Increase of clamps sometimes may reduce the magnitude of clamping force, which helps to reduce stress concentration in the contact zone.

Acknowledgement

This work is supported by the Rolls-Royce plc.

References

[1] Y. F. Wang, Y. S. Wong and J. Y. H. Fuh, 1999, Off-line modelling and planning of optimal clamping forces for an intelligent fixture system. *Int. J. of Machine Tools & Manuf.*, **39**, 253-271

[2] Y. Wu, Y. Rong, W. Ma and S.R. LeClair, 1998, Automated modular fixture planning: Accuracy, clamping and accessibility analysis, *Robotics and Computer-Integrated Manufacturing*, **14**, 17-26

[3] R. O. Mittal, P. H. Cohen and B. J. Gilmore, 1991, Dynamic modelling of fixture workpiece system, *Robotic and Computer-integrated Manufacturing*, **8(4)**, 201-217

[4] J.Y.H. Fuh and A.Y.C. Nee, 1994, Verification and optimisation of workholding schemes for fixture design, *Journal of Design and Manufacturing*, **4**, 307-318

[5] H. Asada and A. By, 1985, Kinematic analysis of workpart fixturing for flexible assembly with automatically reconfigurable fixtures, *IEEE Trans. On Robotics and Automation*, **1(2)**, 86-94

[6] R. A. Marin, P. M. Ferreria, 2001, Kinematic Analysis and Synthesis of Deterministic 3-2-1 Locator Schemes for Machining Fixtures, *Transaction of the ASME*, **123**, 708-719

[7] R. A. Marin and P. M. Ferreira, 2002, Optimal Placement of Fixture Clamps: Minimizing the Maximum Clamping Forces, *Transaction of the ASME*, **124**, 676-685

[8] R. A. Marin and P. M. Ferreira, 2002, Optimal Placement of Fixture Clamps: Maintaining Form Closure and Independent Regions of Form Closure, *Transaction of the ASME*. **124**, 686-730

[9] R. S. Ball, 1900, *A Treatise on the Theory of Screw*, Cambridge University Press.

[10] M. S. Ohwovoriole, 1981, An Extension of Screw Theory, *ASME J. Mech. Des.*, **103**, 725-735

[11] E. C. De Meter, et al, 2001, A model to predict minimum required clamp pre-loads in light of fixture-workpiece compliance, *Int. J. of Machine Tools & Manuf.*, **41**, 1031-1054

[12] J. H. Yeh and F. W. Liou, 1999, Contact condition modelling for machining fixture setup processes, *Int. J. of Machine Tools & Manuf.*, **39**, 787-803

[13] K. Krishmakumar and S. N. Melkote, 2000, Machining fixture layout optimisation using the genetic algorithm, *Int. J. of Machine Tools & Manuf.*, **40**, 589-598

[14] U. Roy and J. Liao, 2002, Fixturing Analysis for Stability Consideration in an Automated Fixture Design System, *Transation of ASME*, **124**, 98-104

[15] B. Li and S. N. Melkote, 1999, Improved workpiece location accuracy through fixture layout optimisation. *Int. J. of Machine Tools & Manuf.*, **39**, 871-883

[16] D. M. Pelinescu and M. Y. Wang, 2002, Multi-objective optimal fixture layout design, *Robotics and Computer Integrated Manufacturing*, **18**, 365-372

[17] Atkinson, A. Doney, 1992, *Optimum experimental designs*, Oxford University Press.

[18] K. Lakshminarayana, 1978, Mechanics of Form Closure, ASME paper 78-DET-32,2-8.

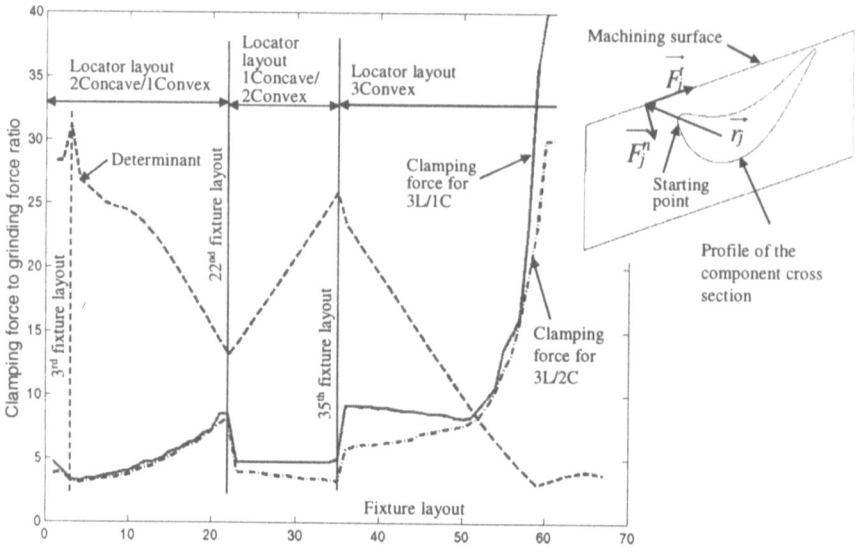

Fig 1 Determinant and minimal clamping force in accordance with locator set **LOC**

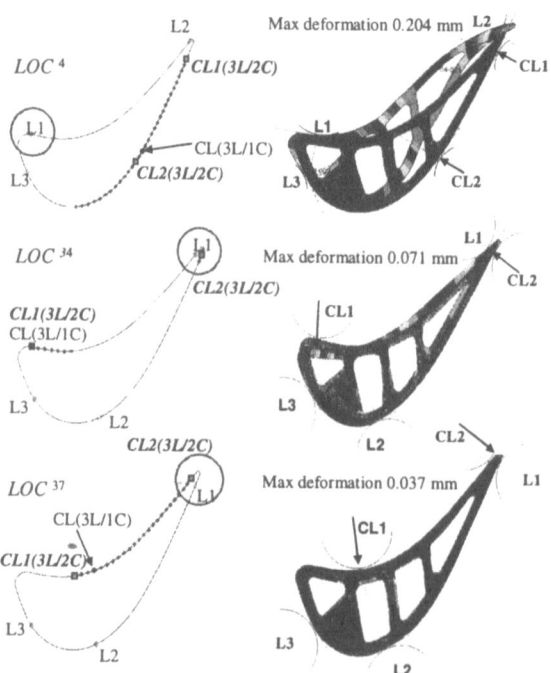

Fig 2 The three types of optimal fixture layouts and their clamping deformation

Web-based Distributed Process Planning for Turning Operations in Small and Medium Enterprises

Ricardo Avila[1], Guillermo Jimenez[2], Ciro Rodriguez[3] *, Luis Canche[3], Olben Falco[1], Miguel Ramirez[4], and Arturo Molina[3]

[1] Centro de Estudios CAD/CAM, Universidad de Holguín, CUBA,
rlar001@yahoo.com, olben@cadcam.uho.edu.cu
[2] Centro de Investigación en Informática, Instituto Tecnológico y de Estudios Superiores de Monterrey (ITESM), MÉXICO, guillermo.jimenez@itesm.mx
[3] Centro de Sistemas Integrados de Manufactura, ITESM, MÉXICO,
ciro.rodriguez@itesm.mx (* corresponding author), al785356@mail.mty.itesm.mx, armolina@itesm.mx,
[4] Departamento de Mecatrónica y Automatización, ITESM, MÉXICO,
miguel.ramirez@itesm.mx

Abstract: Small and medium enterprises (SMEs) in developing countries have limited economic resources and skilled personnel to utilize manufacturing technology, in an environment in which the product development cycle is expected to be short and the quality of the manufactured items has to meet high standards at reduced cost. In this environment, SME's need low cost manufacturing automation in order to compete. This paper describes a framework for low cost integration of distributed engineering applications on the Web, with emphasis on CAPP and CAM functions for turning operations. The proposed approach is based on Web Services technology for CAPP/CAM system integration.

1. Introduction

Manufacturing automation in small and medium enterprises (SMEs) can provide opportunities for higher competitiveness in terms of flexibility, quality consistency and lower cost. This competitiveness can be translated into a better integration of SMEs into the existing supply chains of highly demanding global industries (automobile, home appliance and consumer electronics production).

In the context of the North American Free Trade Agreement (NAFTA), there has been a significant integration of the USA and Mexican manufacturing industries. However, Mexico has a great potential for more SMEs integration into the internationally competitive supply chains. The SMEs barriers more frequently cited for this integration are a) investment capital and b) trained personnel [1]. For many SMEs in Mexico the prospect of introducing automatic machines and support systems (CAD/CAE/CAM/CAPP) into their operations is a significant challenge due to the high cost of the technology and the lack of know-how. Therefore, these SMEs are unable to compete and grow as second-tier and third-tier suppliers because they do not have competitive cost, appropriate delivery time and consistent quality. SMEs across Latin-America have similar restrictions to integrate into global manufacturing as those in Mexico.

One way to break this vicious cycle is to introduce low cost manufacturing automation (machines and CAX systems) to the SMEs in order to generate some of the necessary know-how, confidence and cash flow to invest in more sophisticated equipment. The Inst. Tecnológico y de Estudios Superiores de Monterrey in Mexico and the U. de Holguín in Cuba, together with other R&D institutions, have active programs to develop low cost manufacturing automation such as software-based controllers for machine tools and e-engineering (distributed engineering services for product development and manufacturing).

This paper describes the current efforts associated with bringing low cost CAPP/CAM to SMEs in Latin-America. The system has the following key elements: a) relational database structure for integrated product definition and multi-process planning, b) web-based distributed architecture for low cost access, c) technical terminology in Spanish to facilitate training and usage of this technology.

2. Related Work

Currently, there are a number of research efforts on web-based systems for distributed product development and manufacturing [2] [3]. Some of the specific developments related to distributed CAPP/CAM systems are reviewed next. The early efforts by Kao at the University of South Australia were based on sharing the control of a commercial CAD/CAM system through a UNIX network in order to coedit a CAD model and the CNC programs [4] [5]. A more recent work by Wang, at the Fuzhou University (China), suggested component-based CAD/CAM integration with CORBA over the Internet [6]. At the Institute of Metal Cutting (Poland), Adamczyk has developed a distributed CAD/CAM system over the Internet, integrating the different components with Microsoft Active Platform (ActiveX) [7] [8] [9]. Adamczyk's system is specifically tailored for milling operations in SMEs.

Recently, Wang from the Integrated Manufacturing Technologies Institute (Canada) has proposed a comprehensive process planning architecture based on distributed components over the Internet [10]. Wang has identified the key challenges in the implementation of a distributed process planning system: dynamism, responsiveness, openness, reconfigurability, local optimization of machines, distributed intelligence, seamless integration, high reliability and ease of use [10].

3. Proposed Approach

Previous efforts to develop distributed CAPP/CAM systems have been limited by proprietary or complex Internet integration technology [4] [5] [6] [7] [8] [9]. This paper describes a proposed approach for low-cost dissemination of CAPP/CAM systems in Latin-American job shops through web-based distributed systems. The implementation issues highlighted by Wang are addressed by the key elements of the proposed approach: a) integrated product-process data base (for integration of process planning and CAM functions), b) distributed support modules (for distributed intelligence) and b) Web Services technology for system integration over the Internet (for low-cost reconfiguration and openness due to Web Services simplicity and non-proprietary access).

3.1 General architecture of CAPP/CAM system

The CAPP/CAM system in this work is oriented to support the manufacturing of mechanical components for mining machinery and mineral processing plants, integrating several manufacturing processes (turning, milling, sheet metal cutting, sheet metal rolling, and welding). The general architecture of this system is shown in Figure 1.

The product definition is based on a master CAD model assembly. The system applies a variant process planning for the following processes: sheet metal cutting, sheet metal rolling, and welding. For milling and turning processes, the system applies a combination of generative and variant process planning. For turning operations, the process planning approach is based on multi-constraint cutting tool selection and machining parameters optimization. CNC tool path files are output for turning operations.

Currently the CAPP/CAM system described in Figure 1 has stand alone functionality among SMEs in Cuba. The distributed version of this system will be mainly oriented towards turning operations due to the proliferation of CNC lathes in the region.

3.2 Web Services technology

Web Services technology with Simple Object Access Protocol (SOAP), as the communication layer, has been selected for software component integration. This technology allows the integration of multi-language software components and is available without licensing limitations. The web-based distributed CAPP/CAM allows low cost access by SMEs, as well as agile maintenance and expansion.

Web Services are suggested as the next revolution in application integration [11] [12]. Previous efforts in application integration via the Internet (e.g., CORBA, COM+ and message middleware) are complex or proprietary thus are outside the economic possibilities of small and medium size enterprises (SMEs).

Web services are a light weight and standard based approach to application integration supported by multiple players. The standards are publicly available and many implementations exist. The foundation of Web Services is XML, and many standards use its syntax for messages (SOAP), interface description (WSDL), and publication for potential partners (UDDI) [11] [12].

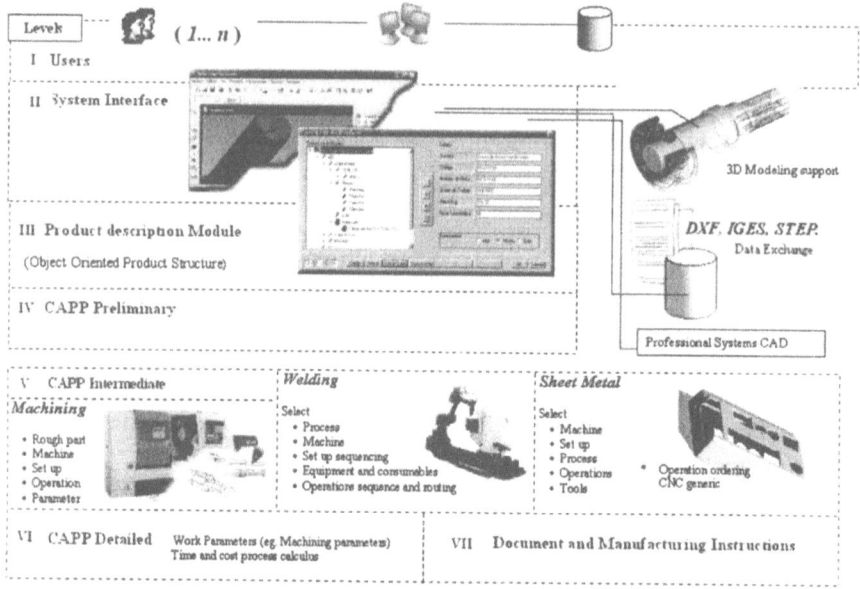

Figure 1. General architecture of the CAPP/CAM system

4. Current Progress

Figure 2 shows the current progress of the proposed CAPP/CAM system. The end user can access the system through the main web page at the Inst. Tecnológico y de Estudios Superiores de Monterrey (ITESM). The major components of the CAPP/CAM, located at U. Holguín, interact with additional components at ITESM. When the processing of the CAD file is completed, CNC programs are returned to the user.

Currently, initial testing of the Web Services technology has been completed, with successful communication between the Mexico and Cuba nodes. In stand alone mode, the CAPP/CAM system is already functional (providing process plan, cutting tool parameters and CNC programs for turning operations). Additional development is required to integrate the current CAPP/CAM system to additional modules and to finish the user interface that receives CAD files.

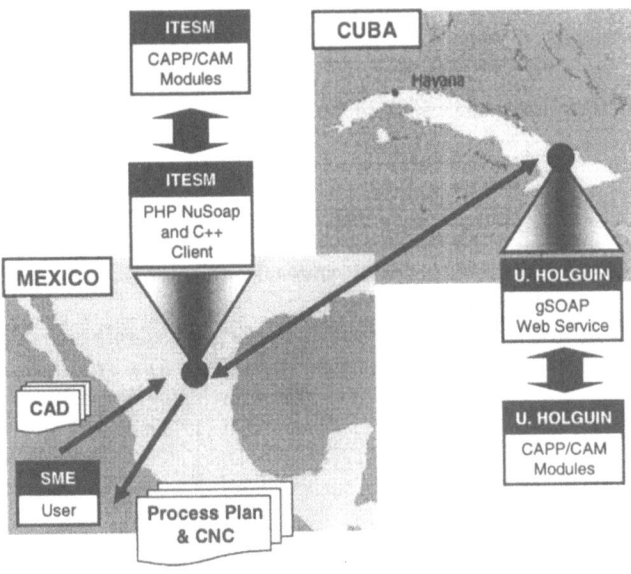

Figure 2. Current progress of the distributed CAPP/CAM system

5. Conclusions

The proposed approach for a distributed CAPP/CAM system based on Web Services technology has the potential of bringing additional low cost automation technology to SME´s in Latin-America. Overall architecture and stand alone testing has been completed. Internet integration and end user interface for the Web are in progress.

The initial testing of the distributed CAPP/CAM system will be in a teaching environment. Additional testing will be conducted with SMEs in Mexico and Latin-America through a network of SMEs that form Virtual Industrial Clusters [13] [14]. Additional modules will be added to expand the system functionality through a network of research institutes and universities in Spain, Portugal and Latin-America (PIBAMAR-Iberoamerican Project for Automation of High Performance Machining Process). Some of the additional functionality will include STEP data exchange standard and virtual machining services, similar to the CyberCut scheme at the U. of California at Berkeley [15] [16].

6. Acknowledgments

The authors would like to acknowledge the support of the Centro de Sistemas Integrados de Manufactura (ITESM), through the Research Chair in Mechatronics, and the Centro de Estudios CAD/CAM (U. Holguín).

This research is funded by a Mexico-Cuba bilateral project (Consejo Nacional de Ciencia y Tecnología, www.conacyt.mx #J200.646/2003), PIBAMAR (CYTED, www.cyted.org, #VII.22) and PYME CREATIVA Subproject: e-Engineering (Banco Interamericano de Desarrollo).

References

[1] Swamidass P.M.; Kotha S.; (1998) "Explaining manufacturing technology use, firm size and performance using a multidimensional view of technology", *Journal of Operations Management*, **17** (1), 23-37

[2] Peng, Qinglin; (2002) "A survey and implementation framework for industrial-oriented Web-based applications", *Integrated Manufacturing System*, **13** (5) 319-327

[3] Yang, H.; Xue, D.; (2003) "Recent research on developing Web-based manufacturing systems: a review", *Int. J. Production Research*, **41** (15), 3601-329

[4] Kao, Yung-Chou; Lin, Grier C. I.; (1998) "Development of a collaborative CAD/CAM system", *Robotics and Computer-Integrated Manufacturing*, **14** (1), 55-68

[5] Kao, Y. C.; Lin, G. C. I.; (1996), "CAD/CAM collaboration and remote machining", *Computer Integrated Manufacturing Systems*, **9** (3), 149-160

[6] Wang, X.K.; (2002) "Research on a dispersed networked CAD/CAM system of motorcycle", *Journal of Materials Processing Technology*, **129** (1), 658-662

[7] Adamczyk, Zbigniew; Kociolek, Krzysztof; (2001) "CAD/CAM technological environment creation as an interactive application on the Web", *J. Materials Processing Technology*, **109** (3), 222-228

[8] Adamczyk Z.; (2000) "A new approach to CAM systems development for small and medium enterprises", *J. Materials Processing Technology*, **107** (1), 173-180

[9] Adamczyk Z.; Malek H.; (1998), "Internet tools supporting creation and management of technological environment of CAD/CAM systems", *J. Materials Processing Technology*, **76** (1), 102-108

[10] Wang, Lihui; Feng, Hsi-Yung; Cai, Ningxu; (2003) "Architecture Design for Distributed Process Planning", *J. Manufacturing Systems*, **22** (2), 99-115

[11] Medjahed, B.; Benatallah, B.; Bouguettaya, A.; Ngu, A. H. H.; Elmagarmid, A. K.; (2003) "Business-to-business interactions: issues and enabling technologies", *International Journal on Very Large Data Bases*, **12**(1), 59-85

[12] Lacroix, E.; St-Denis, R.; (2003) "Web technologies in support of virtual manufacturing environments", *IEEE Conference on Emerging Technologies and Factory Automation*, Proceedings. ETFA '03, Vol. 2, 16-19 Sept., pp. 43-49

[13] Velandia, Marcela; Galeano, Nathalie; Caballero, Daniel; Alvarado, Verónica; Molina, Arturo; (2001) "Creation of Virtual Industrial Clusters" (in Spanish), *XXXI Congreso de Investigación y Extensión del Sistema ITESM*, Monterrey, MÉXICO, January 18-19

[14] Bremer, C.F.; Eversheim, W.; Walz, M.; Molina Gutiérrez, A.; (1999) "Global Virtual Business: A Systematic approach fo Exploiting Business Opportunities in Dynamic Markets", *International Journal of Agile Manufacturing*, **2**(1), 1-11

[15] Zhang, Ynping; Zhang, Chun; Wang, H.P; (2000) "An Internet based STEP data exchange framework for virtual enterprises", *Computers in Industry*, **41**(1), 51-63

[16] Mohole, A.; Wright, P.; Séquin, C.; (2002) "WebCAD: a computer aided design tool constrained with explicit 'design for manufacturability' rules for computer numerical control milling", *P. of the institution of Mechanical Engineers Part B – Journal Engineering Manufacture*, **216**(6), 879-889

Unified Feature Based Integration of Design and Process Planning

G Chen [1], Y-S. Ma [1], G Thimm [2] and S-H Tang [2]

[1]CAD/CAM Lab, School of MPE, Nanyang Technological University, Singapore 639798
mysma@ntu.edu.sg
[2]Design Research Center, Nanyang Technological University, Singapore 639798

Abstract: To realize concurrent engineering, different product development stages must be linked to each other on the basis of an integrated and consistent product information model. Features are used as information units for associating specific sets of geometric entities. In this paper, the semantics of design and machining features are discussed. Furthermore, feature association and unification for information sharing are proposed.

1. Introduction

To achieve concurrent engineering for a mechanical product, many applications are involved. In the product model, the sub-models for each application can be treated as a view. Each view has uniquely defined information entities, such as functions and behaviors in conceptual design, topological entities and form features in detailed design as well as setups and machining operations in process planning. This is why, in reality, different product models are used for different applications. However, the information entities used are essentially associated. Views are linked to each other via associated entities referenced. Any modification in a specific view must ensure the consistency of the whole information model. This aspect has not been fully investigated till now. Identifying information entities, relations and constraints in each view and further generalizing common entities are the first step for developing a consistent product information model. Features are used to associate relevant primitive geometric entities to represent a significant meaning in a specific application. Traditionally, only form features are used. There is a lack of explicitly defined relations between feature attributes and constraints, which together represent feature semantics. These gaps make tracking and managing design or process plan changes difficult. In this paper, semantics of design and machining features are discussed; feature association and unification are proposed.

2. Related Work

Wood *et al.* generalized eleven commonly used features as well as their related fifty three functions in the design of plastic parts [1]. Ranta *et al.* highlighted the information gap between conceptual and detailed design [2], and used an intermediate layer of structural entities/relations to map functional model to feature model. Roy *et al.* pointed out that for a given part function and geometry, the behaviors as well as functional relations between involved geometric entities are uniquely determined [3]. Britton *et al.* proposed a function-environment-behavior-structure model for designing injection mould assembly [4]. A pre-developed mould structure library and an action phenomenon library were used to generate mould structures. Choi *et al.* proposed a hierarchical logic for freeform cavity surface machining [5]. A set of concepts, including freeform features, machining features, unit machining operations and mapping rules between them were proposed. Ye *et al.* proposed an assembly object definition [6]. Assembly objects and primitive/combined constraints were used to determine the position/orientation of parts in mould assemblies. Thimm *et al.* proposed a graph-based method for automatic process plan generation and tolerance analysis for rotational parts [7]. Ma *et al.* proposed the associative feature concept to describe geometric association between feature entities [8]. It highlights that an ideal data structure of a feature definition must be flexible and self-contained.

3. Conceptual Design Features

At the conceptual design stage, combinations of geometric patterns as well as interactions among them realize product functions. Each geometric structure (an assembly or a part) interacts with others to generate a particular behavior for realizing one or more functions. A primitive conceptual design feature is defined as a set of interacting geometric entities, which are critical to realize the required primitive product functions. To support functional reasoning, conceptual design features must include function-related attributes and constraints. Conceptual design feature geometry only includes those geometric entities, which are indispensable for realizing the function. A conceptual design feature definition is given below:

Attributes:	semantics; functions; behaviors (input and output); dimensions; tolerances; and material specifications
Constraints:	spatial constraints; functional constraints
Geometries:	critical geometrical-entities
Methods:	create; edit; check validity; query information

Primitive conceptual design features can be pre-defined and combined to form new ones. A conceptual design feature model represents an assembly that is only partially and vaguely specified in this stage. It consists of a set of primitive or combined conceptual design feature instances. Inter-feature spatial and functional relations are defined among different interacting conceptual design features to realize overall product functions. At this stage, product functions govern the choices of geometric structures and interactions among them. The semantics of a conceptual design feature are represented by the associated attributes of those critical entities.

3. Detailed Design Features

At the detailed design stage, the conceptual design, i.e. critical geometric entities and interactions among them, are further refined into complete product geometries and specifications. A primitive detailed design feature is defined as a set of related geometric entities and has the following definition:

Attributes:	Semantics; patterns; parameters (e.g. diameter); dimensions; tolerances; positions; orientations; material; roughness
Constraints:	geometric constraints; algebraic constraints
Geometries:	parts; assembly; components; features; geometric and topological entities; references; derived entities
Methods:	create; edit; check validity; query information

Primitive detailed design features can be pre-defined; and several related ones can also be defined as a combined feature. A detailed design model is an assembly, which consists of a set of sub-assemblies or components. All sub-assemblies and components are contained (not necessarily in one-to-one manner) in parts. A part consists of a set of primitive or combined detailed design feature instances. Referenced and derived geometrical entities are included in such parts as well. Referencing mechanism enables relations among entities across different parts. Dimensions or tolerances are defined as constraints and attributes in their corresponding levels of parts. Inter-feature geometric or algebraic relations are defined as constraints too. Critical geometric entities of a conceptual design are referred by and associated to the final product model. The required interactions are embedded into appropriate entities' methods, which can manipulate geometric entities, constraints as well as attributes. The associative correspondences between the conceptual and the detailed design models must be managed via a feature manager even though a conceptual design could be realized by multiple detailed designs. The semantics of a detailed design feature are embedded as attributes in different levels. They can be retrieved and modified via supporting methods.

5. Machining Features

Feature-based process planning covers two processes, operation planning and machining passes. Hence, the features involved can be roughly divided into macro machining features (operations) and micro machining features (machining passes). These features have the corresponding methods of generation, aggregation, and sequencing. Machining operations can be defined according to setup or cutter changes. Within each operation, there are one or more machining passes. Machining parameters are determined in the scope of each pass. A primitive machining feature is defined as a set of related geometric entities that represents the volumes removed or faces generated during a machining cut. The primitive machining feature definition is given below:

Attributes:	semantics; machine information; tools; machining parameters; operational and locating datum; dimensions; tolerances and roughness of the machined faces
Constraints:	Machining constraints (power, workspace, etc.); tool constraints (cutter radius, flute length, etc.); geometric constraints
Geometries:	features; geometric and topological entities describing the workpiece before and after the operation or cutting pass
Methods:	create; edit; check validity; query information

The overall process plan of a single part consists of a set of machining features which are sequenced and scheduled according to the nature of precedence of operations and the availability of resources. It can be hierarchically divided into two levels. A part has a set of machining operation features at the first level, each operation feature can be achieved on the same machine with a single setup; it can be machined with one-time location and clamping of the workpiece. Furthermore, each machining operation feature is characterized by an unchanged cutting tool. At the second level, each operation feature can be further derived as a set of primitive machining features, i.e. machining pass features. Some of them can be further decomposed into nested cutting passes, where the machining faces are evolved with incremental changes in both geometry and status attributes. Constraints are defined in the corresponding levels. *Machining allowance allocation constraint*, i.e. depth of each cut, is defined for the micro machining features here to cater for such nested machining passes. *Machining sequence constraints* are defined in part, and operation feature levels, which are further classified as datum constraints, machining constraints (accessibility, rough prior to finish, etc.) and preference constraints (e.g. minimizing tool change). Similarly, the semantics of a machining feature are represented by embedded attributes. The validity of machining features must be checked whenever the design is amended.

6. Feature Association and Unification

From the above analysis, it can be appreciated that different application features have different entities, constraints and semantics. Theoretically, they are linked to each other via some associative relations among the entities used in different applications. The associative entities used by the conceptual design model, and related to the detailed design model, are the critical geometric entities. The required interactions among them represent the behavior of the conceptual features. As long as these critical entities/interactions are not modified, the corresponding product conceptual design (or in turn, the functions) can be regarded as still valid. This characteristic can be used to adjust final product geometry and parameters to lower manufacturing cost or shorten manufacturing time. The associative entities used by the detailed design feature model and related to the machining features are the final product geometry and attributes. These links must be established and maintained for concurrent engineering. However, in reality, communication among different application feature models is difficult due to the different data formats and the lack of association. Feature association can be achieved by a set of relations, including internal references and external references. Internal references relate entities in the same feature, while external references relate entities of different features. Furthermore, different applications link to each other based on associative relations. Hence, feature associations describe feature interactions and are crucial for managing feature validations and invoking necessary feature modifications.

On the other hand, based on the identified generic feature semantics, elements, and relations, a unified feature-modeling scheme has been proposed (to be reported in a separate publication) to integrate applications for information sharing and consistency control. Feature unification means using unified feature concept to represent generic characteristics and methods of different application features.

Application features are derived from unified features. It provides a generic format and common granularities for different applications. Feature association and unification are the two basic elements of the proposed unified feature-modeling scheme. With these two enrichments, features can be used as an interface information type to link different applications to, and hence control consistencies of, the centralized and unified product model. Then the synchronization of detailed geometric modeling and conceptual reasoning processes can be supported.

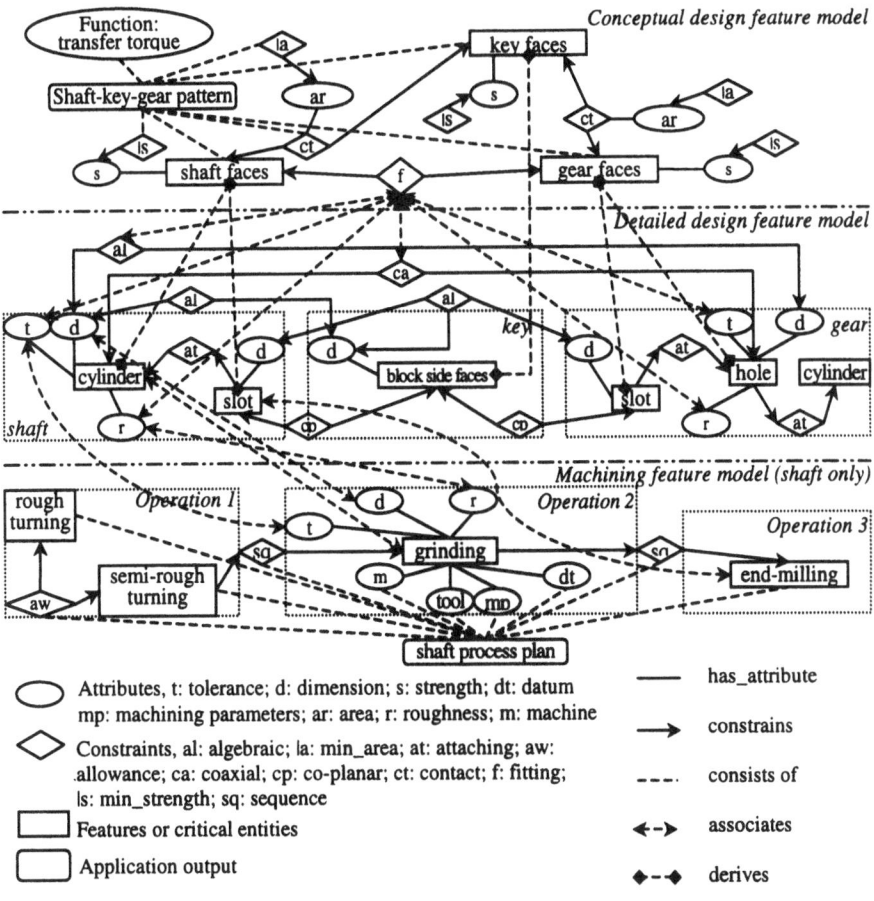

Figure 1 Feature semantic map

7. Example

The development process of a shaft-key-gear mechanism is used to illustrate the unified feature-modeling concept. As shown in Figure 1, in the conceptual design feature, shaft, key and gear faces, are the critical geometric entities. Spatial and functional constraints include fit type, minimum contact area and minimum material strength; they are established under the control of functional reasoning rules. In the detailed design stage, the only assembly includes three parts, shaft, key and gear; they contain six form features. Detailed design features, such as the slot on the shaft, are derived from the critical geometric entities at the conceptual level. Spatial and functional constraints, such as fitting, are transformed and associated with the geometric constraints and attributes in the detailed design. In the process planning stage, four machining features (rough turning, semi-finish turning, grinding and end-milling) are created for the shaft. Machining sequence and machining parameters are generated under the control of process planning application. For clarity, in Figure 1, some entities and relations are left out. This simple example roughly shows the complicated relations between different applications. Different application views are associated with the final product geometry. Such feature association makes the unified feature modeling approach feasible.

8. Conclusion

In this paper, feature-based integration for conceptual design, detailed design and process planning are discussed. Two basic concepts, feature association and unification, are discussed in the context of a unified feature modeling scheme for information sharing and consistency control. It is expected to provide a theoretical basis for the application research in feature-based concurrent engineering.

References

[1] Wood S L and Ullman D G., 1996, *The functions of plastic injection moulding features.* Design Studies, **17**(2), 201-213.

[2] Ranta M, Mantyla M, Umeda Y. et al., 1996. *Integration of functional and feature-based product modeling – the IMS/GNOSIS experience.* Computer-Aided Design, **28**(5), 371-381.

[3] Roy U and Bharadwaj B., 2002, *Design with part behaviors: behavior model, representation and applications.* Computer-Aided Design, **34**(9), 613-636.

[4] Britton G A, Tor S B, Lam Y C, et al., 2001, *Modeling functional design information for injection mould design.* International Journal of Production Research, **39**(12), 2501-2515.

[5] Choi B K, Ko K., 2003, *C-space based CAPP algorithm for freeform die-cavity machining.* Computer-Aided Design, **35**(2), 179-189.

[6] Ye X G, Fuh J Y H, Lee K S., 2000, *Automated assembly modeling for plastic injection moulds.* International Journal of Advanced Manufacturing Technology, **16**(10), 739-747.

[7] Thimm G, Britton G A and Fok S C., 2004, *A graph theoretic approach linking design dimensioning and process planning part 1: design to process planning.* International Journal of Advanced Manufacturing Technology, to appear.

[8] Ma Y S, Tong T., 2003, *Associative feature modeling for concurrent engineering integration.* Computers in Industry, **51**(1), 51-71.

A Graph Theoretic Operations Sequencing Approach for Progressive Die Design

C. Y. Chu[1], S. B. Tor[2] and G. A. Britton[1].

[1]CAD/CAM Lab, School of Mechanical and Production Engineering, Nanyang Technological University, Nanyang Avenue, Singapore 639798.
[2]Singapore-MIT Alliance (SMA) Fellow, SMA-NTU Office, N2-B2C-15, Nanyang Technological University, Nanyang Avenue, Singapore 639798.

Abstract Traditionally strip layout is a manual, experience-based activity. Automation of strip layout is desirable to improve productivity and the quality of design, and to provide computer-aided tools for design. One important, but very difficult, task in automated strip layout design is the determination of a good sequence of stamping operations so that the part can be stamped correctly and efficiently. This paper presents our work to develop a novel, graph-theoretic, operations sequencing method that is capable of generating a stamping sequence automatically, taking into account practical stamping constraints. A graph is used to represent a stamping part and define the relationships between its stamping features. These stamping features are then clustered into workstation sets using a graph colouring algorithm. Next, the sets are ordered to determine an optimal sequence of workstations. The objective function for the optimisation is 'minimisation of the torque difference between two sides of the progressive die'. The proposed approach can speed up the progressive die design process by automating the strip layout design.

1. Introduction

Advances in the field of feature modelling and Artificial Intelligence (AI) mean it is now possible to construct feature- and AI-based systems to solve progressive die design problems, including strip layout design automation. Commercial die design systems are available to help die designers. However, most of these are interactive in nature and usually limited to strip nesting, some simple calculations, and retrieval of catalogue data and compiled "libraries" of die components [1]. Operations sequencing is not automated. In order to reduce the time and cost required at the strip layout design stage, several expert systems have been developed [1,2]. Although these expert systems have made much progress in automatically regenerating design knowledge and constructing the necessary database, there are difficulties in adding new knowledge to existing design rules. For example, the designer's own rules might conflict with the general design rules.

This paper presents our work to develop a novel, graph-theoretic, operations sequencing method that is capable of generating a stamping sequence automatically, taking into account practical stamping constraints. A graph is used to represent a stamping part and define the relationships between its stamping features. These stamping features are then clustered into workstation sets using a graph colouring algorithm. Next, the sets are ordered to determine an optimal sequence of workstations. The objective function for optimisation is 'minimisation of the torque difference between two sides of the progressive die'.

2. Problem Statement

An operation sequence is an ordered set of die operations, which are performed on various workstations. The relationships between stamping features, die operations and workstations are shown in Fig. 1. The transformation of stamping features into die operations is straightforward. The key problems are the mapping of part features to stamping features, assigning die operations to various workstations, and determining the sequence of die workstations in a progressive die given the shape of a part, i.e., operation sequencing.

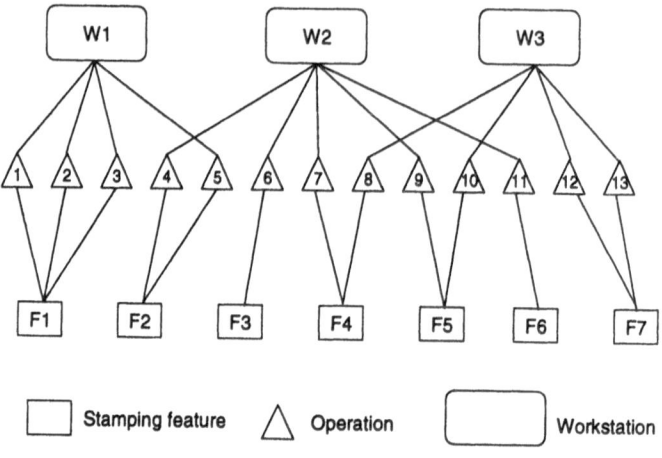

Fig. 1. Relationship between stamping feature, die operation and workstation.

3. Proposed Methodology

Figure 2 shows the proposed method for automated operations sequencing for stamping parts. First, part features are automatically mapped to stamping features. Manufacturability analysis is performed after the mapping. Then a graph model is developed to represent the relationships between the stamping features. After that, the developed graph model is partitioned into sets of mutually independent vertices using a clustering algorithm. Each cluster corresponds to one workstation in the progressive die. Finally, the clustered sets are ordered to give the final sequence of workstations. Each module is briefly described below. For more information on the graph theoretic model refer to [3].

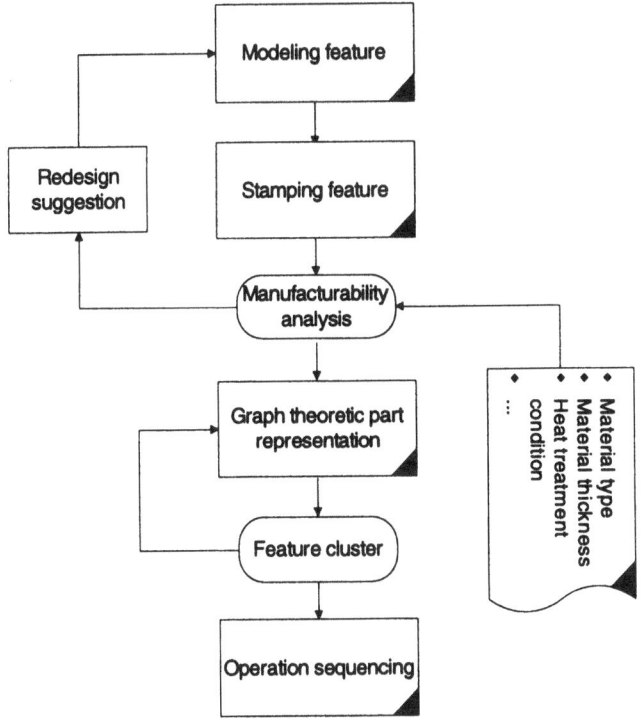

Fig. 2. Proposed method for automated operations sequencing.

3.1 Mapping part features to stamping features module

There is a direct relationship between a stamping feature and the operations required to form it (Figure 1). A stamping feature is produced by one or more operations. By definition a stamping feature contains both geometric information (shape) and manufacturing information (operations required to form it).

The relationship between part geometry and stamping features is more complex. It is not one-to-one. A stamping feature may be composed of one or several geometric features. In some instances, a geometric feature may be decomposed into several stamping features. Mapping rules define these relationships. The rules are based on die design best practices. This module is implemented by a rule-based expert system. The process of mapping a modelling feature to a stamping feature is actually a conversion of a designer's viewpoint to a stamper's viewpoint. Figure 3 shows an example.

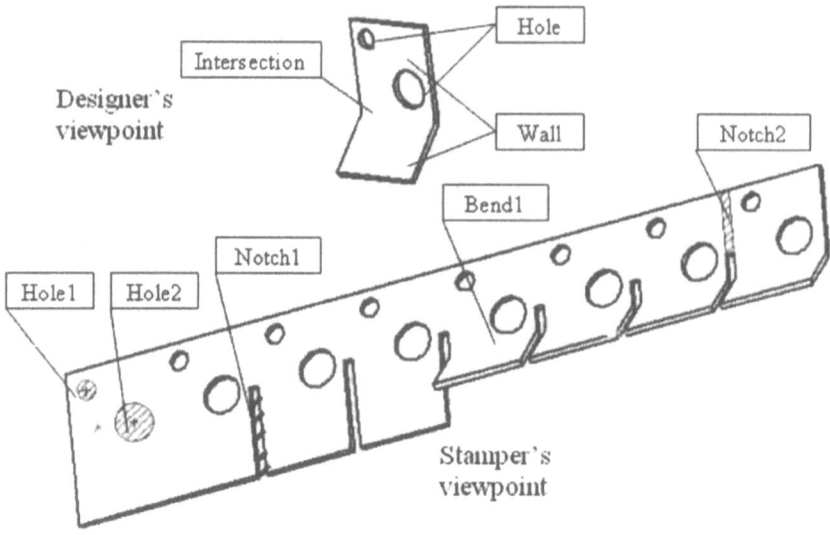

Fig. 3. An example of mapping part features to stamping features.

Stamping features can enable stamping process planning tasks to be performed directly from the geometric model. Stamping features are information carriers that are used to model a stamped part with a set of design and manufacturing information including geometric and non-geometric attributes. Each stamping feature can be manufactured by a specific stamping operation or a combination of stamping operations.

3.2   Manufacturability analysis module

Given a computerized representation of the design, the automated manufacturability analysis determines whether it is difficult or impossible to produce the designed part by stamping. For blanking and piercing, it compares the radius of the holes and the fillet with the criteria in the database. For bending, it compares the distance between the bending line and the internal features with the minimum distance suggested by the database. When a conflict between the actual part and the suggested data from the database is found, the system shows the geometric regions that are infeasible and prompts the designer to redesign the part. This module is implemented using hard coded manufacturability rules.

3.3   Graph-theoretic representation module

In the *graph theoretic representation module*, the part is defined by a graph G (V, E), where V is a set of vertices and E is a set of edges. Each feature of the part is represented as a vertex in the graph. The edges of the graph indicate that the connected features (vertices) are too close to be made at the same workstation. That is, if two stamping features, i and j cannot be stamped at the same station, then vertex i and vertex j are adjacent and an edge $E(i, j)$ links the two vertices. So the graph is an interference graph. Note that if the distance between two vertices is greater than one edge then they can be made at the same workstation. An illustrative example is shown is Figure 4.

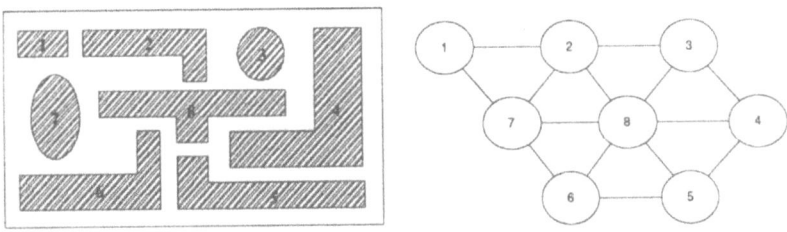

Fig. 4. An example part and its interference graph (Reproduced from [3])

3.4   Clustering of stamping features module

In the *clustering of stamping features module*, the graph is coloured using a graph colouring algorithm. Vertices of the same colour represent features that should be stamped in the same workstation to minimize the number of workstations, while satisfying the requirements of die strength and mounting of punches. For this problem the column generation approach [4] has been found to be suitable.

3.5   Optimum layout of workstations module

After clustering, the workstations are ordered to obtain an optimum layout. A properly designed progressive die should ensure equilibrium between the right- and left-side torques of the die otherwise a torque imbalance occurs, resulting in asymmetric clearances and eventual wear-out of the punches and dies. Hence the aim here is to minimise the torque difference between the left and right sides of the die. The workstation sequence that minimises the torque difference is the optimal sequence. The procedure for determining the optimal sequence is as follows. Different sequences of the clustered sets are formed and the torque difference for each sequence is computed. These values are compared and the sequence with the minimal value is selected.

4. Conclusion

This paper has presented a method for automating stamping operations sequencing. The method is based on graph theory. An interference graph of a part is constructed in which the vertices represent features. The edges link adjacent vertices (features) that cannot be stamped at the same workstation. The graph is partitioned into clustered sets using a graph colouring algorithm. Each set of features in a cluster can be stamped at the same workstation. The clustered sets are ordered using a torque minimisation algorithm. The end result is an optimal layout of workstations that meets all practical die constraints.

The proposed method is being implemented using C++. All modules, except for the mapping module, have been completed. Testing is currently being carried out to verify completeness and robustness of the algorithms. The expert system for the mapping module has been partially implemented in CLIPS.

References

[1] Duffey, M.R. and Sun, Q. *Knowledge-Based Design of Progressive Stamping Dies*. Journal of Materials Processing Technology, 1991, **28**, 221-227.

[2] Lin, Z. C. and Hsu, C. Y. *An investigation of an expert system for shearing cut progressive die design*. International Journal of Advanced Manufacturing Technology, 1996, **11**, 1-11.

[3] Chu, C. Y., Tor, S. B. and Britton, G. A. *A Graph Theoretic Approach for Stamping Operations Sequencing*. Proceedings of the Institution of Mechanical Engineers, B - Journal of Engineering Manufacture, Accepted, 2003.

[4] Mehrotra, A. and Trick, M.A. *A column generation approach for graph colouring*, INFORMS Journal on Computing, 1996, **8**, 344-354.

Data Segmentation Using Implicit Surfaces

Robert J. Cripps and Xiaobo Li

Geometric Modelling Group,
Mechanical and Manufacturing Engineering,
The University of Birmingham, Edgbaston,
Birmingham, B15 2TT, UK.   email: r.cripps@bham.ac.uk

Abstract: An automatic method of data segmentation is presented that uses local surface geometry to characterise the data into coherent sets that define the data segments. In addition, the connectivity of the isolated segments is determined, which is useful information in the follow on activity of surface reconstruction.   The basic ideas behind the characterisation of geometric features using local implicit surface approximation are presented. The segmentation algorithm, which is based on the estimated geometric features, is then outlined. The proposed method is illustrated by segmenting two sets of data. The first is data from a known surface and is used to illustrate the accuracy of the proposed method. The second data set is the well-known benchmark by Bajaj of the blended table corner and illustrates the ability of the method to correctly identify points from a range of surface types, including planes. Some remarks regarding the proposed approach and suggest some areas for future improvements are given in conclusion.

1. Introduction

Since the early 1970's Reverse Engineering (RE) has developed into an important design and manufacturing tool, which enables existing objects with complex structures to be reproduced [1]. The typical sequence of activities in RE include [2] data capture or digitisation of a 3D object (Fig. 1(a)), data filtering to remove gross noise and errors, data segmentation (Fig. 1(b)), surface reconstruction (Fig. 1(c)) and finally some form of surface beautification (Fig. 1(d)). The focus of this article is on data segmentation and in particular, the specification of an automatic segmentation scheme that requires little manual intervention. Since surfaces are expected to be fitted directly to the points digitised from existing objects the recognition of the surface geometry from the scattered points has become more significant [1]. Readers interested in other aspects of the RE processes are referred to the following articles and the references contained within. For digitisation, see [3], for noise reduction see [3], for surface reconstruction see [4] and for surface improvement see [5].

2. Data Segmentation

Complex shaped objects generally require multi-patched surfaces to adequately describe them. When such an object is digitised, it is imperative to detect the potential boundaries of these surfaces from the digitised points. Thus the typical starting point for data segmentation is a grid of 3D digitised points, possibly unordered, which may or may not have been smoothed but which have had gross errors and outliers removed. The less noise in the data, the more reliable the segmentation. The digitised points should be divided into different regions on the basis of the locations of the boundaries or collected into different subsets depending

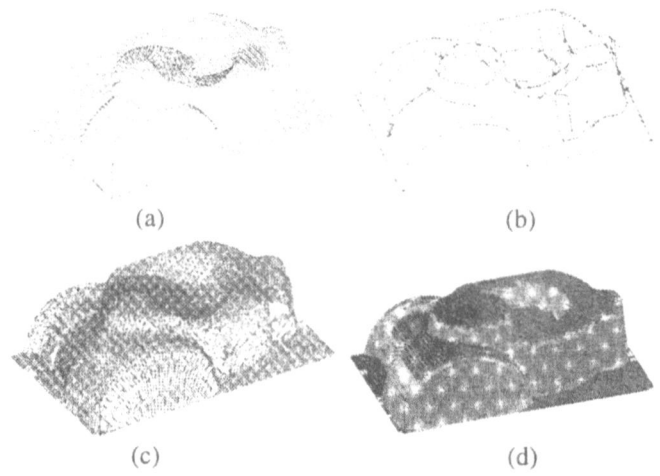

(a)   (b)

(c)   (d)

Figure 1. The reverse engineering process [2].

on geometric features, such as normal, curvatures, and principal directions ([6],[7]). One approach for the recognition of surface geometric features is to locally fit a surface to the scattered points. This local surface is expected to be able to describe a wide range of shapes, but to ensure the surface is local, the procedure should use minimal points in a close neighbourhood. Thus it can be seen that the accuracy of the recognition of surface geometry directly affects the quality of segmentation.

Currently, the most widely used method of characterising the geometry of points is based on the Darboux-frame ([8],[9]). This method estimates the curvatures at a 3D point $\mathbf{P}$ from the scattered point set, by locally fitting a least squares (LS) quadratic surface of the form:

$$z(x,y) = ax^2 + bxy + cy^2$$

where a, b, c are real numbers, to a local neighbourhood, $\wp_k$, which contains $\mathbf{P}$ and at least three points that are closest to $\mathbf{P}$, in terms of their Euclidean distance. The algorithm is easy to implement but has limited geometric description, for example, it can not exactly represent a sphere.

A more accurate method was proposed by Yang and Lee [10] who estimate the principal curvatures by locally fitting a LS parametric quadric of the form:

$$r(u,v) = \sum_{i=0}^{2} u^i v^j \mathbf{Q}_{i,j}$$

to a neighbourhood of sixteen points using chordal parameterisation. This method can describe a wider range of surface shapes than the Darboux-frame. However, the relative higher algebraic degree of the parametric surface makes the estimation remarkably complex and difficult to compute [11]. A rectangular parametric polynomial patch of degree n has algebraic degree $2n^2$, so that this method has algebraic degree 8 [11] as compared to 2 for the Darboux-frame method. In addition, the choice of parameterisation is critical for both surface interpolation and approximation [11]. This method has not been widely used in RE possibly because of these disadvantages. Our motivation is to improve the surface geometry estimation. Hence we need to compromise between the number of data points and the range of shapes during the local fitting, and to compromise between algebraic degree and efficiency. Additionally, we wish to avoid the difficulties of choosing a parameterisation.

3. Implicit Quadratic Surface

Our approach is to use the general implicit quadratic surface, S, represented by:

$$F(x,y,z) = a_1 x^2 + a_2 y^2 + a_3 z^2 + a_4 xy + a_5 yz + a_6 xz +$$
$$a_7 x + a_8 y + a_9 z + a_{10} = 0$$

to approximate the local surface of $\wp_k$. This form is invariant with datum changes and includes the ellipsoid, sphere, cylinder, paraboloid, hyperboloid with a single sheet, hyperboloid with two sheets, hyperbolic paraboloid, cone, two intersecting planes, two parallel planes and two coincident planes. Some of these forms need to be excluded to ensure valid geometric feature estimation. When a valid surface has been found, the coefficients are used to compute the surface normal, curvatures and principal directions, which are used to segment the data.

There are only two assumptions we impose on the data:

- The original surface is connected.
- The point sample is dense enough with respect to the variation in the surface that a sufficiently accurate model can be generated [12].

A general quadratic implicit surface S that approximates $\wp_k$, is fitted to each neighbourhood, $\wp_k$. Before we can use the fitted surface for estimating the local surface geometry we need to check that the surface is admissible. The range of surfaces defined by S includes disconnected surfaces, for example surfaces with multiple sheets. These cannot be used as they would violate our first assumption that the data is connected. We are able to identify all inadmissible cases and hence avoid them. Problems also arise when we have a singular point, at which no geometric features can be estimated, however these points give useful information for boundary detection and are therefore stored for later use.

4. Segmentation

Having determined the characteristics of all data points in the digitised set, we now use the estimated geometric information to assign points to sets. The n unique 3D points, P_0 through P_{n-1}, of digitised data set are stored in an array so points can be referred to by their indices. There are four major steps to the segmentation process.

4.1 Step 1: label points

At each point, P_i, $i \subset [0, n-1]$), estimate the normal vector, n_i and curvatures (Gaussian and mean curvatures), K_i and H_i. By considering the signs of the Gaussian and mean curvatures, each point can be labelled as one of the eight regular surface types, $L_i=[1,8]$, $L_i=9$ iff P_i is singular. We can assign colours to points according to their surface types to give a visual interpretation.

4.2 Step 2: coherent sets

First we collect the neighbouring points that have the same surface type label into the same subset. P_i is assigned to subset R_i, depending on its neighbourhood, $\wp_k$, Gaussian, K_i and mean curvature, H_i. Singular points are not collected into any subset but are used in the next step to determine the nature of boundary points.

4.3 Step 3: boundary identification

Since each subset represents a collection of points belonging to a known surface type, a quadratic surface can be fitted to each subset R_i. The fitting is greatly simplified if we can detect the boundary points in a subset. To detect the boundary points, we first locally project each point in a subset, P_i and its nearest neighbourhoods, $\wp_k$ into the tangent plane of P_i. We next triangulate the projected points using 2D Delaunay triangulation [13]. Then, P_i is a boundary point if and only if a generated angle around it is not less than π. Each point within a subset is labelled as either a boundary point or not. If a singular point is between two coherent subsets, it is considered to be the boundary point of both subsets. Otherwise, it is merged into the nearest subset.

4.4 Step 4: continuity between subsets

It is useful to know how two surfaces should be joined. By considering the estimated normals on adjacent boundaries we can determine the continuity of the subsets implied by the data. We differentiate between edges with normal vector discontinuities and smooth edges. A point, P_i, is on a fold edge if it is a singular point on a boundary of the subset, otherwise it is a smooth edge point.

5. Case Studies

5.1 Simulated data

We apply the generalised segmentation algorithm to a simulated data set with 2500 points shown in Figure 5. This data set is generated from a plane and a part of a cylinder with normal vector continuity. Position errors are added to this simulated data set according to a Gaussian distribution, $x_i \sim N\left\{0,(\delta/150)^2\right\}$, where δ is the smallest step size in each data set.

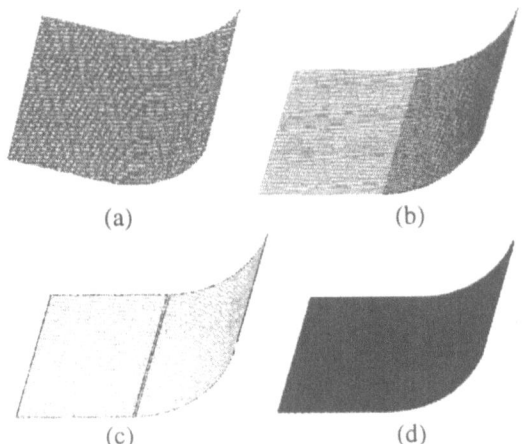

Figure 5. Simulated data. (a) original data points, (b) surface labelling, (c) sub-set
boundaries and (d) final data segmentation.

5.2 Benchmark data

In this Section, we apply the generalised segmentation algorithm to a digitised data
set from Bajaj [14]. The physical model consists of a solid cube where a corner is
blended as a quintic surface, one of the nearest edges to the corner is smoothed out
by a part of a cylinder and the other two nearest edges are smoothed out by parts of
two cones, as shown in Figure 6. This surface model is used as a benchmark in
computer graphics. Figure 6(a) shows the digitised data set from the surfaces of the
model.

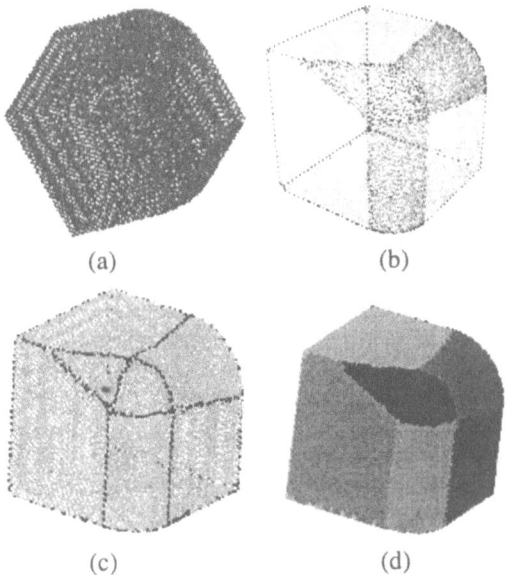

Figure 6. Bajaj data. (a) original data points, (b) surface labelling, (c) sub-set boundaries and
(d) final data segmentation.

6. Conclusions

We have presented a method of data segmentation that requires little manual intervention and has been demonstrated to segment data sets and to give boundary information and segmented points, thus giving the advantages of both existing edge based and surface based methods. The key area of user input is to specify the size of the neighbourhood for local surface fitting. This issue has not been resolved although experimental evidence seems to suggest that a neighbourhood size of 15 gives acceptable results. Another issue is in determining the neighbourhood for each data point within a digitised cloud. This is the most time consuming part of the process and one that we would like to address.

7. Acknowledgements

We are please to be able to acknowledge the support of EPSRC (Grant reference GR/009765) and Delcam, UK.

References

[1] Hoschek, J. and Lasser, D., 1993, *Fundamentals of Computer Aided Geometric Design*, A K Peters, Ltd.

[2] Häfele, K.-H., 1996, POMOS-Point based Modelling system. In Reverse Engineering, Eds. Hoschek, J. and Dankwort, W. B.G. Teubner, Stutgart.

[3] Varady, T., Martin, R. R. and Cox, J., 1997, *Reverse engineering of geometric models - an introduction*. Computer-Aided Design, **29**, 255-268.

[4] Puttre, M., 1994, *Capturing design data with digitised systems*, Mechanical Engineering, **4** (116), 62-65.

[5] Chen, Y. and Wang, Y., 1999, *Genetic algorithms for optimised re-triangulation in the context of Revesre Engineering*. Computer-Aided Design, **31**, 261-271.

[6] Berger, M. and Gostiaux, B., 1988, *Differential Geometry: Manifolds, Curves, and Surfaces*. Springer-Verlag, New York Inc.

[7] Do Carmo, M., 1976, *Differential Geometry of Curves and Surfaces*, Prentice-Hall, Inc, Englewood Cliffs, New Jersey.

[8] Ferrie, F. P., Lagarde, J. and Whait, P., 1993, *Darboux frames, snakes, and super-quadrics: geometry from the bottom up*, IEEE transactions on pattern analysis and machine intelligence, **15** (8), 771-783.

[9] Milroy, M., J., Bradley, C. and Vickers, G. W., 1997, *Segmentation of A Wrap-around Model Using An Active Contour*. Computer-Aided Design, **29**, 299-320.

[10] Yang, M. and Lee, E., 1999, *Segmentation of measured point data using a parametric quadric surface approximation*. Computer-Aided Design, **10**, 449-457.

[11] Pratt, M., 1992, The Virtues of Cyclinds in CAGD. In: *Mathematical Methods in Computer Aided Geometric Design II*, Lyche, T. and Schumaker, L.L., (eds.), Academic Press, Inc.

[12] Floater, M. S., 1996, Mathematical techniques for reverse engineering, SINTEF Report No. STF42 A96019, Oslo, Norway.

[13] Farin, G., 2002, *Curves and Surfaces for CAGD a practical guide*, 5th Edition, Morgan Kaufmann.

[14] Bajaj, C., Ihm, I. and Warren, J., 1993, *Higher-order interpolation and least squares approximation using implicit algebraic surfaces*. ACM Transactions on Graphics. **12** (4), 327-34.

Virtual Knowledge Repository for Intelligent and Distributed Feature-driven Product Realization

G. X. Wang[1], W. Z. Zhang[2] and A. Y. C. Nee[3]

[1]Department of Mechanical Engineering, National University of Singapore, 10 Kent Ridge Crescent, Singapore 119260, engp1724@nus.edu.sg
[2]Institute of High Performance Computing, 1 Science Park Road, #01-01 The Capricorn, Singapore Science Park II, Singapore 117528, zhangwz@ihpc.a-star.edu.sg
[3]Department of Mechanical Engineering, National University of Singapore, 10 Kent Ridge Crescent, Singapore 119260, mpeneeyc@nus.edu.sg

Abstract: The contemporary product design and manufacturing process is knowledge-intensive and collaborative in nature. With the growing popularity of Internet-based technologies, the wide ranges of conventional initiatives involved in improving this process are now seemingly converging on a great concern to create a distributed concurrent and collaborative engineering environment that is able to integrate all the phases of engineering activities together. A central component of this environment is a shared knowledge repository, containing every piece of information relevant to all the sub-processes at different stages assisted with different computer software tools/services which need different knowledge inputs and generate different intermediate models or final engineering outputs. This paper addresses the design and implementation of a virtual knowledge repository supporting a specific collaborative engineering environment, a so-called intelligent and distributed feature-driven product realization environment.

1. Introduction

Design and manufacturing of a product is a complex process which is always subdivided into a set of sub-processes capable of being automated with great speed and precision through the use of task-specific computer-based tools/services. Over the past decades, continuous efforts have been contributed to leveraging diverse technologies, especially the artificial intelligence computations to build up more effective and efficient engineering tools. Simultaneously, many other technical and strategic initiatives, such as CIM (Computer-Integrated Engineering), CE (Concurrent Engineering), PDM (Product Data Management), PLM (Product Lifecycle Management), etc., have also been carried out by researchers and solely or jointly applied to a specific design and manufacturing process. The optimal resolution always tends towards the realization of a functionally and structurally distinct computer-based engineering environment. In this integrated environment, fairly autonomous component systems are organized around modules which share a common database through PDM or PLM and with CE philosophy woven into the internal logic operations with reduced human-machine interfaces.

Recently, this environment has been significantly influenced by the Internet-enabled infrastructure, on which systematic integration and CE-based collaboration are becoming more convenient since sharing and exchanging information among distributed workgroups can be achieved in a simpler and less costly way. One of most useful new concepts is the logically centralized design repository, which is analogous to the product data base in the old-fashioned PDM. For example, the NIST (the National Institute of Standards and Technology) design repository project demonstrated a pilot implementation of a shared source of heterogeneous knowledge and data with Web-based interfaces for creating, browsing and intelligently searching the design repository [7]. Generally put, these reported design repositories are domain-independent and loosely coupled with the real product design and manufacturing environments. In this paper, our concern is to adaptively use this design repository concept to develop a knowledge repository as a built-in central component in a concrete application environment and address relevant implementation issues.

The result of our work is a virtual knowledge repository supporting a so-called intelligent and distributed feature-driven product realization environment. Virtualization means the location-independent view and use of knowledge objects such that one component system can use the intelligence belonging to others within a common framework. Feature-driven product realization is an abstract view of product design and manufacturing processes which can be applied to many practical engineering domains. We first depict this process pattern and its general architecture. The repository design option and some implementation decisions are then made to meet a set of performance requirements.

2. Intelligent and Distributed Feature-driven Product Realization

Feature-driven product realization is a formulized process chain pattern to show how individual component processes within a complete product design and manufacturing process are connected together by feature-based model data flows. It is currently still undergoing development by the authors. The underlying principle is as described in the following. Many product development processes can be viewed as a step-wise process chain, along which each sub-process corresponds to an application tool to fulfil specific tasks using specialized technology knowledge and *modus operandi* for problem solving [1]. One of the most important types of knowledge is how current tasks are dependent on those carried out by its previous process or reflected in data flows, in what extent and in what way the current process data model is dependent on that of the previous process. Feature-based model is the best option to be adopted as the process engineering model. This is because it can promote maximum extent of automation when generating the models using an approach called feature mapping (also called feature conversion or feature transformation): generating the new set of feature instances B from the given one A through knowledge-based reasoning supported by feature mapping knowledge base [1].

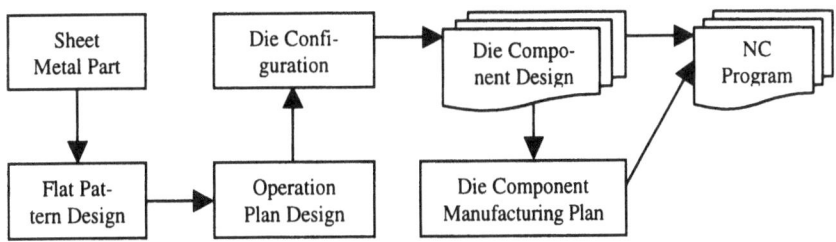

Figure 1: An example feature-driven product realization process

Figure 1 shows an example of the feature-driven product realization process relevant to the development lifecycle of sheet metal products using progressive dies [3-5]. It is important to note that the engineering model in each downstream process cannot be completely and automatically deduced from upstream process model(s). Generating some instance features manually and interactively is still necessary in most cases.

3. Virtual Knowledge Repository as the Centre in the Feature-driven Product Realization Framework

Figure 2 shows the conceptual architecture of the virtual knowledge repository-supported feature-driven product realization framework. It adopts the distributed client-server structure and two types of applications are located on the client side to interface with the users. They are domain neutral tool and a set of engineering tools, each of which corresponds to a process generating a feature-based model. All tools are supported by the process flow management and the knowledge-based services. *A priori* knowledge about the feature-driven product realization process such as that shown in Figure 1 is specified in the process flow management service and the process execution progress is visible to the participating experts who may initiate and inspect its steps. As mentioned in section 2, each engineering tool in a certain process generates its feature model through two combined mechanisms, automatic knowledge-based feature mapping and interactively instantiating a partial set of features. In order to promote task automation when instantiating, knowledge-based approaches may still be used to execute templates or rule-based reasoning, or reuse standard components. The process tools may be non-interactive but require complicated knowledge-based computation, or highly interactive and still need a knowledge-based system support. To gain the best user interactive performance, highly interactive tools are executed on the client side while poorly or non-interactive tools can be executed remotely on the server side. Consequently, the inference engine needed by a tool, as well as the knowledge-base it depends on, may be on the client side or the server side, forming a virtual knowledge repository (the dashed-line enclosed area).

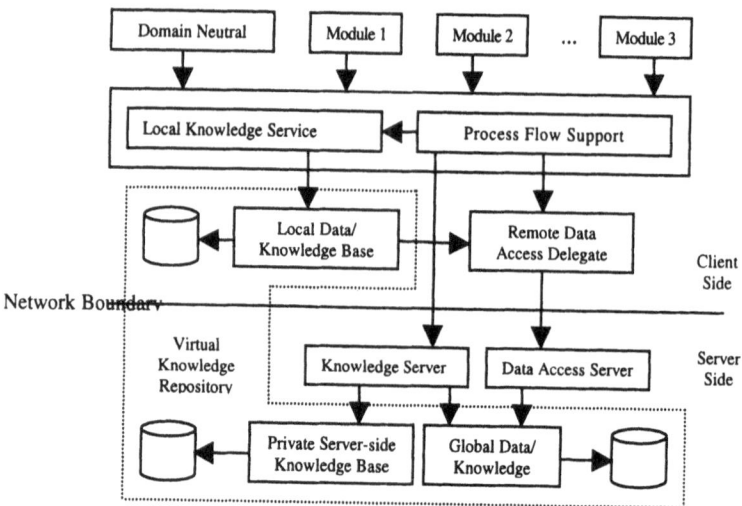

Figure 2: Knowledge repository supported development

4. Distributed Coarse-grain Data Management Service for Virtual Knowledge Repository

To design the knowledge repository, the first thing is to investigate its contents and the required operation behaviour. From section 3, there are collectively two categories of contents encompassed in the knowledge repository, dynamically generated project-dependent feature models and clusters of generic project-independent design and manufacturing knowledge objects. Both of them have a primary copy on the server side for centralized management to maintain coherence and part of them may be migrated to or temporarily cached in the local disk on the client side during run time. Since design of the knowledge repository to accommodate the generic project-independent knowledge is straightforward, the focus is then placed on how knowledge repository design decisions are made to accommodate the dynamically generated project-dependent feature models, which bridge interconnected engineering processes. A key functional component in this information infrastructure is the distributed coarse-grain data management service.

Although the knowledge repository enabled design automation functions operate on the fine-grain, or feature-by-feature level, coarse-grain (one instance feature model including a group of instance features as a unit) data management approach for knowledge repository is still desirable because of its simplicity and flexibility to apply and the necessity to support the enforcement of model dependence (feature mapping) semantics can bypass the feature-driven process management. Coarse-grain data management, or management of groups of files defined by feature model semantics, is similar to conventional approaches adopted by CIM database management. Some enhancements have been taken up in our design based on analysis of the expected repository operation behaviour and subsequently the three general requirements: transparency, efficiency and consistency. As for transparency, the operations on remote files should optimally appear the same on local virtual machines using appropriate mechanisms to hide remote operations. As for efficiency, it is desirable that remote file access delays should be in the order as delays encountered in local access. As for consistency, all file copies in the networked environment to the identical source file should have the same content as expected.

To solve the transparency problem, the remote physical storage device is mapping to a logical name using a protocol, thus file access is either mapped upon local disk access or upon remote file server operations and the user applications are not aware of the distribution. We use Microsoft CIFS (Common Internet File System) [2] as the protocol. The consistency problem is resolved using the CIFS program-resetable opportunistic locks which allow the coarse-grain data management service to safely cache the file data on the client side.

Efficiency problem is especially important because product feature-based models are always very large and Internet bandwidth is limited. To solve this problem, we use a smart design similar to reference [6] that the virtual repository consists of a centralized public area as well as a specific part of the local file system in each machine declared as owned by the repository. The public area stores a complete set of files and the private area is used to store temporal file copies. The logic in the repository management service can control the file access behaviour so that remote access can be minimized while data coherence is held.

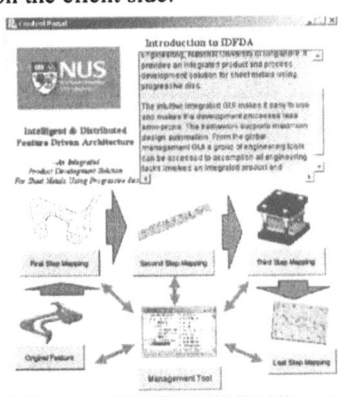

Figure 3: The GUI snapshot of the global process and data flow control tool

5. Preliminary Implementation Experience

The above virtual knowledge repository concept is implemented in a prototype platform, IDFDA (Intelligent and Distributed Feature-Driven Architecture) which provides an integrated product development solution for sheet metals using progressive dies. Figure 3 shows the GUI (Graphical User Interface) snapshot of the global process and data flow control tool through which users initiate and inspect product design and development processes. The knowledge repository management service is activated by this tool and then all the engineering task-specific applications are able to use this service to share or exchange knowledge objects. We have chiefly tested the effectiveness of the coarse-grain data management service. Our preliminary experiment shows that the desired knowledge repository is able to be built up by using the proposed approaches.

6. Conclusions

Knowledge repository is an evolution of traditional centralized design database built in CIM, CE or PDM implementations and flavours the distributed and collaborative engineering process imposed on it with visualized intelligence ability. In this paper, we investigated in detail an intelligent and distributed feature-driven product realization environment, and thus the likely roles allotted to the knowledge repository are analysed. While allowing various types of sharing semantics for a sequence of tightly-associated intelligent applications, the knowledge repository as well as its management service should also satisfy requirements of transparency, efficiency and consistency, etc. This is not an easy task and the knowledge repository design rationale perceived in our research might be useful for many other similar projects.

References

[1]   Zimmermann, J.U., Haasis, S., Van Houten, F.J.A.M., 2002, *ULEO - Universal linking of engineering objects,* CIRP Annals - Manufacturing Technology, Volume 51, Issue 1, 2002, Pages 99-102

[2]   CodeFX publishes, 2001, *CIFS explained,* available at http://www.codefx.com/

[3]   Li, C.Y., Li, J.J., Wen, J.Y., et al, 2001, *HPRODIE: Using feature modelling and feature mapping to speed up progressive die design,* International Journal of Production Research, Volume 39, Issue 18, 15 2001 December, Pages 4133-4151

[4]   Cheok, B.T., Nee, A.Y.C., 1998, *Configuration of progressive dies,* (AI EDAM) Artificial Intelligence for Engineering Design, Analysis and Manufacturing, Volume 12, Issue 5, 1998, Pages 405-418

[5]   Lee, I.B.H., Lim, B.S., Nee, A.Y.C., 1993, *Knowledge-based process planning system for the manufacture of progressive dies,* International Journal of Production Research, Volume 31, Issue 2, February 1993, Pages 251-278

[6]   Oliver, H., Andre, E., 1994, *Design, implementation and evaluation of a distributed file service for collaborative engineering environments,* Proceedings of the 3rd Workshop on Enabling Technologies: Infrastructure for Collaborative Enterprises, 1994, Pages 170-175

[7]   Szykman, S, Sriram, R.D., Regli, C.W., 2001, *The role of knowledge in next-generation product development systems,* The Journal of Computing and Information Science in Engineering, 01 Mar 2001, vol. 1, no. 1, pp. 3-11

CUTTING

High Speed Dry Machining of Aerospace Aluminium Alloys

D.J. Richardson, F. Dailami & J.D. Lanham

RAMP Laboratory, Faculty of CEMS, University of the West of England, Frenchay Campus, Coldharbour Lane, Bristol, BS16 1QY.
David4.Richardson@uwe.ac.uk, Farid.Dailami@uwe.ac.uk

Abstract: This paper explains the industrial drivers towards dry machining and reviews the technical challenges for high speed dry machining of high strength light alloys. Test results are presented that show that less heat is partitioned into the workpiece for high speed dry machining. The understanding and modelling of machining induced temperatures is highlighted as the key issue due to its impact on component performance and its effect on other dry machining parameters.

1. Introduction

Currently commercial machining of aerospace aluminium alloys is accomplished wet, with the introduction of a cutting fluid to remove generated heat and swarf from the working environment. Manufacturing industry would like to embrace dry machining, for both environmental and economic reasons, but the effect of the omission of the cutting fluid is not well researched and understood. Machining is particularly important for the European aerospace industry where a large proportion of material is machined away from a solid billet to produce monolithic structures. This paper describes the technical challenges for dry machining, explains why temperature is the major issue and provides test results that show that less heat is conducted into the workpiece for high speed dry machining.

2. Dry Machining

Cuttings fluids have traditionally been used in machining to reduce friction between the cutting tool and the work, to conduct heat away from the cutting zone and to aid in the removal of swarf [1]. Cutting fluids are considered to be detrimental to the environment and represent an increasingly significant proportion of cutting costs [1, 2]. A survey by Darmstadt University of Technology indicates that 8% to17% of all production costs are associated with cutting fluids [3], and this is confirmed by other studies in Spain and the U.S.A. [4-6]. Environmental regulations concerning airborne cutting fluid vapour are likely to become stricter in the near future [7] and more stringent environmental regulations in much of Europe have increased the disposal costs for spent fluids dramatically.

Thus, there are two key drivers for change to dry machining; the rising direct and indirect costs associated with the use of cutting fluids and the threat of new tighter legislation. Industry is aware that the solution to both of these problems is to machine dry or to minimise the quantities of cutting fluid used.

3. Monolithic Structures

In the European aerospace industry most structural components are manufactured by machining up to 95% of the material away from a solid billet to produce light weight structurally optimised monolithic structures. Wing spars and ribs are typical examples of such structures. Figure 1 shows a photograph of a monolithic structure demonstrator.

Wing ribs are thin shear webs that are attached between the front and rear wing spars and the skins to form part of the wings interior skeleton. The new Airbus A380 has a total of 124 ribs, the largest being 3.1 by 2.0 m. Airbus migrated from conventional machining to high speed machining of wing ribs in 1995 due to the established advantages of improved material removal rates, lower cutting forces and improved surface finish [8, 9]. Conventional cutting speeds were 300m/min and high speed machining is performed at more than ten times these speeds. This has enabled Airbus to produce ribs more quickly and to a lower cost.

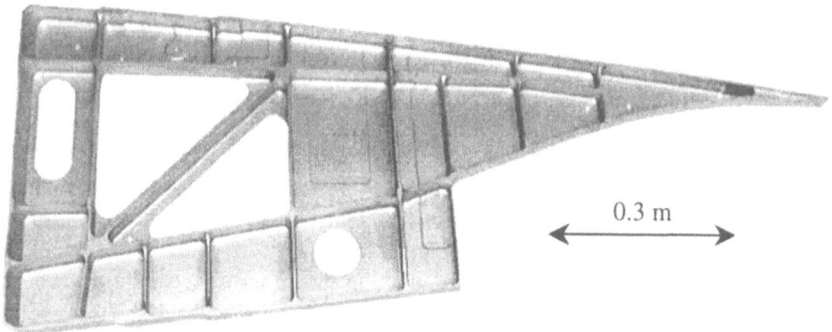

0.3 m

Figure 1, Photograph of monolithic structure demonstrator.

4. Technical Challenges for High Speed Dry Machining

Cutting fluid performs three key functions; it cools the cutting tool, workpiece and machine, it helps to clear the swarf from the cutting area and it lubricates the cutting process therefore providing improved tool life and surface finish. If the cutting fluid is omitted from the production process then each of its functions must be addressed.

4.1 Lubrication

An important task of a cutting fluid is to lubricate the cutting process. However, in the high speed machining of aluminium, high cutting speeds, centrifugal forces and contact pressures reduce the effectiveness of the fluid as a lubricant [10, 11]. To verify this, a single experiment was undertaken where two high speed slotting test cuts were performed under wet and dry cutting conditions. Cutting forces were measured on a dynamometer. The tests were conducted at a cutting speed of 3000 m/min (24,000 rpm) with a 0.2 mm tooth loading.

The test results in figure 2 show that the cutting forces are some 10% higher for dry machining, and that there is more friction as the tooth goes in to and out of the cut. This shows that the cutting fluid is providing a limited degree of lubrication under these cutting conditions.

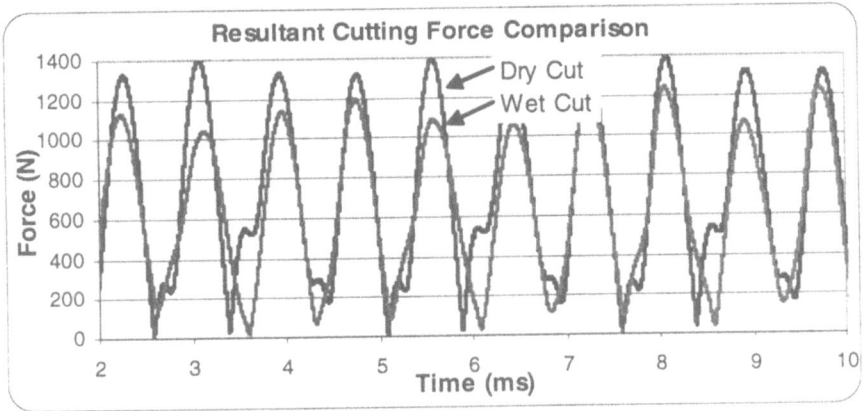

Figure 2, Cutting forces when machining under wet and dry cutting conditions.

Lubrication in dry machining can be addressed by using tool coatings, which lower the coefficient of friction between the tool and the chip. DaimlerChrysler Aerospace [12] demonstrated that diamond and carbon coatings offered reduced friction and eliminated the build-up of material on the cutting edge. Other researchers [13] state that build-up on the tool is reduced or even eliminated under high speed machining conditions; this will be advantageous in high speed dry machining. Lubrication in dry machining is very important and its effects on machining efficiency and cutting temperatures will be analysed in future test programmes.

4.2 Swarf evacuation

It is important to remove swarf quickly and efficiently in order to prevent it from being re-cut and damaging the cutting tool and the workpiece [12] and to minimise heat transfer from the hot swarf to the machine tool. Cutting fluid has traditionally been used for swarf evacuation because it was readily available on the machines. However, some companies [14] suggest that forced air is effective at expelling chips from the cutting area.

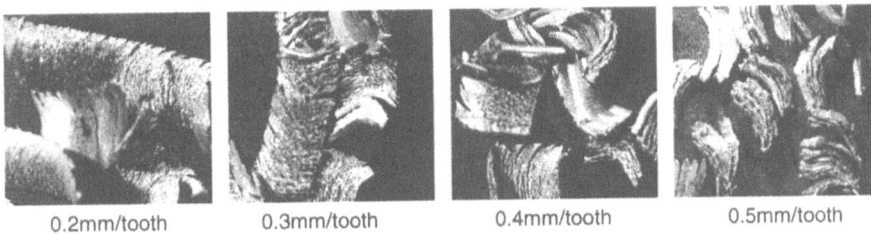

   0.2mm/tooth   0.3mm/tooth   0.4mm/tooth   0.5mm/tooth

Figure 3, Types of swarf produced by varying tooth loadings

Figure 3 shows a photograph of the types of swarf produced when dry machining at varying tooth loadings. The highly segmented swarf produced at a 0.5mm tooth loading is small, light and not curled and should therefore be easily evacuated with compressed air.

4.3 Cooling

It is generally agreed that cutting fluid cools the workpiece, swarf and machine tool but does not reduce the actual tool/work interface temperatures because it can not act directly on the heat source due to the high cutting speeds and forces [11]. The temperature of the cutting environment can be controlled with chilled compressed air or liquid nitrogen. However, it is not presently clear how much cooling is required.

A well-publicised advantage of high speed machining is that the majority of the heat generated in the cutting process is removed in the swarf. This is stated to be because the heat is being removed in the swarf before it has time to conduct into the workpiece [11]. The various authors that have conducted research into this [11, 15] state that approximately 80% of the heat goes into the swarf with the remaining heat distributed into the tool and the workpiece.

Figure 4 shows the results of testing that was carried out to verify this. The test was designed to measure the amount of heat conducted into the workpiece as 0.5 litres of material was machined away from a 1 litre block. The thermocouples were mounted 5mm below the final machined surface. This test was performed dry under both conventional and high speed machining conditions.

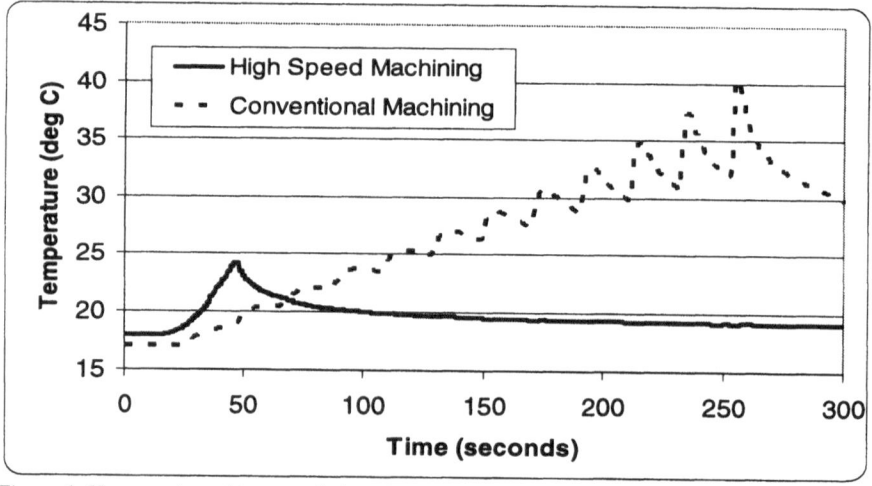

Figure 4, Heat conducted into workpiece in conventional and high speed machining.

Figure 4 clearly shows that the proportion of the heat that is conducted into the workpiece is dramatically reduced for high speed dry machining. At conventional cutting speeds of 300 m/min the workpiece temperature rose by 23.3 °C and at 3000 m/min the temperature only rose by 6.8 °C.

5. Research Direction – Temperature Modelling

The understanding of temperatures induced in the machining process and the effect that they have on the workpiece material under different machining conditions is the key to successful high speed dry milling of light alloys. Cutting temperatures affect machining efficiency, product integrity, product accuracy and tool life. Figure 5 highlights the thermal influences, limits and measurement methods in high speed dry machining.

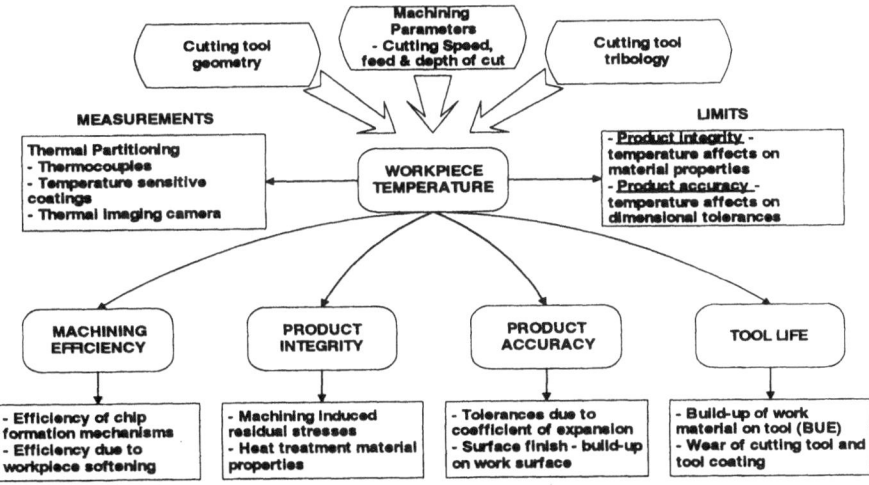

Figure 5, Influences and limits of machining temperatures in dry machining.

There are limits to the magnitude and distribution of heat in the workpiece. Excessive temperatures will affect the materials properties and introduce detrimental residual stresses [16]. A rise in the bulk workpiece temperature will affect machined dimensional tolerances. Aluminium has a high coefficient of thermal expansion of 24µm/m-°C; therefore a 3m long rib would change length by 1.4mm for every 20°C change of temperature.

Machine induced residual stresses are caused by extreme forces and temperatures. Residual stresses are critical in the aerospace industry because they affect both dimensional tolerances and fatigue life. Brinksmeier [16] states that detrimental tensile residual stresses are caused by thermal impacts. In high speed dry machining the temperature effects are likely to be more severe and hence produce higher detrimental residual stresses.

Ongoing research will analyse the hypothesis that the majority of the heat is removed in the swarf for high speed machining. Testing will be carried out to derive information for numerical models. The models will then be validated through comparison with test results. The thermal model developed will be used to optimise machining strategies for high speed dry machining of aerospace component forms by minimising the thermal effects of the machining process on the component.

6. Conclusions

There are three key challenges to the high speed dry machining of light alloys; these are tool life, swarf evacuation and temperature. Tool life can be addressed with tool coatings and swarf evacuation issues can be managed with compressed air and controlled swarf formation. Temperature influences and effects are much more difficult to assess and control. Temperatures induced in the machining process affect product integrity, product accuracy, machining efficiency and tool life. If machining induced temperatures are known and can be modelled then machining strategies can be developed to compensate for these temperature effects. Any requirement for cooling gasses can be ascertained and economics for dry machining can be analysed.

7. Acknowledgements

EPSRC and Airbus are supporting this work through an Industrial Case Award.

References

[1] Sreejith P.S. & Ngoi B.K.A. Dry Machining: Machining of the future, Journal of Materials Processing Technology, 101 (2000), pages 287-291.
[2] Shaw C.F & Mitsuro H. Cost and Process Information Modelling for Dry Machining, Proc. of the Int. Workshop for Environment Conscious Manufacturing–ICEM, Sept 2000.
[3] Cut and Dried Solutions, The Engineer, 21 May 1999, V288, no.7451-7453, pages 21-2+.
[4] Lopez de Lacalle L N et al, Cutting conditions and tool optimization in the high-speed milling of aluminium alloys, Proc. of the IMechE, Vol 215, Part B, 2001, pages 1257-1269.
[5] Cselle T. (Guhring Inc.), The 10 commandments of dry high-speed machining. American Machinist, May 1998, page 66 – 74.
[6] Brinksmeier E. et al. Aspects of cooling lubrication reduction in machining advanced materials, Proceedings of the Inst. of Mechanical Engineers, Vol 213, Part B, pages 769-778.
[7] Littlefair G, Cutting Fluids: Trends in Cutting Technology, Metalworking Production, December 2000, 15(Dec), pages 26-27.
[8] Schultz H, High-Speed Milling of Aluminium Alloys, High Speed Machining – Winter Annual Meeting of ASME, 1984, pages 241-244.
[9] Sahm A, Fiedler U, On the Cutting Edge of High Speed Machining, 4th International Conference on Metal Cutting and HSM, 2003.
[10] Dumitrescu M. et al, Mist Coolant Applications in High Speed Machining of Advanced Materials, Metal Cutting & High Speed Machining, 2002.
[11] Trent E.M. & Wright, P.K, Metal Cutting, ISBN:075067069.
[12] Lahres M, Muller-Hummel P & Doerfel O, Applicability of different hard coatings in dry milling of aluminium alloys, Surface and Coatings Technology 91 (1997) 116-121.
[13] Balkrishna R. & Shin Y.C, Analysis on high-speed face milling of 7075-T6 aluminium, Int.Jrn of Machine Tools & Manufacture 41 (2001) 1763-1781.
[14] Zelinski P, Where Dry Milling Makes Sense, Modern Machine Shop, Oct 2000, 82-87.
[15] Luer K, High Speed Machining of Aluminium for Use in Aerospace Applications, Proceedings from Manufacturing Processes for the 21st Century, 18-19th May, 1998.
[16] Brinksmeier E, Minke E, Nowag L, Residual Stresses in Precision Components, Proc. 5th Int. Conf. on Industrial Tooling, 10-11 September 2003, Southampton, UK

Selected Aspects of High Speed Milling Process

Mr. Pavel Zeman[1], Mr. Jiří Šafek[2] and Prof. Jan Mádl[3]

[1] Research Center of Manufacturing Technology, Horská 3, 128 00 Prague 2, The Czech Republic, p.zeman@rcmt.cvut.cz
[2] Research Center of Manufacturing Technology, Horská 3, 128 00 Prague 2, The Czech Republic, j.safek@rcmt.cvut.cz
[3] Czech Technical University in Praque – Department of Manufacturing Technology, Technická 4, 166 07 Prague 6, The Czech Republic, madl@fsid.cvut.cz

Abstract: The paper describes some results of the research into plastic deformation and cutting temperature in high speed milling of aluminium alloy 2024-T351. The effect of cutting speed and feed on selected parameters in chip formation has been studied.

The experimental and simulation research was carried out by with sintered carbide inserts. It was focused on chip formation, plastic deformation and the cutting temperature. The plastic deformation was studied by means of the chip roots, which were obtained with a quick stop device. The cutting temperature was measured by the tool-workpiece thermocouple. The AdvantEdge 4.2 simulation software, based on the Finite Element Method (FEM) and on the material modelling, was used for the simulation study.

1. Introduction

High speed machining (HSM), was first reported in 1931 by Salomon [1]. One definition is that the process involves machining at considerably higher cutting speeds and feed rates than those used in conventional machining. However, it is most commonly used to describe end milling at high rotation speeds. The process has been adapted to a wide range of applications. In the aerospace sector, HSM is used to remove large volumes of aluminium quickly and to produce thin walled sections in wings [2].

Carl Salomon proposed that there was a peak cutting temperature at an intermediate cutting speed and that when cutting speed was increased from this point, there was a reduction in temperature. Since this claim most of the literature has concluded that there is no corresponding reduction in temperature at higher cutting speeds. McGee [3] suggested that temperature increases with cutting speed up to a maximum that was equal to the melting point of the workpiece. No temperature reduction occurred at higher cutting speeds [3, 4, 5]. This explains why there is no fixed limit to the cutting speed when machining aluminium alloys (other than that imposed by machine tool considerations).

The plastic deformation in the chip formation process leads to changes in the mechanical properties of the work material, e.g. the strength of the material, the hardness of the material and the toughness of the material. Surface finish and integrity is primarily affected by the fact that the primary deformation zone usually extends below the machined surface level. Thus, the plastic deformation parameters can indicate the character of the cutting process and also the character of the machined surface [6, 7]. These parameters always depend on tool and machine tool parameters, cutting conditions, working conditions and workpiece properties. We can determine the effect of each of these cutting process components.

2. Cutting Process Analysis

The effect of cutting conditions on the cutting temperature and the character and intensity of plastic deformation in the chip formation process is presented in this paper. The remaining effects (tool, machine tool, and workpiece properties) did not vary. This paper combines the experimental and simulation investigation of the cutting process. Both methods were performed for up-milling of aluminium alloy 2024-T351 using a sintered carbide tool tip - APKX-M (type P – coated). The experimental part of the study was performed on the high-speed machine tool MCFV 5050 LN as a dry machining process. The cutting conditions used for the cutting temperature measurement and the plastic deformation studies are shown in Tables 1 and 2.

Table 1: The cutting conditions in the temperature measurement.

Cutting parameters	Value	
	Roughing	Finishing
Cutting speed v_c [m.min^{-1}]	From 800 to 2200 after 200	
Feed per tooth f_t [mm]	0.25	0.1
Cutting depth a_p [mm]	4	1
Cutting width a_e [mm]	33	33

Table 2: The cutting conditions in the plastic deformation study.

Cutting parameters	Value					
Cutting speed v_c [m.min^{-1}]	226	281	356	452	563	703
Feed per tooth f_t [mm]	0.88					
Cutting depth a_p [mm]	2					
Cutting width a_e [mm]	2					

The following characteristics of plastic deformation were determined as shown in Figure 1 and Equations (1) and (2).

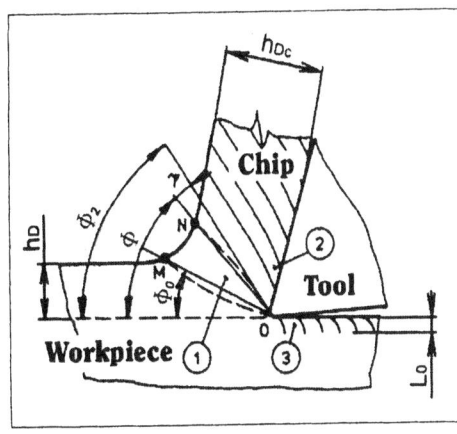

Φ – shear angle [°];

Φ_0 – angle of the start of the primary shear zone [°];

$\Phi - \Phi_0$ – size of the primary shear zone [°];

γ – shear strain [–];

Λ_h – chip-thickness ratio [–];

l_0 – deformation depth in the machined surface layer [mm];

h_{Dc} – chip thickness [mm];

h_D – undeformed chip thickness [mm].

Figure 1: Determination of some characteristics in chip formation process.

$$\gamma = \text{cotg } \Phi - \text{c otg } (\Phi + \delta_o) \qquad (1)$$

$$h = \frac{h_{Dc}}{h_D} \qquad (2)$$

where: δ_o – cutting-wedge angle [°].

3. Experimental Work

The plastic deformation of the workpiece in cutting was investigated by a finished changes study [4]. The real processes in chip formation were caught in chip roots. These were obtained with the use of a quick stop device for milling. The quick stop device function can be seen in Figure 2. A part of the specimen together with the chip root (position 4) is separated from the specimen (position 3) by the cutting force. This is done by reducing the cross section of the specimen in a particular place. Thus, the part of the specimen together with the chip root is removed very rapidly by the tool tip (position 1). The slot was milled out before performing the experiment. A thin lamina made of copper (position 5) was placed into the slot to prevent distortion. The specimen is turned through an angle of 7° in relation to the sliding direction of the tool.

The cutting temperature measurement was done through the natural thermocouple. The natural thermocouple is created by the tool material and the workpiece material. The measurement was focused on determination of the thermoelectric voltage which was created by the temperature change in the tool-wokpiece interface. The set up of the cutting temperature measurement is shown in Figure 3. The workpiece has to be insulated from the table of the machine to prevent the creation of the parasitic thermocouple. The contact on the workpiece was created by the constantan wire which was griped between the workpiece and the insulating washer. It was very difficult to create the contact on the rotating tool. Thus, the contact was created on the rotating spindle. This contact was created by a thin copper lamina, which was pressed on the rotating spindle. The copper lamina was fixed in the magnetic stand, which was mounted on the machine block. A constantan wire was soldered at the end of the lamina, which served as the transfer of the thermoelectric voltage. The thermoelectric voltage was scanned by the data switching exchange. The calibration curve was used for transferring the thermoelectric voltage values to the cutting temperature values.

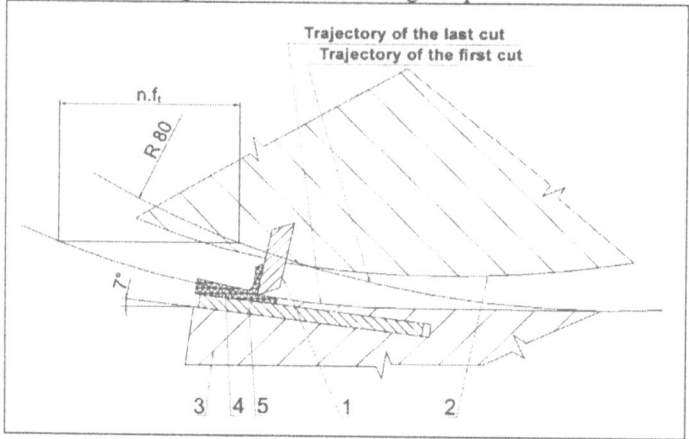

Figure 2: The method of operation of the quick stop device.

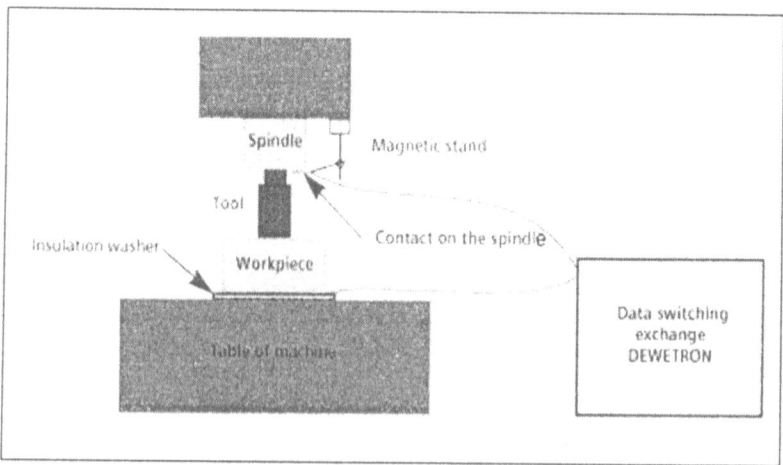

Figure 3: The cutting temperature measurement set up.

3. Simulation Work

The AdvantEdge 4.2 software was used for the simulation of the cutting process. This software is based on simulation through the Finite Element Method (FEM) together with material modelling [8, 9, 10, 11]. It was possible to predict a plastic strain distribution, temperature, and strain rate and maximum shear stress in the workpiece and in the tool, with the help of simulation output data.

The input data of the simulation were the same as the experimental data.

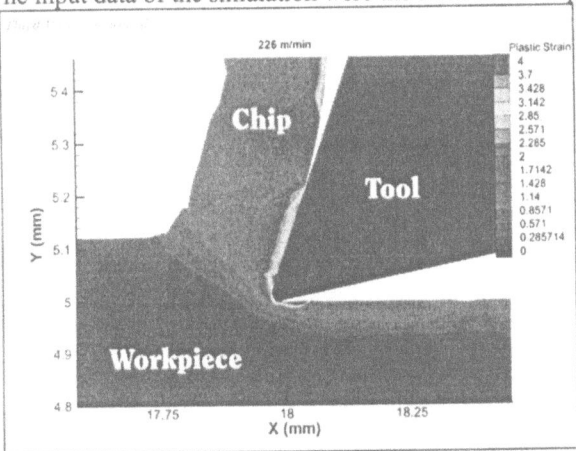

Figure 4: An output of the AdvantEdge software.

4. Results

The parameters Φ_0, Φ slightly increase and $\Phi-\Phi_0$, γ and Λ_h decrease with cutting speed in the experimental results. Since the course of $l_0 = f(v_c)$ is decreasing, the machined surface is less affected by plastic deformation at a higher cutting speed. The simulation results show that courses of parameters Φ_0, Φ and $\Phi-\Phi_0$ vary in a complicated way with cutting speed. The parameters first decrease and then increase with cutting speed. Thus, it is impossible to generalize the effect of cutting speed on these parameters. The parameter γ first increases and then decreases with cutting speed. Nevertheless, it is clear that the effect of parameter γ is very weak (ranging from 1,238 to 1,368), similar to parameter Φ_0 (ranging from 24° to 27°). The dependence $\Lambda_h = f(v_c)$ shows that the chip thickness (h_{Dc}) decreases as v_c increases. The dependence $l_0 = f(v_c)$ shows the increase of l_0 with cutting speed in the simulation. Thus, the machined surface is more affected by plastic deformation at higher cutting speeds.

The measured and simulated values of the cutting temperature are shown in Figure 6. It was found out that the cutting temperature is slightly increasing with the increase in cutting speed for the finishing experimental conditions. The temperature was 305°C at a speed of 800 m.min^{-1} and 345°C at 2200 m.min^{-1}. It was found out that the cutting temperature was also slightly increasing with the increase in cutting speed for the roughing experimental conditions. The temperature range for these cutting conditions is from 402°C to 453°C. For the cutting temperature obtained by the simulation it was found out that the cutting temperature is slightly increasing with the increase in cutting speed. The comparison between the experimental and simulation results shows that the values from the simulation are lower than the measured values for the roughing conditions.

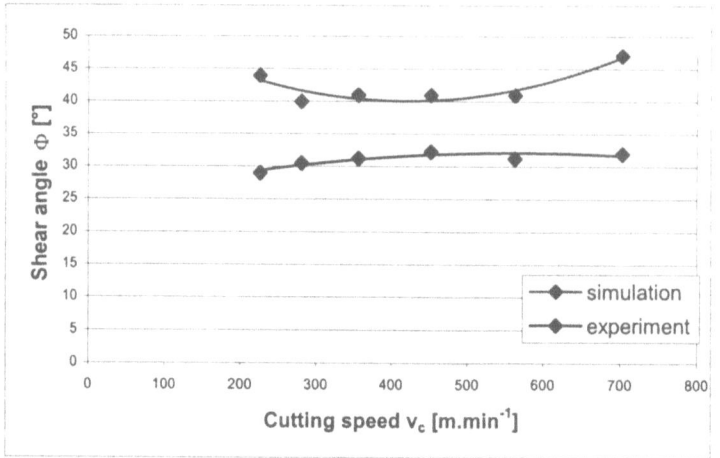

Figure 5: The shear angle as a function of cutting speed.

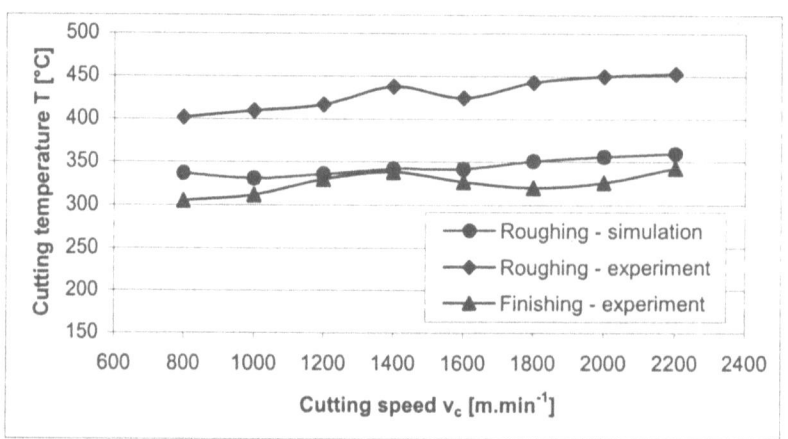

Figure 6: The cutting temperature as a function of cutting speed for different cutting conditions.

6. Conclusions

The experimental results of the plastic deformation study are closer to the theoretical presumptions than the simulation ones. Generally, a rise in cutting speed causes narrowing of the primary shear zone. The primary shear zone extends less below the machined surface level at higher cutting speeds. Thus, the machined surface is less affected by plastic deformation at higher cutting speeds. When compared with the theoretical presumptions, our study of the parameters of plastic deformation indicates that the experimental results are more precise than the simulation results. The simulation results have a somewhat different course than the theoretical presumptions (except dependence $\Lambda_h = f(v_c)$).

The measured and simulated values of the cutting temperature are slightly increasing with the increase in cutting speed. In the comparison of measured values in roughing and finishing it was found out that the values in roughing are higher than the values in finishing. The comparison of the experimental and simulation results shows that the values obtained by the simulation are lower than the measured values for the roughing conditions.

The differences between the experimental and simulation results are probably caused by other effects than cutting conditions. These effects occurred mainly in the simulation investigation. It seems that other influences in simulation (inaccuracy of the simulation model, the simplification and limitations of the simulation software) have a great effect on the simulation results. The greatest difference is observed between the simulation and experimental results for parameter l_0.

Acknowledgements

The results of the project LN00B128 were financially supported by the Ministry of Education of the Czech Republic.

References

[1] Salomon C., 1931, *Verfahren zur bearbeitung von metallen oder bei einer bearbeitung durch schneidende werkzeuge sich ahnlich verhaltenden werkstoffen*, German Patent no. 523594.

[2] Dewes R.C., Ng E., Chua K.S. et al, 1999, *Temperature measurement when high speed machining hardened mould/die steel.* Journal of Material Processing Technology.

[3] McGee F.J., 1979, *High speed machining – study: methods for aluminium workpieces*, American Machinist.

[4] Mádl J., 1988, *Experimentální metody v teorii obrábění.* ČVUT.

[5] Trent E.M., 1991, *Metal cutting*, Butterworth, London.

[6] Mádl J., 2001, *Theoretical Aspect of Precise Machining.* Proceedings ICPM 2001.

[7] Mádl J., 2002, *Dry Machining Versus Cutting With Cutting Fluids.* Manufacturing Technology.

[8] Marusich TD, Ortiz M., 1995, *Modeling and Simulation of High-Speed Machining.* Int. J. Numer. Meth. Eng.

[9] Özel T, 2003, *Modeling of Hard Part Machining: Effect of Insert Edge Preparation in CBN Cutting Tools.* Journal of Materials Processing Technology.

[10] Kumbera T.G., Cherukuri H.P., Patten J.A., Brand C.J., Marusich T.D., 2001, *Numerical Simulation of Ductile Machining of Silicon Nitride with a Cutting Tool of Defined Geometry*, Proceedings of 4th CIRP International workshop on Modeling of machining operations.

[11] http://www.thirdwavesys.com.

Wear and Failure of High-Speed Steel Bimetal Bandsaws When Cutting Ball-bearing Steel

Mohammed Sarwar [1], Martin Persson [2], Håkan Hellbergh [3]

[1,2]Northumbria University, Newcastle upon Tyne NE1 8ST,
mohammed.sarwar@unn.ac.uk
[3]Bahco Metal Saws AB, Box 833, S-53118 Lidköping,Sweden.
hakan.hellbergh@bahco.com

Abstract: This paper reports experimental data on the wear and failure modes of high-speed steel bimetal bandsaw blades cutting annealed ball-bearing steel circular bars. Several different methods of assessing the wear modes and mechanisms are evaluated; Cutting and thrust force components, Set width, "Out-of-square" cutting, Wear land area and geometry,; and, Surface characteristics of cut-off workpieces.

The failure mode established in the current work when bandsawing ball-bearing steel with a bimetal blade is "out-of-square cutting". There is a linear increase in the thrust force component as the bandsaw teeth wear, whereas the cutting force component experiences a non-linear increase. The out-of-square failure mode appears to be caused by unbalanced corner wear to the set teeth. The corner wear also influences the total set width and the quality of cut surfaces. The out-of-square failure mode occurred before the thrust force and wear land area reached the tertiary stage of wear. This work should be of great interest to the production engineer associated with cutting-off processes.

1. Introduction

Generally, bandsawing is an important operation in a variety of industries, particularly steel suppliers, which need to cut-off to size raw material for secondary processes. Owen [1] states that bandsawing is growing in popularity as a cutting-off method compared to power hack-sawing, circular sawing etc. Bandsawing offers the advantage of high automation possibilities, high cutting rate, low kerf loss, straightness of cut, good surface finish and long tool life. The bandsawing operation is fairly well understood today, owing to recent scientific work carried out by several researchers [2,3,4,5,6,7,8]. This paper reports experimental data on the wear and failure modes of bandsaw blades cutting ball-bearing steel round bars. Several different methods of evaluating the wear modes and mechanisms have been assessed.

Much research has been carried out for the wear of traditional cutting tools, such as milling, drilling and turning tools. The development of the single cutting edge tools has introduced advanced tool materials, such as Tungsten carbide, Ceramics, Cubic Boron Nitride and Diamond. It has also brought surface engineered tools with advanced coating materials via enabling technologies (PVD, CVD etc.). Despite this, high-speed steel cutting edges still play an important role for machining operations such as bandsawing.

Bandsawing differs from many machining methods and has four important distinctive characteristics;

- The depth of cut per tooth is small (5 μm to 50 μm), leading to a complex combination of chip formation modes [9].
- The bandsaw tooth cutting action is intermittent.
- The number of active cutting edges in contact with the workpiece at any instant can vary.
- The chip formed has to be accommodated within the gullet.

Although there has been previous research in the area of bi-metal bandsaw wear [3,4,5,6,7,8], recent improvements and developments of the bandsaw blades and processes show that the previous research can be further developed to achieve a better understanding of bandsaw wear, and hence gain greater tool life.

2. Test Methodology

2.1 Bandsaw blade

A bimetal bandsaw blade with conventional tooth setting (alternating straight–left–right) was selected for the experimental work. The blades were fully examined with regards to the nominal and actual properties according to tables 1 and 2. The workpiece material was also characterised with regard to its microstructure and hardness, shown in tables 3 and 4.

Table 1. Selected bandsaw product, nominal properties

Product	Bahco 3851-54-1.6-PSG-2/3
Production method	Ground teeth
Tooth pitch	Variable pitch, 2/3 teeth per inch
Tooth height configuration	Equal tooth height
Backing thickness	1.6 mm
Bandsaw blade width	54 mm
Bandsaw loop length	8800 mm
Tooth set pattern	Straight-Right-Left sequence
Tooth rake angle	+10°
Tooth clearance angle	32°

Table 2. Measured properties

Back to tooth tip height	*Average values (mm);* Neutral teeth: 54.25 ± 0.007* Right-set teeth: 54.27 ± 0.005* Left-set teeth: 54.31 ± 0.011*
Tooth edge radius	Average value 7 ± 0.99* μm
Tooth hardness (M42 high-speed steel)	Average 887, Vickers 5kg
Backing hardness (Spring steel AISI D6A)	Average 490, Vickers 5kg

* 95% confidence interval

Table 3. Ball-bearing steel round bar

Material trade name	Ovako 803J
Bar dimension	Ø 120 mm
Hardness	Average value 208 ± 5*, Vickers 5kg
Microstructure	Ferrite and cementite with globular carbides finely precipitated in matrix
Condition	Surface oxide, fully annealed
Corresponding standards	SAE 52100, B.S. 534A99(En31), DIN 100Cr6

* 95% confidence interval

Table 4. Ball-bearing steel, chemical composition in weight %

C	Si	Mn	Cr	Ni	Mo	Cu	V
1.00	0.26	0.28	1.42	0.16	0.05	0.21	0.01

2.2 Bandsaw machine features

The bandsaw machine (Behringer HBP650/850A/CNC) has been instrumented in order to measure cutting force and thrust force components, cutting speed and feed rate. The bandsaw machine characteristics together with its instrumentation and data acquisition system were calibrated before the test programme was commenced. The machine was dynamically calibrated using a 3-axis Kistler dynamometer and relevant transducers.

2.3 Cutting conditions used in wear testing

Initial 'mapping' sawing tests were carried out in order to establish cutting speeds and feed rate values which would be suitable for the wear test programme. These are shown in table 5.

Table 5. Selected cutting conditions for ball-bearing steel

Work-piece material	Horizontal cutting speed (m/min)	Vertical feed rate (mm/min)	Average depth of cut per tooth (µm)
Ball-bearing steel	80	32	4.7

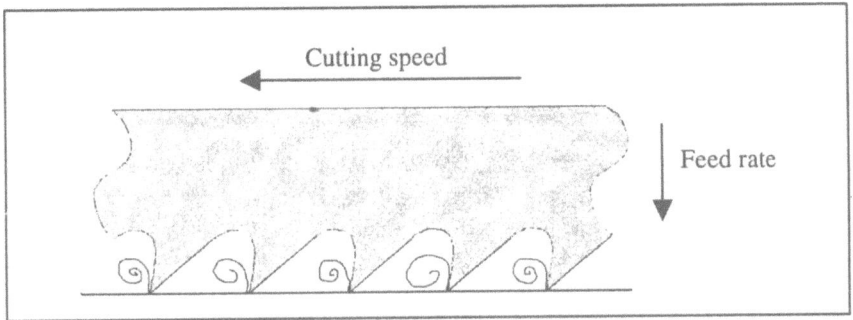

Figure 1. Cutting speed and feed rate

3. Evaluation Methodologies for Wear Tests

The wear testing was carried out by using 3 bandsaw loops to cut sections of the work-piece. Wear related properties were simultaneously recorded at set intervals. Each bandsaw loop was run until failure. The following measurements were taken;
• Cutting and thrust force components
• Set width
• "Out-of-square" cutting, blade run-out
• Wear land area and geometry
• Surface characteristics of cut-off workpieces (waviness)

In order to produce meaningful data, the parameters were clearly defined. The cutting and thrust force components were measured at the maximum diameter of the workpiece, resulting in the maximum value for the cutting-off of a round section. The 'set' width was measured at several points around the bandsaw loop. The "out-of-square" cutting (run-out) was defined as the measured difference in lateral position between the start and the end of the cut. The wear land was optically measured using a stereo microscope and associated computer software. The surface characteristic of the cut-off sections was, in this case, the macroscopic waviness of the surface, an average value of 3 measurements.

4. Results

4.1 Failure mode

The first and perhaps most important information of the wear test is the resulting failure mode. The failure mode criteria used was run-out, as this was the first failure mode to occur. All 3 bandsaw samples failed in the same mode, run-out, as stated in table 6. The variation in result is large, multi-point tools can have a large variation due to a large number of factors.

Table 6. Out-of-square failure, run-out criteria of 0.5 mm.

Bandsaw sample number	Number of sections cut before out-of-square cutting failure
1	857
2	644
3	979

4.2 Cutting and thrust force components

Figures 2 to 4 show the force results together with the out-of-square value from the 3 tested bandsaw loops. There appears to be a correlation between the out-of-square cutting and the increase in cutting force. In the end of product life there is a more rapid increase in cutting force whereas the thrust force continues with a constant wear rate. The reason for the increase in cutting force can be increased friction to the sides of teeth as the band starts to lean to one side and cut out-of-square. The thrust force does not appear to be affected by the out-of-square cutting.

There ought to be a visible difference in corner-wear on the teeth that are set to the left and right. This is investigated and reported later in the paper, section 4.4.

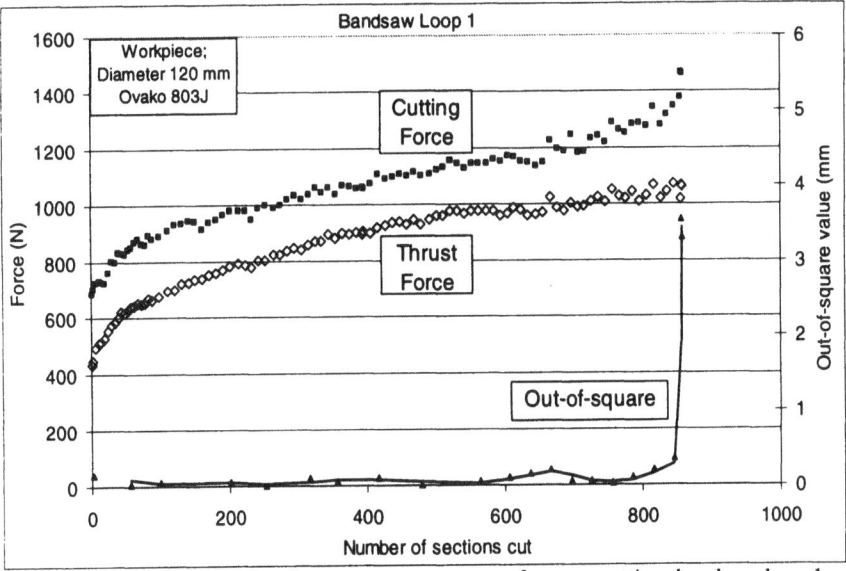

Figure 2. Thrust and cutting force components, out-of-square cutting, bandsaw loop 1

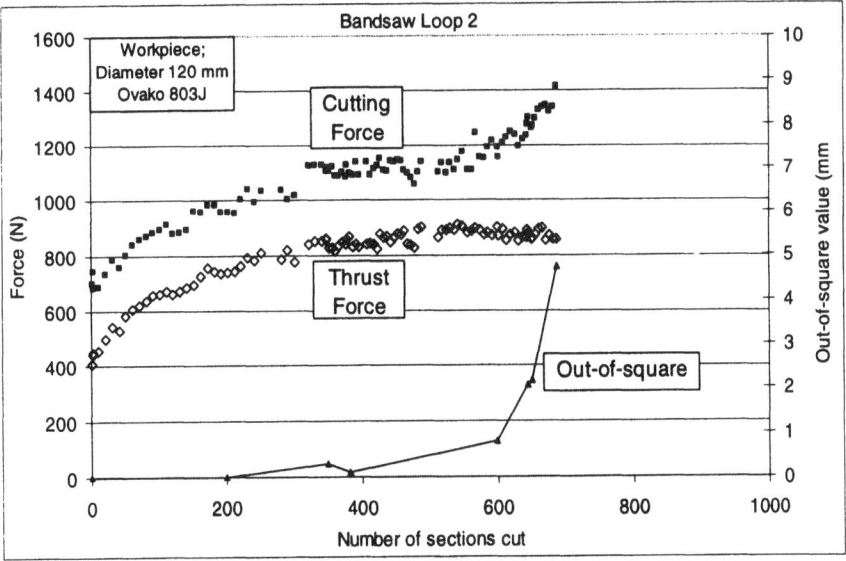

Figure 3. Thrust and cutting force components, out-of-square cutting, bandsaw loop 2

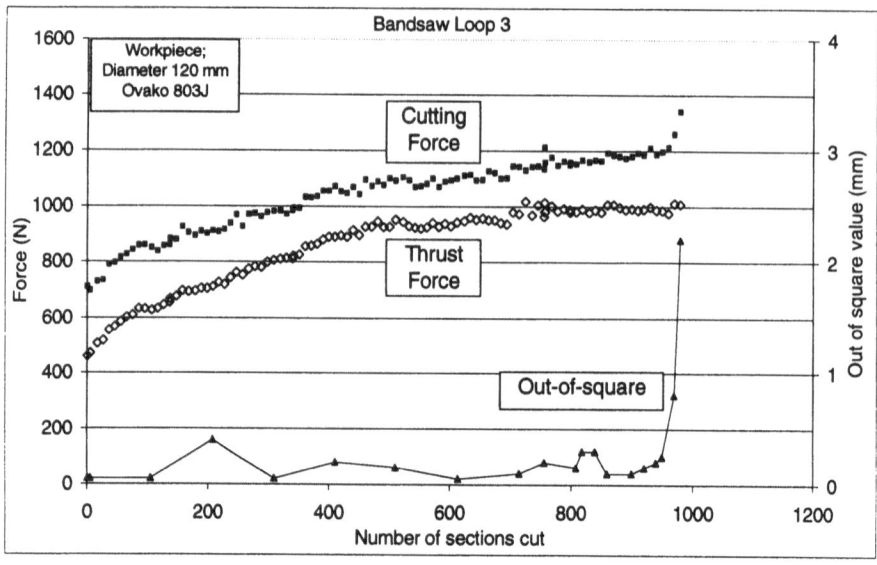

Figure 4. Thrust and cutting force components, out-of-square cutting, bandsaw loop 3

4.3 Set width results

The 'set' measurements of the blades in new condition are shown in table 7, followed by the loss of set shown in table 8. Since the bandsaw loops fail due to out-of-square cutting, set wear measurements were taken in order to establish the influence of set-wear on run-out. The results show a marginally greater loss of set on one side for all bandsaw loops, indicating that set loss (corner wear) could have caused the out-of-square cutting. However, the difference in set loss is very small and is therefore not a very good indication of the level of corner-wear. Instead, corner wear will be presented in section 4.4.

Table 7. Set magnitude values for new bandsaw samples

	Set magnitude right (mm)	Set magnitude left (mm)
Bandsaw sample 1	0.416 ± 0.01*	0.415 ± 0.01*
Bandsaw sample 2	0.44 ± 0.01*	0.415 ± 0.01*
Bandsaw sample 3	0.396 ± 0.01*	0.365 ± 0.01*

* 95% confidence interval

Table 8. Loss of set magnitude, ball-bearing steel test

	Loss of set, right-set teeth, average (mm)	Loss of set, left-set teeth, average (mm)
Bandsaw sample 1	0.09 ± 0.01*	0.05 ± 0.01*
Bandsaw sample 2	0.05 ± 0.01*	0.07 ± 0.01*
Bandsaw sample 3	0.07 ± 0.01*	0.05 ± 0.01*

* 95% confidence interval

4.4 Wear land results

Since the wear on the tooth edges will finally determine the failure mode of the bandsaw blade, measurements were taken of the wear land of the cutting edges.
There wear land area (not presented) showed a linear increase, this measurement did not show correlation to the out-of-square cutting failure mode.

However, the measurement of worn corners, figure 5, showed correlation to the out-of-square cutting values. The length of the wear land was measured for the right/left-set tooth corners, representing the level of corner wear. There appears to be a correlation between the increasing wear to the left-set teeth and the increasing out-of-square cutting value. Increasing corner wear can possibly lead to a higher lateral force acting on one corner, forcing the bandsaw blade to cut out-of-square. This measurement of corner wear is a stronger indication of the failure than the measurement of set-wear (in section 4.3), it provides a larger difference between the right- and left-set teeth.

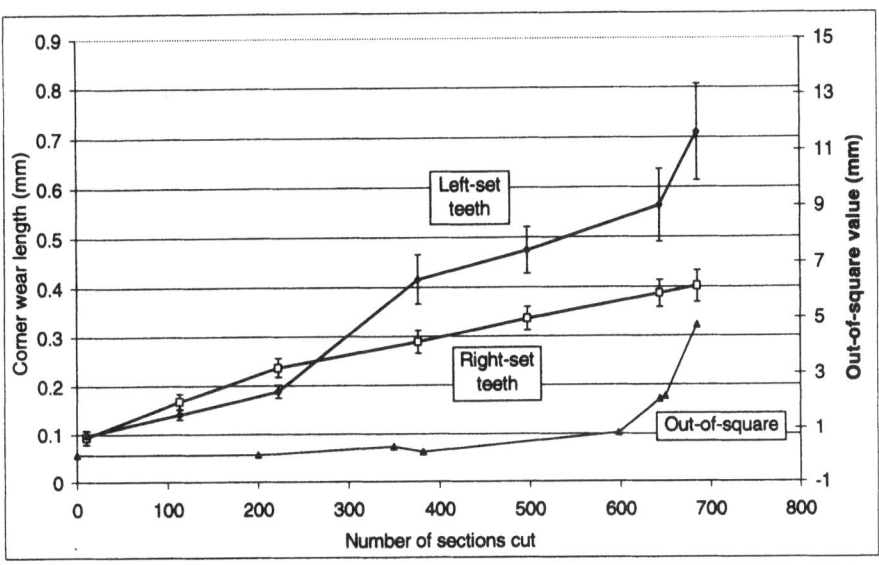

Figure 5. Corner wear length, 95% confidence interval, bandsaw loop 2

4.5 Surface characteristics results

The surface waviness value appears to have a correlation to the amount of out-of-square cutting, as shown in figure 6. It is indeed the outer corners of teeth that produce the resulting cut-off surface, therefore this result is yet one more indication of the significance of the corner wear level.

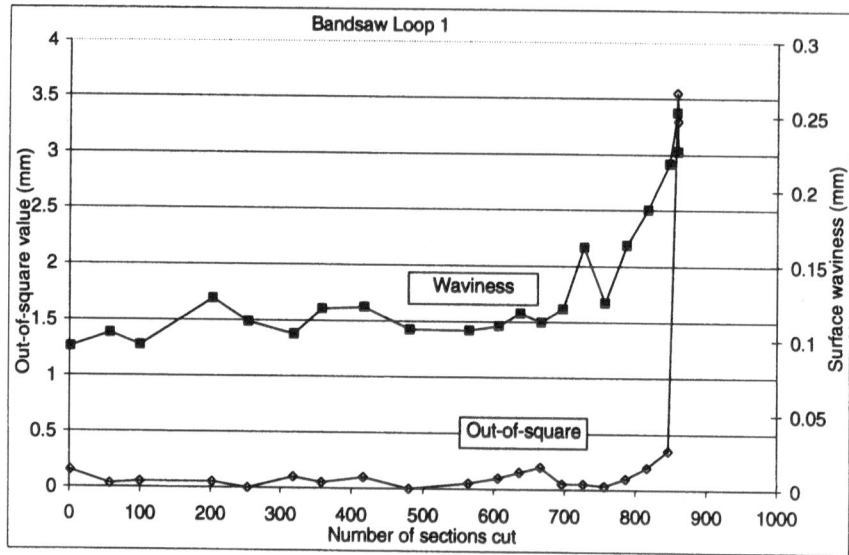

Figure 6. Out-of-square cutting and surface waviness, bandsaw loop 1

5. Conclusions

The failure mode established in the current work when bandsawing ball-bearing steel with a bimetal blade is "out-of-square cutting". There is a linear increase in thrust force as the bandsaw teeth wear, whereas the cutting force experiences a non-linear increase. The out-of-square failure mode appears to be caused by unbalanced corner wear to the set teeth. The corner wear also influences the total set width and the quality of cut surfaces. The out-of-square failure mode occurred before the thrust force and wear land area reached the tertiary stage of accelerated wear. This work should be of great interest to the production engineer associated with cutting-off processes and the tool designer associated with bandsaws.

References

[1] J. V. Owen, Manufacturing Engineering, Feb. 1997, 28-38
[2] M. Sarwar, P.J. Thompson, International MTDR Conference, Manchester, (1981) 295-303
[3] P. Wallén, S. Jacobson, S. Hogmark, Report UPTEC 87 104 R, Uppsala University, Sweden, 1987
[4] C. Andersson, J.-E. Ståhl, H. Hellbergh, International Journal of Machine Tools and Manufacture, 41 (2001) 237-253
[5] S. Elanayar, Y.C. Shin, J. of Manuf. Sc. and Eng., 118 (1996) 359-366
[6] D.W. Smithey, S.G. Kapoor, R.E. DeVor, International Journal of Machine Tools and Manufacture, 40 (2000) 1929-1950
[7] P.M. Archer, S.R. Bradbury, M. Sarwar, 5:th National Conference on Production Research, Huddersfield Polytechnic, (1989) 443-451
[8] M.M. Ahmad, B. Hogan, E. Goode, International Journal of Machine Tools and Manufacture, 29 (1989) 173-183
[9] M. Sarwar, P. J. Thompson, Production Engineer, June 1974

Tool Wear Behaviour of Micro-Tools in High Speed CNC Machining

BT Hang Tuah Baharudin[1], N. Dimou[2] and K.K.B. Hon[3]

Department of Engineering, University of Liverpool, L69 3GH, UK.
[1]btht@liv.ac.uk, [2]ndimou@liv.ac.uk, [3]hon@liv.ac.uk

Abstract: Cutting tool life is one of the most important economic considerations in metal cutting. In roughing operations the various tool angles, cutting speeds and feed rates are usually chosen in order to achieve economic tool life. In micro-machining, unpredictable tool life and premature tool failure are major problems. Furthermore, it is impractical to determine the tool life of micro end-mills with a diameter in the region of 1mm using the standard criterion as given in the ISO 8688-2:1989 'Tool Life Testing in Milling'. In this investigation, the tool life criteria of micro tools were evaluated for the machining of H13 tool steels and Titanium Alloy Ti6Al4V. The correlation between tool wear and the cutting forces were also studied. Inspections of the tool edges by Scanning Electron Microscopy (SEM) revealed the progression of tool wear, and in several other cases, sudden tool breakages were also observed.

1. Introduction

Manufacturers increasingly adopt rapid prototyping and manufacturing techniques to maintain a high level of performance in the present competitive environment. The global trend today is moving towards miniaturising components which many believe as the gateway to increase substantially the value of their products and market share. Since the trend is to make products more compact, greater attention is given by many manufacturers on micro machining in their operations.

Previous research by Tansel [1] studied the wear and breakage of micro end-milling. Wear and tool failure mechanisms were found to be complicated in micro machining. Micro end-mills with less than 1mm diameter have a very short and unpredictable tool life in metal cutting and tool life is only acceptable at low feed rate. Cutting forces were investigated by Pandit [2] and Gygax [3], while Takata [4], Lan [5], Atlantis [6] and many others concentrated on tool failure detection.

Cutting force characteristics of micro end-milling operations are almost the same as those of conventional milling. However, the dominant wear and breakage mechanisms are very different. The cutting edges of conventional end mills wear out during machining thus resulting in gradual increase of cutting forces. This increases is highly unlikely to cause any catastrophic failure such as breakage unless the tool is used beyond its wear criterion limit. However, in micro end milling, it is the tiny shafts of the micro tools that break when either the cutting edges become dull because of build-up-edge or because of material loss, or chip clogging. Konig [7] showed that the main reason for tool failure in micro drilling operations was chip clogging. Therefore any changes is cutting forces is important for consideration.

Tool deflection is another case which is closely related to micro milling. In this case, the flexibility and deflections within the cutter is beneficial in attenuating the overload in a sudden transient situation, as well as attenuating chatter. It was discovered that the effective diameter of an end mill was equivalent to approximately 80% of its diameter, i.e., a 1 mm diameter cutter has an effective diameter of 0.8 mm [8].

Although a significant number of investigation have been carried out on tool wear and tool life of micro tools, many findings are based on estimations and comparisons between two or several sets of wear data. There is no available data or indicative figure which can determine whether any particular micro tool has reached its tool life after machining operation. It is not practical to apply the same criterion to micro tools with a diameter in the region of 1 mm as given in the ISO 8688-2:1989 'Tool Life Testing in Milling' [9]. Some researchers rely on the micro tools reaching catastrophic failure, others on machining qualities such as surface finish and burr, while some redefine the meaning of 'wear' in micro machining, such as Tansel [1].

2. Research Aim

The aim of this investigation is to study the wear behaviour of micro tools in accordance to ISO 8688-2:1989 'Tool Life Testing in Milling – End Milling – Part 2'. According to ISO Standard, the end of tool life of an end-mill is defined by a flank wear of 0.3 mm. Since this study deals with micro-tools of diameters from 0.5 mm to 1.5 mm, the end of tool life naturally will have to be smaller than stated in ISO 8688-2. For example, 0.3 mm flank wear for an end-mill tool of 0.5 mm diameter implies that 60% of its effective cutting edge has been worn and this clearly is inadmissible. Furthermore, by this point, the tool would probably have reached the catastrophic failure stage beyond the end of normal tool life criterion or incapable of producing the desirable surface qualities.

The reasons for which each tool may be considered to have reached the end of their useful tool life will be different in each case depending on the tool geometry and dimensions, cutting conditions, product qualities, tool material and coatings and the workpiece material. The time at which the micro-tools cease to produce workpieces of the desired size or surface quality will determine the end of their useful life.

As part of this research, the propagation of flank wear with machining time will provide a useful tool life limit or wear criterion for each micro-tool. To produce flank wear graphs, the tool wear has to be measured accurately after each specified cutting interval. The measurements of the tool wear were made by means of Scanning Electron Microscopy (SEM). This repeatable action is only possible by marking the exact side of the cutting tools to be measured and by careful positioning of the tool in the SEM chamber.

3. Experimental Equipment

3.1 Cutting tools

All eight micro cutting tools used in this experiment are advanced solid 2 flute micro grain carbide tools, square end with 30^0 right hand helix, coated with TiAlN. For 1mm and 1.5 mm diameter tools, one tool was used for the experiment but for 0.5 mm diameter tools, two tools were used in order to replicate the experiment for greater data reliability. All tools have a shank diameter of 3 mm and overall length of 38 mm. Table 1 gives the properties of the TiAlN coating of the tools.

Table 1. Properties of Balzers TiAlN Coating

Coating Colours	Violet/Grey
Microhardness	3,500 (HV 0.05)
Dry Coeff. of Friction Against Steel	0.4
Coating Thickness	3-4 μm
Maximum Working Temperature	800^0C
Key Characteristic	Excellent oxidation resistance
Primary Applications	Dry/semi dry machining, workpiece >45HRC

3.2 Workpiece Materials

Two workpiece materials were chosen for this study, i.e., H13 tool steel and Ti-6Al-4V titanium alloy. H13 steel (BS4659:1989 – Tool and Die Steels) is a mid range hardened tool steel. This is generally used as a hot work tool steels which can withstand relatively high working temperature between 315^0C to 650^0C. Some of its applications include hot forging dies, hot turning dies, pressure die casting, extrusion dies and extrusion mandrels. In this study, the workpiece was hardened and tempered to 45-55HRC.

Titanium alloy Ti-6Al-4V (BS 2TA 11:1974) was chosen because it is widely used in the aerospace industry and in bio-medical application. It is generally classified as a difficult to machine material with high chemical reactivity to most cutting tool materials.

3.3 Machine tool, data acquisition system and surface roughness measurement

All machining operations of the present work was carried out on a Mikron HSM 700 high speed milling machine. This machine is capable of 42,000 rpm spindle speed with 12.5kW power, 40m/min rapid feed and 20m/min cutting feed.

Machining Strategist, a Windows based CAM software, was used to generate the part programs for the milling operations. After importing an IGES file of a CAD part, the program then carried out a triangulation function before cutting passes was generated, linked and post-processed, via a DNC link from the CAM software onto the CNC controller of Mikron HSM 700. A Kistler 9265B Quartz, three-component dynamometer was used for measuring the three orthogonal cutting forces during machining. All cutting force signals were conditioned by Kistler charges amplifiers before they were sent to a high-speed data acquisition board for post processing and data analysis.

Surface roughness was measured using a WYCO interferometric microscope surface profiler with nanometer accuracy in the Z direction. The system will generate a 3D surface profile following scanning and this surface data can then be extracted with the software provided. An average reading from eight different sets of data was used to determine the surface roughness of the workpieces at regular intervals.

4. Experimental Technique

The tools mentioned earlier were divided into two equal groups to machine the respective workpiece. Each tool performed a slot milling on the workpiece and the flank wear was measured using SEM after specified cutting intervals. Since the workpiece block was designed to be square with 30 mm each side, the length of each cutting runs was 30 mm. With a known feedrate and depth of cut, wear can be measured against machining time or volume of machined materials. A vegetable based coolant, VASCOMILL MKS 68 was used for the experiment by means of spray mist technique to deliver the coolant to the cutting zone in all cutting operations.

The cutting conditions for each micro tool are given in Table 2 as the same conditions apply to both H13 tool steel and Ti-6Al-4V titanium alloy. The number of runs for each tool, wear and surface roughness measurement is given in the same table. To reduce the overall machining and measuring time, 0.5 mm tools were measured after every two cutting intervals or every 60 mm of cutting operations whilst 1.0 mm and 1.5 mm tools were measured after every four cutting intervals. The two cutting intervals used for 0.5 mm tools because the 0.5 mm tool is smaller and it is believed that any changes in the tool geometry may not be detected if four cutting intervals was used instead.

Table 2. Cutting parameters and measurement intervals

Cutting tool diameter	0.5 mm	1.0 mm	1.5 mm
Depth of cut (mm)	0.01	0.05	0.75
Feedrate (mm/min)	130	130	130
Spindle speed (RPM)	26,000	14,000	9,500
Run intervals (1 run = 30 mm)	2	4	4

5. Results and Discussions

In order to study the overall machining performance with the tool wear progression, an effective recognition of the progressive patterns of tool wear is a prerequisite. Previous experimental work has shown that multilayer TiAlN coated milling cutters perform well when machining H13 Steel and titanium alloys compared to other coated tools.

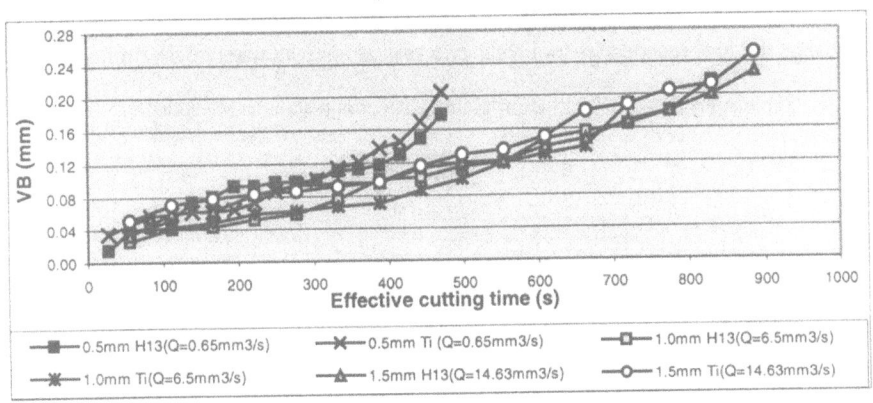

Figure 1. Flank wear progression when machining H13 Steel and Ti-6Al-4V Titanium Alloy

From Figure 1, it is obvious that the tool life for 0.5 mm tools is shorter than 1.0 and 1.5 mm tools. The graph for 0.5 mm tool shows some similarities with conventional size tools as the general wear graphs which have three stages, the primary or initial wear zone, followed by a secondary wear zone or steady state region, where the wear normally increases gradually over a longer time scale, and finally followed by tertiary or accelerated wear zone. It can be seen that after 388 seconds or 0.9 m of machining, the rate of wear increases dramatically. This is normally the point where the tool life criterion of that specific tool cutting specific material is taken. In this case, the tool life criterion is approximately 0.11 mm. For 1.0 and 1.5 mm tools, the initial wear stage is visible. After the first few runs, the wear had a sudden increase indicating the first stage of wear progression. After this stage, both graphs showed a steady increases in flank wear but do not show the typical third stage of the wear progression because one cutting flute for each cutter was broken and no longer produce any effective cutting. The tool life for the 1.0 and 1.5 mm tools were therefore 831 and 886 seconds respectively. Figure 2a shows the flank wear of 0.5 mm tool after 1.02 m of machining and Figure 2b shows the tool breakage of 1.0 mm tools after 1.8 m of machining.

Similarly, when machining Ti-6Al-4V titanium alloy, the 0.5 mm tool have a shorter tool life than 1.0 and 1.5 mm tools. The wear criterion of 0.5 mm tool is approximately 0.14 mm, which is slightly higher compared to machining H13 tool steel although both readings were taken at the same cutting time. It can be seen that the tool life criterion of 1.0 mm is 0.14 mm. This value is less than 0.16 mm obtained with 1.0 mm tools in machining H13 as seen in Table 3. Nevertheless, this value was taken after 12 minutes of machining compared to just over 11 minutes for machining titanium. This suggests that the wear rate in machining titanium is higher than machining H13. Tool wear increases dramatically into the tertiary zone with an extra one minute of machining until the tool breaks. For 1.5 mm tool, the wear in primary zone is slightly higher than the other tools. However, tool wear increases steadily until 12 minutes of cut, beyond which the wear increases sharply. This indicates that the tool has entered the tertiary wear stage, thus the tool reached its tool wear criterion at approximately 0.18 mm.

Table 3. Results of VB Criterion, Cutting Times and Volume Machined

Diameter (mm)	Material	VB Criterion (mm)	VB/Dia	Cutting Time at VB Criterion (minutes)	Total Cutting Time (minute)	Total Volume Machined (mm³)	Failure Mode
0.5	H13	0.11	0.220	6.46	7.85	5.10	Worn tools
0.5	Titanium	0.14	0.280	6.00	7.85	5.10	Worn tools
1	H13	0.16	0.160	12.00	13.85	90.00	Broken tools
1	Titanium	0.14	0.140	11.08	12.00	78.00	Broken tools
1.5	H13	0.17	0.113	12.00	14.77	216.00	Broken tools
1.5	Titanium	0.18	0.120	11.08	14.77	216.00	Worn Tools

The workpiece surface roughness R_a in the feed cutting direction was measured after initial machining. This was carried out after two cutting passes was performed for each tool. These results were then compared with the surface roughness values obtained when the tools reached their end of life. It is expected that the value R_a from the final cutting process would be much higher than initial R_a values. Although chatter or other factors can caused high surface roughness, the increases of surface roughness is normally due to increase in tool wear. Worn, fractured cutting edges or tools that generally have lost their original profile and sharpness due to wear will be unable to produce high quality machined surfaces.

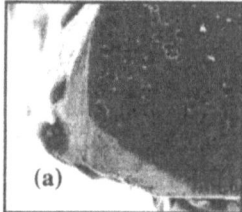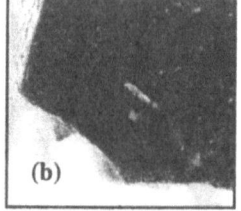

Figure 2. SEM images of some of the worn tools.

For example, if the tool edge was fractured at the early stage of machining, or when the tool reaches its end of life, the tool may end up rubbing the workpiece material generating higher level of force and vibration. The worn flank face will also have higher surface roughness thus increasing the coefficient of friction between the tool and the workpiece. The combination of vibration and increasing coefficient of friction will ultimately generate poor surface quality.

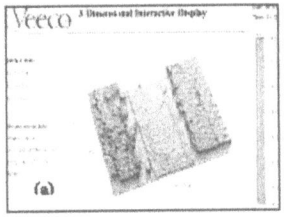
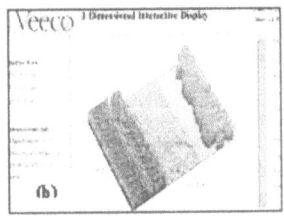

Figure 3. Surface roughness of initial and final machining of Ti6Al4V for the 0.5mm tool.

As flank wear increases the tool-workpiece contact area becomes larger and hence rubbing on the workpiece surface becomes stronger. This results in a poor surface finish as well as high friction forces and temperatures, which may eventually lead to tool breakage as in the case of 0.1 mm tool for machining titanium. Figures 3(a) and 3(b) show the surface roughness of 0.5 mm tools in machining titanium alloy. Initial machining produced a surface roughness of 0.14μm R_a. After 471 seconds, when the tools have reached its end of tool life, the surface roughness was 0.18μm R_a. The deterioration of surface quality can also be used as an indication of tool wear. However, unless an upper limit of surface roughness is specified, this figure will only serve as an indication of tool wear and would not be able to determine clearly if and when a particular micro tools have reached its wear criterion.

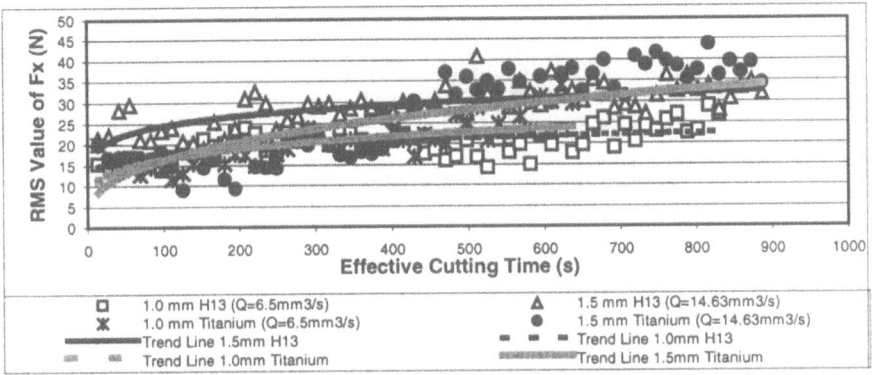

Figure 4. F_x cutting forces when machining H13 Steel and Ti-6Al-4V Titanium Alloy.

A close examination shows that the 0.5 mm tool was unable to produce a consistent and good surface finish towards the end of its life because it had lost a significant percentage of its effective cutting edge compared to its original shape. Similarly as the 1.0 mm tool in machining Titanium alloy had an early nose tip breakage, it could not produce good surface quality. Figure 4 shows the trend of the RMS value of F_x which has a degree of similarity with wear progression. The result suggests that as the flank wear increases, the cutting forces increases.

6. Conclusion

The main objective of this research is to investigate the wear progression and to estimate the wear criterion for micro tools in high speed milling. From the experimental results, the wear progression behaved in a similar manner as conventional sized tools showing primary, secondary, and tertiary wear stages. However, the tool life criterion is always smaller than the standard 0.3 mm flank wear as given in ISO 8688-2:1989. It also shows that in micro machining, tool life is unpredictable because the tools are prone to breakage due to sudden fluctuation in cutting forces. Cutting forces can be used as an indication of tool wear in machining. Further research with more micro tools is needed in order to give a more precise relationship of tool wear criterion for micro machining.

6. Acknowledgement

The authors would like to thank International Tooling Corporation (ITC) UK and Balzers Limited UK for supporting this investigation and to John Curran and Walter Perrie from Liverpool University, for their invaluable technical support.

Refererences

[1] I. Tansel, O. Rodriguez, M. Trujillo, E. Paz, W. Li., 1998, Micro-end milling-I, Wear and breakage, Int. J. Machine Tools and Manufacture 38, 1419-1436

[2] S.M. Pandit., 1982, Frequency decomposition of cutting forces in end milling, Proc. 10th NAMRC, 393- 400.

[3] P.E. Gygax., 1980, Cutting dynamics and process-structure interactions applied to milling, Wear 64, 161-184.

[4] S. Takata, M. Ogawa, P. Bertok, J. Ootsuka, K. Matushima, T. Sata., 1985, Real-time monitoring system of tool breakage using kalman filtering, Robotics and Computer Integrated Manufacturing 2 (1) 33-40.

[5] M.S. Lan, Y.Naerheim., 1986, In-process detection of tool breakage in milling, J. Eng. for Ind., Trans. ASME 108, 191-197.

[6] Y. Altintas, I. Yellowley., 1989, In-process detection of tool failure in milling using cutting force models, J. Eng. for Ind., Trans. ASME 111, 149-157.

[7] W. Konig, K. Kutzner, U. Schehl., 1992, Tool monitoring of small drills with acoustic emission, Int. J. Machine Tools and Manufacture 32, 487-493.

[8] L.Kops, D.T. Vo., 1990, Determination of the equivalent diameter of an end mill based on its compliance, CIRP Annals 39 (1), 93-96.

[9] ISO Standard, 8688-2, Tool life testing in milling-Part 2: End milling (1989).

Chip Morphology and Cutting Forces in High Speed End Milling of Titanium 6Al-4V Alloy

M. El-Houry, N. Driver and P. T. Mativenga[1]

[1]Department of Mechanical Engineering, UMIST, UK, p.mativenga@umist.ac.uk

Abstract: Titanium alloys pose considerable problems in manufacturing, especially when machined at high cutting speeds. Despite great advancements in cutting technology, only limited success has been achieved in improving the high-speed machinability of titanium alloys. In this paper, chip morphology, cutting forces and tool wear are studied under minimum quantity lubrication (MQL) and dry conditions in high speed milling using ball end tools. Predictive models for chip length in end milling are also revisited. For the range of speeds tested, resultant cutting forces and maximum chip lengths were found to reduce with cutting speeds. Dry machining was associated with shorter chips and lower cutting forces when compared to MQL conditions. In addition, the study also establishes a mechanism of chip segmentation in ball nose end milling of titanium at high cutting speeds.

1. Introduction

Studies based on various criteria such as tool life and productivity conducted by Komanduri and Reed [1], Machando and Wallbank [2] and Ezugwu and Wang [3] have shown that titanium and its alloys are difficult to machine. Low thermal conductivity, tendency to pressure weld, low modulus of elasticity, thermal plastic instability and fluctuations in cutting forces are among the characteristics that have been identified as factors behind the poor machinability of titanium alloy.

On the machining of Ti alloys with carbide tools, Drearnley and Grearson [4] and Donachie [5] state that uncoated straight grade cemented carbide (WC-Co) is usually the preferred tool material. This was later verified and confirmed by Jawaid et. al. [6] who found the straight tungsten carbide (WC-Co) grade to be superior to other cutting tool materials such as ceramic, solid lubricant coating (SLC) or even cubic boron nitride (CBN) in the machining of titanium alloys. Several cooling and lubrication techniques such as those employing Liquid Nitrogen (LN$_2$), High Pressure (HP) coolant and Minimum Quantity Lubrication (MQL) have been given great emphasis as potential strategies that would furthermore promote the high speed machining of titanium alloys. Despite the great concern, little research work focusing on chip formation has been done in the past years. Chip formation is influential to a great extent on the entire machining process as it determines the finish of the workpiece surface, cutting temperature, tool wear and cutting forces. In this paper, it is considered necessary to perform chip analysis and establish a theory explaining the formation of chips for high-speed ball-nose end milling of titanium alloys. In addition, cutting forces are measured and related to chip length.

2. Experimental Setup and Conditions

A ball nose end milling cutter was used in dry and MQL high speed machining of titanium 6-4 alloy. The experiment was performed on a Mikron HSM 700 with 12.5 kW of power and maximum rotation speeds of 42,000 rpm. The cutting process parameters used were an axial depth of cut of 0.2 mm, radial width of cut of 0.35 mm, feedrate of 6000 mm/min in up-cut milling mode. Spindle speeds of 7500, 15000, 22500 and 30000 rpm were used in the test. The test material was a 185x70x50 mm Ti 6Al-4V alloy rectangular block. The cutting fluid mist employed was a vegetable-based oil, VASCOMILL MKS 68. Samples of chips were collected for each test run. Cutting forces were measured using a three-component Kistler dynamometer (9265 B) of natural frequency 2.7 kHz in Z and 1.7kHx in X and Y. A two flute cutter was chosen so as to limit the tooth passing frequency to regions clear of the natural frequency of the dynamometer. The milling cutters used were Fraisa micro-grain carbide ball nose end-mills of tool diameter 6 mm, helix angle 30^0 and rake angle 0^0 with TiCN coating.

3. Theoretical Model for Chip Length

The maximum chip length, L_C, is the distance subtended between the 2 extreme corners of the chip. Figure 1 shows a view of an un-deformed chip in a plane perpendicular to the cutter axis.

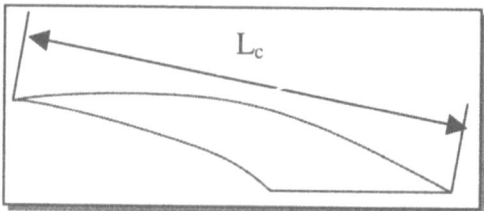

Figure 1 – Chip Length

The chip length can be approximated using various methods, DeVries [7] suggests two methods:

$$L_C \approx r\phi_s \qquad (1)$$

where ϕ_s is the swept angle subtended by the cutting edge during cutting and r is the tool radius.

When $b \ll r$, the following equation can be used:

$$L_C \approx \sqrt{2br} \qquad (2)$$

where b is the radial width of cut,

Problems can arise using these approximations of chip length. When S_Z, the feed per tooth, is comparable to b, the width of cut, as is typical case in HSM finish milling operations, equations (1) and (2) do not give an accurate indication of maximum chip length. Geometrical analysis of Figure 1 was preformed to obtain an equation for maximum outer profile or chip length, which also accounts for the feed motion. A more accurate method of estimating the chip length can be calculated by taking into account the feed motion of the cutter. From Figure 2, the following relationships can be deduced:

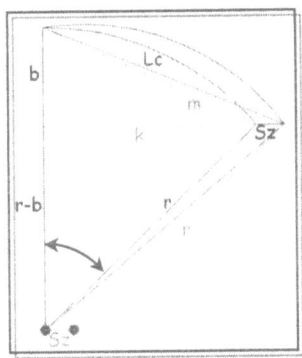

$$L_C = \sqrt{b^2 + (r\sin\theta + S_z)^2} \qquad (3)$$

This can then be rearranged to give:

$$L_C = \sqrt{2br + 2S_z\sqrt{2br - b^2} + S_z^{\,2}} \qquad (4)$$

For the case when b << r, equation (4) can be simplified to:

Figure 2 –*Cutting Kinematics* $$L_C \approx \sqrt{2br} + S_z \qquad (5)$$

4. Experimental Analysis: Results and Discussion

Chip samples were examined under a microscope. A basic understanding of the mechanism underpinning chip formation was established through measurement and comparison to theoretical models. Generally, chips can be sorted into two categories, segmented or serrated chips. Segmented chips were predominant throughout these experiments, nevertheless quite a significant number of serrated chips were also observed.

Serrated chips resulted from both dry and vegetable oil mist machining environments. Most chips suffered breaks at the base, a part of the chip where tremendous tearing had occurred. Tears caused sister chips to split. Figure 3 depicts the chip morphology: (a) serrated chip; (b) evidence of chip segmentation; (c) splitting mode; (d) proposed splitting theory.

The chips resulting from MQL machining environments were longer and wider than those obtained from dry machining as shown in Figure 4 and 5. The chip length predicted by the classical and proposed model was larger than the collected chips. This showed that chip spitting was a dominant occurrence in the cutting tests. In general, the size of the chips reduced with spindle speed. This could be attributable to enhanced chip splitting and reduced chip load as model by the proposed Equation 4.

Serration was a fairly commonly repeated phenomenon, with the segmented chips appearing to be resulting from the splitting of the sister chips. Measurements for individual chips extracted from the serration presented were done and recalling the twin chips for the same cutting conditions, it could be noted that the width of the base of one chip was almost twice as long as that of any of the chips here, whilst the length was slightly longer. Furthermore, when comparing the twin chips to the theoretical chip it was also observed that the width of the latter is roughly twice as that of any of the sister twins. Parts of chips were gathered and arranged thus giving the outline of a chip with dimensions close to those of the theoretical chip.

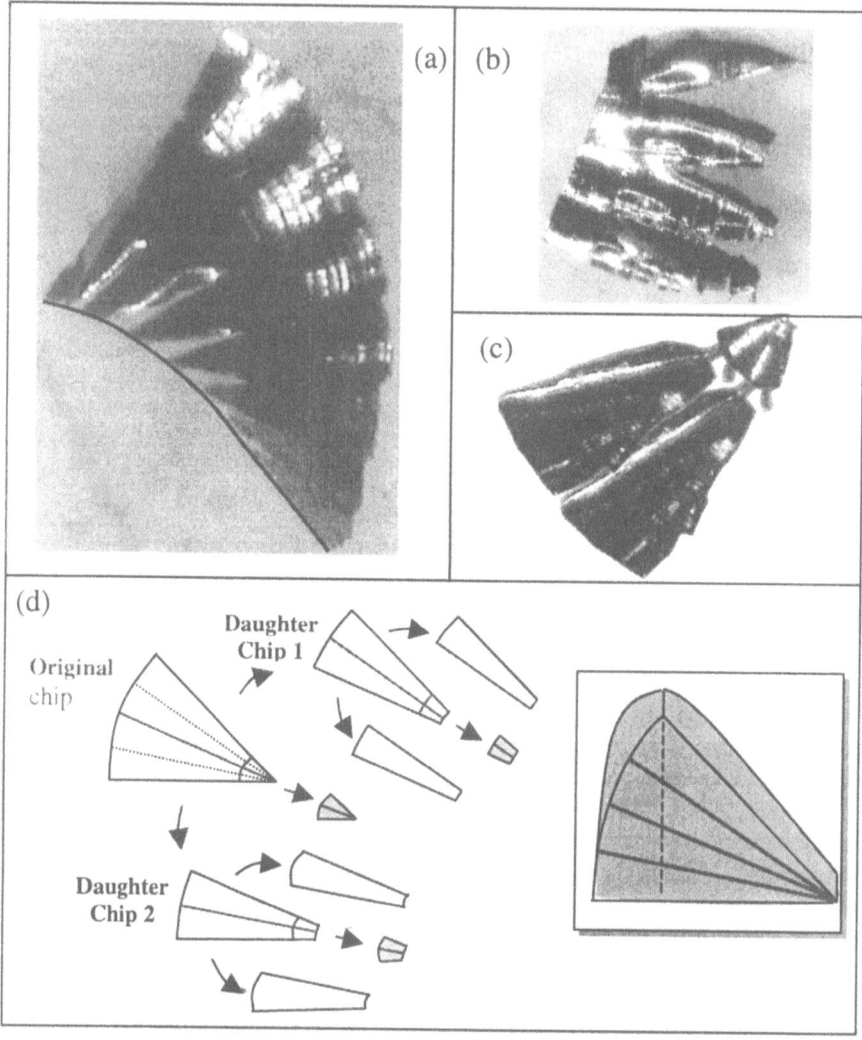

Figure 3 – Chip Serration & Segmentation

The resultant cutting forces for maximum and mean forces respectively are shown in Figure 6. This investigation revealed a decrease in cutting force with increasing cutting speed. This trend was observed for both segmented and serrated chip formation. Dry machining performed better than wet in this experiment with respect to lowering cutting forces.

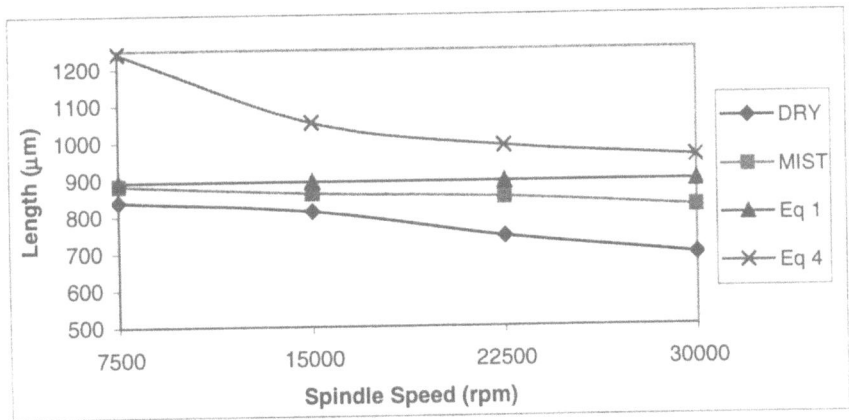

Figure 4 – Maximum Chip Length

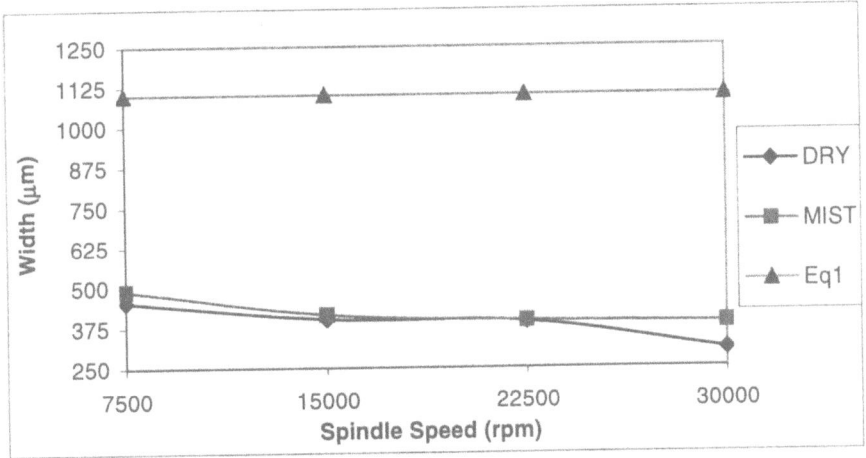

Figure 5 – Maximum Chip Width

In the dry machining of titanium alloys, cutting temperatures are expected to be higher than under MQL conditions. The high interface temperatures could account for the lower cutting forces observed under dry machining conditions. This study shows that there is a greater tendency for chips to segment under dry machining and this could be attributed to differences in temperatures at the tool-chip interface. In the cutting tests, the base of the chip always broke away from the parent chip. In ball end milling, cutting forces and strain rates are a function of the depth of cut due to the variations in chip load and of the cutting velocity respectively. The chip thickness variation for ball-nose end milling is given in Equation (6):

$$h = S_z \sin(\theta).\sin(\alpha) \qquad (6)$$

where h is the chip thickness, θ is the tool rotation angle, and α is the depth of cut emersion angle. The existence of chip segmentation in the width dimension in ball nose end milling is promoted by the differential strain and cutting force profile in the depth direction. Higher interface temperatures as in dry machining promoted segmentation as well.

As expected, mean values of forces, shown in figure 6, follow the same decreasing trend where both plots reveal a drop in root mean square forces at higher speeds, with the dry cutting forces experiencing a greater drop.

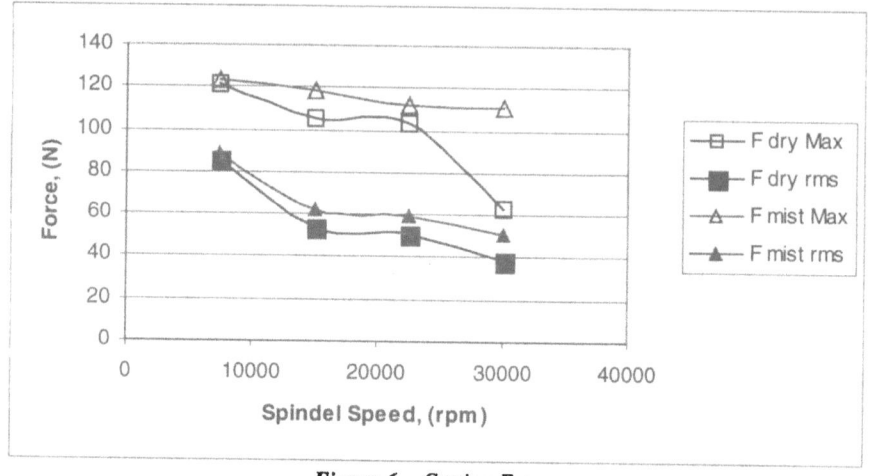

Figure 6 – Cutting Forces

5. Conclusions

Chip segmentation occurred in all cases covering the whole range of spindle speeds used. The combined effect of low thermal conductivity of titanium and high heat production led to higher local temperatures within shear bands. This paved the way for the formation of segmented chips. Splitting occurred twice and in each time the chip tended to split into two main daughter chips, which were roughly of equal sizes. Mist machining generated larger chips compared to dry machining. Forces in machining with vegetable oil mist were higher than dry machining and this correlated well with the findings reported above regarding chip geometry.

7. Acknowledgements

The authors wish to acknowledge the cooperation of Prof. K K B Hon from the University of Liverpool for the access to a machining centre for this investigation.

References

[1] Komanduri R and Reed W.R, 'Evaluation of carbide grades and a new cutting geometry for machining titanium alloys", *Wear*, Vol. 92, pp 113-123, (1983)

[2] Machado A.R and Wallbank J, "Machining of titanium and its alloys - a review", *Proceedings of the Institution of Mechanical Engineers*, Vol. 204, pp 53-59, (1990)

[3] Ezugwu E.O and Wang Z.M, "Titanium alloys and their machinability – a review", *Journal of Material Processing Technology*, Vol. 68, pp262-274, (1997)

[4] Dreamley P.A and Grearson A.N, "Evaluation of principal wear mechanisms of cemented carbides and ceramics used for machining titanium alloy IMI 318", *Materials Science and Technology*, Jan., Vol. 2, pp 47-58, (1986)

[5] Donachie M.J, "Titanium a technical guide", ASM International, OH, ISBN 0871703092, (1988)

[6] A. Jawaid, C.H. Che-Haron, and A. Abdullah, Tool wear characteristics in turning of titanium alloy Ti-6246, J. Mater. Process. Technol. 92-93 329-334, (1999)

[7] "Analysis of Material Removal Processes", W. R. DeVries Springer-Verlag, 135-138, (1992)

On Effect of Cryogenic Cooling on Tool Wear in Turning Ti-6Al-4V Alloy

K. A. Venugopal [1], S. Paul [2] and A. B. Chattopadhyay [3]

[1] Malnad College of Engineering
Hassan – 573 201, INDIA
kav@mech.iitkgp.ernet.in
[2] Indian Institute of Technology Kharagpur
Kharagpur – 721 302, INDIA
spaul@mech.iitkgp.ernet.in
[3] Indian Institute of Technology Kharagpur
Kharagpur – 721 302, INDIA
abcme@mech.iitkgp.ernet.in

Abstract: Low productivity due to rapid tool wear is the main reason for considering Ti-alloys as difficult to machine. Unlike in machining steels, crater wear in machining of Ti-alloys is equally significant as flank wear. In the present investigation, the maximum crater depth and flank wear have been measured while turning Ti-6Al-4V alloy with uncoated tungsten carbide inserts under dry, wet and cryogenic cooling environments. Compared to dry and wet machining environments, cryogenic cooling in the form of liquid nitrogen (LN_2) jets reduces the crater wear. Significant improvement in tool life also has been obtained under cryogenic cooling. However, SEM-EDAX analysis of the worn tool crater shows presence of Ti under all environments though in varying degrees.

1. Introduction

Rapid tool wear due to fluctuating stresses and thermal load, high cutting temperatures concentrated at the tool tip and chemical reactivity of Ti with all the tool and coating materials, puts Ti-alloys in the class of "difficult to machine" materials [1-3]. For economical machining of Ti-alloys, control of high cutting zone temperature through effective and proper cooling technique is very essential.

Cryogenic cooling using liquid nitrogen jets have yielded substantial technological benefits along with environmental friendliness in machining of alloy steels [4]. Such cryogenic cooling has also been initiated in machining titanium alloys [5-7].

The main objective of the present work is to investigate the relative role of cryogenic cooling over dry and wet conditions on tool wear in turning Ti-6Al-4V alloy by uncoated carbide inserts.

2. Experimental Procedure

Ti-6Al-4V rods of size ϕ 150 × 600 mm have been plain turned in a 11 kW centre lathe by ISO K20 SNMA 120408 type uncoated carbide inserts at cutting velocity of 85 m/min, feed of 0.2 mm/rev and 2.0 mm depth of cut under dry, wet (soluble

128

oil) and cryogenic cooling. Liquid nitrogen was impinged in the form of two jets through a specially designed nozzle towards the face and flank of the tool tip as shown in Fig.1

(a) Front view of the liquid nitrogen delivery nozzle

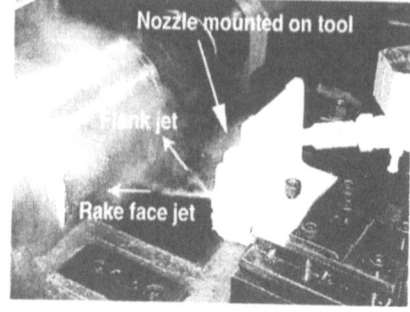

(b) Liquid nitrogen jets discharging from the nozzle

Fig.1 Liquid nitrogen delivery nozzle

At regular intervals the tool was withdrawn for studying its condition and measuring its crater and flank wear developed under different environments. Finally the worn out tools were observed under SEM.

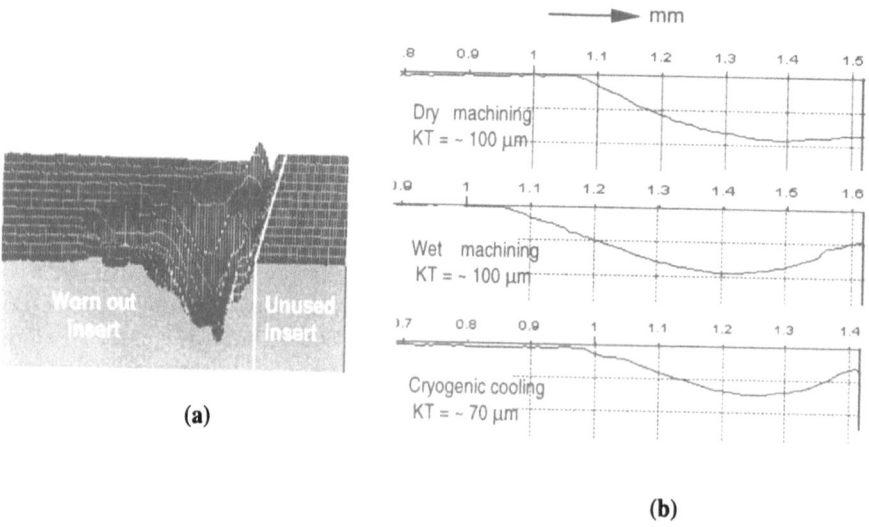

Fig.2 (a) Pattern of crater wear (b) Crater depth profile after 3 min. of machining Ti-6Al- 4V

3. Experimental Results

In machining materials like Ti-alloys, the crater wear is not only intensive but also irregular in depth with adherent deposits of work material. This warrants scanning the entire crater area to determine the deepest point and measure the maximum crater depth at that point. In the present study, the crater depth has been measured using Taylor Hobson NCS Profilometer. The crater area was scanned and mapped in 3D on a computer using Taly μultra software. Fig.2 (a) shows one such mapping while Fig.2 (b) typically shows the crater profile at the maximum crater depth region under different machining environments.

The pattern and extent of the tool wear attained under the three environments were observed in SEM. Fig.3 typically shows the SEM views. The progresses of the average flank wear (V_B), maximum flank wear (V_M), nose wear (V_S) and the edge depression (E) along with the crater depth with machining time are depicted in Fig.4

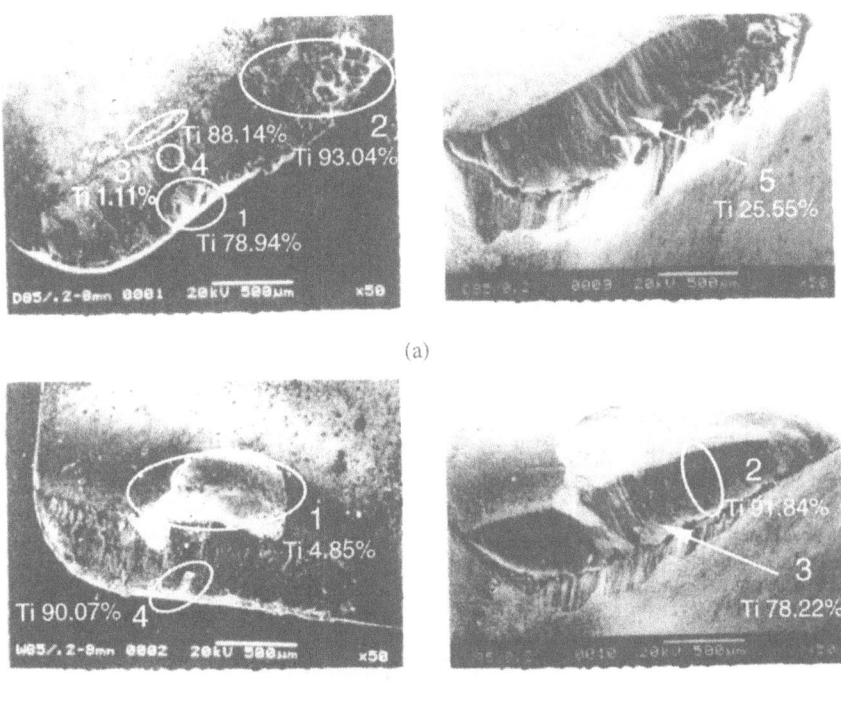

(a)

(b)

Fig.3 Tool conditions after 8 minutes of machining under (a) dry (b) wet and (c) cryogenic cooling conditions (views in the first column show the rake surface and those in the second column show the bird's eye view of the worn inserts. EDAX analysis was carried out at the encircled and numbered locations showing percentage of Ti) *(contd.)*

130

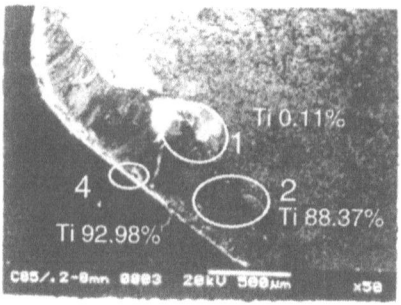 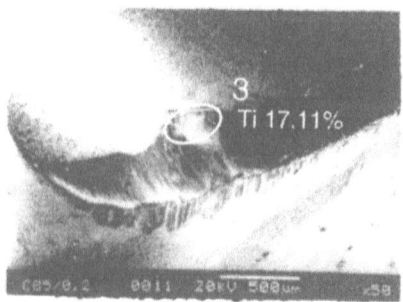

(c)

Fig.3 *(contd.)*

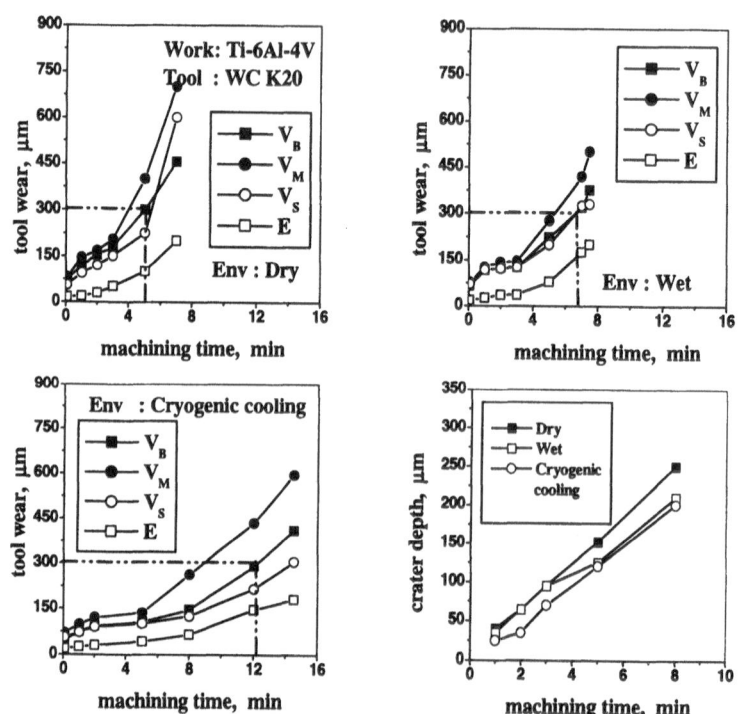

Fig.4 Growth of the various tool wear features in turning Ti-6Al-4V alloy
under different environments

4. Discussion

Most of the commonly used cutting tool materials reportedly suffer from chemical
instability and rapid tool wear when used in machining Ti and its alloys [3].
Further, titanium from the chip adheres to the tool preventing relative sliding at the
tool-chip interface. This leads to the formation of a boundary layer of titanium at

the interface and the relative motion between the chip and the tool is generated internally by shear within the chip. Repeated tearing and transport of this titanium layer by the chip underside will also result in tool material being pulled away causing increased crater wear. This phenomenon has been noticed in the present study as can be seen in Figs.2 and 3. The deep cratering noticed in Fig.2 and presence of Ti on the tool rake surface as noted in Fig.3 indicate strong adhesion of chip material on the tool rake face and rapid crater formation by adhesion wear and the grain pull out from the rake surface. This has been evidenced even under wet and cryogenic machining of this alloy as can be seen in Fig.3.

Fig.4 visualises both crater and flank wear which expectedly increased with machining time under all the three machining environments, the depth of crater being deepest under dry machining. Up to a machining time of 3 minutes, the crater depths under dry and wet machining were more or less equal and deeper (~ 100 µm) than under cryogenic machining (~ 70µm). However, at the end of 5 minutes, the crater depths under wet and cryogenic machining were almost equal (~ 130 µm) while it was ~ 150 µm under dry machining. Continued machining up to 8 minutes resulted in a crater depth of ~ 250 µm under dry machining and ~ 200 µm under wet and cryogenic machining.

Also, it is evident that the material loss from the tool face is more under dry and wet machining than cryogenic machining, implying more intense chip-tool interaction in the former case than the latter. This point is confirmed in the SEM views of the worn inserts. It is interesting to note that chunks of tool material have flaked off the tool face under wet and cryogenic, possibly due to thermal shocks/thermal stresses. Flaking has been more severe under wet machining. This phenomenon is absent in dry machining.

The SEM views in Fig.3 reveal adherent deposits of titanium from the chip on the crater and more profusely near the outer end of the cutting edge engaged, irrespective of the machining environments, though in varying amounts. The EDAX analysis carried out at different locations on the crater (encircled and numbered regions) confirms this fact. Piled up work/chip material at the depth of cut can be seen under dry machining. Such piled up deposits are notably absent under wet and cryogenic machining.

Though significant depression of cutting edge occurred under all the three machining environments, such detrimental effect was found to be minimum during cryogenic machining.

Examination of the worn flank in the SEM views reveals the relative beneficial effects of cryogenic machining, where the flank has undergone fairly uniform and lesser wear. It is more intensive and highly irregular under dry machining. It can also be observed that, under dry machining there is considerable auxiliary flank wear. Worn regions indicate severe abrasive marks compared to the crater where the surface appears smooth. At the end of 8 minutes of machining, the crater depths under wet and cryogenic machining were almost same. This indicates the inability of the cooling mediums to penetrate the chip-tool interface and control the high temperature. For a set tool life criteria of average flank wear, $V_B = 300$ µm (ISO 3685:1993), the tool lives obtained have been approximately 5 minutes and 6.5 minutes under dry and wet environments respectively while under cryogenic cooling, the tool life increased to 12 minutes, almost a hundred percent increase as depicted in Fig.4. This emphasises the fact that, proper application of cryogenic cooling offers substantial technological benefits along with clean environment in manufacturing industries.

5. Conclusion

a) Compared to steels, Ti-6Al-4V alloy, when machined causes much rapid and severe damage of the uncoated carbide inserts seemingly due to high chemical reactivity of that alloy resulting in intensive tool wear and grain pull out.

b) In machining of Ti-6Al-4V alloy by carbide tools, the strong chip-tool bondage prevents desired cooling effect even while cryogenic cooling by liquid nitrogen jets.

c) Unlike crater wear, the flank wear of the carbide tools significantly decreased by application of liquid nitrogen jets especially due to more effective cooling at the flank surface.

d) In machining Ti-6Al-4V alloy by carbide tools, the tool life assessed based on limiting average flank wear has significantly improved by proper cryogenic cooling.

Acknowledgement

The authors are thankful to the SERC division of Department of Science and Technology (DST), Government of India for the financial support [Sanction III.5 (104)/2000-ET dated 20.04.2001]

References

[1] Ezugwu E. O. and Wang Z. M., 1997 *Titanium alloys and their machinability,* International Journal of Material Processing Technology, **68**, 262-274

[2] Dearnley P. A.and Grearson A. N., 1986 *Evaluation of principal wear mechanisms of cemented carbides and ceramics used for machining titanium alloy IMI 318*, Material Science and Technology, **2**, 47-58

[3] Hartung P. D. and . Kramer M., 1982, *Tool wear in titanium machining*, Annals of CIRP, **31** (1), 75-80

[4] Dhar N. R, Paul S. and Chattopadhyay A. B., 2001, *Influence of cryogenic cooling on tool wear, dimensional accuracy and surface roughness in turning AISI 1040 and E 434DC steel*, Wear, **249**, 932-942

[5] Wang Z. Y, Rajurkar K. P and Fan J., 1996, *Turning Ti –6Al –4V alloy with cryogenic cooling*, Transactions of NAMRI/SME, **XXIV**, 3-8

[6] Hong S. Y, Ding Y, Jeong W., 2001, Friction *and cutting forces in cryogenic machining of Ti – 6Al – 4V*, International Journal of Machine Tools & Manufacture, **41**, 2271-2285

[7] Venugopal K. A, Tawade R., Prashanth P. G. et. al, 2003, *Turning of titanium alloy withTiB₂ –coated carbides under cryogenic cooling*, Proc. Instn. Mech. Engrs. **217** Part B: J. Engineering Manufacture, 1697-1707

Wear Characteristics of PCD when Machining Wood Based Composites

Philbin, P [1*], Gordon, S [2] and Pretorius, N[3]

[1, 2] Department of Manufacturing and Operations Engineering, University of Limerick, Plassey, Limerick, Ireland. paul.philbin@ul.ie
[3] Market Support Centre, Element Six, Shannon, Co. Clare, Ireland.

Keywords: Machining, Composites, Polycrystalline diamond.

Abstract: Wood based composite materials are used extensively for interior and exterior construction applications. These materials are typically pressed into sheet form and then machined to the required size and profile. Polycrystalline Diamond (PCD) is the cutting tool material increasingly used to machine these materials. The major benefit of using PCD is extended tool life resulting from its superior properties over traditional tool materials.

A test rig was designed to investigate PCD tool wear modes when sawing wood based composite materials. In order to save time the testing was done using a panel saw fitted with a replaceable tooth insert saw-blade adapted to single PCD edge format. Exploratory tests were carried out in order to establish the general wear modes and underlying wear mechanisms at a micron and sub micron scale. Abrasive and chipping wear modes were found for sawing different materials. The underlying wear mechanisms appeared to be dislodgement of PCD grains and micro fracture of the PCD respectively.

1. Introduction

A composite material consists of two or more base materials, combining the favourable characteristics of each [1]. Wood based composite materials are increasingly used in manufacturing applications. There is a broad range of these materials, ranging from engineered wood boards to inorganic-bonded wood composites [2, 3]. Tool wear, dimensional accuracy and the machined surface finish are important considerations in the machining of these materials [4].

Cutting tools used in traditional wood machining such as saws, drills and routers are now also used in the machining of wood based composites. The main cutting tool materials used in woodworking are stellites, cemented carbides, and polycrystalline diamond (PCD). While each cutting tool material has found its own niche of applications, the extreme hardness of diamond, its high thermal conductivity and low coefficient of friction make it an ideal material for cutting tools [5]. These properties are the reason for the superior performance of PCD and diamond-coated carbides over cemented carbide tools, such as increased wear resistance, improved ability to machine to closer tolerances and reduced acoustic emissions [6].

1.1 PCD as a cutting tool material

PCD tools are increasingly used in high volume machining of wood based composites due to their superior tool life and associated cost savings resulting from reduced downtime necessary for tool changing [7-13]. PCD is a synthesised, extremely tough inter grown mass of randomly oriented diamond crystals bonded to a tungsten carbide substrate. It is manufactured by sintering together micron sized diamond grains at high pressure and temperature in the presence of a solvent/catalyst metal, usually cobalt or cobalt/nickel alloy [5]. During the sintering process, the voids between PCD grains are filled the cobalt binder. Unlike cemented tungsten carbide however, individual diamond grains actually bond to one another [14]. The result is a tough, hard product that will retain its shape and strength if some of the metal matrix is removed. In the case of cemented carbide, when the binder phase is removed the tungsten carbide grains break away from the parent material and from each other. This difference was readily observed by looking at fracture surfaces of PCD and of cemented tungsten carbide. Fracture surfaces of PCD show brittle fracture of the diamond crystals, whereas in the case of cemented carbide, ductile fracture of the binder phase and inter-granular fracture of the carbide phase is predominant [7]. This explains why the wear of cemented tungsten carbide woodworking tools occurs by erosion of the matrix and subsequent loss of carbide grains [15, 16]. In the case of PCD, if the matrix is eroded grains are still held together by the bonding between them.

1.1.1 Wear modes of PCD in woodworking.

Micro chipping, abrasive wear (edge rounding) and gross tool fracture are the main PCD tool wear modes in woodworking. Gross tool fracture results in a sudden catastrophic failure of the cutting edge, usually in the early stages of cutting. Foreign materials in the workpiece such as hard inclusions in chipboard are believed to cause tool fracture [11]. PCD is more vulnerable to fracture than tougher tungsten carbide tools [7]. Once a tool is fractured it is immediately removed from service because the resulting workpiece surface finish becomes poor.

All other wear modes are gradual and do not lead to instant failure of the cutting edge. Micro chipping of PCD can occur when machining hard materials at aggressive feed rates, e.g. 0.4 mm/tooth and above. Chipping occurs where the fracture toughness of the PCD is exceeded in a local area at the cutting edge. Research shows that chipping wear occurs when machining inhomogeneous wood-based materials [17]. Fine grained PCD is less susceptible to edge chipping in the machining of wood based composites [18]. The degree to which chipping will determine the life of the tool depends primarily on the quality of the finish required on the work piece [15]. Abrasive wear occurs when the cutting edge recedes uniformly. It is thought that abrasive wear occurs on the PCD edge when machining homogenous wood-based materials [17]. Abrasive wear will increase steadily with the amount of material machined, until a point is reached where the tool cannot cut effectively or produce the desired surface finish. The tool is then replaced and removed from service for disposal or regrinding where possible.

In recent investigations into PCD tool wear modes in woodworking it was proposed that the PCD wear mechanism occurs by the initiation of micro-cracks. These cracks were assumed to have occurred due to external impact loadings contributing to cleavage fracture within PCD grains, thus weakening the tool edge and leading to wear. Inter-granular wear, grain cleavage, peeling and spalling of grains have been proposed as possible wear mechanisms of the PCD [17, 19].

2. Equipment

Due to its superior properties over traditional cutting tool materials, massive amounts of wood based composite have been machined with PCD without any significant tool wear [20-23]. In order to investigate the wear modes of PCD when sawing wood based composites it was necessary that a test rig be developed to carry out a time compression test using the most abrasive wood based composites available. Design of the test rig used in this work has been reported previously [24].

The test materials selected were Fibre Cement Board (FCB) and High Pressure Laminated Flooring (HPL). FCB consists of wood chips and Portland cement. The cement contains Silica which is abrasive to machine. HPL consists of a high-density fibreboard with a ceramic overlay and backing layer which are also abrasive to machine. Machining these abrasive materials caused rapid wear so that it was possible to investigate the initial wear of the PCD cutting tools.

The saw teeth used in the sawing tests were prepared with PCD by the industrial partner. A 10-micron grain size, general purpose PCD grade was used in all of the machining tests. The PCD segments were wire EDM cut to size and then brazed on to a saw tooth. Initial tools were mechanically ground using an EWAG diamond-grinding machine. A slight chamfer was also ground on the PCD corners where the relief faces meet. This was done to prevent chipping of the PCD at these edges. The tools used had an effective rake angle of $6°$ and an effective relief angle of $18°$ resulting in a cutting tool included angle of $66°$.

3. Experimental Procedure

Sawing tests were carried out using the test rig and workpiece materials described above. A single board was sawn for FCB tests and three stacked boards were sawn for HPL tests. The width of cut was 3mm, half the width of the PCD cutter. A constant cutting speed of 2826 m/min (3000 rev/min) was used across all tests. Feed rates were varied across the tests, between 0.05 and 0.2 mm/tooth. The feed rate values were selected to match feed per tooth values commonly used for PCD sawing. Larger sliding distances resulted from using lower feed rates.

When testing was complete the cutters were first examined using a Meiji Stereo Optical Microscope at 15X and 45X magnification to identify tool wear modes. The wear was initially classified either as abrasive or chipping wear. Abrasive wear was thought to be uniform and to increase linearly with the distance cut. The severity of the abrasive wear was determined by measuring the maximum and average wear land (VB) values on the relief face. Chipping wear was thought to be stochastic however, so it may not be repeatable at a given feed rate. Therefore it was important to measure the concentration of the chipping wear, i.e. the number of chips on cutting edge and use this as a measure of repeatability. The severity of chipping wear, that is the depth of the chips on the relief face was also measured.

Later a Scanning Electron Microscope (SEM) was used to investigate the possible underlying tool wear mechanisms on a micron and sub-micron scale.

4. Results and Discussion

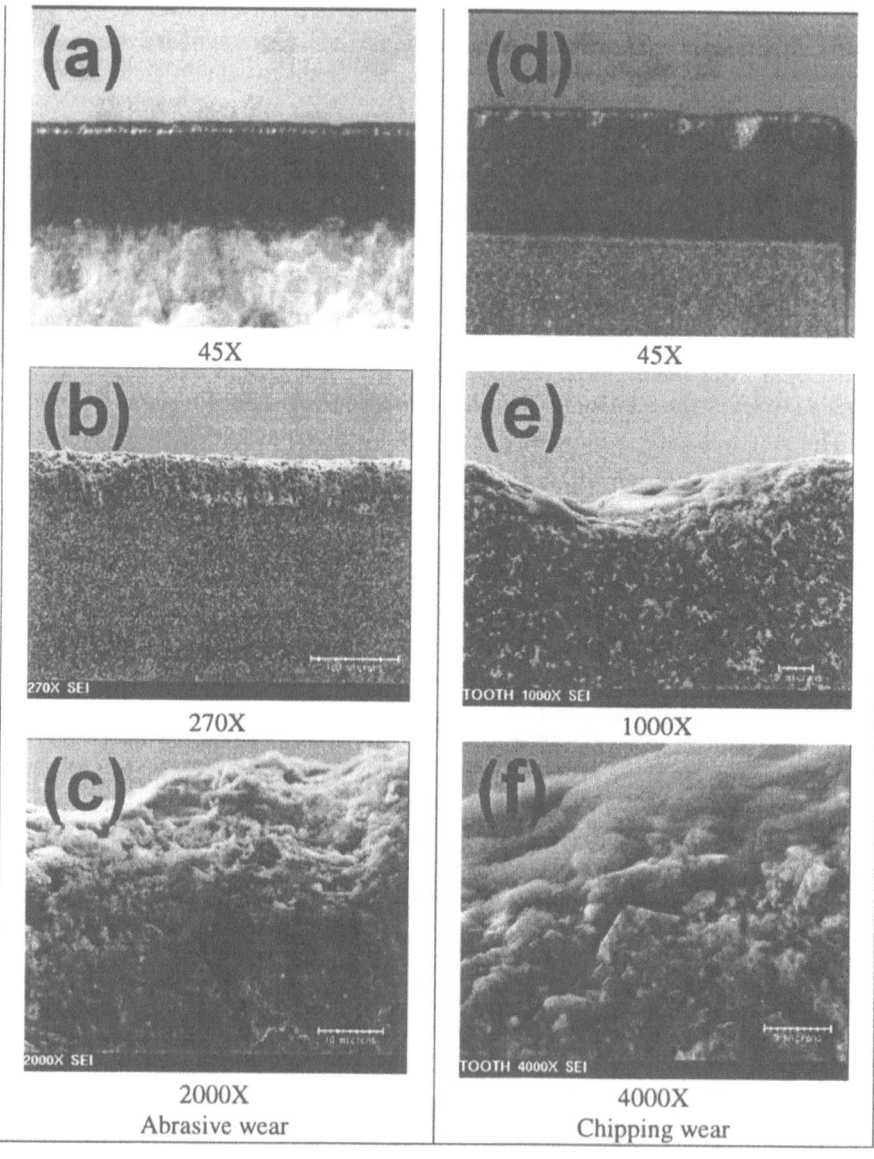

Figure 1: Tool wear pictures

Where FCB was machined abrasive wear was the dominant wear mode. The average maximum wear land value was 0.1 mm. Small voids, around 10 microns in size were noted in the wear surface when viewed under optical microscope (Figure 1(a)) and SEM (Figure 1(b)). The voids seem to indicate locations where diamond grains were dislodged from the cutting edge. A close up view revealed the damaged surface (Figure 1(c)). It is possible that diamond grains became weakened after abrasion of the cobalt binder around them, and then became dislodged from the tool edge when subjected to the intermittent cutting force in the panel sawing operation.

Where HPL was machined a small amount of edge recession was noted, but chipping wear was the dominant wear mode. Chips of up to 160 microns in size were observed (Figure 1(d)). The average chip size was 100 microns and the average number of chips on an edge was two. Closer examination of chips in the SEM revealed that the chipped surface is much more clearly defined (Figure 1(e)). The surface resembles a fracture surface of PCD, with diamond grains clearly evident (Figure 1(f)). It is thought that the whole chip surface was produced in one instance of fracture, when the strength of the cutting edge was exceeded. Intermittent cutting through the hard ceramic layer of the HPL seems to be the reason for chipping of the PCD edge.

While micro cracks have been reported previously for PCD machining tests on wood based composites [17, 19], no evidence of micro cracks was noted in this study.

5. Conclusions

1. A test rig for the panel sawing of wood based composites was successfully implemented. The single edge blade allowed for some time compression of wear testing (sawing) on the work materials described.

2. Tests implemented were only considered to go into the very initial region of tool wear for PCD when machining wood based composites. No deterioration in the quality of the sawn edge was observed for these tests.

3. Abrasive wear was the dominant PCD wear mode when sawing the relatively homogenous FCB material, especially at lower feed rates where sliding distance was greater. Further investigation of this wear mode using SEM observations indicated that dislodging of diamond grains after binder removal is the probable wear mechanism.

4. Chipping wear was the dominant PCD wear mode when sawing the inhomogeneous HPL material. It is proposed that chips are produced in one instance of fracture, when the strength of the cutting edge was exceeded.

References

[1] Kaw, A.K., *Mechanics of Composite Materials.* 1997: Boca Raton : CRC Press.

[2] Smulski, S., ed. *Engineered Wood Products: A Guide for Specifiers, Designers and users.* 1997, PFS Research Foundation.

138

[3]   Dietz, T. and K. Bohnemann. *Calcium Silicate Hydrate in Fiber Cement Sheets and Autoclaved Aerated Concrete*. in *Inorganic bonded wood and fiber composite materials*. 2000.

[4]   Gordon, S. and M.T. Hillery, *A review of the cutting of composite materials*. Proceedings of the I MECH E Part L Journal of Materials:Design and Applications, 2003. **217**(1): p. 35-45(11).

[5]   Wilks, J., *Properties and applications of diamond*.

[6]   Clark, I.E., *Polycrystalline diamond tooling in the woodworking industry*, in *Finer points*. 1995. p. 6-20.

[7]   Prakash, L. *Comparison of Polycrystalline and cemented carbide tools in woodworking*. in *Superabrasives 1991*. 1991.

[8]   Jennings, M., *PCD plays its part at Galtee*, in *Industrial diamond review*. 2000. p. 256-258.

[9]   Hayes, D., *Low noise PCD tooling from Japan*, in *Industrial diamond review*. 1999. p. 62-64.

[10]   Gittel, H.J., *Cutting tool materials for high performance machining*, in *Industrial diamond review*. 2001. p. 17-21.

[11]   Gladu, J. *Developments in applications of PCD in woodworking*. in *Superabrasives 1991*. 1991.

[12]   Bai, Q., Y. Yao, and S. Chen, *Research and development of polycrystalline diamond woodworking tools*. International Journal of Refractory Metals and Hard Materials, 2002. **20**(5-6): p. 395-400.

[13]   Cook, M.W. and P.K. Bossom, *Trends and recent developments in the material manufacture and cutting tool application of polycrystalline diamond and polycrystalline cubic boron nitride*. International Journal of Refractory Metals and Hard Materials, 2000. **18**(2-3): p. 147-152.

[14]   Anthony, D.J. *Concepts of Polycrystalline Diamond Tooling for the Woodworking Industry*. in *First international symposium on tooling for the wood industry*. 1990. Raleigh, North Carolina.

[15]   Steinmetz, K. *PCD Polycrystalline Diamond - Tailor made for woodworking applications*. in *Superabrasives 1991*. 1991.

[16]   Sheikh-Ahmad, J.Y. and J.A. Bailey, *The wear characteristics of some cemented tungsten carbides in machining particleboard*. Wear, 1999. **225-229**(1): p. 256-266.

[17]   Miklaszewski, S., M. Zurek, P. Beer, et al., *Micromechanism of polycrystalline cemented diamond tool wear during milling of wood-based materials*. Diamond and Related Materials, 2000. **9**(3-6): p. 1125-1128.

[18]   *Syndite Product Range*. 2004, Element six.

[19]   Bai, Q.S., Y.X. Yao, P. Bex, et al., *Study on wear mechanisms and grain effects of PCD tool in machining laminated flooring*. International Journal of Refractory Metals and Hard Materials. **In Press, Corrected Proof**.

[20]   Sen, P.K., M.W. Cook, and R.D. Achilles. *Various diamond cutting tool materials for the machining of HPL wood flooring*. in *Ultrahard materials technical conference*. 1998. Windsor, Ontario, Canada.

[21]   Steinmetz, K. and P.J. Heath, *SYNDITE CTC002 - A PCD material for woodworking*, in *Industrial Diamond Review*. 1989.

[22]   Jennings, M., *Medite uses SYNDITE*, in *Industrial Diamond Review*. 1997.

[23]   Collins, J.L., M.W. Cook, and T. Ninnis, *New developments in ultrahard machining of wood*, in *Industrial Diamond review*. 2001.

[24]   Philbin, P. and S. Gordon. *Machining of wood based composites using a polycrystalline diamond (PCD) saw blade*. in *20th International Manufacturing Conference*. 2003. Cork institute of Technology, Cork, Ireland: CIT Press.

ELECTRO-PHYSICAL AND CHEMICAL PROCESSES

System Identification and Controller Design for the Tool Position Control System of an Electro Chemical Discharge Machining (ECDM) Machine

T.K.K.R. Mediliyegedara[1], A.K.M De Silva[2], D.K. Harrison[3], J.A McGeough[4]

[1]School of Engineering, Science and Design, Glasgow Caledonian University, Glasgow, G4 0BA, U.K., k.mediliyegedara@gcal.ac.uk
[2]School of Engineering, Science and Design, Glasgow Caledonian University, Glasgow, G4 0BA,
[3]School of Engineering, Science and Design, Glasgow Caledonian University, Glasgow, G4 0BA,
[4]School of Engineering and Electronics, The University of Edinburgh, U.K., Alrick Building, The King's Buildings, Mayfield Road, Edinburgh, EH9 3JL,.

Abstract: The performance of an Electro Chemical Discharge Machining (ECDM) machine, in terms of surface finish and rate of machining is significantly influenced by the performance of the tool position control system, process controller, and the pulse generator. This investigation concentrates on the tool position control system and the paper presents a system identification for the controller design.

Closed-loop identification is carried out in this work. A variety of system identification models were utilised for the model estimation. These estimated models were then simulated and the best fit model was identified and selected. A model verification process was carried out to verify the accuracy of the estimated model. Further, a controller was designed for the tool position control system of an ECDM machine.

1. Introduction

1.1 Electro Chemical Discharge Machining (ECDM)

ECDM is a manufacturing process which combines features of Electro Chemical Machining (ECM) and Electro Discharge Machining (EDM), by having a cathodic tool and an anodic workpiece, which are separated by a gap filled with an electrolyte, and pulsed DC power applied between them. This leads to discharges in the electrolytes, thus achieving both electrochemical dissolution and electro-discharge erosion of the workpiece [1]. One of the major advantages of ECDM, over ECM or EDM, is that the combined metal removal mechanisms in ECDM, yields a much higher machining rate [2].

1.2 System identification

There are three main ways to identify the dynamics of a system. One approach is to develop a mathematical model of the system using the laws of physical sciences. The second approach is based on the experimental data. The third approach is combination both approaches mentioned above. The process of constructing models from experimental data is called system identification.

1.3 Closed loop identification

In closed loop system identification, the identification data are collected from a closed loop test where the underlying process is fully or partly under feedback control. There are many advantages of closed loop tests over open loop tests such as it is easy to carry out because manual control action will be reduced considerably or eliminated; it produces better model for control; under the same variance or amplitude constraints of the outputs, the control performance of the model will be higher, because input amplitude can be larger in a closed loop test [3]. In this work, a closed loop test was carried out due to the above mentioned advantages.

2. Selection of System Identification Model

The following models are common in identification literature: Finite Impulse Response (FIR) model, Auto Regressive with eXternal input (ARX) model also known as least-square error model, Output Error (OE) model, Auto Regressive Moving Average with eXternal input (ARMAX) model, Box-Jenkins (BJ) model. ARX ,OE, ARMAX and BJ models are called parametric models and the FIR model is called a non-parametric model. For closed loop identification, the choice of a model structure depends on three often conflicting issues: The compactness of the model; The numerical complexity in parameter estimation; The consistency of the model in closed loop identification. Table 1 compares the advantages and disadvantages of various model structures. From Table 1, it is clear that only possible identification model structures are ARX model, ARMAX model and Box-Jenkins model for a closed loop identification test.

Table 1: Comparison of various model structures for the system identification [3]

Model structure or estimation method	Numerical difficulty	Compactness	Consistency in closed-loop test
FIR	Low	Low	No
ARX	Low	Medium	Yes
OE	High	Highest	No
ARMAX	High	High	Yes
BJ	High	Highest	Yes

3. Design of the System Identification Experiment

3.1 Tool position control system

The tool position control system of the ECDM machine consists of a DC brushless servo motor, a brake, a ball screw, a linear encoder and a tool holder. Figure 1 shows a schematic diagram of the tool position control system. A brushless type DC servo motor was selected as an actuator for the tool position control system. A dual shaft motor was selected, since it is required to mount a brake on one side of the motor and a ball screw on the opposite side. A ball screw is used to convert a rotary motion into a linear motion. Since it is possible to reduce the backslash, a ball screw was selected for the design. A linear encoder is used to obtain the position feedback of the tool holder. The tool holder is used to clamp a tool.

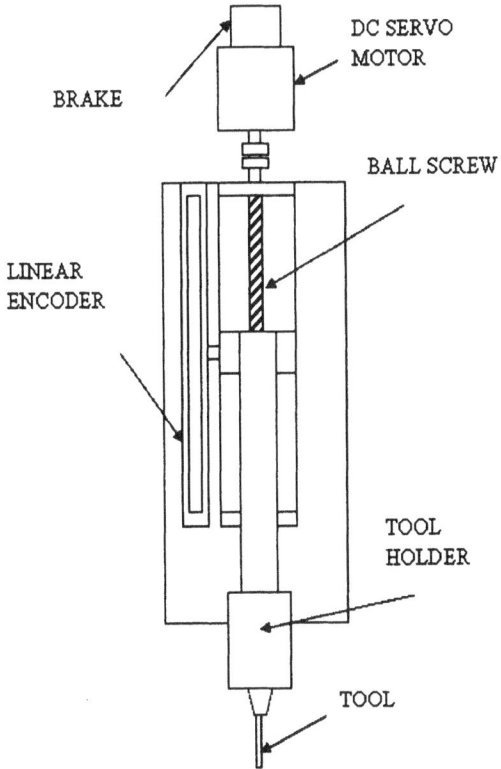

Figure 1: A schematic diagram of the tool position control system

3.2 Experimental setup

Figure 2 show a block diagram of the system identification experimental setup. A motion control board, which is capable of producing an analogue output and capable of obtaining an encoder input, is fixed to a Personal Computer (PC). A linear optical encoder, the resolution of which is 1 μm, was used as the linear position feedback device.

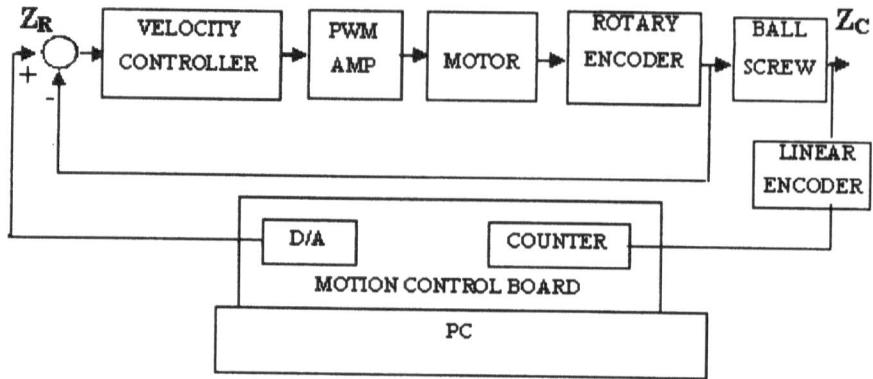

Figure 2: A block diagram of the system identification experimental setup

A motion control algorithm, which is used to count the number of encoder pulses and to produce a reference signal (Z_r) to the tool position control system, was programmed with the LabVIEW 7.0 software package. The MATLAB 6 software package was used to estimate the parameters and validate the model.

3.3 Selection of the input signals [4]

A Discrete Interval Binary Noise (DIBN) was employed to excite the system persistently. The sampling time of the tool position control system is estimated to be 10ms. Therefore, the sampling time of the identification experiment was selected as 10 ms. The switch time of the DIBN sequence is usually set to about 1/3 of the settling time of the plant. The settling time of the system was calculated by using an experimental step response. The settling time of the system was found to be 300 ms. Therefore, the switch time of the DIBN was selected as 100 ms. The amplitude of the DIBN sequence was selected as 10000 pulses. The length of the DIBN sequence was selected to be as 53. Therefore, the total identification time is 5.3 s.

4. Parameter Estimation

4.1 Estimation of the transfer function and model verification

The MATLAB 6 software package was used to estimate the model. The number of delays (nk) was obtained from an experimental step response which is shown in Figure 3. The magnitude of the step response is 10000 pulses. ARX, ARMAX, BJ models were obtained. It was found that an ARX model gives a consistent and accurate model when the order of the polynomials A, B, and C are 2, 2, and 1 respectively. Step signal having 10000 encoder pulses were used to simulate the estimated model. A simulated step response is shown in Figure 4.

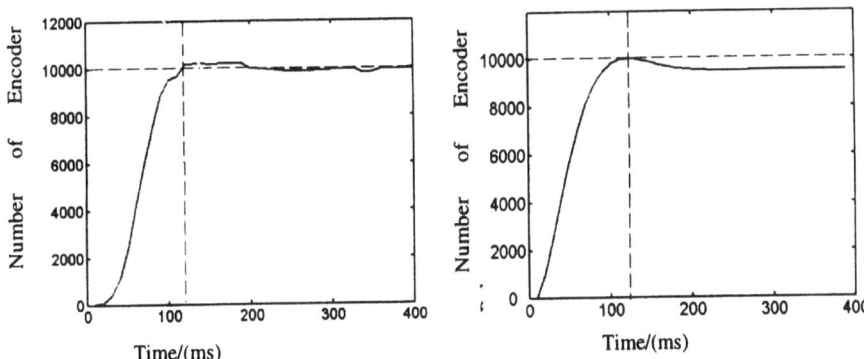

Figure 3: An experimental step response   Figure 4: A simulated step response

4.2 ARX model

An ARX model was selected, since it gives a simple transfer function with required accuracy. The general form of an ARX model can be written as follows.

$$y(t) = G(q)u(t) + H(q)e(t) \qquad [1]$$
$$G(q) = q^{-nk} B(q)/A(q) \qquad [2]$$
$$H(q) = C(q)/A(q) \qquad [3]$$
$$A(q) = 1 + a_1 q^{-1} + \cdots + a_{na} q^{-na} \qquad [4]$$
$$B(q) = b_1 + b_2 q^{-1} + \cdots + b_{nb} q^{-nb+1} \qquad [5]$$
$$C(q) = 1 + c_1 q^{-1} + \ldots c_{nc} q^{-nc} \qquad [6]$$

Where, A, B and C are the polynomials of the delay operator q^{-1}. The constants na, nb and nc are the orders of the polynomials A, B and C respectively. The number nk is the number of delays.

5. Controller Design and Future Work

As far as the ECDM process is concerned, it is necessary to have a zero overshoot in the tool position control system. To achieve this requirement a Proportional-Integral-Derivative (PID) controller was designed. Rise time, settling time and maximum overshoot of the tool position control system were 5ms, 50ms and zero percent respectively.

Strategies for the process control of EDM and ECM have been studied in the past. But there remains a need for an effective and efficient gap width control system for ECDM. Some researchers have employed fuzzy controllers for EDM process control [5]. The future studies will be directed toward the designing of a fuzzy controller to control the gap width of the ECDM machine. Further, the above estimated model and the parameters will be used to design the ECDM process controller.

6. Conclusions

A closed loop system identification experiment was carried out to identify the dynamics and to develop a model of a tool position control system for an ECDM machine. An ARX model was developed. The model was verified by using experimental step response data. A PID controller was designed. A closed loop test is easy to carry out, as it reduces the influence of disturbances to the process operation. The future work will be directed towards the design and implementation of an intelligent process controller for the ECDM process.

References

[1]   De Silva A.K.M, Khayry A.B. and McGeough J.A., 1995, *Process Monitoring and Control of Electroerosion-dissolution Machining*, IMechE Conference Transactions, 11th International Conference on Computer-Aided Production Engineering, pp. 73-78.

[2]   De Silva A.K., 1988, Process *Developments in Electrochemical Arc Machining*, PhD Thesis, University of Edinburgh.

[3]   Zhu, Y.C. and Butoyi, 2002, *Case studies on closed-loop identification for MPC*, Control Engineering Practice, Vol.10, pp. 404-411.

[4]   Pathirana, S.D., 1989, *Self – tuning   tracking and compliance control of a manipulator actuated by pneumatic artificial muscles*, PhD Thesis, University of Tokyo.

[5]   Yan, M.T., Li, H.P. and Liang J.F., 1999, *The Application of Fuzzy Control Strategy in Servo Feed Control of Wire Electrical Discharge Machining*, International Journal of Advance Manufacturing Technology, vol 15, pp. 780-784.

Mathematical Modelling for Wire Electrical Discharge Machining of Aluminum-Silicon Carbide Composites

Samy Ebeid[1], Raouf Fahmy[2] and Sameh Habib[2]

[1] Faculty of Engineering, Ain Shams University, 11517 Cairo, Egypt,
drsami@menanet.net
[2] Shoubra Faculty of Engineering, Zagazig University, Cairo, Egypt.

Abstract: The increased role of advanced materials in engineering design optimization has enhanced the development of new material types. Metal matrix composites now allow the designer to select needed material properties. A key in the use of these materials is the cost effective manufacturing and dimensional control of the designed parts. Composites are a new class of engineering materials which have a high potential for many industrial applications. The traditional methods of machining these materials are very slow and expensive, and can also cause strength degradation due to the formation of subsurface cracks or other defects. The WEDM process is one of the best alternatives, for machining an ever-increasing number of high-strength, non-corrosive and wear resistant materials such as Aluminum Silicon Carbide (AlSiC).

In this work, an attempt has been made to develop mathematical models for optimizing wire electrical discharge machining characteristics such as the material removal rate, cutting speed and the surface roughness. The process parameters taken into consideration are the average machining voltage, the pulse frequency, the workpiece height, the kerf size and the percentage volume fraction of SiC present in the aluminum matrix.

1. Introduction

Composite materials are formed by the combination of two or more materials to achieve properties that are superior to those of its constituents. The main components of composite materials are fibres and matrix. The fibres provide most of the stiffness and strength, and the matrix binds the fibres together. Other substances are used to improve specific properties. The demand for materials having high mechanical performance has led to the development of a new generation of materials, such as AlSiC alloy composites. The mechanical properties of AlSiC offer a combination of higher values of hardness, strength and stiffness. Another important feature of the material and process combination is that the designers can specify mechanical strength criteria previously not feasible.

The presence of the hard reinforcing ceramic makes these materials difficult for conventional machining. Amongst various non-conventional machining methods, wire electrical discharge machining (WEDM) is one of the most versatile and useful technological processes for machining intricate and complex shapes as it is independent of material hardness [1,2,3].

In wire electrical discharge machining, a thin 50-300 µm diameter wire is used as the electrode. Material is eroded ahead of the traveling wire by spark discharges. Either the workpiece or the wire can be moved to cause the wire to cut similarly to a band saw. There is no mechanical contact between the wire and the workpiece in wire EDM [4].

The main objectives of this research are: to study the effects of the volume percentage of SiC in the composite, pulse frequency and average machining voltage on the metal removal rate, cutting speed and the surface roughness. In order to improve the performance of this process, optimal cutting conditions have to be determined and so mathematical models need to be established. Thus, the present work develops the mathematical models necessary to predict the MRR, CS and the surface roughness within the operating region.

2. Experimental Work

The experiments are performed on a high precision 5 axis CNC wire electrical discharge machine (Robofil 300), manufactured by Charmilles Technologies Corporation. The Robofil 300 allows the operator to choose input parameters according to the material and height of the workpiece and tool material from a manual given by the WEDM manufacturer. In this work, the wire used is hard brass 250 μm diameter. The wire travelling speed ranges between 8-12 m/min. In wire EDM, the cutting speed is defined as the area cut by the electrode wire in unit time (mm²/min) [5]. The cutting speed and material removal rate have been calculated [6] and the kerf size was estimated as twice the offset value. To calculate the pulse frequency, the time between two pulses (B) in μs and duration of pulse (A) in μs were measured. The pulse frequency was then estimated from the following relation:

$$F = 1 / (A + B) \tag{1}$$

The surface roughness (Ra) of each machined workpiece was measured using the TAYLOR-HOBSON Taylsurf. Each experiment was replicated twice for better results.

For the purpose of building mathematical modelling for the material removal rate, cutting speed and surface roughness, some factors have been considered such as pulse frequency (F), average machining voltage (V), kerf size (K), percentage volume of SiC (P), workpiece height (H) and machining feed rate (V_f).

The volume fraction of SiC present in the aluminum matrix used during the experiments ranged from 6% to 25%. The selected workpiece height ranged from 5 to 25mm. Also, three levels of surface roughness (rough, semi fine and fine) were selected for this work. Based on these settings a total of 60 experiments each having a combination of different levels of factors were carried out.

The mathematical models for MRR, CS and surface roughness are built by using the LAB Fit program for data fitting. The program builds up the best fitting equation with a regression value of not less than 0.98.

3. Results and Analysis

The purpose of developing the mathematical models relating the output parameters (MRR, CS and Ra) with their process parameters (F, V, K, P, H and V_f) is to facilitate the optimization of machining metal matrix composites by WEDM. The mathematical models are thus represented by:

$$MRR = 0.5362\ F - 0.00174\ F^2 - 0.289\ V + 0.00511\ V^2 - 848.979\ K$$

$$+ 2525.624\ K^2 - 0.7638\ P + 0.0151\ P^2 + 2.635\ H - 0.0439\ H^2 \qquad (2)$$

$$CS = 1.3912\ F - 0.00425\ F^2 - 1.5656\ V + 0.016\ V^2 - 1672.171\ K$$

$$+ 5262.145\ K^2 - 2.5612\ P + 0.0486\ P^2 + 8.8565\ H - 0.1567\ H^2 \qquad (3)$$

$$Ra = -0.01422\ F + 0.0000432\ F^2 - 0.1564\ V + 0.001014\ V^2 + 48.58\ K$$

$$- 47.558\ K^2 - 0.0208\ P + 0.0003\ P^2 - 0.1609\ V_f + 0.00321\ V_f{}^2 \qquad (4)$$

Figures 1 to 8 show the interactive effects of the pulse frequency, average machining voltage, kerf size, workpiece height and machining feed rate versus the percentage volume of SiC on material removal rate and surface roughness. From Fig. 1 it is evident that, with the increase in pulse frequency for all percentage volumes of SiC, the MRR decreases non linearly: however, the increase in percentage volume decreases the MRR. This is due to the presence of a large amount of SiC particles.

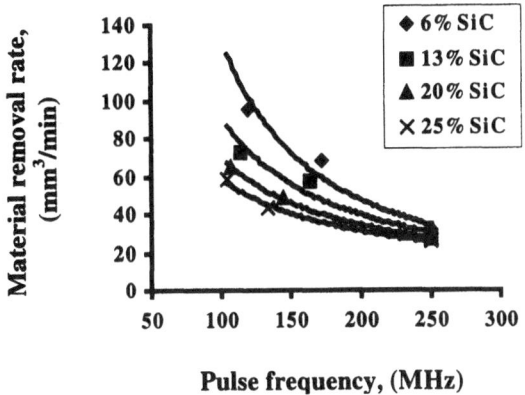

Fig. 1 Relationship between material removal rate and pulse frequency.

Fig. 2 shows that the material removal rate decreases with the increase of average machining voltage for all percentage volumes of SiC, probably due to a decrease in conductivity of the work material with refractory dispersions.

The relationships between material removal rate with kerf size and workpiece height are shown in Figs. 3 and 4. As the kerf size and workpiece thickness increase the MRR increases directly for all percentage volumes of SiC.

From Figs. 5 and 6 it is clear that with an increase of both pulse frequency and average machining voltage, the surface roughness decreases for all percentage volumes of SiC. This is because the surface roughness value is proportional to the MRR which later decreases with an increase of pulse frequency and average machining voltage. With an increase in the volume fraction of SiC, the voids leave very small cavities on the surface thus causing an increase in the surface roughness.

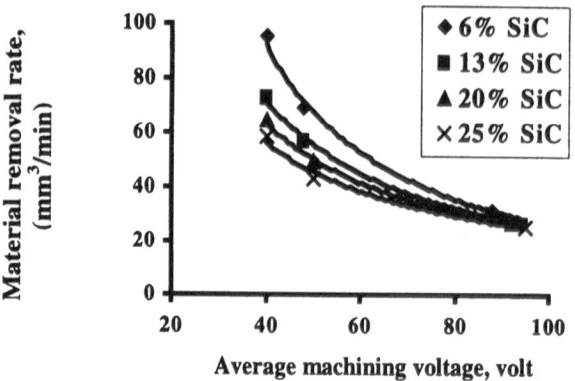

Fig. 2 Relationship between material removal rate and average machining voltage.

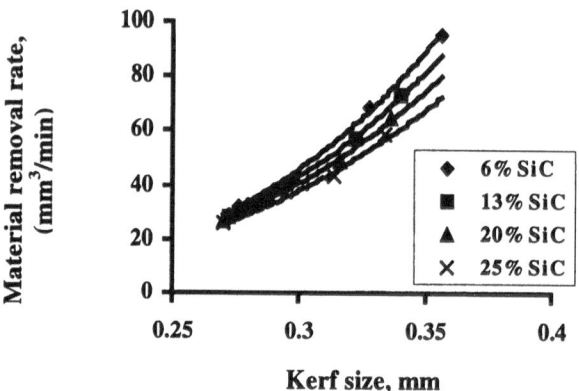

Fig. 3 Relationship between material removal rate and kerf size.

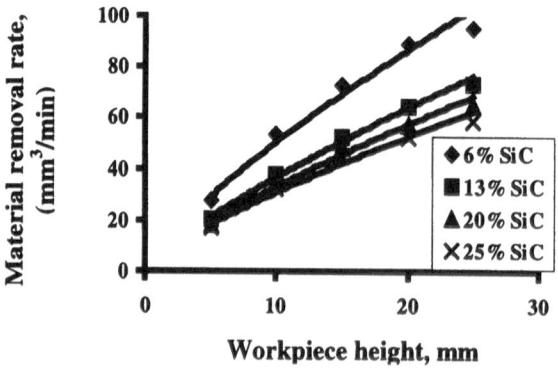

Fig. 4 Relationship between material removal rate and workpiece height.

Figures 7 and 8 show that the increase of kerf size and machining feed rate result in the increase of surface roughness for all percentage volumes of SiC.

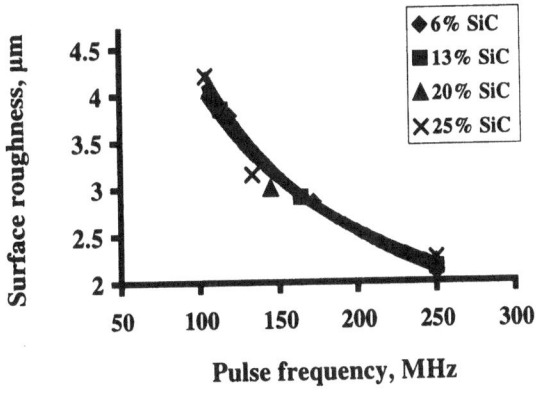

Fig. 5 Relationship between surface roughness and pulse frequency.

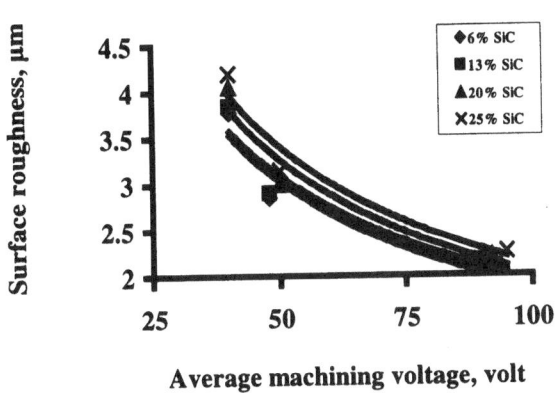

Fig. 6 Relationship between surface roughness and average machining voltage.

4. Conclusions

The analysis of the experimental observations show that the material removal rate and surface roughness are greatly influenced by the presence of SiC present in the metal matrix composite. The material removal rate and surface roughness decrease with the increase of percentage volume of SiC, pulse frequency and average machining voltage. MRR increases with the increase of both kerf size and workpiece height. Surface roughness increases with the increase of both kerf size and machining feed rate. Moreover, an optimal combination of the various process parameters can be made in such a way that the surface roughness can be minimized

152

and the material removal rate and cutting speed can be maximized, using equations (2 to 4).

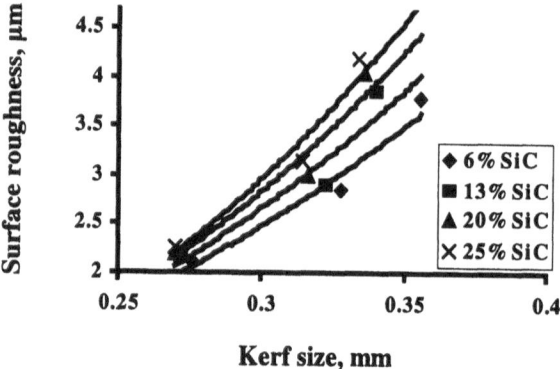

Fig. 7 Relationship between surface roughness and kerf size.

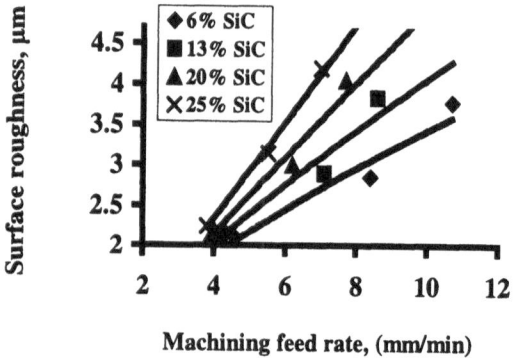

Fig. 8 Relationship between surface roughness and machining feed rate.

References

[1] Lindberg R. A., 1990, Processes and Materials of Manufacture, Published by Allyn and Bacon, Boston.

[2] Kumar B., 1981, Introduction to Manufacturing Technology, Khanna Publishers, Delhi.

[3] Karthikeyan R. et al, 1999, Mathematical modelling for electric discharge machining of aluminum-silicon carbide particulate composites, Journal of Materials Processing Technology, Vol. 87, pp. 59-63.

[4] Rajurkar K. P., Equipment used in the wire electrical discharge machining process, http://www.unl.edu/nmrc/raju1/.html.

[5] Luo Y. F., 1995, An energy-distribution strategy in fast-cutting wire EDM, Journal of Materials Processing Technology, Vol. 55, No. 3-4, pp. 380-390.

[6] Ebeid S., Fahmy R. and Habib S., 2003, An operating and diagnostic knowledge-based system for wire EDM, 7th International Conference, KES, Oxford, UK, pp. 691-698.

A Fuzzy Logic Control Approach to Electrochemical Machining (ECM)

V. J. Keasberry, A. W. Labib, J. Atkinson and H. W. Frost

UMIST, PO Box 88, Manchester M60 1QD, v.keasberry@student.umist.ac.uk

Abstract: Electrochemical machining (ECM) is a manufacturing process that offers a number of advantages (e.g. no mechanical stress) over its nearest competitors as certain trends in production move towards the micro scale. Maintaining optimum ECM process conditions ensures higher machining efficiency and performance. This paper presents the development of a fuzzy logic controller to add intelligence to the ECM process. An experimental ECM drilling rig, at UMIST, was improved through the integration of a fuzzy logic controller into the existing control system. Matlab (Fuzzy Logic Toolbox) was used to build a fuzzy logic controller system, which control the feed rate of the tool and the flow rate of the electrolyte. The objective of the fuzzy logic controller was to improve machining performance and accuracy by controlling the ECM process variables. The results serve to introduce innovative possibilities and provide potential for future applications of FLC in ECM. Hybrid controllers that integrate fuzzy logic into the control system allow for intelligence to be incorporated into ECM controllers. As the future of ECM moves into electrochemical micromachining (EMM), the need for process uncertainty control in this area may be met by FLC, which has advantages over conventional methods of process control.

1. Introduction (Background)

1.1 Electrochemical and electrochemical micro machining

ECM is an electrochemical anodic dissolution process [1]. Direct current is passed between a workpiece (the anode) and a pre-shaped tool (the cathode), with electrolyte flowing through the gap to complete the circuit. The tool shape is copied (formed) into the anodic workpiece surface through the metal being dissolved into metallic ions by electrolytic action. So ECM changes the shape of the workpiece by removing metal through electrolytic action. EMM has evolved from ECM due to miniaturisation developing as a future trend in production demands. Bhattacharyya et al comment that because of the inherent machining advantages ECM has over other processes, ECM appears to be very promising as a future micromachining technique [2]. These authors state that, "In EMM the inter-electrode gap control plays an important role." It follows that the inter-electrode gap has a major role in the accuracy of the machining process [2-5]. In conclusion Bhattacharyya et al [6] comment that further work is required into EMM inter-electrode gap control and machining parameter control. Bhattacharyya et al continued their work into EMM and presented further findings from their experimental EMM rig set-up. They discuss features such as inter-electrode gap control, which consists of a Boolean logic-based control system that compares power input signals to a reference voltage.

The performance of ECM is affected by many interlinking machining process variables. Optimum process conditions allow for high ECM performance in terms of surface finish and rate of machining. It follows that optimum process conditions are desirable in order to ensure optimum machining conditions and performance. Process deterioration is identified through indirect interpretation of a process variable moving out of the acceptable operational range. The interlinking variables are highly nonlinear and complex in nature and it is therefore very difficult to develop an exact mathematical model to control the machining process. The result is that conventional linear control strategies become extremely complex to construct and may still not adequately model the ECM process. Based upon the present authors' research findings, fuzzy logic control (FLC) is ideally suited to ECM, as FLC is an effective control strategy for nonlinear/uncertain process applications. In recent times, published work relating to the application of fuzzy logic to ECM is limited. However, FLC has been applied to other members of the non-conventional machining processes family, e.g. electrodischarge machining [7, 8]. Surmann. H and Huser. J [9] applied a fuzzy logic controller to electropolishing of cobalt chromium dental cast alloys. They presented a FLC with 16 fuzzy rules, which completely automated the polishing process. The FLC approach may be used to extract key interlinking machining process features, capturing this knowledge in a fuzzy rule base to be used as the control system's inference engine. The addition of FLC to ECM provides expert real-time operational intelligence, which may result in more consistent levels of machining performance.

1.2 Fuzzy logic interlude

Zadeh, L. introduced fuzzy set theory in the 1960's [10]. He recognised that many shades of grey found in the real world were not covered by Boolean logic. In fuzzy logic instead of something being 100% true or false it deals in degrees of membership ranging from zero to one, and so something can be partially true and false at the same time. Classical Boolean logic prevailed before fuzzy logic, the idea being the whole universe could be either A or not-A and so everything is accounted for wholly either by one group or another. However, it was proven by Kosko, that Classical Boolean logic is just a special case of fuzzy logic [11]. Fuzzy logic allows nonlinear functions to be modelled; it is easy to use, to understand and to implement. The following part of this paper presents the design and implementation methodology for the developed ECM FLC created.

2. Methodology

An experimental ECM drilling rig, was improved through the integration of a fuzzy logic controller into the existing PC based, in-house Visual Basic (VB) control system. Previously, machining decisions were subject to interpretation from the machine operator, as the machine operator managed the VB controller and rig manually through indirect interpretation of process variable measurements (e.g. If the flow rate of the electrolyte reduces, then the manual flow valve is opened). The fuzzy logic controller was created in Matlab (Fuzzy Logic Toolbox) [12]. The fuzzy logic controller, consisting of two inputs and two outputs was created to monitor and control both the feed rate and the flow rate in the ECM process. The fuzzy logic controller created serves two purposes: (1) to investigate the concept of integrating fuzzy logic into the UMIST experimental ECM drilling rig and, (2) to investigate the potential of FLC to ensure higher levels of machining performance.

In order to control the ECM drilling process, the following input variables were identified: Voltage, current, measured flow rate, electrolyte conductivity and electrolyte temperature. However, as this was the first attempt at integrating FLC into the experimental ECM drilling rig, a simple fuzzy logic controller was created. This controller takes two inputs: current and measured flow rate; mapping them to two outputs: tool feed rate and flow rate (see Figure 1). De Silva et al [3] discuss the need for adequate electrolyte flow in ECM. Also, it is well documented that in order to achieve high machining performance the electrolyte condition must be in an optimum state, in terms of temperature, conductivity and stage of life. As the current ECM drilling rig does not have the capability to treat spent electrolyte solution; it was accepted that after every machining run, that the electrolyte condition would deteriorate, i.e. increasing iron-hydroxide levels even though a simple filter system is present. Experimental results are thus presented with the understanding that results may be subject to electrolyte deterioration factors. This is acceptable in the light of the above two FLC research purposes. The fuzzy logic controller created aims to provide gap and electrolyte flow control, so even if the electrolyte condition deteriorates, FLC of the inter-electrode gap would benefit, as the controller can then be observed to react to the undesired process conditions if they were to have an adverse effect on machining performance.

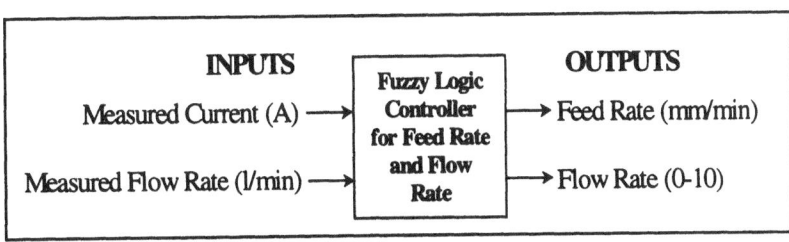

Figure 1: The fuzzy logic controller

The major reason for inter-electrode gap and electrolyte flow rate control is to prevent the onset of sparking. Sparking is an ECM process occurrence that can lead to damaged electrodes, poor surface finish and loss of machining accuracy. By controlling the inter-electrode gap, sparking reduction can be managed and kept to a minimum. Also, adequate electrolyte flow ensures that machining debris, which may otherwise cause sparking, is flushed out from the inter-electrode gap.

156

2.1 Design of the fuzzy logic controller

During electrochemical drilling the behaviour within the inter-electrode gap is extremely nonlinear. The experience and knowledge of the machine operator are critical sources of information used to identify process deterioration, through interpretation of interlinking machining process variables (e.g. electrolyte condition and measured current). FLC allows the experience and knowledge of the machine operator to be elicited as fuzzy rules that map input space to output space, trading off precision with significance. What is created is a means to provide the ECM controller with decision-making attributes that can be classified as artificial intelligence. The following part of this paper presents the fuzzy rules (figure 2) and membership functions (MF) (figures 3 and 4) created and used to control the drilling rig.

1.	IF Current is below 15 AND Measured Flow Rate is Operating THEN Flow Rate is 5.
2.	IF Current is 15-20 AND Measured Flow Rate is Operating THEN Flow Rate is 5.
3.	IF Current is 20-25 AND Measured Flow Rate is 4 THEN Flow Rate is 6.
4.	IF Current is 25-30 AND Measured Flow Rate is 4 THEN Flow Rate is 7.
5.	IF Current is 30-35 AND Measured Flow Rate is 3 THEN Flow Rate is 8.
6.	IF Current is 35+ AND Measured Flow Rate is 3 THEN Flow Rate is 9.
7.	IF Current is 35+ AND Measured Flow Rate is below 2 THEN Flow Rate is 10.
8.	IF Current is Dangerous THEN Feed Rate is Reverse Feed Rate.
9.	IF Current is Too High THEN Feed Rate is Very Slow.
10.	IF Current is Operational AND Measured Flow Rate is Operating THEN Feed Rate is Optimal.
11.	IF Current is Too Low THEN Feed Rate is Optimal.
12.	IF Measured Flow Rate is Critically Low THEN Feed Rate is Reverse Feed Rate.
13.	IF Current is Operational THEN Feed Rate is Optimal.

Figure 2: The fuzzy rules.

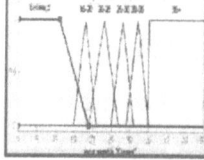 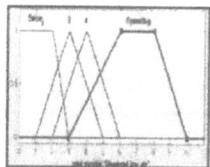 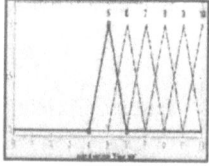

Figure 3: Membership Functions for flow rate input (first two) and output variables

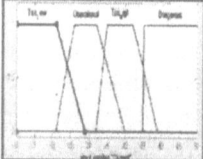 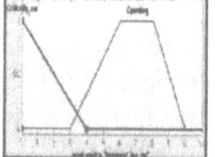 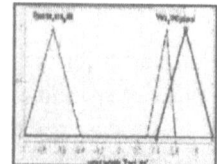

Figure 4: Membership Functions for feed rate input (first two) and output variables

Figure 2, shows the 13 rules generated, of which the first 7 manage the electrolyte flow rate through the control of the rig's valve/actuator set. The valve actuator is controlled through a 0-10V input range and so the universe of discourse has been set from 0-10V. Rules 8-13, control the tool feed rate, which accounts for the tool moving in either a positive (away from the workpiece) or negative (towards the workpiece) direction. All rules were generated through operator experience and then fine-tuned to provide the desired effect, which accounts for the number of rules being 13 and not 46. The universe of discourse of the MF, for each variable, was defined through experiments and operator knowledge about the rig (e.g. as in the example of flow rate output above). For both output variables, the same two input variables (current and measured flow rate), have been mapped separately to them. This is because when the operator relates the inputs to the output, in each case, the linguistic terminology used is different. This implies that there are different MF for the same input variables when mapped to different outputs.

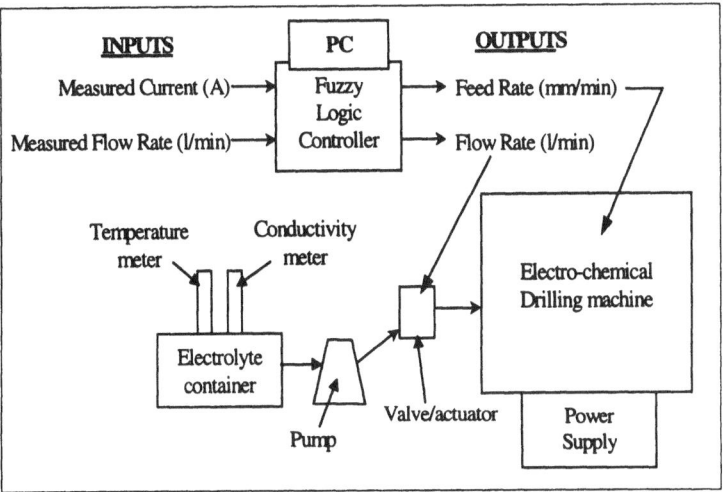

Figure 5: System schematic diagram

The existing controller consisted of in-house software developed in VB to control the tool feed rate. The introduction of the new valve/actuator set allowed for the electrolyte flow rate to be controlled using an upgraded version of the VB controller. To integrate FLC into the existing VB controller look-up tables were created from the output control surfaces created in Matlab (Fuzzy Logic Tool box). The look-up tables serve as the controller intelligence (inference engine), as the VB program was upgraded to compare process readings with values in the look-up tables to decide what tool feed rate and electrolyte flow rate to be set as the output. Figure 5, shows a system schematic diagram with the inclusion of the fuzzy logic controller. Various process readings are taken, but the fuzzy logic controller only uses two of the inputs, as shown.

3. Results and Discussion

The two output surfaces used to create the look-up tables in the VB controller can be seen in Figure 6.

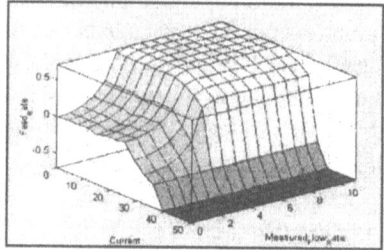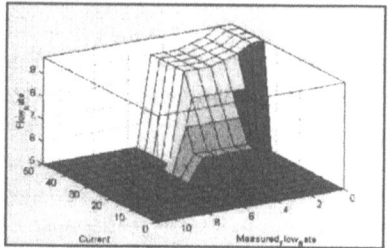

Figure 6: Output Control Surfaces for flow rate (left) and feed rate (right)

Figure 7, presents a picture of the experimental ECM drilling rig and also an axial section through a 10mm depth hole machined using the fuzzy logic controller. Throughout the machining run the operator did not exercise any decision-making control. The graph in Figure 8, presents all the measurements from a machining run where the operator purposely forced the tool to short circuit with the workpiece. In this case current rises and the flow rate falls, an undesirable machining occurrence. FLC was then engaged to see how the fuzzy controller performed. On three occasions the fuzzy controller brought the machining process back to optimum conditions, without any operator intervention. Critical conditions triggered the tool feed rate to reverse and also the flow rate to increase. The process changes are in line with what the operator would have done in such a situation. However, the test was an artificial situation; in normal machining conditions the fuzzy controller would never allow the machining process to reach such critical conditions.

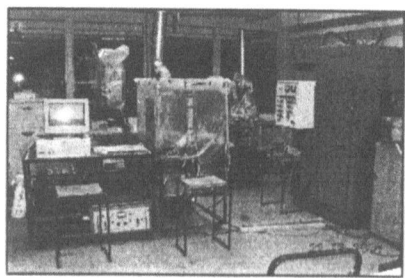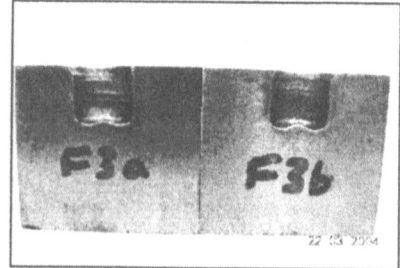

Figure 7: ECM rig (left) and fuzzy logic machined hole (right)

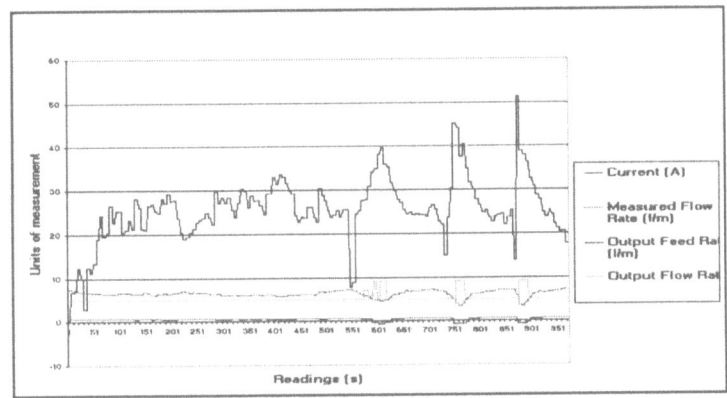

Figure 8: Graph showing the fuzzy controller correcting a critical machining situation

The drop in current towards the middle of the graph represents the end of a machining run and then the start of the critical testing machining run. Optimal current measurements, expected for this rig, would be between 19-25 A. Although the fuzzy logic controller slightly over-compensates for the critical current occurrence, this can be adjusted through 'fine tuning' of the MF. What these results serve to prove is that FLC can be used to gradually control the ECM process. Other studies apply Boolean logic-control, which entails turning off the power and reversing the electrode when a short circuit is detected [2]. Their findings suggest that further process optimisation is needed. Perhaps the use of FLC will improve EMM accuracy, due to the process being given 'intelligence'.

4. Conclusions

This research addressed the two aims stated above. FLC was integrated into the UMIST experimental ECM drilling rig. The MF would still need further refinement to ensure finer control of the process. The fuzzy controller gradually controlled the ECM process and corrected critical situations. The encouraging results suggest that the application of FLC to ECM has potential. Further developments could entail the use of FLC to manage the whole ECM system, e.g. from electrolyte maintenance to power pulse management during the machining process. If micro-spark management in EMM can be realised better machining accuracy should result. A form of Hybrid Controller (HC) is proposed as a possible future for ECM/EMM process control. A HC would incorporate FLC to provide real-time process intelligence, acting as the inference engine within the system controller.

References

[1]   De Barr, A.E., Oliver, D. A., *Electrochemical Machining*. 1975: Macdonald & Co Ltd.

[2]   Bhattacharyya, B., Mitra, S., Boro, A. K., *Electrochemical machining: new possibilities for micromachining*. Robotics and Computer Integrated Manufacturing, 2002. **18**: p. 283-289.

[3]   De Silva, A.K.M., McGeough, J. A., *Process monitoring of electrochemical micromachining*. Journal of Materials Processing Technology, 1998. **76**: p. 165-169.

[4]   Yong, L., Yunfei, Z., Guang, Y., Liangqiang, P., *Localized electrochemical micromachining with gap control*. Sensors and Actuators A Physical, 2003. **108**: p. 144-148.

[5]   Bhattacharyya, B., Munda, J., *Experimental investigation on the influence of electrochemical machining parameters on machining rate and accuracy in micromachining domain*. International Journal of Machine Tools & Manufacture Design Research and Application, 2003. **43**: p. 1301-1310.

[6]   Bhattacharyya, B., Munda, J., *Experimental investigation into electrochemical micromachining (EMM) process*. Journal of Material Processing Technology, 2003. **140**: p. 287-291.

[7]   Zhang, J.H., Zhang, H., Su, D. S., Qin, Y., Huo, M. Y., Zhang, Q. H., Wang, L., *Adaptive fuzzy control system of a servomechanism for electro-discharge machining combined with ultrasonic vibration*. Journal of Materials Processing Technology, 2002. **129**: p. 45-49.

[8]   Lin, C.T., Chung, I. F., Huang, S. Y., *Improvement of machining accuracy by fuzzy logic at corner parts for wire-EDM*. Fuzzy Set and Systems, 2001. **122**: p. 499-511.

[9]   Surmann, H., Huser, J., *Automatic electropolishing of cobalt chromium dental cast alloys with a fuzzy logic controller*. Computers Chemical Engineering, 1998. **22**(7-8): p. 1099-1111.

[10]   Zadeh, L.A., *Fuzzy sets*. Information and Control, 1965. **8**: p. 338-353.

[11]   Kosko, B., *The new science of fuzzy logic: Fuzzy Thinking*. 1994: Flamingo.

[12]   Mathworks, *Fuzzy Logic Toolbox: documentation*. 2004 (http://www.mathworks.com/access/helpdesk/help/toolbox/fuzzy/fuzzy.html, accessed 5/4/2004)

A Fuzzy Logic Approach for the Pulse Classification of Electro Chemical Discharge Machining (ECDM)

T.K.K.R Mediliyegedara[1], A.K.M De Silva[2], D.K. Harrison[3], J.A McGeough[4]

[1]School of Engineering, Science and Design, Glasgow Caledonian University, Glasgow, G4 0BA, U.K., k.mediliyegedara@gcal.ac.uk
[2]School of Engineering, Science and Design, Glasgow Caledonian University, Glasgow, G4 0BA, U.K., A.DeSilva@gcal.ac.uk
[3]School of Engineering, Science and Design, Glasgow Caledonian University, Glasgow, G4 0BA, U.K., D.K.Harrison@gcal.ac.uk
[4]School of Engineering and Electronics, The University of Edinburgh, U.K., Alrick Building, The King's Buildings, Mayfield Road, Edinburgh, EH9 3JL, U.K., J.McGeough@ed.ac.uk

Abstract : This paper presents the pulse classification of the ECDM process using Fuzzy Logic (FL). An Electro Discharge Machining (EDM) machine was modified by incorporating an electrolyte system and by modifying the control system. Gap voltage and working current waveforms were obtained. A fuzzy rule base was designed and simulated to classify pulses. Different ECDM process variables were considered to design the pulse classification rule base. The performances were compared by changing the process variables and by changing rules in the fuzzy rule base. The classification accuracy was measured in the classification system. A quantitative analysis was performed to evaluate the fuzzy pulse classifier with different process variables and with different fuzzy rules for the ECDM machine.

1. Introduction

ECDM is a manufacturing process which combines features of Electro Chemical Machining (ECM) and EDM, by having a cathodic tool and an anodic workpiece, which are separated by a gap filled with electrolyte, and pulsed DC power applied between them. This leads to discharges in electrolytes, thus achieving both electrochemical dissolution and electro-discharge erosion of the workpiece [1]. One of the major advantages of ECDM, over ECM or EDM, is that the combined metal removal mechanisms in ECDM, yields a much higher machining rate [2].

The performance of ECDM, in terms of surface finish and rate of machining, is affected by many factors. Relationships between these factors and machining performance are highly non linear and complex in nature. Therefore, it is very difficult to develop a relationship between those factors and the machining performance with conventional mathematical modelling. This fact makes it very hard to formulate control strategies for the process control of ECDM. Pulse classification plays a vital role in the formulation of control strategies. Strategies for pulse classification in EDM have been studied in the past but there remains a need for an effective and efficient pulse classification system for ECDM.

In a fuzzy gap width controller, the control signal of the ECDM process is generated directly by the linguistic rules acquired from the knowledge of experts and expressed through the theory of fuzzy sets. In the traditional control approach, the controller is designed based on mathematical models of the plant. Since the ECDM process is very complex to model mathematically, the utilization of a traditional control approach is limited. A Fuzzy Controller (FC) does not require a mathematical model and it is less sensitive to parameter variations and noise disturbance when compared to traditional model based controllers.

Zheng et. al, [3] has employed a fuzzy controller to control the EDM process. They have shown that extracting fuzzy rules from dynamic data on an off-line basis makes fuzzy rules more practical than creating fuzzy rules based on the experts experience. Zhang et al [4] has utilized an adaptive fuzzy control system for EDM.

2. Pulse Types of in the ECDM Process

It is possible to identify five distinct types of pulses in the ECDM process such as Electro Chemical Pulse (ECP), Electro Chemical Discharge Pulse (ECDP), Spark Pulse (SP), Arc Pulse (AP) and Short Circuit Pulse (SCP). SP and AP also known as Normal Discharge Pulse and Abnormal Discharge Pulse respectively. Open Circuits Pulse (OCP) is not present in ECDM as some electrochemical current flows even with a larger gap [2]. De Silva (1988) [2] has presented a detailed analysis of various pulses in ECDM.

3. Experimental Procedure

An EDM machine was modified by incorporating an electrolyte system and changing the control system. NaNO3 was used as the electrolyte. A mild steel work piece and copper electrode were used. The duty ratio and pulse duration were set to be 50% and 100µs respectively. The above mentioned five types of pulses were acquired using a storage oscilloscope at a sampling frequency of 1 MHz. The MATLAB 6 software package was used to model and to simulate the FC.

4. Fuzzy Pulse Classification System (FPCS)

4.1 Selection of input and output variables

A block diagram of a FPCS is shown in Figure 1. The FPCS is composed of three calculation steps: fuzzification, fuzzy inference, and defuzzification. The linguistic rule base of the classifier, implements the classification strategy. There are four input variables such as I_1, I_2, I_3 and I_4. Normalised values of Peak Voltage (PV), Average Voltage (AV), Peak Current (PC) and Average Current (AC) of a pulse were used as I_1, I_2, I_3 and I_4 respectively. There are five outputs such as O_1, O_2, O_3, O_4 and O_5 corresponding to ECP, ECDP, SP, AP and SCP.

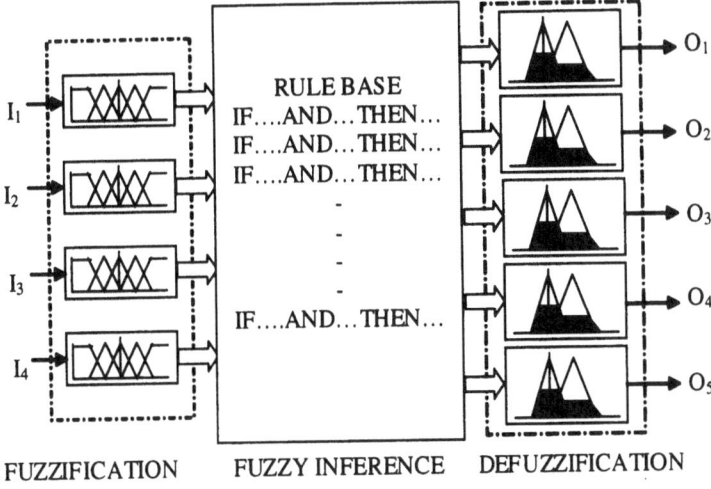

Figure 1: Block diagram of the fuzzy pulse classification system (FPCS)

4.2 Fuzzification

Five linguistic terms are used to create the input linguistic variables such as Very Small (VS), Small (S), Medium (M), Large (L) and Very Large (VL). Membership functions of the normalised peack voltage (I_1) is shown in Figure 2. Similarly, membership functions for normalised average voltage (I_2), normalised peak current (I_3) and normalised average current (I_4) were defined.

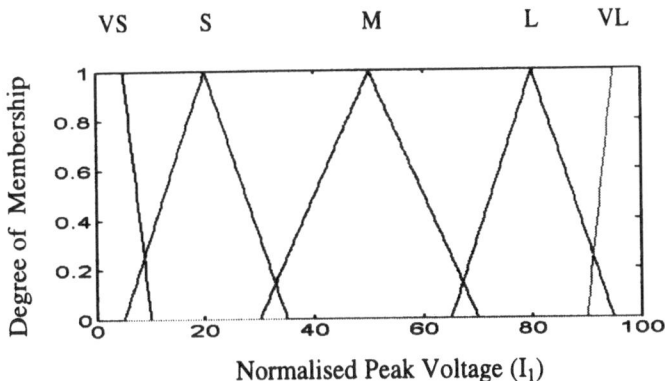

Figure 2: Membership functions of the normalised peak voltage (I_1)

4.3 Fuzzy rule base

The fuzzy rule base contains the experts knowledge necessary to classify pulses. In the pulse classification ECDM, there are four input variables each having five linguistic terms. Therefore, the number of possible rules is $5^5 = 3125$ but, some rules do not exist in the normal operation. So it is possible to identify fuzzy rules which are relevant to the normal operation conditions. If PV, AV, PC, and PV are S, L, M and S respectively then the corresponding pulse is ECDP and so on.

4.4 Defuzzification

The mean of the maximum defuzzification method was employed. Two linguistic terms were used to create the output linguistic variables such as 'Yes' (Y) and 'No' (N). Membership functions of all output variables (O_1, O_2, O_3, O_4 and O_5) are in the same shape. Figure 3 shows the membership functions of the output corresponding to ECP (O_1) of the fuzzy pulse classification system.

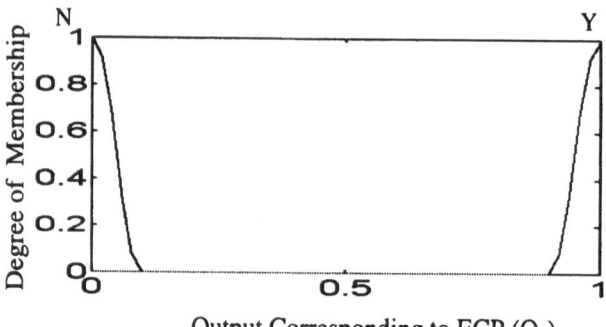

Output Corresponding to ECP (O_3)

Figure 3: Membership functions of the output corresponding to ECP (O_1) of the fuzzy pulse classification system

5. Measuring the Performance of FPCS

Classification Accuracy (CA) is introduced to compare the performance of the classification system. The CA of a pulse type was defined as follows.

$$\text{CA of 'X' Type Pulses} = \left\{ \frac{\Sigma X_i}{n_x} - \frac{\Sigma Y_i}{n_y} \right\} \times 100\% \qquad [1]$$

Where,

X_i - Simulated output value from 'X' output node of ANN for i^{th} pulse when the input values are corresponding to 'X' type pulses,

Y_i - Simulated output value from other output nodes of ANN for i^{th} pulse when the input values are corresponding to X 'type pulses,

n_x - Number of 'X' type pulses used in the test data set,

n_y - Number of other types of pulses used in the test data set .

Average Classification Accuracy (ACA) is introduced to measure the overall accuracy of the classification system. The ACA is defined as follows.

$$ACA = (CA_{ECP} + CA_{ECDP} + CA_{SP} + CA_{SP} + CA_{SCP})/5 \qquad [2]$$

Where, CA_{ECP}, CA_{ECDP}, CA_{SP}, CA_{SP} and CA_{SCP} are the classification accuracy of ECP, ECDP, SP, AP, and SCP respectively.

6. Preparation of theTest Data Set

Thirty pulses were selected from each pulse type. The normalised values of Peak Voltage (PV), Average Voltage (AV), Peak Current (PC) and Average Current (AC) were calculated. PV, AV, PC and AC were used as the inputs for FPCS. The test data set, the structure of which is shown in Table 1, consists of 150 data points.

Table 1: Structure of the test data set

Data Points	1-30	31-60	61-90	91-120	121-150
Pulse Type	ECP	ECDP	SP	AP	SCP

7. Results and Discussion

The simulated results are shown in Figure 4. The vertical axis (Y) of the graph indicates the output values of the FPCS. In the ideal situation, if a pulse is a ECP, then the value of the output O_1 should be equal to '1'. Other output values (O_2, O_3, O_4 and O_5) should be equal to '0'. Figure 4 shows that the simulated values from the output corresponding to ECP (O_1) is nearly equal to '1', for the first 30 pulses. That means the first 30 pulses have been classified as ECP by the FPCS. Table 2 summarises the classification accuracies for the five different pulse types mentioned above. The average classification accuracy of the proposed FPCS is about 88 %.

Table 2: Classification Accuracies

	ECP	ECDP	SP	AP	SCP	Average
Accuracy (%)	92.50	78.56	89.99	86.11	93.60	88.15

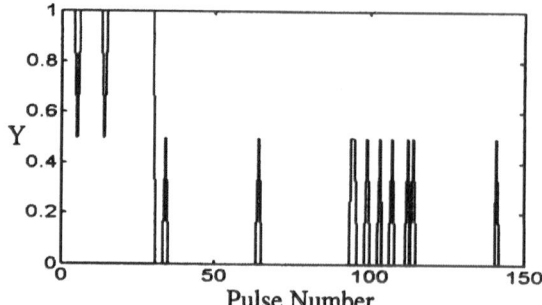

Figure 4: The simulated values of the fuzzy pulse classification system from the output corresponding to ECP (O_1)

8. Conclusions

In this paper, an FPCS for pulse classification in the ECDM process has been established and analysed based on the ECDM process variables. Fuzzification of the input variables, designing of the rule base and the defuzzification stage are described. Four features such as PV, AV, PC and AC were used successfully for classification of pulses in the ECDM process. The term "Classification Accuracy" has been defined to measure the accuracy of a pulse classification system. Simulation results showed that the proposed pulse classification system can be successfully used in the pulse classification of the ECDM process.

References

[1]   De Silva A.K.M, Khayry A.B. and McGeough J.A., 1995, *Process Monitoring and Control of Electroerosion-dissolution Machining*, IMechE Conference Transactions, 11th International Conference on Computer-Aided Production Engineering, pp. 73-78.

[2]   De Silva A.K., 1988, Process *Developments in Electrochemical Arc Machining*, PhD Thesis, University of Edinburgh.

[3]   Zheng H., Liu Q.B., Gou Y.F., et al, 1998, *A Fuzzy Controller for EDM Process*, 12th International Symposium for Electromachining (ISEM-XII), Aachen, Germany, pp. 185-191.

[4]   Zhang J.H., Zhang H., Su D.S., et al, 2002, *Adaptive fuzzy control system of a servomechanism for electro-discharge machining combined with ultrasonic vibration*, Journal of Material Processing Technology, 129, pp. 45-49.

[5]   Tarng, Y.S., Tseng C.M. and Chung, L.K., 1997, *A Fuzzy Pulse Discriminating System for Electrical Discharge Machining*, Int. J. Mech. Tool Manufacture., Vol. 37, No. 4, pp. 511-522.

FORMING

Principles for on-line Monitoring of Tool Wear During Sheet Metal Punching

W. Klingenberg[1], U.P. Singh[2]

[1]University of Groningen, WSN 745, PO Box 800, 9700 AV Groningen, The Netherlands, w.klingenberg@bdk.rug.nl
[2]University of Ulster at Jordanstown, Shore Road, Newtownabbey, Co. Antrim BT37 0QB, Northern Ireland, up.singh@ulster.ac.uk

Abstract: Significant progress was made in experimental, analytical and numerical modelling of the process of punching/blanking of sheet metal. These models, however, may still not be sufficiently adequate to accurately predict the process in all circumstances. Therefore, researchers have continued to search for methods to monitor and control some key aspects in real time. After an overview of existing literature, some principles of the punching/blanking process are discussed. Detail is added to these principles by the presentation of experimental results, indicating that progressive tool wear can be recognised from progressive changes to the shape of the force-displacement graph of the process. The potential use of such information as input to a monitoring system is demonstrated here.

1. Introduction

Research of process planning and the application of in-process control play an increasingly important role in the quest for low cost flexible manufacturing operations. In recent years, the first tangible concepts, which could serve as bases for in-process control solutions for sheet metal operations, started to emerge, combining an understanding of the principles of in-process control of manufacturing processes and the technological aspects of the process in question [1-3]. For manufacturing processes outside of sheet metal forming, for example turning, on-line monitoring of tool wear and in-process control principles were reported by other researchers [4,5].

Punching/blanking is among the most important sheet metal manufacturing processes in mass production of metal parts and components in a great variety of industries. Punching and blanking are probably as old as any sheet metal forming operation, which is said to date back to at least 250 BC. However, the operation is still not fully understood or captured by any comprehensive process model, although significant progress was made, through experimental modelling [2,3,6,7] analytical modelling [1,2,8,9] and numerical (Finite Element) modelling [1,2,10,11] of the punching/blanking process.

These models, however, may still not be sufficiently adequate to accurately predict the process in all circumstances. Therefore, researchers have started to search for methods to monitor and control some key aspects of the punching/blanking process in real time. Some of these research efforts focus on Artificial Neural Networks (ANN), for example by training an ANN with input from the results of a large number of Finite Element simulations [11]. Others propose to use pattern recognition techniques to monitor and control tool wear, for example from measurements of the peak force [12]. Pattern recognition and

Statistical Process Control (SPC) solutions [13] in combination with analytical and experimental models may serve as the basis for future progress in this field.

It is known from other investigators that it is fair to assume tool wear causes the cutting edges of the punch to be rounded [6,7,15]. It is also known that an increase in the tool radius causes an increase in burr height [10], and in [2] some measurements confirming these known effects are presented. Less well known is the behaviour of the process in the form of the Force-Displacement graph. Some work in this area was reported by Choy and Balendra [7]. Noteworthy is also the early research by Archard [14].

2. Force-displacement Graphs During Progressive Tool Wear

The stages of the process can be recognised from the force-displacement graph. At macro-level the process can be divided into five stages, which can also be recognised from the graphs in Figure 1 [2,6]. These are elastic deformation, indicated by a straight line, plastic bending, shear deformation, which causes the contact area between the product and the blank to diminish, causing a gradual decrease in the force, and finally ductile shear fracture. A possible tail in the graph after the point of ductile fracture is caused by friction: the work done due to friction is dissipated when forcing (pushing) the slug through the die hole.

Typical tool wear consists of wear on the side of the tool, due to burnishing of the blank, and damage to the cutting edge of the tool. The tool wear creating most of the dimensional inaccuracies is due to the breakdown of the corners of the tool tips, which increases the burr height. As mentioned, other researchers [7,10] have demonstrated that the geometry of the tool after tool wear may be approximated by a rounded shape, creating a punch tip radius.

It seems reasonable to suggest that the introduction of a radius on the tip of the tool causes additional plastic deformation of the blank, since blank material is forced to obey the shape of the tool and flows around the radius of the punch. The radius at the tip of the punch therefore causes a delay in the start of shear cutting, extending the amount of plastic deformation of the blank. The effect of the radius is to reduce the stress concentration in both the tool and the material. Figure 1 shows the Force-Displacement curves for the punching/blanking process using cold rolled steel with the initial blank thickness $H_0 = 1.6$ mm. The material characteristics of this cold rolled steel are: proof stress $\sigma_{0.2} = 209$ MPa, ultimate tensile strength $\sigma_B = 304$ MPa, elongation at break $e_{break} = 36\%$, work hardening factor $C = 489$ MPa, work hardening exponent $n = 0.17$. Clearance is defined as $(D_d-D_p)/2H_0$, with D_d the diameter of the die, D_p the diameter of the punch and H_0 the sheet thickness.

The measurements in this section were taken as part of a comprehensive investigation into characteristics of the punching/blanking process [2]. For these measurements, a C-frame press was used with a capacity of 60 tonnes. The press was fitted with a four pillar die set, configured for punching, as well as a load cell and a displacement transducer [2].

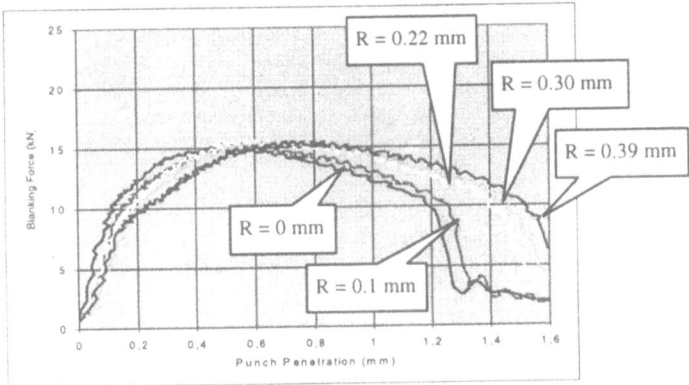

Figure 1. Force-Displacement curves for increasing punch radii.
Cold rolled sheet steel, $H_0 = 1.6$ mm, clearance = 5%, $D_p = 9.84$ mm.

Figure 1 shows, as expected [12], an increase in the maximum punching force as a result of an increased punch radius. It also shows a consistently increasing delay in the start of shear cutting, represented by the peak force. It is revealed that the final rupture of the blank, indicated by the sudden drop of the punching force, is suppressed by the introduction of a radius. This observation is consistent with an increasing radius. Also the reduction of the punching force during shearing, indicated by the downward slope, is affected by the tool radius.

There appear to be several opportunities to monitor the punching/blanking process. The punching force seems to increase with increasing radius. However, the increase is marginal and within 5% of the original force for this particular material. In any measuring environment, it may be difficult to reliably identify this increase as the onset of tool wear. A better way, directly related to the introduction of a tool radius, seems to be to analyse the work done up to the point of maximum force together with the punch penetration at the point of maximum force. It was observed that these values increase consistently and significantly with the tool radius, i.e. with tool wear. Figure 2a shows increases in work done to F_{max} of up to 25% with the introduction of a radius of 0.22mm. Figure 2b illustrates that an introduction of a punch radius of 0.22 mm again causes an increase of up to approximately 25% in punch penetration at the point of maximum force. It is noted that the punch penetration prior to the start of shear cutting will also be dependent on the clearance between the tool and the die. These and other phenomena were also investigated and reported in [2].

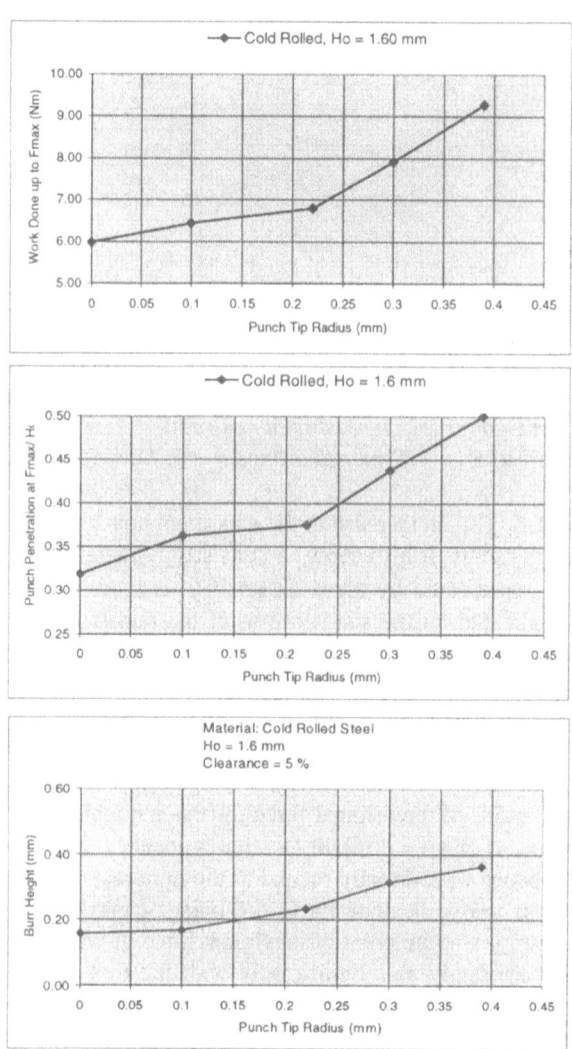

Figure 2.   Introduction of a punch tip radius
    (a)   Work done up to F_{max}
    (b)   Punch penetration at F_{max}
    (c)   Burr height

Figure 2c shows measurement of the burr height with increasing punch radius. The trend of the graph, an increase in burr height due to an increase in product radius, is consistent with literature values [6]. It is noted that for this particular material, the increase in the burr height was marginal, probably negligible, with the introduction of a radius of 0.1 mm on the tip of the punch. Only at tool radius values of 0.2 mm and greater does the result become significant in terms of an increased burr height. A comparison with Figure 2a shows a significant increase in the work done during the stages of elastic and plastic deformation of the blank material. This increase is consistent for both materials and could be used, possibly together with other features in the force-displacement graph, as input information to monitor the punching/blanking process.

It should be noted that measurements are presented here for a single type of sheet steel. As part of a wider programme [2], also measurements were taken using annealed steel with thickness $H_0 = 2.95$ mm. Note that in this case, if the same punch and die are used, clearance is smaller, since clearance between the punch and the die is usually expressed as a percentage of the sheet thickness. The experiments with the annealed steel produced the same trends. Again, the work done increased significantly with the punch tip radius [2]. Since tests were carried out for only two materials, generalisations of the above findings may not be possible from the presented observations. These findings can, however, serve as a justification for further research into the possibilities to monitor tool wear in punching/blanking from ongoing analysis of the shape of the force-displacement graph. To provide a suitable view of these possibilities, it is necessary to briefly review possible further considerations and consequences of such work.

3. Further Research

An automated monitoring and diagnosis system (or process control system) is of great importance in a Computer Integrated Manufacturing (CIM) environment. Such a system enables unmanned production and the manufacturing of high quality products. An effective process control system can be developed if useful data can be gathered during the process. This paper sketches some principles which could serve in developing an in-process control application for the on-line monitoring of tool wear. The authors wish to note that the principle demonstrated here covers only one among a number of possible ways to monitor the punching process [1-3]. In their view, a more extended report on the current subject should include the following considerations.

In order to appreciate any possible inconsistencies in experimental results and indeed to be able to minimise these, one should be aware of known scatter effects in sheet metal materials [2]. In fact, the stability of the Force-Displacement graph for particular punching/blanking characteristics is an issue that needs to be addressed. It is likely that machine and material characteristics are not stable and all kinds of error patterns may exist with respect to the measurement of the work done and the punch penetration at the point of maximum force. Indeed, further work should be aimed at possible ways to separate error patterns from process signals applicable in an in-process control environment. First indications suggest that a possible solution may lie in the use of artificial neural networks to monitor these error patterns in an automated way. Also the variety of operations at a punching machine, during the life of a punching tool, complicates the development of an automated monitoring and diagnosis system. Possibly, the relative stability of the peak force (Figure 1) may be of some use in on-line analysis. The present authors propose and intend to conduct further work in an effort to progress the integration of punching/blanking in Computer Integrated Manufacturing environments.

References

[1]   W. Klingenberg, U.P. Singh, 2003, *Finite Element simulation of the punching/blanking process using in-process characterisation of mild steel*, J. of Materials Processing Technology 134(3) 296-302.

[2]   W. Klingenberg, 2000, *Numerical modelling of the punching/blanking process using in-process characterisation of steel*, DPhil. Dissertation, University of Ulster.

[3]   W. Klingenberg, U.P. Singh, 2004, *Design and Optimisation of Punching/Blanking systems, aided by experimental modelling*, accepted for publication, Int. J. of Vehicle Design.

[4]   S.K. Choudhury, K.K. Kishore, 2000, *Tool wear measurement in turning using force ratio*, Int. J. of Machine Tools & Manufacture, 40 899-909.

[5]   T. Yandayan, M. Burdekin, 1997, *In-process dimensional measurements and control of workpiece accuracy*, Int. J. of Machine Tools & Manufacture 37(10) 1423-1439.

[6]   K. Lange, 1985, *Handbook of Metal Forming*, McGraw-Hill, New York.

[7]   C.M. Choy, R. Balendra, 1996, *Experimental analysis of parameters influencing sheared-edge profiles*, Proc. 4th Int. Conf. on Sheet Metal, University of Twente, The Netherlands, Vol. II, pp. 101-110.

[8]   A.G. Atkins, 1980, *On Cropping and Related Processes*, Int. J. of Mechanical Sciences 22 215-231.

[9]   D.C. Ko, B.M. Kim, 2000, *Development of an analytical scheme to predict the need for tool regrinding in shearing processes*, Int. J. of Machine Tools & Manufacture 40 1329-1349.

[10] R. Hambli, 2002, *Prediction of burr height formation in blanking processes using neural network*, Int. J. of Mechanical Sciences 44 2089-2102.

[11] R. Hambli, 2001, *Blanking tool wear modeling using the Finite Element Method*, Int. J. of Machine Tools & Manufacture 41(12) 1815-1829.

[12] W.B. Lee, C.F. Cheung, W.M. Chiu, L.K. Chan, 1997, *Automatic supervision of blanking tool wear using pattern recognition analysis*, Int. J. of Machine Tools & Manufacture 37(8) 1079-1095.

[13] R. Burman, 1995, *Manufacturing Management Principles and Systems*, McGraw-Hill, New York.

[14] J.F. Archard, 1953, *Contact and rubbing of flat surfaces*, Journal of Appl. Phys. 24 981-988.

Effects of Hot Rolling Conditions on Warm Formability of Cast Magnesium Alloy Sheets Manufactured by Continuous Strip Casting Process

H. Watari.[1], K. Davey[2], M. T. Rasgado[2], L. D. Clark[2], T. Haga[3], N. Koga[4]

[1]Oyama National College of Technology, 771 Nakakuki, Oyama, Tochigi, Japan
watari@oyama-ct.ac.jp

[2]UMIST, PO BOX   88, Sackville street, Manchester, M60 1QD,UK
keith.davey@umist.ac.uk, maria.rasgado@umist.ac.uk, dane.clark@umist.ac.uk

[3]Osaka Institute of Technology, 5-16-1 Omiya, Asahi-ku, Osaka, 535-8585 Japan
haga@ med.oit.ac.jp

[4]Nippon Institute of Technology, 4-1 Gakuen-dai, Miyashiro-machi, Minami-saitama-gun, Saitama-ken, 345-850 Japan, koga@ nit.ac.jp

Abstract: This paper is concerned with the development of a continuous strip casting technology to facilitate the manufacture of magnesium sheet alloys economically whilst maintaining high quality.   Established in the paper is warm formability of cast magnesium alloy sheets after being hot rolled by continuous casting process. Effects of manufacturing condition on possible forming were clarified in terms of casting temperatures and roll speeds. It has been found that magnesium sheet with 3.0 to 4.0 mm thickness could be produced at a speed of 25m/min.   The temperatures of the molten magnesium, and the roll speeds were varied to find an appropriate manufacturing condition.   Rolling and heat treatment conditions were also changed to examine which condition would be appropriate for producing wrought magnesium alloys with good formability.   Microscopic observation of the crystals of the manufactured wrought magnesium alloys was performed.   It has been found that a limiting drawing ratio of 2.7 was possible in a warm deep drawing test of the cast magnesium alloy sheets after being hot rolled.

1. Introduction

Magnesium alloys are expected to play an important role in next-generation materials, which have possibilities of contributing to lighten the total product weight when magnesium products can be used to replace aluminum and mild steel products.   The specific density of magnesium alloy is about 2/3 that of aluminum and 1/4 of that of iron. When alloyed, magnesium has the highest strength-to-weight ratio of all the structural metals.   Moreover, because of the ease of recycling of metallic materials, magnesium has received global attention from the standpoint of environmental preservation. Utilization of magnesium alloys has mainly depended on casting technology, for instance, thixso-forming because of its less workable characteristics due to the crystal structure of the hexagonal close-packed lattice.   Recently, demands have been raised in automotive and electronics industries to reduce the total product weight [1]. Unfortunately, the major barrier to greatly increased magnesium alloy use is still primarily high manufacturing cost. One of the keys to solving this problem is to develop semi-solid roll strip casting technology to manufacture magnesium sheet alloys economically while maintaining high quality.

The authors, therefore, have investigated the effectiveness of continuous casting process for magnesium alloys [2]. This paper describes the forming characteristics of the cast magnesium alloy sheets after being hot-rolled in a warm deep drawing test. Established in the paper are the appropriate manufacturing conditions for the production of high quality strip using a purpose built semi-solid roll strip casting mill. Influence of process parameters such as melt temperature, roll speeds are ascertained. A warm deep drawing test of the cast magnesium sheets after being hot rolled was performed to demonstrate the formability of the magnesium alloy sheets produced by semi-solid roll strip casting process. Microscopic observation of the crystals of the manufactured wrought alloy sheets were performed to investigate effects of the hot rolling and heat treatment conditions on crystal growth in the products.

2. Experimental Procedures

2.1 Twin roll strip caster and experimental conditions

Figure 1 indicates the twin-roll strip caster for the horizontal casting direction. It includes a source of molten metal that feeds into the space between a pair of counter-rotating, internally cooled rolls. The principle dimensions of the strip caster and tundish are presented in Table1. Illustrated in Table 2 are the experimental conditions to investigate an appropriate manufacturing formation to successfully produce magnesium alloy sheets by twin-roll strip casting. Casting temperatures were changed from 620°C to 630°C to keep the molten metal in the tundish in a semi-solid state. Temperatures of the molten magnesium in the melting pot and tundish were measured by thermo-couples. Roll casting speeds were varied from 5 m/min to 30 m/min to examine which roll speeds are appropriate for solidifying the molten magnesium. The roll gap between the upper and lower rolls was set to 1.9mm to 3.3mm.

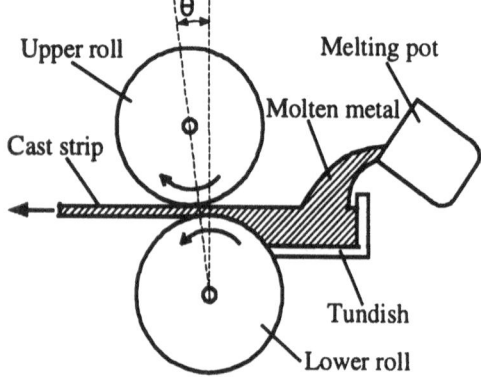

Fig.1. Schematic illustration of horizontal twin roll caster.

Table 1. Dimensions of horizontal twin roll caster and tundish.

Rolls	
Materials	Copper alloy
Upper roll	Φ300mm*100mm
Lower roll	Φ300mm*100mm
Roll speed	0-150m/min (Max.)
Inclination angle θ	$0°$
Tundish	
Material	Insulator
Volume	$20.0*10^5 mm^3$

Table 2. Experimental conditions.

Temperatures $(°C)$	620,625,630
Roll speeds (m/min)	5,10,15,20,25,30
Roll clearance (mm)	1.9-3.3
Shield gas	CO_2, N_2, CO_2+SF_6, No shield

2.2 Material

AZ31B Magnesium alloy was used in the experiment. The physical properties of the material are listed in Table 3. Magnesium ingots are heated to $650°C$ in a melting pot with an electric furnace. In the magnesium melting process, magnesium oxide and other suspended nonmetallic matter were removed with flux that preferentially wets the impurities and carries them to the bottom as sludge. After the refining process, the molten magnesium metal in the melting pot was carried to the strip caster, and poured into the tundish to manufacture magnesium strip.

Table 3. Physical properties of material.

Density	$(kg/m^3 \cdot 10^3)$	1.78
Liquidus temperature	$(°C)$	630
Solidus temperature	$(°C)$	575
Specific heat	$(J/kg \cdot °C)$	1040
Therma-conductivity	$(W/m \cdot °C)$	96

2.3 Hot rolling process

Hot rolling process was performed to obtain wrought magnesium alloy sheets with globular and fine microstructures. The cast strip sheets were milled to obtain the sheets with 2.0mm thick to remove oxide film. In the hot rolling process, the temperature of the cast strip was elevated to 400°C by heaters, and the ground sheet with elevated temperature was rolled by several rolling pass schedules until the sheet thickness became 0.8mm. Next, the 0.8mm thick sheet was annealed at 250-400°C for 2 hours. The annealed sheet was rolled again until the sheet thickness became 0.5mm. Finally, the rolled magnesium sheet was annealed at 250-400°C for 2 hours, and cooled in an electric furnace. In the hot rolling process, a 9m/min roll speeds was chosen. Under the condition of 350°C, cracks were seen during hot rolling process even though the reduction was less than 30 %. The 400°C temperature in hot rolling was chosen to keep cast products from cracking.

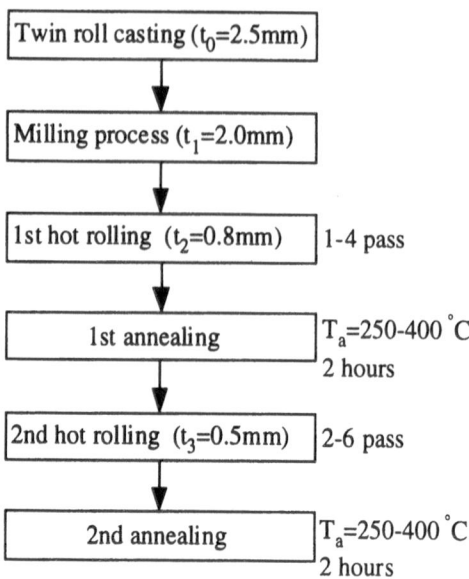

Fig.2. Flow chart of manufacturing process of wrought magnesium alloy sheets.

2.4 Warm deep drawing test

A warm deep drawing test of the cast magnesium sheets after hot rolling was performed to examine the formability of the magnesium alloy sheets produced by semi-solid strip casting. The forming conditions in the test, and dimensions of the deep-drawing tool are described in Table 4. The diameter of the punch is 28.8mm and it is cooled by water flowing through inner of the punch. A solution type lubricant was used as a lubricant in the deep-drawing test. The limiting drawing ratio was investigated by the deep-drawing test at 250°C, in terms of rolling and heat treatment conditions. Two drawing speeds, 30mm/s and 2.5 mm/s, were chosen in the test.

Table 4. Dimensions of deep drawing tool and forming conditions.

Punch diameter D_p	(mm)	28.8
Radius of punch R_p	(mm)	3.0
Die diameter D_d	(mm)	30.0
Radius of die R_d	(mm)	2.0
Drawing speed V_d	(mm/s)	2.5, 30
Forming temperature T_f	(°C)	250

3 Results and Discussion

3.1 Rolling condition and warm formability

The cast strips manufactured by continuous casting process were hot rolled to gain fine and globular microstructures. It has been found that rolling condition and annealing condition during the hot rolling affect formability of the hot rolled magnesium alloy sheets. In this experiment, firstly, a 60 % rolling reduction was chosen to gain 0.8mm thick sheets from 2.0mm thick sheets during the first hot rolling process. In the second hot rolling process, less than 10 % rolling reductions were chosen in four times rolling passes when 0.8mm thick sheets became 0.5mm thick sheets. Figure 3 shows the relation between annealing temperature and limiting drawing ratios of the hot rolled AZ31B magnesium sheets. At relatively high drawing speed (V_d=30mm/s), a limiting drawing ration of 2.7 in the case of 350 °C was possible in a warm deep drawing test. At relatively low drawing speed (V_d=2.5mm/s), a limiting drawing ration of 3.4 was also obtained in the case of 350 °C. These results suggest that the obtained wrought magnesium alloy sheets that were hot rolled after the strip casting process, have equivalent levels of plastic forming ability compared with the wrought magnesium alloy sheets obtained by conventional slab process.

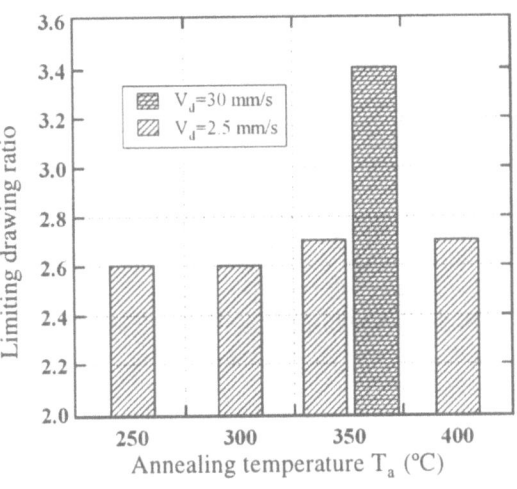

Fig.3 Relation between annealing temperature and limiting drawing ratio.

3.2 Microstructure of manufactured alloy sheets

Figures 4 present photographs of the micro crystals of AZ31B alloy sheets being hot rolled after the roll strip casting process. The photograph shown in Fig.4 (a) was annealed at 250°C, the photograph shown in Fig.4 (b) was annealed at 350°C. We could not recognize clear differences between the Fig. 4 (a) and Fig. 4 (b), however the mean grain size of Fig. 5 (a) was slightly smaller than that of Fig. 4 (b). We can also see that the grain sizes of the crystal in Fig. 4 (a) and 4 (b) are

180

less than 10 micrometers. Before hot rolling, the mean gain size of the cast magnesium alloy sheets was 30 micrometers. Although the experimental results obtained suggest a key to the relation between the crystal grain size of products and plastic formability in elevated temperature, more detailed experimental research will be required.

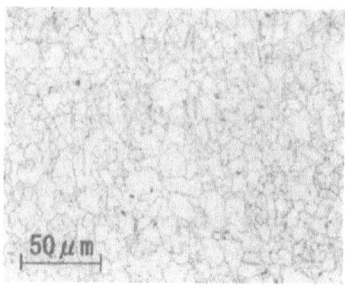

Fig. 4 (a) Crystal (T_a=250°C)

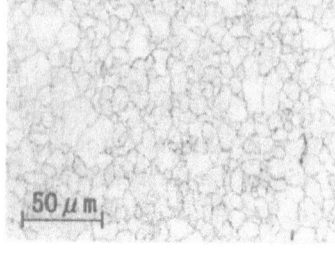

Fig. 4 (b) Crystal (T_a=350°C)

4 Conclusions

Effects of hot rolling conditions on limiting drawing ratio along with microstructure of manufactured alloy sheets were investigated. The conclusions obtained are as follows:

1) It was found that appropriate roll speeds are from 5 to 25 meters per minutes at 625 °C, and from 5 to 10 meters per minutes at 630 °C in the present experiment.
2) In the hot rolling process, a temperature of 400°C was necessary to avoid cracks occurring in cast magnesium alloys.
3) It was found that appropriate annealing condition was 350°C for two hours in the hot rolling process.
4) The mean grain size of the manufactured wrought magnesium alloys sheets was less than 10 micrometers.

5. Acknowledgements

This research was partially supported by the Japanese Ministry of Education, Science, Sports and Culture, Grant-in-Aid for Scientific Research (B) 15360076, 2003.

References

[1] S. Yoshihara et. al. : J. Mater. Process. Technol. Vol. 143-144 (2003), p. 612-p.615.
[2] H. Watari et. al. : Materials Science Forum, Vols. 426-432, (2003), p. 617-p.622.

The Application of Thermal Hydroforming in the Automotive Industry

Michael Keigler [1], David K. Harrison [2], Anjali K. M. De Silva [3], Herbert Bauer [4]

1 Aalen University of Applied Sciences, Aalen, Germany, michael.keigler@fh-aalen.de
2 Glasgow Caledonian University, Glasgow, UK, D.K.Harrison@gcal.ac.uk
3 Glasgow Caledonian University, Glasgow, UK, ade@gcal.ac.uk
4 Aalen University of Applied Sciences, Aalen, Germany, herbert.bauer@fh-aalen.de

Abstract: Thermal hydroforming process was developed to overcome the problems of forming aluminium alloy sheets or tubes, which are used extensively in the automotive industry for weight reduction and corrosion resistance. FEA simulation was used to predict the thermal hydroforming process and was verified by experimental analysis. Strains of up to 130 % were reached during the experiments, which corresponds to five times higher strains than that present in AlMg alloy at room temperatures. The FEA model can be used to assess the feasibility of new components.

1. Introduction

The increased demand by the automotive industry for lightweight, corrosion resistant materials has seen a growth in the use of aluminium alloys. One major disadvantage of aluminium alloys however is its poor formability compared to that of steel. To this end, new forming methods are being developed. One such method is "Thermal Hydroforming" – essentially a hydroforming process that uses global or local heating of aluminium sheets or tubes to enhance the formability [1].

Research work carried out since about the 1920s on the improved formability of materials with increased temperature has lead to the development of superplastic forming today. Superplastic forming is mainly used for limit-lot productions in the aircraft and spacecraft industry as it requires high temperatures (0.5 x melting point), low strain-rates (10^{-5} - 10^{-1} 1/s) and the use of specific materials (Grain size < 10µm). Thermal hydroforming differs from superplastic forming as it can be used for materials which do not posses superplasticity. These are mainly aluminium alloys from the groups 5xxx (AlMgSi) and 6xxx (AlMg), which have the ability to be formed at lower temperatures (<450°C) and higher strain-rates. Therefore thermal hydroforming is suitable for large scale series production.

Since thermal hydroforming is relatively a new technology very few publications can be found on it. This paper investigates the developments of a FEA model to predict the feasibility of new applications on thermal hydroforming, particularly with regard to automotive industry. Preliminary experiments were performed in order to obtain data for formulating the FEA model. The produced components were measured (strains and wall-thicknesses) and with the findings a FEA- Model was developed. This FEA- Model is important in order to be able to give statements about the accuracy of the calculation and to carry out further investigations theoretically. From hydroforming at room temperature it is known that the FEA-simulation is an important and helpful tool during the feasibility investigation of components. Similarly, FEA- simulation can be used for thermal hydroforming to optimise the process before tool production.

2. The Experimental Tool

Preliminary investigations were carried out to assess the formability with increasing temperature for commonly used automotive aluminium alloys. For alloys of the AlMgSi group no significant improvements in its formability was observed, whereas AlMg alloys showed enhanced formability with increasing temperature. This is evidenced by the stress/ strain curves vs. temperature as shown in Figure 1.

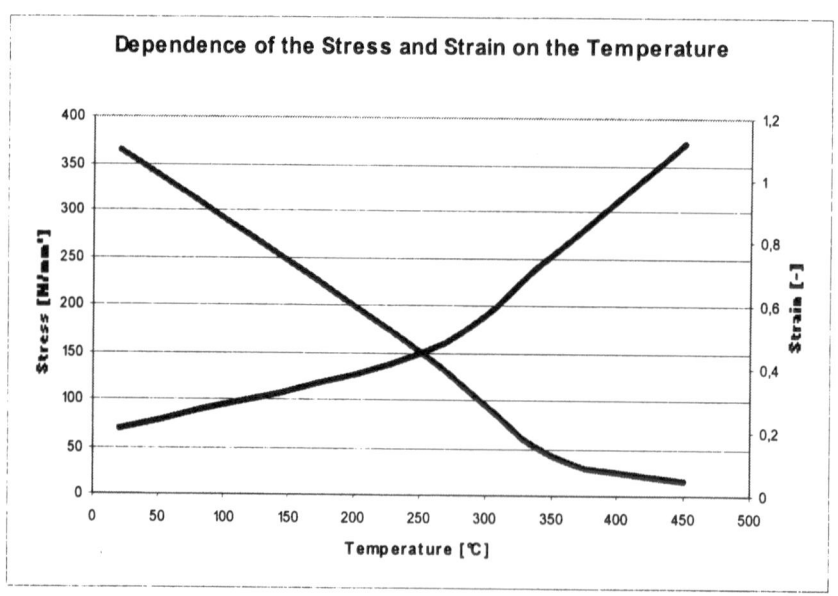

Figure 1: Dependence of the stress and strain of a AlMg alloy on the forming temperature

The improved formability with increase in temperature can be attributed to two factors. Firstly, the yield strength of the material decreases with increasing temperature, thereby forming loads are greatly reduced. This means that lower internal pressures, lighter tools and smaller machines are , which reduces the cost of the process [2] and also facilitates a reduction in cycle time. Secondly, the increase in the deformation-degree at higher temperatures facilitates more complex components to be realised and improves the possibility of integration of secondary components (e.g. rips, part connections). Therefore, from simple single parts a high-complexity component with high functionality can be produced, further reducing the costs.

Figure 2 shows the experimental setup that was developed to investigate the effects of the temperature rise on the formability of materials.

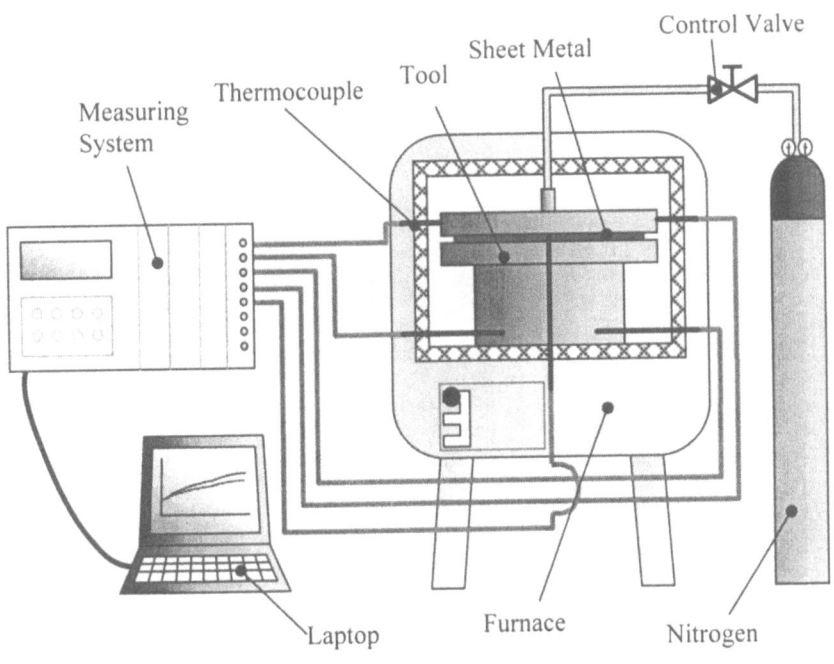

Figure 2: Experimental Setup for the thermal hydroforming

Nitrogen was used as forming-medium as it, has among other things, the advantages that it is cheaper than Argon, and does not form an explosive mixture (like air and oil at higher temperatures). A furnace is used for warming the tool and the aluminium alloy workpiece. The furnace control is used for temperature regulation and with thermocouples the temperature is controlled during the experiments and recorded for subsequent evaluation. A circular sheet workpiece of diameter 200 mm and a thickness 4 mm was chosen as the initial geometry. From this a cup is formed at 450 °C which has a diameter of 150 mm and a depth of 48 mm. The best results were achieved at an internal pressure of 40 bar. In this case the radii formed were good, the forming depth was reached and the cup wall fits to the tool wall (Figure 3). The actual forming time of these experiments was 1 minute and 30 seconds. However, difficulties were experienced when removing the formed part off the tool, indicating the presence of high frictional forces during forming.

The tool was constructed in such a way that no material flow-in occurs from the edge of the initial sheet. Thus it can be assumed that forming occurs completely as a result of the strains in the aluminium alloy. The strains were measured using a measuring grid on the sheet. A maximal extension of 130 % was realised with a thinning of wall-thickness up to 1 mm.

Figure 3: Aluminium alloy cup formed at 450 °C (diameter 150 mm, depth 48 mm)

3. FEA- Simulation of the Thermal Hydroforming

Deformation processes are non-linear, that means at every point of time another stress and deformation state occurs. The nonlinearity of the forming processes is characterized by:

- Nonlinearity of the material
- Nonlinearity of the boundary conditions
- Geometrical nonlinearity

For the simulation of the thermal hydroforming process the explicit FEA-Calculation program LS-DYNA software from Livermore Software Technology Corporation (LSTC) was used. LS-DYNA is a FEA- Program for high-grade non-linear, dynamic problems. It is used particularly in the crash, occupant and forming simulation [3], however, it is also used for the simulation of drop-tests or for ultimate load calculations.

The thermal hydroforming experiment of a cup described previously was simulated with the aid of LS-DYNA software package. In this case the axes-symmetry of the component was utilised. The necessary material parameters (flow-curve) were determined from crush tests. Other necessary material attributes are the density and the Young's modulus. The unknown parameter during the simulation and during the experiments was friction. In order to account for the friction Coulomb - or Kinematic model can be used. Thus for the Simulation of the thermal hydroforming the Coulomb's friction model of LS-DYNA was used:

Coulomb's friction model:

$$\tau = \mu \cdot \sigma_n$$

The tangential-contact-stress τ is calculated with the material parameter μ (friction coefficient) and the contact pressure σ_n, which is the strength component acting orthographically to the friction surface. The variable μ is largely unknown and was investigated with the aid of the FEA model. Using the FEA simulation model the best analogy with the reality was achieved for a friction coefficient of 0.12. The maximum difference between simulation and experiment was in this case 0.15 mm. However, on average the difference is less than 0.1mm (Figure 4).

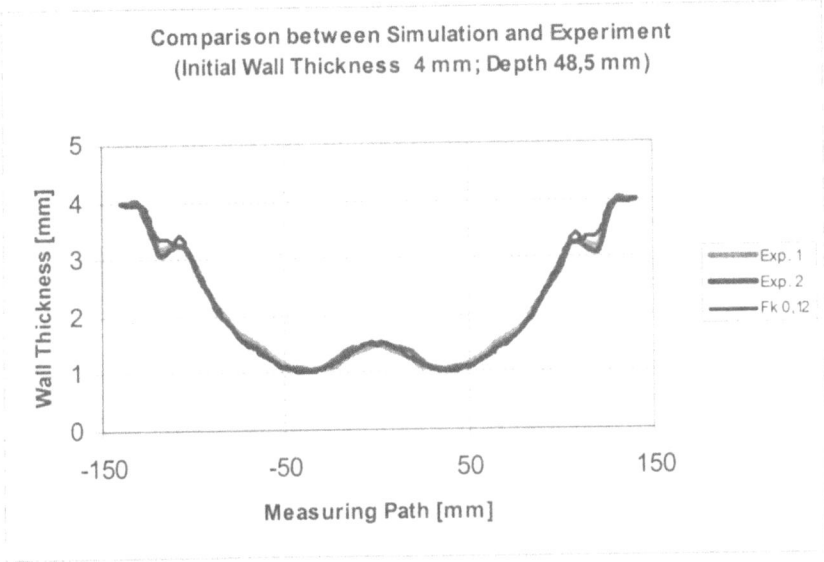

Figure 4: Comparison between the FEA simulation and the experimental wall thickness

Further experiments with different sheet thicknesses and aluminium alloys were carried out. These tests confirmed the close correlation between the FEA simulation model the experimental results.

From these results it can be concluded that:

- The friction is greater for thermal hydroforming than for hydroforming at room temperature (μ=0.06 – 0.08)
- The thermal hydroforming process is computable with the created FEA-Model and thus feasibility investigations can be carried out

4. Summary

From the preliminary experiments performed the potential of the thermal hydroforming process was established. Strains of 130 % were able to be reached during the experiments, which corresponds to four- to five times higher strains than that present in AlMg alloy at room temperature. The simulation model created showed very close correlation to the experimental results. Thus it enables the feasibility study of new components. The necessary material parameters can be obtained by the crush test or the cup testing method.

Further investigations are being performed to ascertain techniques for reducing friction during thermal hydroforming. This is important in order to improve the flow of material and to obtain a homogeneous wall-thickness distribution. In order to improve the efficiency of the thermal hydroforming it is necessary to work on possibilities for the reduction of the cycle time.

References

[1] Wieser, Keller, Brünger (2002): "Non Heat-Treatable Aluminum Materials for Automotive Application ", In Papers of the Interantional Conference "New Developments in Sheet Metal Forming", Juni 4-5, Fellbach/Stuttgart, Germany, MAT-INFO Werkstoff-Informationsgesellschaft mbH, pp. 365-383, ISBN 3-8355-305-0

[2] Treude (2002): "Reduce the Cost of Hydroformed Parts" , Metalforming Magazine, Volume 36, Number 12, December, 2002, pp. 30-33

[3] Haas, Bauer, Hall, Mihsein (2001): "State of the Art in the use of (LS-DYNA) forming simulation in hydroforming and preceding processes", Third European LS-DYNA Conference 2001, June 18-19, Paris, France

Dieless Drawing for Flexible Processing of Microstructure and Mechanical Properties.

Klaus von Eynatten [1], J N Reissner [2]

Institute of Virtual Manufacturing, Technoparkstr. 1, CH - 8001 Zürich
[1] kve@ivp.bepr.ethz.ch
[2] reissner@ivp.bepr.ethz.ch

Abstract:
This paper deals with the potential of the Dieless Drawing technique; esp. in means of flexible processing of microstructure and resulting mechanical properties. Therefore, the heat-treatable steel C35E was evaluated and drawn up to a reduction in area of $R_D = 45\%$ with good accuracy (h11). The variation in diameter was studied depending on the deformation temperature T_D and total strain φ. A wide variety of microstructures and properties was realised on the same as-received material, e.g. from a ductile ferritic/pearlitic microstructure ($\sigma_{UTS} = 705$ MPa, $\varepsilon = 23\%$) to a high strength martensitic microstructure ($\sigma_{UTS} = 1600$ MPa, $\varepsilon = 4.8\%$) or a surface hardened material with a ductile, machineable core.

1. Introduction

The idea of Dieless Drawing (DLD) was first published by Weiss and Kott [1] in 1969. However, information about industrial application of this process is rare [2], although many researchers all over the world have contributed work to evaluate this novel wire drawing technique [1-7].

The Dieless Drawing technique is a non-contact, flexible metal forming process without the use of a conventional wiredrawing die. The die is replaced by a heating, measuring and cooling (HMC) device (*Figure 1*).

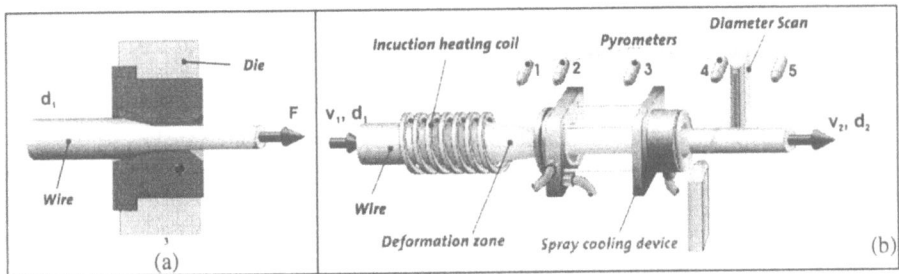

Figure 1: Concept of (a) conventional wire drawing and (b) Dieless Drawing

Due to the contact-free deformation a number of advantages compared to the conventional wire drawing emerge
    - high degree of freedom of the process parameters (T, φ, v_2, etc.)
    - flexible processing of shape, microstructure and mechanical properties

- high reduction of area per draft
- possibility of thermo-mechanical processing
- drawing of high strength alloys
- low costs (no die, wear, lubricant, maintenance, set-up time)
- etc.,

outnumbering the disadvantages like

- complicated interdependence of the process parameters
- process sensitivity, precision
- low productivity.

1.1   Principle of dieless drawing

An axially loaded wire (difference in feeding speed v_1 and drawing speed v_2, $v_1 < v_2$) is heated (induction coil) to an elevated temperature to initiate plastic deformation. The deformation is restricted by subsequent cooling (spray cooling device). Deformation takes place in the high temperature range between induction coil and cooling device, which approximately describes the deformation zone L_D. So far, the DLD-process has been theoretically analysed in detail by various authors [e.g. 8-10].

The resulting diameter d_2 after Dieless Drawing is determined only by considering the law of volume constancy

$$d_1^{\,2} \cdot v_1 = d_2^{\,2} v_2 . \tag{1}$$

The total strain is given only by the ratio between v_1 and v_2

$$\varphi = \ln\left(\frac{v_2}{v_1}\right). \tag{2}$$

The reduction in area R_d is given by

$$R_d = 1 - \frac{v_1}{v_2}. \tag{3}$$

Since DLD is a thermo-mechanical process the microstructure and, therefore, the mechanical properties of the wire drawn can be controlled by the heating and cooling conditions. They basically depend on the total strain φ, the strain rate $\dot{\varphi}$, the maximum temperature T_D, the temperature distribution $\dot{T}_X$, $\dot{T}_r$ (axial and radial) and the deformation end temperature (T_E).

In this paper the results of an experimental Dieless Drawing program carried out concerning the microstrucural evolution of a heat-treatable steel (C35E) are presented. The dependence of the process stability (variation in diameter ΔD) towards the deformation temperature and total strain is also covered in this work.

2. Experimental Procedure

In order to evaluate the DLD-technology a 'quasi' continuous Dieless Drawing pilot plant has been designed and operated at the Swiss Federal Institute of Technology.

The machine is supplied with a 20kW heating generator and a translucent spray cooling device. It is possible to draw wires with an initial diameter ranging from $d_1 = 10\text{-}20mm$. Due to the limited heating power the maximum drawing speed is e.g. 60mm/s for $d_1 = 12mm$ and $R_d = 45\%$. The machine is supplied with a diameter scan, 5 pyrometers and two flow meters for water and air cooling to guarantee reproducible drawing conditions.

The examination of the flexible processing of mechanical properties has been carried out on the heat treatable steel C35E. The chemical composition is given in _Table 1_. The as-received wire diameter was $d_1 = 15^{-0.03}mm$.

Table 1: Chemical composition

Material	Element wt-%								
	C	Si	Mn	P	S	Cr	N	Ni	Cu
C35E	0.34	0.22	0.67	0.007	0.026	0.17	0.005	0.06	0.101

The process stability in means of diameter variation of the DLD-drawn material was examined at different deformation temperatures $T_D = 750°C - 950°C$. The total strain was varied from $\varphi = 0.3 - 0.9$ ($R_d = 25\% - 60\%$).

Common techniques were applied to analyse the microstructural evolution and mechanical properties.

3. Results and Discussion

3.1 Influence of deformation temperature and total strain

In Dieless Drawing temperature and plastic deformation affect each other resulting in a complicated interdependence of the process parameters. The effect of deformation Temperature T_D and total strain φ on the variation in diameter ΔD is shown in _Figure 2_.

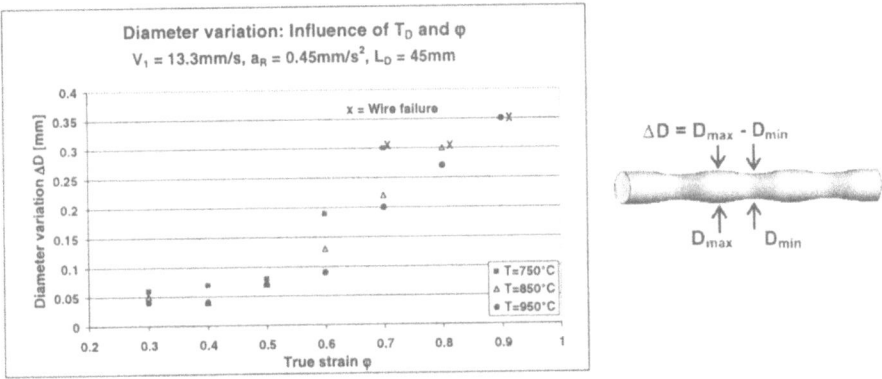

Figure 2: Diameter variation ΔD vs deformation temperature T_D and total strain φ during Dieless Drawing of C35E heat-treatable steel.

With higher deformation temperature the axial temperature gradient along the deformation zone L_D also increases, resulting in a more uniform deformation. Therefore, the process stability increases with the deformation temperature, which is known to be the most sensitive process parameter in Dieless Drawing [5].

Up to a total strain of $\varphi = 0.6$ ($R_D = 45\%$) high precision in diameter ($\Delta D < 0.1mm$) was obtained. This is in the range of cold drawn processing (h11). Increasing the total strain ($\varphi > 0.6$) results in a more non-uniform deformation. However, with carefully set parameters C35E has successfully been drawn to a reduction of area $R_D = 65\%$ without fracture.

3.2 Flexible processing of microstructure and mechanical properties

With different Dieless Drawing strategies a wide variety of microstructures and mechanical properties was obtained. The stress-strain curves of the different tensile testing specimens with their mechanical properties (yield strength σ_Y, tensile strength σ_{UTS}, maximum elongation ε) are given in *Figure 3*. All specimens exceed the properties of the as-received material.

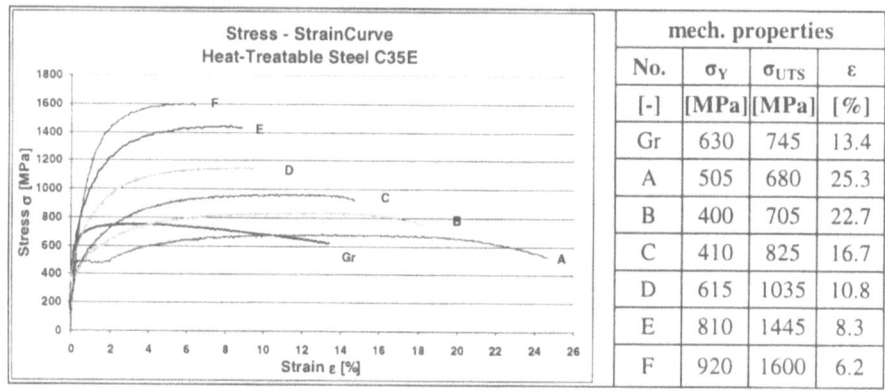

No.	σ_Y	σ_{UTS}	ε
[-]	[MPa]	[MPa]	[%]
Gr	630	745	13.4
A	505	680	25.3
B	400	705	22.7
C	410	825	16.7
D	615	1035	10.8
E	810	1445	8.3
F	920	1600	6.2

Figure 3: Mechanical properties of C35E in as-received condition (Gr) and after Dieless Drawing (A-F).

The observed microstructures are shown in *Figure 4*. The as-received typical cold drawn microstructure (Sample Gr) is characterised by a banded structure of ferrite and pearlite grains. After Dieless Drawing the banded structure was successfully eliminated except for Sample A. This is due to the deformation temperature T_D being lower than the ferrite to austenite ($\alpha \rightarrow \gamma$) transformation temperature A_{r3}, $T_D < A_{r3}$. Since grain refining occurs during Dieless Drawing the drawn specimens experienced high elongation during tensile testing with grain sizes $D < 10\mu m$. The multi-phased microstructure of specimen B - D results in relatively low yield strength and a strong work hardening effect with high tensile strength. These grades qualify for subsequent cold bulk forming processes without intermediate annealing for the production of e.g. fasteners. Specimen E and F developed high strength with moderate elongation and work hardening for high performance applications.

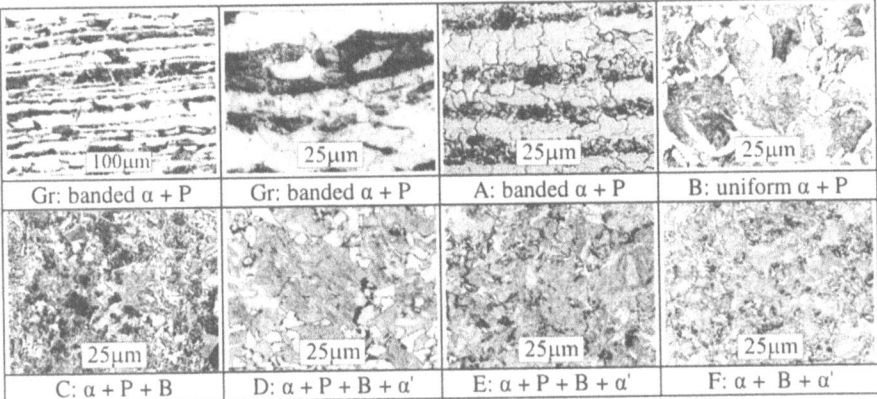

Figure 4: C35E: Microstructures in the as-received condition and after Dieless Drawing.
(α = Ferrite, P = Pearlite, B = Bainite, α' = Martensite)

3.3 Surface hardening treatment

The limiting factor for the wire cooling during Dieless Drawing is the thermal conductivity of the material. By exposing the wire to an effective spray cooling for a short time a surface hardening treatment can be implied. The result is a very hard surface (α' + B) with a ductile core (α + P) without any alignment of grains, as seen in *Figure 5*. This grade qualifies for e.g. drilling long holes along the longitudinal axis.

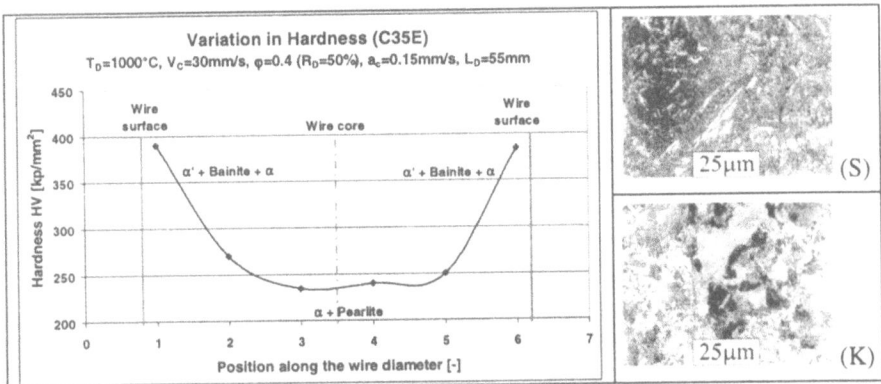

Figure 5: Surface hardened wire with hard surface (S) and ductile core (K)

4. Conclusions

It was found that the process stability during Dieless Drawing improves with deformation temperature due to an increase of the axial temperature gradient along the deformation zone. Up to a reduction in area of $R_D = 45\%$ high precision in diameter ($\Delta D < 0.1$mm, h11) was achieved.

192

The mechanical properties of the drawn wire were found to vary considerably with the deformation conditions, especially heating temperature and cooling conditions. With a heat-treatable steel (C35E) a wide variety of microstructures and hence mechanical properties was produced. Therefore, the Dieless Drawing technique can be used for thermo-mechanical treatment. DLD is very prospective in tailoring properties since the cooling conditions can be varied in a certain range during processing. A customer tailored wire with a combination of e.g. hardened surface and ductile core, full hardened, soft annealed or uniform ferritic/pearlitic microstructure can be dieless drawn (*Figure 6*).

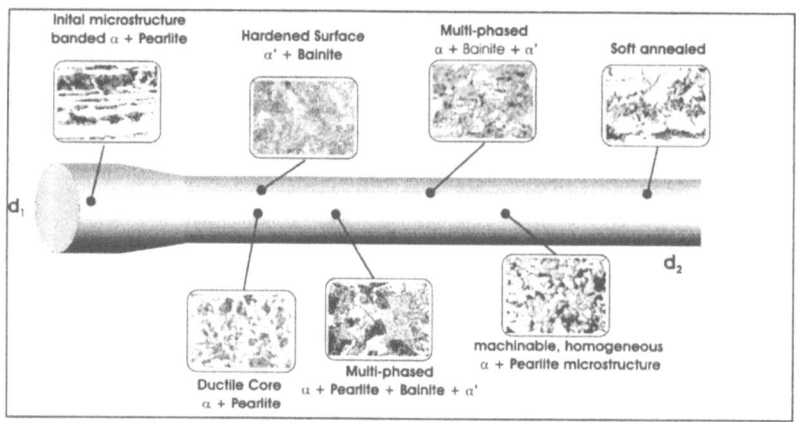

Figure 6: Potential of Dieless Drawing for the production of e.g. a tailored wire with the respective properties.

Acknowledgements

The authors would like to thank Steeltec AG for their cooperation and the Commission for Technology and Innovation (KT I/CTI),Switzerland, for the sponsoring of this research project.

References

[1] Weiss V., Kott R.A., WIre Journal, Sept. (1969), pp 182
[2] Sekiguchi, H., Kobatake,K., Advanced Technology of Plasticity, 1, (1987), pp 347
[3] Kawaguchi, Y., Katsube, K., Mujrahashi, M. et al, Wire J. Int., Vol 12 (1991), pp 53
[4] Tiernan, P., Hillery, M.T., Proc. Inst. Mech. Engi., Vol 216 Part L (2002), pp 167
[5] Wengenroth, W., Pawelski, O., Rasp, W., Steel Research 72 (2001), pp 402
[6] Gliga, M., Canta, T., Wire Industry, Vol 6 (1999),pp 294
[7] Wang, Z.T., Zhang, S.H., Xu, Y. et al, J. Mater. Pro. Tech., Vol 120 (2002), pp 90
[8] Fortunier, R.,Sassoulas, H.: Int. J. Mech. Sci. Vol 39, No 5 (1997), pp. 615
[9] Pawelski, O., Kolling, A., Steel Research 66 (1995), No 2, pp 50
[10] Sekiguchi, H., Kobatake, K., Osakada, K.: Pro. 15th MTDR Conference (1974), pp 539

An Intelligent Multiagent Approach for Friction Surfacing Applications

V. I. Vitanov[1], N. Javaid[1], G. Sapundgiev[2]

[1]Cranfield University, School of Industrial and Manufacturing Science, MK430AL.
[2]Technical University of Sofia, Faculty of Automation, Bulgaria.

Abstract: This paper discusses a multi agent approach for the selection of major friction surfacing parameters. With the intention to overcome the reasoning limitations of the single approaches by combining the strength of the employed agents. The agents approach has been used in order to emphasise the need for more general understanding of the complex interaction between the friction surfacing equipment and the processing environment. The agents include fuzzy logic reasoning as an integration environment, case based reasoning reflecting the experience from around 3000 experiments, FEA reasoning to analyse the relationship between the critical process parameters and for synthetic modelling as well as forecasting techniques to analyse and design new experiments.

1. Introduction

Surface engineering derives its importance and usage from the fast growing industrial requirements, typically operations at high temperature and pressure resulting in greater wear and corrosion and often the inadequacy of conventional materials to deliver the required operational and performance standards, thereby urging the need for composites to meet the criteria.

A wide range of surface improving techniques has been developed and successfully employed in industry but most of them come with the inherent process defects of porosity, slag inclusion, excessive dilution, cracking etc. To overcome some of the above undesirable effects friction has been exploited to deliver enough thermomechanical energy, as required for successful joining or deposition of wide variety of materials. Friction surfacing is considered to be a derivative of friction welding, retains all the features of its source - solid phase (without melting), forged micro-structure, negligible dilution, narrow heat affected zone, higher deposition rates and strong bonds. The process involves rotating a consumable rod (MechtrodeTM) of required coating material to a desired speed and then bringing in contact with the substrate (stationary initially) under an axial pressure, the heat from friction is sufficient to initiate plastic flow in the Mechtrode. During the transient process frictional interface starts to move up from the substrate surface into the rod with peak temperatures approaching the melting point of the depositing material. The substrate is then traversed across the face of the Mechtrode at a preset velocity, which causes the material to deposit downwards from the hot frictional interface onto the substrate surface. The achieved bond strength is a function of the applied force, rpm, traverse movement velocity and direction. On completion of linear traverse, depending upon the desired length of coating, the axial pressure is released and momentarily the rotating Mechtrode is withdrawn. The width, thickness, HAZ and bond strength are dependent on process parameters, material properties and dimensions/geometry of Mechtrode and substrate. The process is environmentally clean, efficient with no fumes, spatter or high intensity light and noise emissions.

During the past decade friction surfacing process has been established to an extent when it can be exploited commercially. The research [1 to 9] however has revealed that the input process parameters (force, RPM, and traverse velocity) are of vital importance to the quality outcome. Never the less existing models of the relationships between critical parameters and the process quality attribute are still generic and of limited practical value. An extensive review of the literature, technical reports and interviews with practitioners reveal that the relations have not been examined rigorously using appropriate analytical, numerical methods and DOE techniques. There is no reliable engineering methodology at present that allows for the selection of optimal process parameters for a particular material family, geometry and equipment.

A review of the literature sources [1, 2, 3, and 4] has revealed that the recommended process parameters for a range of material families vary considerably. For instance, reported results for depositing Stainless Steel on Mild Steel (consumable rod of Stainless Steel and substrate of Mild Steel) are shown in table 1, which clearly are indicative of the fuzzy nature of the reported results for optimal conditions. Further conclusion that can be drawn from these results and also as observed by the authors, is that an outcome of friction surfacing process is not only dependent on the input process parameters but also on the type of material used, diameter of the consumable, direction of the traverse movement and substrate geometry. Because the relationships between these variables are not known, some claims concerning the selection of optimal conditions for the process becomes questionable.

Table 1. Intervals for RPM, axial pressure and traverse speed

S/No.	Coating Material	Substrate Material	Consumable Diameter(mm)	Rotational Speed(RPM)	Axial Pressure(MPa)	Traverse (mmlmin)
01	SS	MS	12.7	1500 - 3000	5.5-19.4	156 - 500
02	SS	MS	20.0	300 - 2400	30 - 93	60 - 240
03	SS	MS	25.0	550	101.9	300
04	SS	MS	25.0	550	101.9	318
05	SS	MS	25.0	500 - 700	79.5 - 101.9	120 -240

In an attempt to model process conditions for a particular material family and to cater for the non-linearity and uncertainty of the process, fuzzy logic based agent seemed a viable option. Which can be later on integrated with the DOE and statistical techniques to fine tune the system. Fuzzy models and controllers have found particular applications in complex industrial systems that contain linguistic variables or data ofuncertain nature and cannot be modelled successfully using traditional approaches [9 to 13]. More specifically this paper considers the development of methodology for process modelling, based on the relationships that exist between the following process variables;

1) Vz (vertical downward velocity of the Mechtrode), resulting Force developed during the process for a specific material family. This agent is specifically useful for the conversion of parameters optimised under laboratory conditions to the ones that can be used in an industrial set-up.

2) F (axial Force applied on the Mechtrode), V (substrate traverse velocity), N (revolutions per minute of the Mechtrode) and the resulting Bond strength

3) F (axial Force applied on the Mechtrode), V (substrate traverse velocity), N (revolutions per minute of the Mechtrode) and the resulting thickness and width of the deposit for a specific material family.

Systematic collection and storage of large quantities of data from laboratory and industrial experimentation has resulted in a database containing the results from over three thousand experiments

2. Relationships between Process Parameters

This database has formed the foundation for the initial analysis to establish the above-mentioned relationships between process parameters and to produce Fuzzy models.

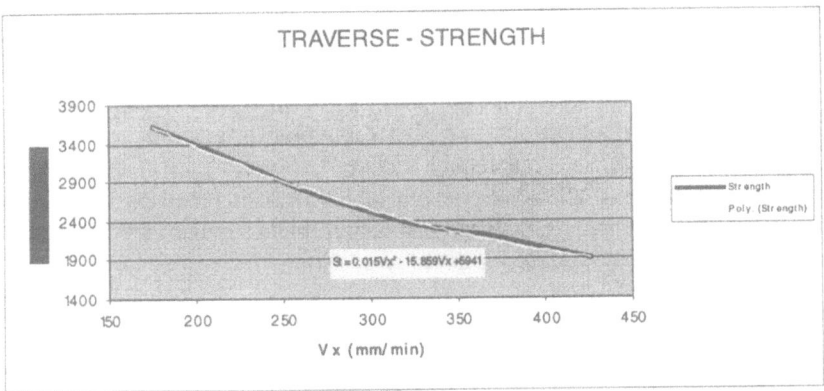

Figure 1 Relationship between traverse of the substrate and the bond strength

Similar relationships, as in Figure 1, were established between major process parameters and was followed by regression analysis to define the type of functions. This helped in developing an understanding about the behaviour of the process to the range of input parameters and their outcome. And the knowledge gained was utilized in defining the range of data, shape of membership functions and most importantly in building rules for the Fuzzy Inference Engine rules based on the deposition efficiency and the cold laps were also formulated, since these are not incorporated into any of the models, they have not been discussed.

3. Multiagent Approach for Friction Surfacing Process

Since the machine opted for experimental work was a CNC Vertical Machining Centre with a coordinate based system, which applies the preset feed rate to all the three axes so it was practically not possible to set force as an input parameter but measured during the process using a 'Kistler' force and torque sensor. From the feed rate the downwards velocity vector of the Mechtrode was calculated, initially a relationship was developed between this velocity vector Vz and the measured force, shown in Figure 2.

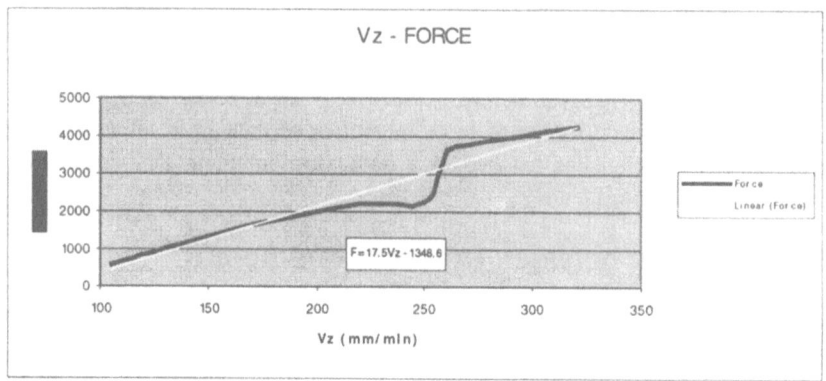

Figure 2. Relationship between velocity V_Z and the measured force

The basic structure of agents is shown in Figure 3, which serves as an expert system for selecting optimum process conditions for a given range of input parameters for a specific material family. Figure 4 shows a part of the inference mechanism for predicting the bond strength of the deposit based on three inputs (force, traverse and rotational speed) and one output (bond strength) and the figure 6 shows the overall response surface for the same system.

Clearly, first order polynomial was used to approximate the relationship between velocity vector 'Vz' and the measured force 'F', which as per assumption didn't give the exact values over the entire domain. To improve the accuracy, agent I was prepared, which evaluated the force from the feed rate of Mechtrode. Triangular shape membership functions were employed to describe the fuzzy sets and an overlap of 50% between the adjacent sets, for the predetermined minimum and maximum values per set over the whole range of input and output universe.

Force as an output from the agent 1 was fed as an input to agent 2 and agent 3 along with the rotational speed of the Mechtrode and traverse velocity of the substrate. Due to the complexity of the agent 2 with the number of associated rules, which where quite high, agent for bond strength was built separately. Agent 3 predicts the width and thickness of the deposited coating, measured torque during the process was not taken as an input parameter, just for the purpose of implementing and validating

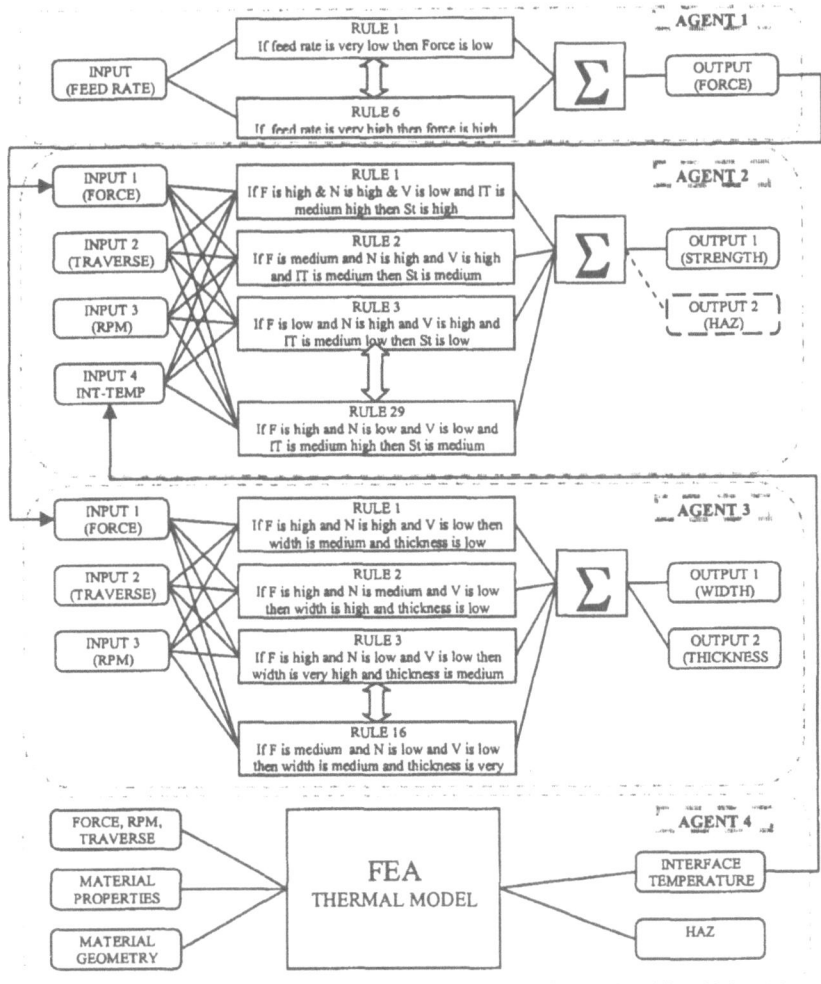

Figure 3. Schematic of the Multiagent approach

The value of this system is in reducing the lead time and hence, cost for the determining the optimum parameters for a given coating material on a given substrate geometry. This is an important feature when developing the process for new applications because the optimum conditions depend on the thermal system, which will vary when the materials, Mechtrode diameter and the substrate geometry are changed.

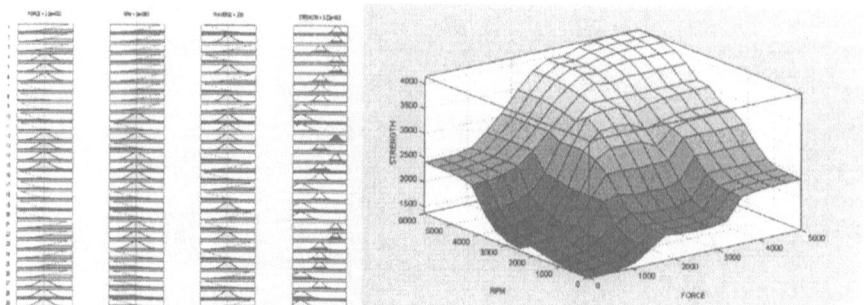

Figure 4. Inference mechanism and response surface of FS fuzzy inference system

4. Conclusions

- The relationship between the velocity vector V_z and the measured force can be described and evaluated by Fuzzy logic (agent 1)
- Reverse modelling (agents 2 & 3) for the process parameters and the quality state variables showed very good correlation compared to the experimental results
- Multiagent methodology can successfully be employed in the selection of optimal process conditions for a range of material families.

References

[1] W. Batchelor, S. Jana, C. P. Koh, C. S. Tan, The effect of metal type and multi-layering on friction surfacing, J. Mat. Proc. Tech. 57 (1996) 172 - 181.

[2] E. D. Nicholas, W. M. Thomas, Metal deposition by friction welding, Welding J. August (1986) 17 - 27

[3] E. D. Nicholas, Friction Surfacing, ASM Handbook Vol. 6,1993, pp 321 - 323.

[4] T. Shinoda, Q. Li, Y. Katoh, T. Yashiro, Effect of process parameters during friction coating on properties of non-dilution coating layers, Surf. Eng. 14 (3) (1998) 211-216

[5] G. M. Bedford, A. Davies, J. R. Sharp, Micro-friction surfacing in the manufacture and repair of gas turbine blades, Third International Charles Parsons Turbine Conference, Newcastle, (1995) 683 - 693.

[6] G. M. Bedford, V. I. Vitanov, I. I. Voutchkov, Decision support system for the frictec (friction surfacing) process, Proceedings of the Thirteenth National Conference on Manufacturing, Glasgow, UK, 9 - 11 September 1997, (1997) 580 -584.

[7] G. M. Bedford, L. J. Ward, P. J. Tooley, B. J. Wilson, R. J. Sharp, Large scale friction surfacing, EUROMAT'95, Proceedings of the Fourth European Conference on Advance Materials and Processes, Italy, Sept. 1995, (1995) 441- 444.

[8] I.I. Voutchkov, V. I. Vitanov, G. M. Bedford, Neurofuzzy model based selection of process parameters for friction surfacing applications, Proceedings of the Thirteenth National Conference on Manufacturing, Glasgow, UK, 9- 11 Sept 1997, (1997) 491-495

[9] T. I. Liu, X. M. Yang, G. 1. Kalambur, Design for machining using expert system and fuzzy logic approach, J. Mater> Eng. Perform. 4 (5) (1995) 599 - 609.

[10] R. X. Du, M. A. Elbestrawi, S. Li, Tool condition monitoring in turning using fuzzy set theory, Int. J. Mach. Tool Manufact. 32 (1992) 781 -796.

An Unequal Diameter Twin Roll Caster for Aluminum Alloys Casting

Toshio Haga[1], Masaaki Ikawa[2], Hisaki Watari[3], Shinji Kumai[4]

[1] Department of Mechanical Engineering, Osaka Institute of Technology,
5-16-1, Omiya, Asahiku, Osaka city, 535-8585, Japan   *haga@med.oit.ac.jp*
[2] Graduate School of Osaka Institute of Technology, Japan
[3] Department of Mechanical Engineering, Oyama national collage of Technology,
771, Nakakuki, Oyama city, Tochigi, 323-0806, Japan
[4]Department of Materials Science and Engineering, Tokyo Institute of Technology,
4259, Nagatsuda, Midoriku, Yokohama city, Kanagawa, 226-8502, Japan

Abstract: An unequal diameter twin roll caster was adopted for the strip casting of the aluminum alloys. The diameter of the upper roll was 250 mm, and that of the lower roll was 1000 mm. The width of the rolls was 100 mm. 6111 aluminum alloy strip was-cast at speeds of 5, 10 and 20 m/min. This caster could cast the strip at the speeds higher than the conventional twin roll caster. The start of casting was very easy. The strip did not stick to the roll without the lubricant. This was the effect of the low separating force. The microstructure of the 6111 was not a columnar structure but near to the equiaxed structure. This caster was adopted with closed top nozzle and side dam plate and cooling slope. Lower superheat casting was carried out using the cooling slope.

1. Introduction

The conventional twin roll caster for aluminum alloys (CTRCA) has some advantages. For example, rapid solidification, low equipment cost, low running cost, etc. However, the CTRCA has disadvantages, too. The start of casting is not easy. Casting speed is slow. Sticking of the strip to the roll occurs. In the present study, an unequal diameter twin roll caster (UEDTRC) was adopted for the strip casting of aluminum alloys in order to improve the disadvantages [1]. 6111 aluminum alloy was-cast to investigate properties of UEDTRC.

2. Twin Roll Caster of the Present Study

Figure 1 shows a photograph of the UEDTRC, and schematic illustration of the UEDTRC is shown in Figure 2. A lower roll was 4 times larger than an upper roll. The diameter of the upper roll was 250 mm. The upper and lower rolls were made from mild steel, and were water cooled. When the molten metal was poured in the nozzle, the rolls were rotating at intended speed and roll gap was set at intended gauge. The unsolidified metal did not overflow through the roll gap. This was the effect of the roll position. The base of the UEDTRC was a melt drag single roll caster [2-4]. A small upper roll was added to the melt drag twin roll caster. There are two effects of the small upper roll. One is improvement of the upper surface of the strip. The other is the improvement of the heat transfer between the roll and the strip. The load exerted by the rolls makes the heat resistance small. The top of the lower roll was higher than the meniscus of the molten metal. The unsolidified metal did not flow out to the rotating direction, but solidified metal was dragged. Control of roll speed and roll gap and use of a dummy plate were not needed at start of the casting. The nozzle was shaped at the curvature of the roll in the melt drag single roll caster and previous UEDTRC [1-4]. The nozzles of the UEDTRC of the previous and present study are shown in Figure 3. In the previous UEDTRC, a one-piece nozzle was used. In the present study, an assemble type nozzle was used. The clearance between the nozzle and roll, the side dam plate and the roll could be controlled. It is easy to prevent leak of the molten metal from the clearance between the roll surface and the nozzle in the nozzle of the present study. The assembled nozzle is superior in terms of the cost and ease of use.

Figure 1: Photograph of the UEDTRC of the present study

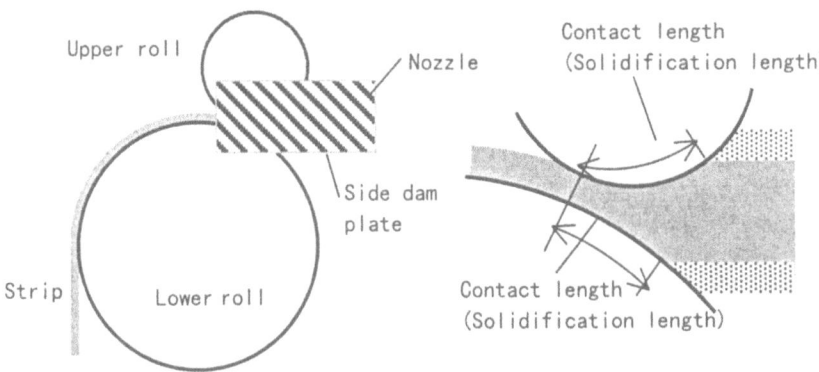

Figure 2: Schematic illustration unequal diameter twin roll caster

Table 1: Specification of a caster and experimental conditions

Upper roll	Mild steel, water cooling, Diameter 250 mm, width 100 mm Position 15 degrees
Lower roll	Mild steel, water cooling, Diameter 1000 mm, width 100 mm
lubricant	No use
Roll speed	5, 10, 20 m/min
Contact length between the roll and melt (solidification length)	Upper roll 50 mm Lower roll 60mm
Load	0.06KN／(mm per width)
Specimen	A611, 3 kg
Casting temperature	680 °C

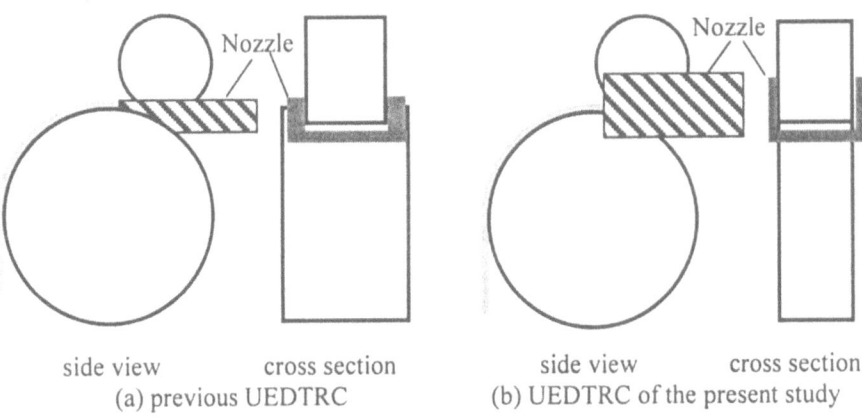

Figure 3: Shape of the nozzle of UEDTRC of previous and present study

3. Experimental Conditions

Table 1 shows the experimental conditions. 6111 of Al-Mg-Si alloy was cast into strip. The strip was cast at speeds of 5, 10 and 20 m/min. These casting speeds were higher than that of the CTRCA. The strip was homogenized at 540°C for 6 hours and cold rolled until 1mm thick. The 1mm-thick strip was annealed at 540°C for 2 hours, and cold rolled until 0.5 mm thick. T6 heat treatment was operated to 0.5mm-thick strip. The condition of T6 heat treatment was as below. Solution temperature was 540°C, holding time was 2 hours, and the specimen was immediately water quenched. Aging temperature was 160°C and holding time was 6 hours. Mechanical properties were investigated with a tension test and a 180 degrees bending test.

4. Results and Discussion

4.1 Casting using an equal diameter twin roll caster

6111 strip could be cast continuously at speeds up to 20 m/min by the UEDTRC. The sticking of the strip to the roll did not occur in the condition of no lubricant. The as-cast strip was not straight but had curvature at the roll speeds of 5 and 10 m/min. The curvature was as the same as the curvature of the lower roll. The small separating load was one of the reasons for the curvature. The lower solidification layer was thicker than the upper solidification layer. This different thickness of the solidification layer may be another reason for the curvature of the strip.

The relationship between the roll speed and strip thickness is shown in Figure 4. The strip thickness became thinner as the roll speed became higher. When the roll speed was 20 m/min, the strip was thinner than 3 mm. This shows that UEDTRC can cast thinner strip than that cast by the CTRCA. The solidification length shown in Figure 2 was constant in the present study. The solidification length influences the strip thickness. When the solidification length becomes longer, the strip becomes thicker.

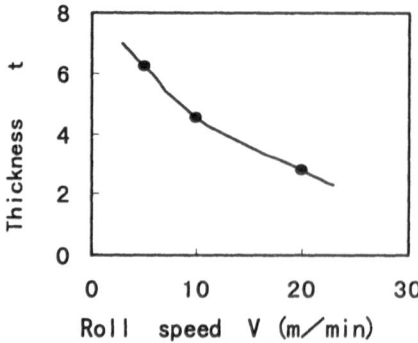

Figure 4: Relationship between the roll speed and strip thickness.

4.2 Surface and thickness

The surfaces of strip are shown in Figure 5. The upper side of the strip was not same as the lower side surface. This may be the influence of the contacting condition between the molten metal and the roll. The surface could be improved by cold rolling. The difference between the surfaces was not observed after cold rolling. A 20% reduction of cold rolling was enough to improve the surface.

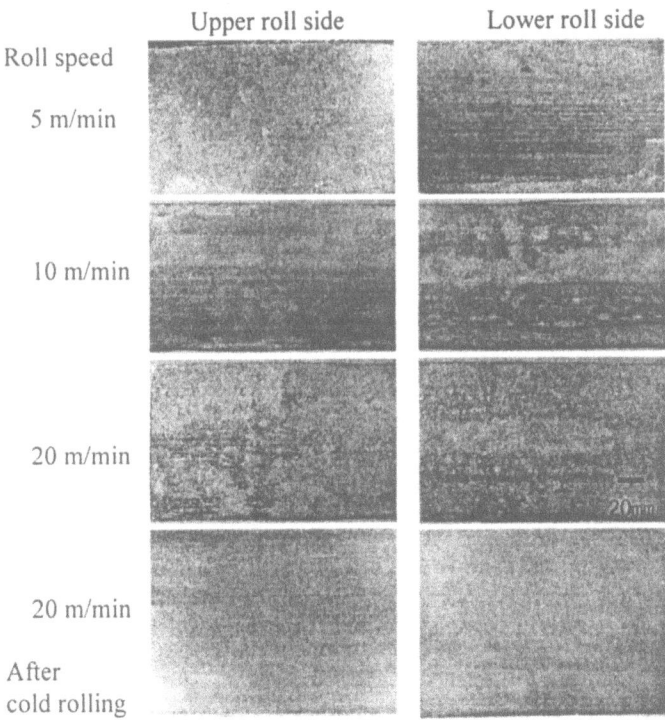

Figure 5:   Surface of as-cast and cold rolled 6111 strip

4.3 Microstructure

The Microstructure of the as-cast strip by the CTRCA was usually a columnar structure. However, the microstructure of the as-cast strip by UEDTRC of the present study was near to an equiaxed structure. The microstructure is shown in Figure 6. The microstructure of as-cast strip was not uniform in the thickness direction. However, the microstructure became uniform in the thickness direction after some mechanical and heat treatment.

4.4 Mechanical properties

Results of the tension test of a T6 heat treatment test-piece are shown in Figure 7. Roll speed did not affect the mechanical properties. The mechanical properties of the roll cast strip were almost the same as those of the ingot metal. Figure 8 shows the result of the 180 degrees bending test of a T4 heat treatment test-piece. The cracks did not occur at outer surface. This shows that the strip cast by the UEDTRC has good ductility. The microstructure of roll cast strip was very fine. This has a good effect on the ductility of the roll cast strip.

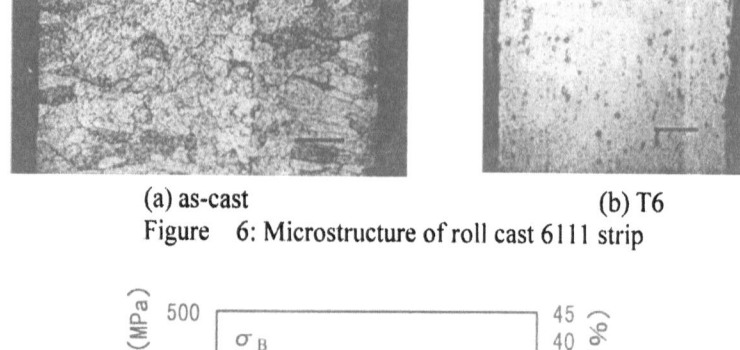

(a) as-cast   (b) T6

Figure   6: Microstructure of roll cast 6111 strip

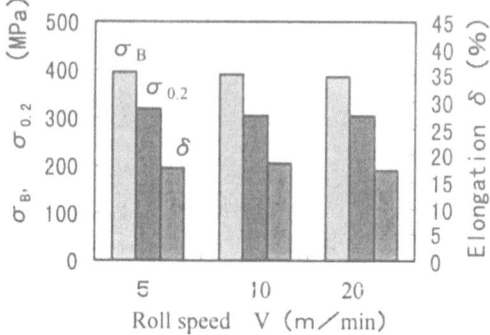

Figure 7: Result of the tension test; tensile stress, proof stress and elongation

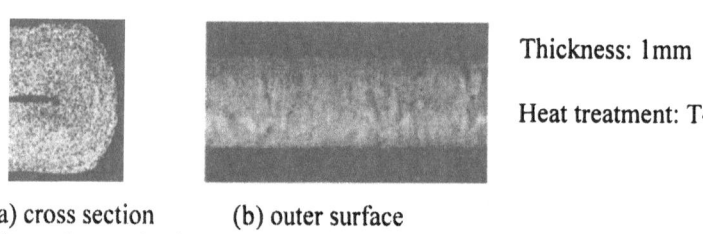

Thickness: 1mm

Heat treatment: T4

(a) cross section   (b) outer surface

Figure 8: Result of the 180 degrees bending test.

References

[1]   N Toyama, H Aho, H Arai and H Yoshimura, (1986) *Direct casting of stainless sheet by unequal-diametered twin roll method*, Tetsu-to-Hagane, **71** A245.

[2]   D B Love and J D Nauman, (1988) *Controlling the physical and mechanical properties of cast stainless steel band*. TMS Proceedings of an International Symposium on Casting of Near Net Shape Products, 597.

[3]   L E Hackman and T A Gaspar, (1986) *Producing strip by melt over flow*, Ind Heat, 153 36.

Optimisation of the Joining Process Selection in the Early Phase of the Automotive Development Process

V.I. Vitanov[1], J.W. Stahl[1], M. Ganser[2]

[1] School of Industrial and Manufacturing Science, University of Cranfield, Bedfordshire, England, V.Vitanov@cranfield.ac.uk, JanStahl@compuserve.com
[2] Research & Development Centre Munich, BMW Group, Munich, Germany, Markus.Ganser@bmw.de

Abstract: Cost validation at the end of the design stage is of critical importance for the automotive industry. This paper focuses on the development of a methodology for cost analysis of new design concepts in the preliminary phase of the automotive Product Development Process (PDP). The development of a generic Optimisation System to facilitate the design process will be discussed. The Optimisation System is aimed at providing the Design Department with the option to analyze the relative cost of new design concepts and to assist with the cost reduction in the Early Phase of product development.

1. Introduction

Cost management plays an ever-increasing role in the globalisation of the automotive industry and has led to an increase in competition in the quality car-manufacturing niche. In particular, the development of new designs accounts for a great percentage of the expenses incurred during the development process [1]. A portion of these expenses is due to the fact that, at the end of the design process of a particular subsystem, the designer does not have a tool to help with the comparison of future manufacturing costs of different design options. At present, this is a lengthy procedure that can take up to several months and requires input from three different departments. The optimisation procedure described in this paper aims to improve the interaction between the Design, Planning and the Finance Departments involved in the process. The intended main user of the optimisation tool is a designer that may require information about manufacturing costs. The proposed approach makes it possible to optimise the investment in specific technologies subject to manufacturing costs. The precise cost can be calculated only after obtaining process plans and information concerning the operation of the manufacturing facility. The aim of the proposed approach is to help with the cost analysis of different manufacturing options and to focus the efforts of experts on the most promising ones, saving time, funds and increasing the overall quality of the product.

2. Analysis of the Existing Process

2.1 Product Development Process (PDP)

To successfully manage the entire vehicle development process, automotive companies use the Product Development Process (PDP). PDP allows for the overall synchronisation and control of the development processes inside a specified timeframe, as is shown in Figure 1. The PDP is subdivided into certain phases starting with the Preliminary Phase, so called Early Phase of development, and finishing with the Start of Production (SOP) Phase.

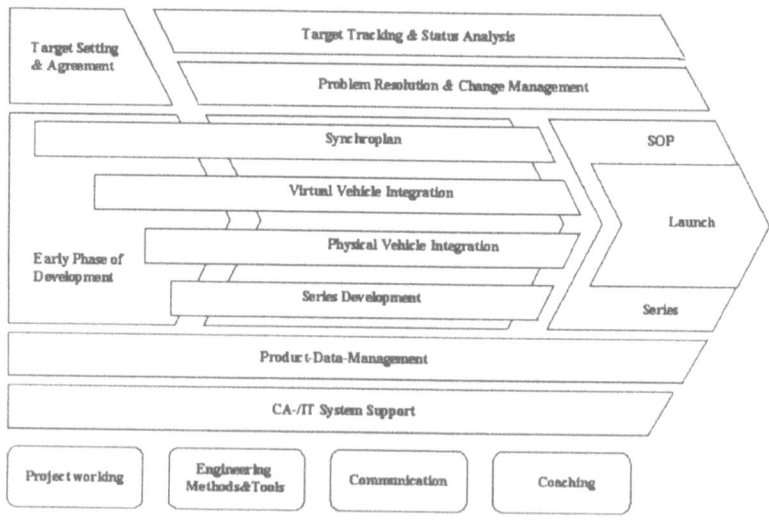

Figure 1: Product Development Process.

An important part of the Early Phase of the PDP is the Concept Phase. Within the Concept Phase, that is, the first stage of the Early Phase, a Design trend definition on the basis of an initial design model can be found. The problem in this part of the PDP is the lack of data and information concerning future costs of new materials, manufacturing processes, joining techniques, etc. Under such circumstances a designer cannot reach a proper decision about the future manufacturing costs of the new design concepts without extra input from Process Planners and Plant Managers, which is a lengthy process [3]. Almost 90 % of the future product costs will be defined in the Early Phase of the PDP by the Design and Planning Department inside of the Concept Phase [4].

2.2 Drawbacks and opportunities inside of body-in-white design process

The Design Department assumes the responsibility for developing coherent technical design concepts. This comprises the configuration of the vehicle, design characteristics and design trends, innovation concepts, component concepts and system concepts. At the preliminary phase of the automotive PDP, body-in-white (BIW) design is one of the most complicated and extensive areas. This is one of the core competences of the automotive industry and is not normally outsourced. The analysis of the BIW manufacturing processes and the integration of the activities of the Design, Planning and Finance departments in the structuring of the work sequences and working contents are of critical importance. The idea behind the proposed approach is to represent the design concept using working contents, which are further transformed into joining techniques, as shown in Figure 2.

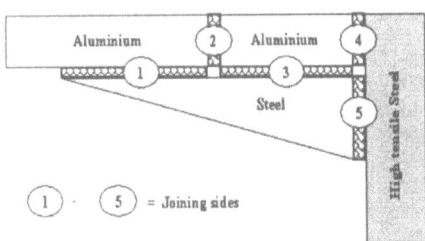

		Material	Length [mm]
1 Joining side		Aluminium - Steel	520
2 Joining side		Aluminium - Aluminium	275
3 Joining side		Aluminium - Steel	460
4 Joining side		Alu - High tensile Steel	275
5 Joining side		Steel - High tensile Steel	350

Figure 2: Schematic of a wheelhouse and joining sides.

It is expected that the BIW designer has the expertise to apply joining techniques for new design concepts but is not familiar with the cost decision models and the corresponding optimisation procedures. Working contents can serve as a common denominator in this situation between the above-mentioned departments. At the next stage of the optimisation process, 'joining technique matrices' are prepared in collaboration with the Planning and Finance departments. Inside of these matrices cost and material data that are relevant in initiating the optimisation of the Joining Process Selection method are compiled from the actual running process. These cost data include the cost rates for the fixed investment and the operational costs for the joining techniques with its machinery.

The discrete nature of the joining process suggests that dynamic programming can be used for process optimisation purposes. A designer has to make a decision regarding the optimal combination of joining techniques in respect of the manufacturing cost incurred. The optimisation procedure begins with the identification of the sides that have to be joined together, their joining length, and the material of the participating parts - as shown in Figure 2 - followed by selection of the appropriate joining techniques for each joining side, as indicated in Figure 3.

3. Optimisation of the Joining Process Selection

For illustrative purposes the proposed optimisation method is explained below on the basis of a generic joining techniques case study.

3.1 Algorithm inside of the joining process

Considering the process of joining selection separately at every joining side, a designer has to choose among several alternative technically applicable joining technologies. Knowing the material required on each joining side, a list of joining techniques that are authorized for a particular joining side can be obtained from the matrix shown in Figure 2. Figure 3 demonstrates one possible combination of the process of allocation of joining technologies to the joining sides.

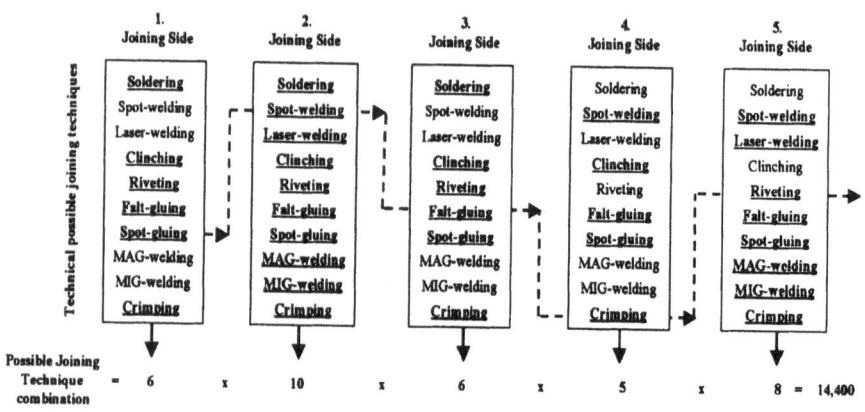

Figure 3: Allocation of joining techniques to the joining sides.

Considering the example shown in Figure 2 representing a wheelhouse with 5 joining sides and its material diversity, the allocation process can be illustrated as is shown in Figure 3. Taking into account the number of authorized joining techniques for every joining side, the possible number of joining technique combinations has been calculated as 14,400. Even from this oversimplified example it is apparent that it is irrational to obtain a global optimum subject to manufacturing costs via exhaustive enumeration. To emphasise this conclusion it is worth mentioning that, in reality, in an automotive BIW wheelhouse design the expectation is for approximately 200 joining sides, each with an average of 4 appropriate joining techniques - which increases the number of possible combination up to 4^{200}. To find a global optimum on the basis of such a great number of possible joining technique combinations is practical only if an optimisation technique can be used to avoid the enumeration problem. The optimisation procedure used in this paper is based on the dynamic programming method.

3.2 Optimisation procedure for allocation of joining techniques to joining sides

Dynamic programming (DP) has been selected as a mathematical procedure to improve the computational efficiency of the allocation of joining techniques to joining-sides problem. The idea behind DP is to decompose the global optimisation problem into smaller, and hence computationally simpler, sub-problems called stages. In the example discussed above, the stages represent the joining sides. Dynamic programming typically solves the optimisation problem in stages. The computations in the different stages are linked through recursive computations in a manner that yields a feasible optimal solution to the entire problem.

After breaking down the problem into sub-problems, the computations are carried out in stages. Each sub-problem is then considered separately with the objective of reducing the volume and complexity of computations. A stage in DP is defined as the portion of the problem that possesses a set of mutually exclusive alternatives from which the best alternative is to be selected.

		Stage 1		Stage 2		Stage 3		Stage 4		Stage 5	
Joining side No.:		1.		2.		3.		4.		5.	
Joining side Length [mm]:		520		275		460		275		350	
Joining side Material:		Aluminium - Steel		Aluminium - Aluminium		Aluminium - Steel		Aluminium – High tensile Steel		Steel - High tensile Steel	
No.	Joining Technique / Costs	c_1	M_1	c_2	M_2	c_3	M_3	c_4	M_4	c_5	M_5
1	Soldering	25	15	15	10	30	20	--	--	--	--
2	Spot-welding	--	--	10	15	--	--	15	20	20	30
3	Laser-welding	--	--	30	20	--	--	--	--	15	10
4	Clinching	15	40	10	20	15	30	10	15	--	--
5	Riveting	20	30	10	25	20	10	--	--	10	25
6	Falt-gluing	35	25	30	15	25	15	30	20	25	20
7	Spot-gluing	30	20	40	10	40	15	25	10	30	15
8	MAG-welding	--	--	15	15	--	--	--	--	20	25
9	MIG-welding	--	--	20	10	--	--	--	--	15	20
10	Crimping	25	20	30	20	25	25	30	20	15	15

c_j = Investment costs for the joining technology, M_j = Manufacturing costs of the joining technology of joining side j

Figure 4: Joining techniques - joining sides starting matrix.

In terms of the example in Figure 2, each joining side defines a stage, with the first up to the fifth stages having up to ten alternatives respectively, as shown in Figure 4. These stages are interdependent because all five joining sides must compete for a *limited* investment costs budget. The aim of the above DP problem is the optimisation of the sum of manufacturing costs: $M_{opt.} = M_1 + ... + M_5$ under the compliance of the constraint: $c_1+ \ldots +c_5 \leq 90$. In the present example, utilised data are deduced from real world sources.

The accomplishment of the DP algorithm results in the joining technique combinations assigned in Table 1, which achieve the optimal solution for manufacturing costs of 60 and investment costs of 85.

Five iterations were needed to achieve an optimal solution, which is a significant improvement in comparison with the enumeration approach that requires valuation of the entire set of 14,400 joining technique combinations.

Table 1: Optimal solution for manufacturing costs.

Permissible Solutions		1. Joining side	2. Joining side	3. Joining side	4. Joining side	5. Joining side	Investment costs	Manufacturing costs
1. Solution	[1,1,5,4,3]	Soldering	Soldering	Riveting	Clinching	Laser-welding	85	60
2. Solution	[1,9,5,4,3]	Soldering	MIG-welding	Riveting	Clinching	Laser-welding	90	60

The importance of using an appropriate optimisation model increases with the increase in number of joining sides.

4. Conclusions

Optimal selection of joining technology options can be achieved at the preliminary phase of the design development process.

The introduced DP optimisation procedure is applicable to a variety of optimisation and constraint criteria, such as the optimisation of the reliability, weight, energy, running costs etc. of the joining techniques process.

Optimisation at an early stage in the design and development process likewise helps the Planning and Finance Departments in proffering the most appropriate design concepts, subject to selected manufacturing criteria.

5. Acknowledgements

This research has been supported by the Research & Development Centre of the BMW Group, Germany, and the School of Industrial and Manufacturing Science, Cranfield University, UK. The authors would like to express their sincere thanks to Mr. Appold and Mr. Schrödinger for their attention and useful comments.

References

[1] Nitschke F., 1998, Markt- und prozessorientiertes Kostenmanagement von Entwicklungsvorhaben im Automobilbau. Dr. Kovač Verlag Hamburg.
[2] Ehrlenspiel K., 1998, Kostengünstig Entwickeln und Konstruieren, Springer Verlag Berlin Heidelberg.
[3] Beitz W., Birkhofer H. and Pahl G., 1992: Konstruktionsmethodik in der Praxis. In: Konstruktion, 44 (1992) 12, pp. 391-397.
[4] Botta V., 1996, Mitlaufende Kalkulation für eine frühzeitiges Kostenmanagement. KRP Kostenrechnungspraxis, Sonderheft1.

New Bending Technologies for the Automobile Manufacturing Industry

Peter Gantner[1], David K. Harrison[2], Anjali K. M. De Silva[3] and Herbert Bauer[4]

[1] Aalen University of Applied Sciences, Aalen, Germany, Peter.Gantner@fh-aalen.de
[2] Glasgow Caledonian University, Glasgow, UK, D.K.Harrison@gcal.ac.uk
[3] Glasgow Caledonian University, Glasgow, UK, A.DeSilva@gcal.ac.uk
[4] Aalen University of Applied Sciences, Aalen, Germany, Herbert.Bauer@fh-aalen.de

Abstract: This paper is concerned with a relatively new bending technique, known as Free-Bending. Its possibilities to realise geometries with almost arbitrary bending radii, with freely definable bending angles, which can be in several planes make this technique very interesting for an application in the automobile industry, especially in the field of car chassis and space frames. Therefore the paper shows some practical examples of geometries which were investigated with bending tests and examined by means of the finite element method. Furthermore the extended possibilities and the restrictions of this new technology are discussed.

1. Introduction

In the last few years spot welded sheet metal parts in chassis and space frames have increasingly been replaced by hollow profiles due to their the better stiffness characteristics and reduced weight. The tubes or profiles used are mostly bent and often additionally hydroformed. However, the traditional bending processes like Rotary Draw Bending or Roll Bending are often unsuitable for these geometries as the component may be pre-damaged from the clamping which can lead to failures during the down stream hydroforming process. Thus, it is necessary to investigate the characteristics of the new bending technologies. One of these is Free-Bending a technique which offers new potential and also allows extended design possibilities in comparison to techniques like Rotary Draw Bending. Especially in concert with the hydroforming process pre-bent geometries are required which have bends with different bending radii, bend-in-bends or spline bends. For the production of such "special" geometries the Free-Bending technique is very suitable, since this technique allows the bending of splines, bends with a bending angle more than 180° and bend-in-bends with a fast bending speed and without re-clamping.

However, it is necessary to be able to simulate this bending technique with a close-to-reality simulation model in order to secure the producibility of the component. Also, regarding the simulation of the whole production process and concerning the process security, it is essential to be able to simulate this bending process together with the down stream process steps like hydroforming, because only by regarding the complete process may a safe prediction about the feasibility be possible [1].

In the following sections the new bending technology, Free-Bending, is introduced and its functional principle is described. Furthermore, a simulation model is shown and the results of the simulation are compared with the results from the bending tests accomplished. To demonstrate the extended design possibilities a component is shown which was bent on the one side with "normal" bends and on the other side with a bend-in-bend and a spline bend.

2. Free-Bending

The Free-Bending technique allows the bending of almost any required geometry. The special concept allows the possibility to create almost arbitrary bending angles and bends over 180° together with spiral forms. The bending radii are freely-definable and the bends can flow in a transition-less way together in several planes [2]. Also, it is not required to re-clamp the tube during the bending and for different radii no tool changes are necessary.

The bending tests which are shown in this paper were done with a NISSIN CNC Tube Bender. This Free-Bending machine can bend tubes with a bending speed up to 350 mm/s which allows a very good cycle time regarding the application of this technology in series production [3].

2.1 The Free-Bending functional principle

The bending concept of Free-Bending can also be referred to as push bending with a die which is moveable in four axes. The work piece which can be a tube or profile is pushed along the z-axis through a guide by the pusher die that also restrains the component against rotation. This means that the bending is accomplished by pushing the work piece through the guide and through the movable bend die. The bending zone is thereby between the guide and the bend die. Thus there is no die with a pre-defined bending-geometry but the geometry is defined by the motion curve of the bend die. The Figure 1 shows exemplarily the bending dies and the tube.

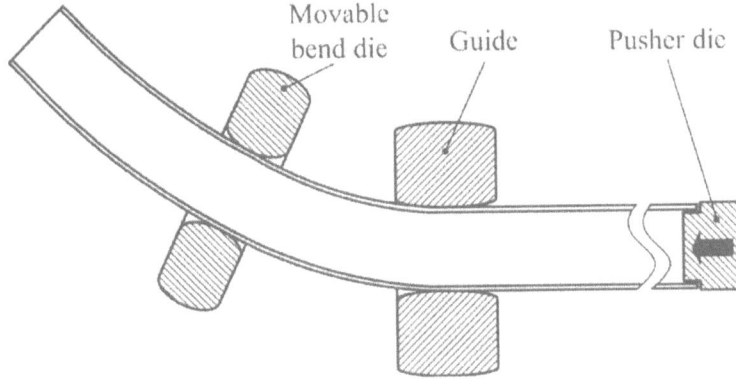

Figure 1: An exemplary representation of the Free-Bending principle

2.2 Possibilities and restrictions of Free-Bending

The advantages of this bending technique regarding the freely definable radii and its possibility to create bend-in-bends or spline geometries make this technology suitable for pre-bent hydroformed components or special geometries which are optimised with regard to minimised rotational flow. Due to re-clamping not being required this bending technique is particularly suitable for components with multi-bends. The more bends a component has the greater is the efficiency of this bending technique. Also, there is only one die necessary for different bending radii which can save considerable costs in comparison to conventional Rotary Draw Bending. A further advantage is the possibility to combine the bending with the torsion of profiles. Figure 2 represents an example of a geometry which was bent with "normal" bends and with a bend-in-bend and a spline bend.

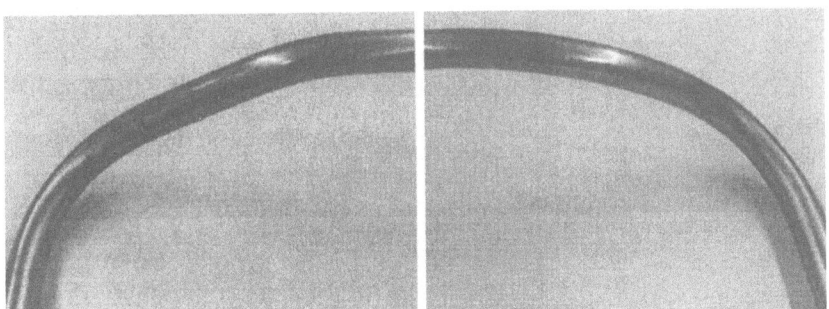

Figure 2: Geometry with "normal" bends (left side) and geometry with a bend-in-bends and a spline bend (right side)

However, this technology also has some restrictions. One of them is the minimal bending radius, since the cross-section of the tube becomes oval in the area of the bending zone when the bending radius is too small, and then it is not possible to push the tube through the bend die. Thin-walled tubes can also be a problem, because the buckling strength of a tube limits the bending radii. The risk of buckling of a tube or profile increases with the decreasing ratio of s_0/D_0 (wall-thickness to profile or tube diameter). Some bending tests have also shown that the repetition accuracy depends very strongly on the lubrication between the tube and the dies. The slightest changes of the lubricant film can cause significant deviations of the geometry.

3. The Simulation of Free-Bending

For the simulation of Free-Bending a LS-DYNA® simulation model was created which is exactly designed like the real bending dies of the NISSIN CNC tube bender: the guide which fixes the tube in the X- and Y-direction, the bend die which is spherically seated on the guide and the ring which is driven in the X- and Y-direction and which is also seated spherically on the bend die. The pusher die is not necessary, since the axial feed was directly applied on the last node row of the meshed tube. All dies are meshed with shell elements and are defined as rigid bodies and between the connected dies contacts are defined. The tube is meshed with 4-node fully integrated shell elements and the flow behaviour of the material of the tube is defined with a stress-strain curve. Contacts are defined between the guide and the tube as well as between the bend die and the tube. The process parameters for the control of the ring and the axial feed are taken from the CNC controller of the bending machine. The complete FE-model of the NISSIN CNC tube bender is represented in Figure 3.

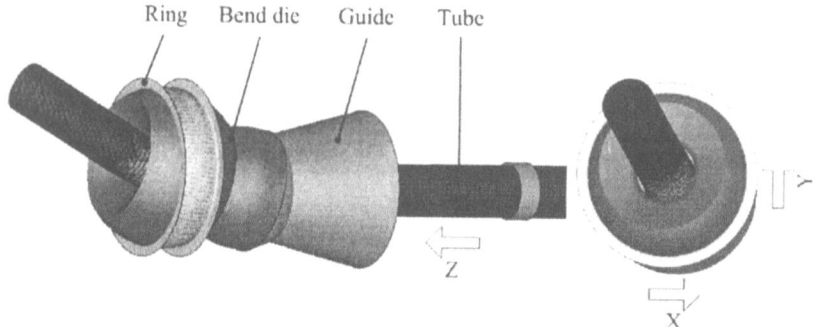

Figure 3: The simulation model of the NISSIN CNC tube bender

4. The Accomplished Bending Tests

The geometry of the first bending tests was a tube with six bends with different bending radii and bending angles (Figure 4). For the comparison of the simulation results with the measured values from the bent tubes the results of the simulation were reconstructed into Pro/ENGINEER® and the bending lines were computed by means of a special program [4]. This comparison shows strong deviations of the bending angles. The Table 1 represents the results of the bending angles from simulation in comparison with the measured angles from the bending tests (cf. [5]).

Table 1: Comparison of the first bending test with the simulation results

	Bend 1	Bend 2	Bend 3	Bend 4	Bend 5	Bend 6
Measured angle	54.7°	95°	33.1°	Spline	Spline	54.7°
Simulated angle	63.6°	102.8°	38.4°	-	-	61.6°

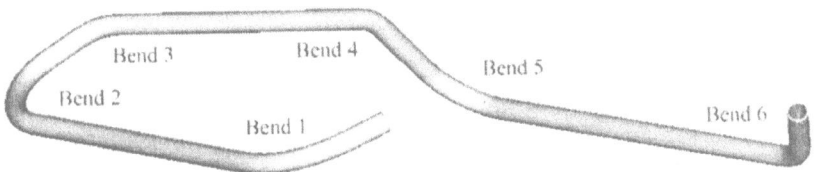

Figure 4: Bent steel tube (25 mm diameter, 2 mm wall-thickness) with 5 bends

To get a close-to-reality simulation model, some further bending tests were accomplished. Therefore, steel tubes with a 25 mm diameter and a 2 mm wall-thickness were bent. The bends have a 100 mm bending radius and different bending angles: 22.7°, 45.0°, 67.5°, 90.0°, 112.5° and 135°. The bent tubes were measured with a coordinate measuring machine, in order to be able to determine the bending line. The evaluation of the bending lines in comparison to the simulation results shows that the deviations increase linearly with the bending angle (Figure 5).

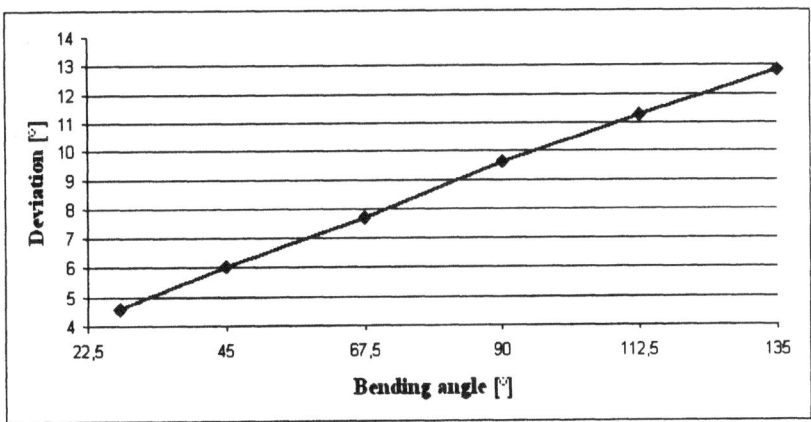

Figure 5: The deviations of the comparison of the bending angles from the simulation results and from the measured tubes

5. Summary and Future Work

In this paper a new bending technology was presented. Free-Bending offers a high degree of capability concerning the bending geometry and bending speed. Particularly, components with a high number of bends with different bending radii and bend-in-bends are suitable for manufacturing with this technique (e.g. fuel feed lines). With its possibility to bend splines the technique is also applicable for geometries which are optimised regarding characteristics such as flow. However, the minimal bending radius of the NISSIN CNC tube bender which is approximately 2.5 x outside diameter leads to restrictions concerning the part spectrum.

The simulation results of the developed finite element model have shown significant deviations regarding the bending angle. However, the comparison of the further bending tests with the simulation results have shown that the deviation increases linearly with respect to the bending angle. Thus a correction factor for the bending angle is possible to get close-to-reality simulation results. With such a simulation model, further investigations regarding the wall thickness distribution and the strains are possible. It is also intended to accomplish further investigations with the developed simulation model to define the limits of this technology.

References

[1] Bauer H., Haas A., Lerch I., et al (1999): "FEA-Process simulations of Hydroforming" Hydorforming of Tubes, Extrusions and Sheet Metals, Volume 1, pp. 253-261; International Conference on Hydroforming 1999, Fellbach/Stuttgart, Germany, October 1999; ISBN 3-88355-284-4

[2] Murata M., Kato T. (2003): "Highly Improved Function and Productivity for Tube Bending by CNC Bender"; TubeNet – The Site For Tube And Pipe Industries, http://www.tubenet.org [accessed March 2003]

[3] Neu J. GmbH (2003): "Technical Data Sheet"; http://www.neu-gmbh.de [accessed March 2003]

[4] Gantner P., Harrison D. K., De Silva A. K. M., et al (2003): "Forming Process Optimisation Through Surface Reconstruction", Proceedings of the AMPT (Advances in Materials and Processing Technologies) 2003, July 8-11, 2003, Dublin, Ireland. A. G. Olabi, M. S. J. Hashmi. pp. 1630-1633, ISBN 1-872327397

[5] Gantner P., Harrison D. K., De Silva A. K. M., et al (2003): "Investigation of Free-Bending – a new Bending technique", Proceedings of the First International Conference on Manufacturing Research (ICMR2003), Advances in Manufacturing Technology – XVII, September 2003, Glasgow, United Kingdom, Y Qin, N. Juster. pp. 69-73, ISBN 1 86058 412 8

Deep Cryogenic Treatment Electrodes Life and Microstructure for Spot Welding Hot Dip Galvanized Steel

Cuirong Liu[1], Xinhua Yang[2], Zhisheng Wu[3] , Ping Shan[4]

[1,2]Department of Materials Science and Engineering of Taiyuan Heavy Mechinery Institute, Taiyuan,P.R.China, lcr@tyhmi.edu.cn,
[3,4]Material Science and Engineering School of Tianjin University, Tianjin,P.R China,wuzs@eyou.com

Abstract: In this paper, the way of deep cryogenic treatment of electrodes for spot welding hot dip galvanized steel is put forward firstly to improve electrode life. Electrodes are treated by deep cryogenic treatment way with different deep cryogenic treatment parameters. Deep cryogenic treatment electrodes life experiment is carried out. The electrical conductivities of the deep cryogenic treatment electrodes and the non-deep cryogenic treatment electrodes are tested. The SEM area scanning and the SEM back scattering of the electrodes are carried out. The grain degree of deep cryogenic treatment electrodes and the non-deep cryogenic treatment electrodes are tested by X-Ray Diffraction. The experimental results show that deep cryogenic treatment makes Cr, Zr in deep cryogenic treatment electrodes emanate dispersedly and makes the grain of deep cryogenic treatment electrodes smaller than non-cryogenic treatment ones, so that the electrical conductivity and the thermal conductivity of deep cryogenic treatment electrodes are improved very much , which make electrode life for spot welding hot dip galvanized steel improve obviously.
Key Words: Spot welding; hot dip galvanized steel; Electrode life; Cryogenic treatment ; Microstructure

1. Introduction

One of the outstanding bargains in sheet metal fabrication in general , and automotive manufacturing in particular, has to be resistance spot welding. The demand being placed on spot welding are rising constantly, primarily because of the automotive industry increasing usage of Zinc-coated or galvanized sheet steels to enhance vehicle corrosion protection. However , the welding of Zinc-coated steels is tremendously different than welding the bar or uncoated versions of those same steel products. One of the key differences is that the coating plates onto and also alloys with the copper alloy welding electrodes , resulting in greatly accelerated electrode wear. The wear is believed to be due to differences in melting temperatures and resistivity of the coatings, combined with the differing current levels required.

To improve the spot welding quality of Zinc-coated steel plate , many researchers have been looking for varied kinds of ways improving the thermal physical properties and electrical conductivity of electrodes, such as developing new copper alloy materials ,changing the properties of electrode surface and developing the high –quality complex material etc..But all of them is limited to improve electrode properties.

Deep cryogenic treatment process has not been used in improving the electrode properties for spot welding zinc-coated steel plate. In this paper, the deep cryogenic treatment technology is first applied to treatment of electrodes for spot welding hot dip galvanized steel plate to improve electrode life.

2. Material and Equipment

2. 1 Material

Hot dip galvanized steel plate of 0.8mm thickness and Cr-Zr-Cu alloy electrodes were used. The properties of the galvanized steel are listed in Table 1 and the properties of the electrode are listed in Table 2.

Table 1: Mechanical properties of galvanized steel plate

Material	Coating thickness(g/m^2)	σ_s(Mpa)	δ (%)
Hot-deep galvanized steel	100	270-380	$\geqslant 24$

Table 2: Components and properties of electrode

Material	Chemical component (%)	Brinell hardness (HRB)	Conductivity (IACS)	Softing temperature(℃)
Cr-Zr-Cu	Zr 0.08-0.15 Cr 0.25-0.4	64	75%	550

2.2 Equipment

A Panasonic YR-700CM2 single phase resistant spot welding, MIYACHIMM-326B current measuring device and a deep cryogenic treatment unit are used in this study.

2. Deep Cryogenic Treatment of Electrodes

A gas deep cryogenic treatment method is employed for treatment of electrodes. Electrodes are put into the deep cryogenic treatment room; liquid nitrogen from an exit nozzle is directly gasified in deep cryogenic treatment room and the temperature of the room becomes lower and lower with absorbing heat of low temperature nitrogen and gasification of liquid nitrogen. The rate of descending temperature and duration of treatment temperature are controlled by the volume of liquid nitrogen flow and the self-adjustment of temperature can be reached.

In the paper, the rate of descending temperature is 6℃/min, treatment temperature is -150 ℃ and -170℃ respectively and the duration of the temperature is two hours and four hours. Finally, the temperature of the deep cryogenic treatment room is naturally gone up to ambient temperature.

The material of the spot welding electrode, which is conical, is Cr-Zr copper alloy. The diameter of the electrode is 20mm and the length is 60mm. There are three groups of deep cryogenic treatment process parameter and three electrodes are treated with each group of parameters.

3. Electrodes Life Experiment

The parameters of spot welding process experiment are as follow: welding current is 9.3kA, welding time is 8 circles, closed time is 20 circles, pre-pressing time is 99circles, electrode pressure is 3724N, pressure duration is 7 circles and the volume of cooling water flow is 3L/min. Deep cryogenic treatment electrodes life and non- deep cryogenic treatment electrodes life are shown in Figure1. The experimental results show that deep cryogenic treatment electrodes life for spot welding hot dip galvanized steel is improved obviously, electrode-to-workpiece sticking and weld spattering are very serious and the color of weld surface is brass when the hot dip galvanized steel is welded by the electrode without deep cryogenic treatment, otherwise weld spattering phenomena doesn't happen during the welding process and there are little brass color on the weld surface, tendency of the Cu-Zn alloying is little after deep cryogenic treatment of electrodes.

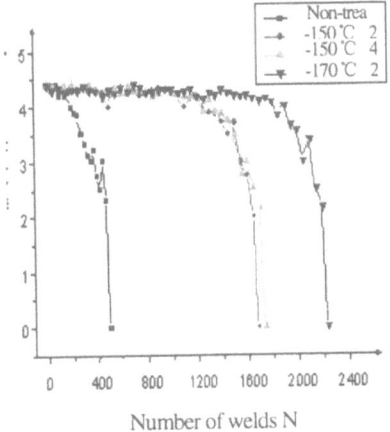

Fig.1   deep cryogenic treatment   electrode life

4. Observation and Analysis of Microstructure of Spot Welding Electrodes

4.1 Observation and analysis of microstructure with scanning electrical microscope

The back scattering by a scanning electrical microscope(SEM) for spot welding electrode before and after deep cryogenic treatment is shown in Figure 2, and the SEM plane scanning is shown in Figure 3and Figure 4.

From Fig.2, it can be seen that the soundness of the basal body before deep cryogenic treatment is lower and there are lots of micro cavities, which destroy the lattice structure and continuity of the material. However, after deep cryogenic treatment, the micro cavities in the basal body reduce significantly and the soundness of the basal body is obviously increased.

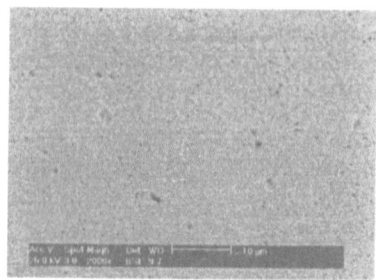

(a) Before deep cryogenic treatment   (b) After deep cryogenic treatment

Fig.2. SEM back scattering for Cr-Zr-Cu alloy structure.

(a) Before deep cryogenic treatment   (b) After deep cryogenic treatment

Fig.3. The distribution of Cr element in Cr-Zr-Cu alloy

(a) Before deep cryogenic treatment   (b) After deep cryogenic treatment

Fig.4   The distribution of Zr element in Cr-Zr-Cu alloy

It can be found that the distribution of element Cr and Zr in the sample has obvious change. There are a lot of diffusive Cr and Zr particles in the copper basal body because dissolvability of Cr and Zr in copper is rapidly decreased and supersaturated Cr and Zr are educed due to deep cryogenic treatment.

4.2 Observation and analysis by X-ray diffraction

The grain size of electrode sample before and after deep cryogenic treatment is measured by X-ray diffraction. The results are shown in Figure 4 and grain sizes of Cr-Zr-Cu alloy before and after deep cryogenic treatment are shown in Table 3.

Table 3: Analysis report of the grain sizes of X-ray diffraction.

Testing parameter	Spot welding electrode Before deep cryogenic treatment	Spot welding electrode after deep cryogenic treatment
Testing Sample half-wave Width (cm)	4.0	4.5
Testing Sample $2\theta(°)$	43.28	43.36
Standard sample half-wave Width (cm)	3.0	3.0
Standard sample 2θ (°)	26.6	26.6
Average grain size of 111-crystal plane/nm	159.7	82.9

The change of grain size can be measured through measuring the width change of x-ray diffraction spectrum before and after deep cryogenic treatment. The table3 shows that average grain size before deep cryogenic treatment is 159.7nm and one after deep cryogenic treatment is 82.9nm. The grain size is obviously refined.

5. Resistivity Testing and Analyzing of Deep Cryogenic Treatment Electrodes

Resistivity before and after deep cryogenic treatment is tested by the DC double arms bridge and the testing data are shown in Table 4.

Table 4: Testing data of resistivity of the electrode.

Treatment process	Non-treatment	-150°C2h	-150°C4h	-170°C2h	-170°C4h
Resistivity ($\Omega\,mm^2/m$)	1.801063	1.362078	1.425138	1.049448	0.976894

The data in Table 4 show that the resistivity has obvious change. The resistivity of the electrode decreases and the thermal conductivity greatly increases when the temperature of deep cryogenic treatment decreases and temperature duration enlongs.

6. Relation Between Microstructure and Properties of Deep Cryogenic Electrodes

The free electrons which form current continuously collide with positive ions during the directional movement and their movement is resisted, so electrical resistance is produced. If metal contains a little impurity and forms solid solution, its resistivity increases. The reason is that the impurity atoms make the metal structure distort, which sets off extra electron scatter. The uneven solid solution is formed as the micro-region distribution of solute and solvent atoms is not even, that is to say, there are atoms offsetting regions in solid solution, where its components are different from the average components of solid solution, or there are short distance ordered regions in solid solution. All above situations can strongly scatter electrons, which makes the uneven solid solution have high resistance.

The heat conductivity process is the transporting process of energy in the material. The flaws, such as elastic distortion, dislocation and lattice flaw ,are resulted from the solute atoms in solid solution and can lead to the electrons scatter and make the heat conductivity coefficient decrease. when the ordered structure is formed in the solid solution alloy, the average free distance which can transmit electron is increased. so the heat conductivity coefficient will be obviously bigger than that of disordered .In addition, if there is gas in the metal or there are diffusive distribution holes in the metal ,the material's heat conductivity coefficient will be influenced and decreased because gas isn't the good heat carrier.

The SEM analysis results of the electrode copper alloy show that there are many micro cavities in the electrode basal body before deep cryogenic treatment, which make the electron which moves directly suffer the sharply scattering function so that the material's capacity of heat transmission and electrical conductivity are decrease . However, the quantity of the micro cavities in material after deep cryogenic treatment is obviously less than that before deep cryogenic treatment. The soundness of the basal body is obviously increased so that electrode capacity of heat transmission and electrical conductivity are improved.

The electrode material belongs to low-concentration solid solution because the content of Cr and Zr in copper alloy is very small. The solid solution's resistance in low-concentration solid solution submits to Matthissen ruler[1]. Deep cryogenic treatment makes the solid solution Cr and Zr in electrode copper alloy precipitate from the basal body so as to decrease the concentration of solute atoms in the basal body , increase the purity of the basal body , descend the additional resistance produced by solute elements and improve the electrode copper alloy's capacity of heat transmission and electrical conductivity[2,3].

The copper alloy's grains size after deep cryogenic treatment decreases from 159 nm to 82.9nm ,which is half of that before deep cryogenic treatment. In addition, many Cr and Zr particles whose distribution is diffusive appear in basal body after deep cryogenic treatment . Since there is the linear relation between the metal material's yield strength and the reciprocal of the grain diameter's square root and the more grain interfaces there are in basal body, the more fine grain is , the strengthened effect is larger , the electrode's strength is increased[4] and the spot welding process performance of the galvanized steel is improved by deep cryogenic treatment.

7. Conclusion

The deep cryogenic treatment makes Cr-Zr-Cu alloy electrodes soundness of spot welding galvanized steel plate higher, many Cr and Zr particles whose distribution is diffusive appear in basal body, the copper alloy's grains more fine than that before deep cryogenic treatment, so the electrode's capacity of heat transmission and electrical conductivity are improved, the electrode's strength is increased and the spot welding process performance of the galvanized steel is improved by deep cryogenic treatment.

8. Acknowledgment

We would like to express appreciation to our two sponsors for this project namely: Nation Natural Science Foundation of China (Project No.50175080) and Science and Technology Development Foundation of   Education committ of Shanxi Province (Project No.200262).

References

[1] Xu Jingjuan,Deng Zhiyu, Zhang Tongjun. The Analysis of Metal's Physical Property. Shanghai : Shanghai Science and Technology Press.1988:36

[2] Qu Guangpu. A New Dry Refrigerating Method of Improving Wear Resistant and Other Properties of Metal Parts. Cryogenic Refrigerating Technology Applyment Shanghai: Shanghai Jiaotong University Press. 1993:22

[3] Wang Yi, Cong Jiyuan. Improving the Electrical Property of Cu-Cr vacuum Contact Material. Low Voltage Electrical Equipment. 1993.(3):43

[4] [4] Cui Zhongxi. Metal Materials Science and Heat Treatment. China Machine Press. 1989:182

LASER PROCESSING

Effect of High Power Diode Laser Surface Melting on Corrosion Resistance of Magnesium Alloys

G. Abbas[1,2], Lin Li[2] and Zhu Liu[3]

[1]Laser Processing Research Centre, Department of Mechanical, Aerospace and Manufacturing Engg, UMIST, Manchester, M60 1QD, UK, Lin.li@umist.ac.uk
[2]National University of Sciences and Technology (NUST), Pakistan, gabbas5@yahoo.com
[3]Corrosion and Protection Centre, UMIST, Manchester, M60 1QD, UK, zhu.liu@umist.ac.uk

Abstract: This paper reports the effect of laser surface melting on corrosion performance of magnesium alloys AZ31 and AZ61. A 1.5 kW high power diode laser (HPDL) was used to melt about 1 mm thick surface layers of the Mg alloy samples. With careful selection of adequate laser processing parameters, it was possible to obtain sound and crack-free overlapping melt tracks. Scanning electron microscopy was used to examine microstructure and distribution of phases in the matrix region of the melted layer. The laser-melted samples were subjected to immersion test, using 5% NaCl solution of pH10.5, for ten days. The results of weight loss measurement revealed a marked improvement in the corrosion resistance of both alloys studied. Fine microstructures were observed in the melt zone with an average grain size of less than 5 μm.

1. Introduction

High strength to weight ratio of Mg alloys make them a promising alternative to aluminium and steel alloys used in the aerospace and automotive industries where weight reduction plays a significant role in the economy of fuel consumption. Unfortunately, Mg alloys could not yet be accepted widely in these industries due to their inferior wear and corrosion properties. A substantial amount of work has been carried out in the last decade to improve the corrosion and wear properties of Mg alloys using different techniques [1-16].

Laser surface melting has been found a reliable means of improving the surface properties of Mg alloys due to higher power densities and short interaction times associated with laser materials processing. High power CO_2 and Nd:YAG lasers have been widely used for this purpose [17-19]. It has been reported in some recent studies that high power diode lasers (HPDLs) can be used as an economical substitute of CO_2 and Nd:YAG lasers due to some intrinsic characteristics such as higher energy conversion efficiency, small physical size, lightweight and a long lifetime [20-24]. However, HPDLs are not yet used for the surface melting of magnesium alloys.

The present paper reports surface melting of magnesium alloys AZ31 and AZ61 for improvement of corrosion properties. This is a continuation of the authors' previous work on the subject using CO_2 laser [25]. However, in the present work a high power diode laser (HPDL) was used as a heat-generating source. Although the economical aspects of the process were not studied in the present work, considering the quality of melt surface, corrosion test results and the process cost of both the

CO_2 and HPDL, it can be envisaged that high power diode lasers can be a viable alternate to CO_2 lasers both on small scale and on industrial basis.

2. Experimental Procedure

2.1 Materials used

Two types of Mg alloy plates were used as substrate in the present study for laser melting. The nominal composition of each alloy and the substrate dimensions are given in table 1.

Table 1

Nominal composition and substrate dimensions of Mg alloys used

Sr No.	Mg Alloy	%Mg	%Al	%Zn	%Mn	Dimensions (mm)
1.	AZ-31	96.0	3.0	1.0	0.3	25 x 50 x 5
2.	AZ-61	93.0	6.0	1.0	-	25 x 50 x 8

2.2 Laser processing

A 1.5 kW high power diode laser (HPDL) was used as a heat-generating source for producing 50% overlapping melt tracks. The laser melt pool was shrouded by argon gas to prevent excessive oxidation of the substrate. The specimens were melted using a constant laser power of 1.5 kW and substrate traverse speed of 50 mm/s (transverse to the laser beam direction) with a rectangular beam of 3.5 x 2.0 mm dimension.

2.3 Microscopy

The general features of laser-melted specimens and the surfaces of corroded samples were observed using optical and scanning electron microscopes (AMRAY 1810 electron microscope). Transverse sections were cut from the overlapping laser melted tracks and mounted in Bakelite. The samples were then polished and etched using 5% aqueous acetic acid solution.

2.4 Corrosion test

Immersion test was carried out to study the corrosion performance of the laser-melted samples. One square centimetre sample of each as-received alloy and overlapping laser melted surface was cut and mounted in bakelite. 5wt% NaCl solution was prepared in ionized water. Small amount of $Mg(OH)_2$ was added to stabilize the pH of solution at 10.5 for the entire duration of the test. All the samples were polished (on 1200 emery paper), thoroughly washed and weighed before subjecting to the immersion test. The samples were individually immersed in the salt solution using a separate 250 ml glass beaker for each sample. The test was carried out for consecutive ten days. The samples were then taken out of the salt solution and soaked in a different solution, containing 200 g/l CrO_3 + 10 g/l Ag_2CrO_3, for removing the corrosion product. The weight loss was measured by comparing the initial and final weight of the samples. Corrosion rate was then calculated as weight loss in mg/cm^2/day.

3. Results and Discussion

3.1 Microstructure of laser melted surfaces

The microstructure of the substrate material of both the alloys AZ31 and AZ61 used was studied to correlate with the microstructures of the laser-melted samples. These alloys contain Al as a major alloying element whereas small amounts of Zn and Mn are also present. In additions, there are a number of impurity elements like Fe, Cu and Ni. The initial microstructure of both the alloys was coarse grain cellular type structure with grain size in the range of 30-100 μm as shown in Figs. 1 and 2. This type of Mg-Al alloys are solidified as two phase alloys consisting of α–matrix (Mg solid solution) large cells surrounded by β-phase which is essentially $Mg_{17}Al_{12}$ type particles. Some of the alloying elements and the impurity elements form intermetallic particles and the most important are those of the form $Al_xMn_yFe_z$ where x, y and z vary with the composition of the alloy After laser treatment, both the alloys were resolidified as fine dendrites of α–matrix surrounded by β–phase. The mean secondary dendrites arm spacing of laser-melted samples was less than 5 μm as shown in Figs. 3 and 4.

3.2 Corrosion results

Four different types of samples (two substrate and two laser melted) were subjected to immersion test for ten days. The immersion test results showed distinct corrosion behaviour of different samples. Laser melted Alloy AZ61 showed the lowest whereas Alloy AZ31 substrate showed the highest corrosion rate in this test. However, the corrosion resistance of both the alloys used was improved to some extent by the laser surface melting as depicted in Fig. 5. The corrosion resistance in the laser-melted surfaces is improved mainly due to the refinement of microstructure. The β-phase and the intermetallic particles are more finely distributed in the laser-melted region and surround the α–matrix region. Both the β-phase and these intermetallic particles are nobler than the matrix. They are also more or less good catalysts for the cathode reaction. The anode reaction, which occurs mainly in the matrix, is initiated in the vicinity of β-phase and the intermetallic particles.

The β-phase has a dual effect on the corrosion reaction. In addition, to being an initiator, it also acts as a barrier to corrosion growth in the matrix. The β-phase forms a network around the α- matrix with many corrosion barriers. The corrosion resistance of magnesium alloys decreases with decreasing aluminium content as illustrated in Fig. 5. The reason is that when the concentration of aluminium decreases, the amount of β-phase decreases. As a consequence, there are still many sites for corrosion initiation, but less corrosion barriers due to the lower content of β-phase. These results are in agreement with a recent study on the subject where it has been reported that the corrosion resistance of Mg alloys is improved by the presence of β-phase in the matrix [26].

3.3 Corroded surface morphologies

In order to determine the corrosion behavior of Mg alloys, a detailed analysis of the morphology of the corroded surfaces was done. Figs. 6 (Alloy AZ31 substrate), 8 (Alloy AZ61 substrate), 10 (Laser melted surface of Alloy AZ31) and 12 (Laser melted surface of Alloy AZ61) show the optical micrographs of the corroded surfaces of as-received and laser surface melted samples at lower magnifications, X

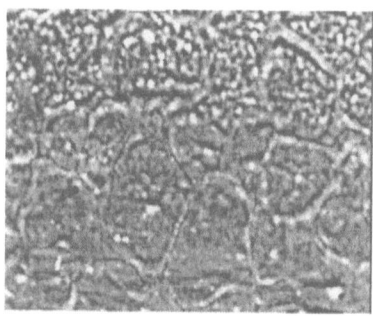

Fig.1 SEM micrograph of melt-substrate interface of Alloy AZ31 (X 500).

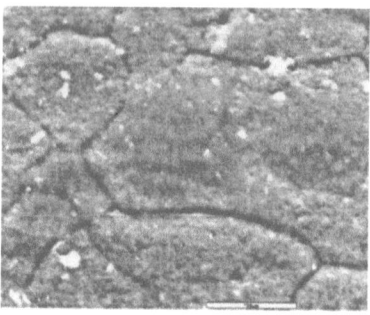

Fig.2 SEM micrograph of Alloy AZ61substrate (X 1200)

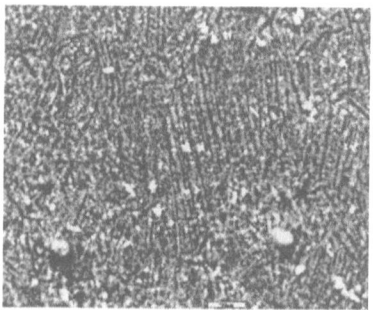

Fig.3 SEM micrograph showing laser melted area of Alloy AZ31 (X 500).

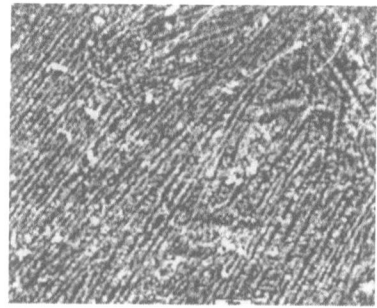

Fig.4 SEM micrograph showing laser melted area of Alloy AZ61 (X 500).

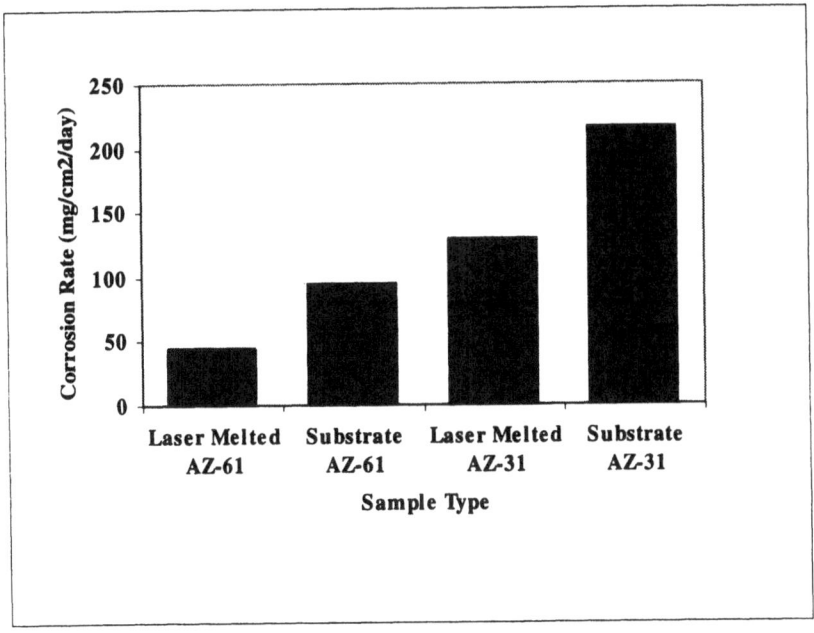

Fig. 5 Corrosion rate of substrate and laser melted samples of Alloy AZ31 and AZ61.

5. SEM micrographs of the same surfaces at higher magnification are shown in Figs. 7, 9, 11 and 13 respectively. All these samples show the corrosion of top surfaces after immersion in 5 % NaCl solution of 10.5 pH value for consecutive ten days.

It is apparent from the surface morphologies of the corroded samples that the corrosion in the as-received substrates (Figs. 6-9) is more severe as compared with the laser-melted samples (Figs. 10-13). The corrosion in as-received substrates is almost uniform through out the entire surface of the sample. However, corrosion in the laser-melted samples was found to be localized in distinctly small regions. As discussed in the previous section, the grain size and volume fraction of the α–matrix (Mg solid solution) was small in the laser melted samples as compared to the as-received substrates. Since the α–matrix is more susceptible to the corrosion attack, the corrosion areas in the laser treated samples are small as compared to the as-received substrates. Also, the corrosion propagation in laser treated samples is restricted by the presence of large volume of β–phase and the intermetallic particles, which act as a barrier to corrosion, in the laser-melted samples.

The conclusions made on the basis of the surface morphologies of the corroded sample are in agreement with the most of the previous studies on the subject and also supportive to the results obtained in the NaCl solution immersion test as shown in the Fig. 5.

232

Fig. 6 Optical micrograph showing corroded surface of Alloy AZ31 substrate (X 5)

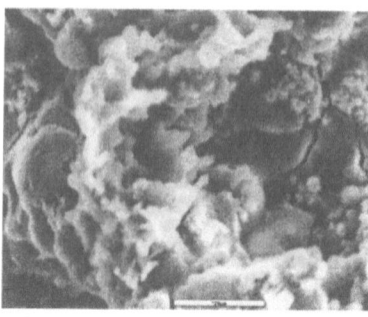

Fig. 7 SEM micrograph showing corroded surface of Alloy AZ31 substrate (X 1200)

Fig. 8 Optical micrograph showing corroded surface of Alloy AZ61 substrate (X 5)

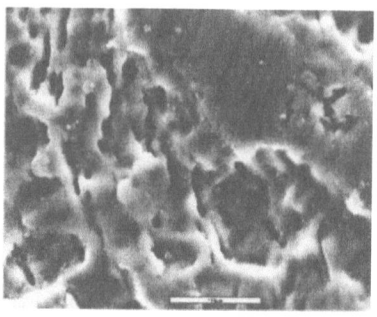

Fig. 9 SEM micrograph showing corroded surface of Alloy AZ61 substrate (X 1200)

Fig. 10 Optical micrograph showing corroded surface of laser melted Alloy AZ31 (X 5)

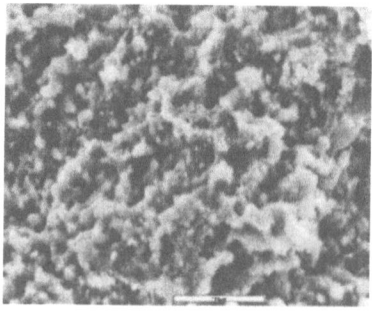

Fig. 11 SEM micrograph showing corroded surface of laser melted Alloy AZ31 (X 1200)

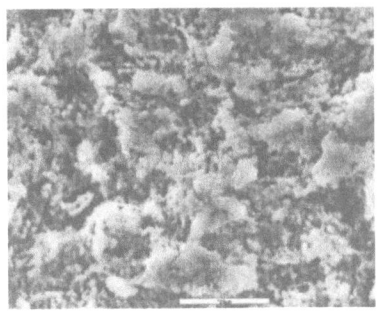

Fig. 12 Optical micrograph showing corroded surface of laser melted Alloy AZ61 (X 5)

Fig. 13 SEM micrograph showing corroded surface of laser melted Alloy AZ61 (X 1200)

4. Conclusions

High power diode laser has been found a suitable source of heat for melting the magnesium alloys. Corrosion resistance of magnesium alloys is affected by the chemical composition and the microstructure. Laser surface melting of magnesium alloys produced fine dendritic microstructures. Laser surface melting of magnesium alloys AZ31 and AZ61 improved their corrosion resistance markedly. Corrosion resistance of Alloy AZ61 was greater than that of Alloy AZ31 due to higher Al content in the former.

5. Acknowledgements

One of the authors, G. Abbas, would like to acknowledge the financial support of Higher Education Commission of Pakistan to carry out this research work.

References

[1]   Li, Pei Yong, Hai Jun Yu, Shen Chuan Chen et al, Factors Affecting the Corrosion Resistance of Cast Magnesium Alloys, Magnesium Technology 2003, TMS, pp. 51-58.

[2]   Skar, Jan Ivar and Darryl Albright, Emerging Trends In Corrosion Protection of Magnesium Die-Castings, Magnesium Technology 2002, TMS, pp. 255-262.

[3]   Shahin, G. E., Corrosion and Wear Resistance of Electroless Nickel on Magnesium Alloys, Magnesium Technology 2002, TMS, pp. 263-268.

[4]   Shrestha, S., A. Sturgeon, P. Shashkov et al, Improved Corrosion Performance of AZ91d Magnesium Alloy Coated with the Keronite Process, Magnesium Technology 2002, TMS (The Minerals, Metals & Materials Society), pp. 283-288.

[5]   Mathieu, S., C. Rapin, J. Hazan et al, Corrosion behaviour of high pressure die-cast and semi-solid cast AZ91D alloys, Corrosion Science, Vol. 44, 2002, pp. 2737-2756.

[6]   Gray, J. E. and B. Luan, Protective coatings on magnesium and its alloys — a critical review, Journal of Alloys and Compounds, Vol. 336, Issues 1-2, 2002, pp. 88-113.

234

[7] Bonora, P. L., M. Andrei, A. Eliezer et al, Corrosion behaviour of stressed magnesium alloys, Corrosion Science, Vol. 44, Issue 4, April 2002, pp. 729-749.

[8] Yamamoto, A., A. Watanabe et al, Improvement of corrosion resistance of magnesium alloys by vapour deposition, Scripta Materialia, Vol. 44, 2001, pp. 1039-1042.

[9] Peter, J., A. Blau et al, Sliding friction and wear of magnesium alloy AZ91D produced by two different methods, Tribology International, Vol. 33, 2000, pp. 573–579.

[10] Hoche, H., H. Scheerer, D. Probst et al, Plasma anodisation as an environmental harmless method for the corrosion protection of magnesium alloys, Surface and Coatings Technology, Volumes 174-175, September-October 2003, pp. 1002-1007.

[11] Hoche, H., H. Scheerer, D. Probst et al, Development of a plasma surface treatment for magnesium alloys to ensure sufficient wear and corrosion resistance, Surface and Coatings Technology, Volumes 174-175, September-October 2003, pp. 1018-1023.

[12] Lee, M. H., I. Y. Bae, Formation mechanism of new corrosion resistance magnesium thin films by PVD method, Surface & Coatings Tech, Vol. 169-170, 2003, pp. 670-674.

[13] Angelini, E., S. Grassini, F. Rosalbino et al, Electrochemical impedance spectroscopy evaluation of the corrosion behaviour of Mg alloy coated with PECVD organosilicon thin film, Progress in Organic Coatings, Volume 46, Issue 2, March 2003, pp. 107-111.

[14] Han, E. H., W. Zhou et al, Corrosion and protection of magnesium alloy AZ31D by a new conversion coating, Materials Science Forum, Vol. 419-422, 2003, pp. 879-882.

[15] Mori, K., Z. Kang, J. Oravec et al, Corrosion resistance of polymer-plated magnesium alloys, Materials Science Forum, Vol. 419-422, 2003, pp. 889-896.

[16] Chiu, L. H., H. A. Lin, et al, Effect of aluminium coatings on corrosion properties of AZ31 magnesium alloy, Materials Science Forum, Vol. 419-422, 2003, pp. 909-914.

[17] Murayama, K., A. Suzuki et al, Surface modification of magnesium alloys by laser alloying using Si powder, Materials Science Forum, Vol. 419-422, 2003, pp. 969-974.

[18] Yue, T. M., A. H. Wang and H. C. Man, Corrosion resistance enhancement of magnesium ZK60/SiC composite by Nd:YAG laser cladding, Scripta Materialia, Vol. 40, Issue 3, January 1999, pp. 303-311.

[19] Yue, T. M., A. H. Wang and H. C. Man, Improvement in the corrosion resistance of magnesium ZK60/SiC composite by excimer laser surface treatment, Scripta Materialia, Vol. 38, Issue 2, December 1997, pp. 191-198.

[20] Li, Lin, The advances and characteristics of high-power diode laser materials processing, Optics and Lasers in Engineering 34 (2000), pp. 231-253.

[21] Pashby, I. R., S. Barnes et al, Surface hardening of steel using a high power diode laser, Journal of Materials Processing Technology 139 (2003) pp. 585–588.

[22] Bachmann, Friedrich, Industrial applications of high-power diode lasers in materials processing, Applied Surface Science 208-209 (2003), pp. 125-136.

[23] Barnes, S., N. Timms, B. Bryden et al, High power diode laser cladding, Journal of Materials Processing Technology 138 (2003), pp. 411–416.

[24] Szweda, Roy, Diode Laser market pauses to regroup, The Advanced Semiconductor Magazine Vol15 - NO 6 - August 2002, pp. 42-44.

[25] Abbas, G., Zhu Liu, P. Skeldon and Lin Li, Corrosion behaviour of CO_2 laser melted magnesium alloys, Paper to be presented at E-MRS Spring Meeting, France, May 2004.

[26] Song, G., Amanda L. Bowles et al, Corrosion resistance of aged die cast magnesium alloy AZ91D, Materials Science and Engineering A, Volume 366, 2004, pp. 74-86.

Shell Assisted Layer Manufacturing (SALM)

A K Egodawatta[1], D K Harrison[2], A K M De Silva[3] and G Haritos[4]

[1] Glasgow Caledonian University, G4 0BA, UK, aeg@gcal.ac.uk
[2] Glasgow Caledonian University, G4 0BA, UK, D.K.Harrison@gcal.ac.uk
[3] Glasgow Caledonian University, G4 0BA, UK, ade@gcal.ac.uk
[4] Softnet Consultancy Ltd, Birmingham, B29 6RT, UK, george.haritos@conteyor.com

Abstract: This paper presents the technology and some experiments results of a novel Rapid Prototyping/Tooling/Manufacture (RP/RT/RM) process, Shell Assisted Layer Manufacturing (SALM), which is based on layer manufacturing technology (LMT). The SALM process executes two essential steps to develop one layer of the part. Initially, it develops the outer shell (boundaries) of the particular layer of the part using the Fused Deposition Modelling (FDM) technique and then the shell is filled with UV (Ultraviolet) curable resin. The deposited resin layer is then cured using a UV light source. This procedure is repeated until the complete part is built. The process has the potential to give improved quality, productivity and process speed.

1.   Introduction

During the last 18 years, many different types of RP processes have been developed to meet the demand for reducing the "time-to-market" of a product. Some of these have been more successful than the others. For a example, Stereolithography (SLA), Fused Deposition Modelling (FDM) are widely used in industry today due to their continuous system improvements through research and development. However other processes such as Solid Ground Curing (SGC) and Light Sculpting (LS) did not survive in the industry. Their failure to survive can be attributed to limitations of dimensional accuracy, reliability, build envelope and process speeds, complexity of the process, machine the high cost and the greater amount of post processing involved.

The latest development in RP includes two new processes: Perfactory (Envisiontec GmbH, Germany) [1, 2] and Objet Eden (Objet Geometries Limited, Israel) [3, 4, 5], which are at the early stages of commercialisation. Table 1 summarises some key features these two new processes.

Table 1: Comparison of Perfactory and Objet Eden RP processes

Process	Manufacturer	Development Methodology	Layer thickness μm	Accuracy μm	Maximum Build size mm
Perfactory	Envisiontec GmbH, Germany	UV curing of Photopolymer resin	20–150	25	190x160x230
Objet Eden	Objet Geometries Ltd, Israel	UV curing of Photopolymer resin	16 (0.0006")	N/A	340x330x200

There is still an urgent need for new and affordable techniques with better performances in order to enhance the benefits and proper use of rapid processes in product development. Shell Assisted Layer Manufacturing (SALM) is aimed at developing a fast, reliable, cheap, simple, and accurate process with better surface finish and a flexible build envelope. Currently, there is no published literature available for RP/RT/RM processes that has similar operating principles to SALM. This paper presents experimental investigations on concept realisation and feasibility testing of SALM.

2. The Experiments

The SALM process is based on layer manufacturing technology. The system is working with UV curable resin as the model (part) material and P400 SR (Soluble Release, Acrylic copolymer thermoplastic, FDM water soluble support structure material) [6] plastics as the shell material that can be easily removed after building the part. Alternatively, if the part is to be built with the shell, then the shell is built with ABS (Acrylonitrile Butadiene Styrene) [7] plastic using the FDM machine (Figure 1). Fundamental experiments were carried out using a FDM 3000 machine for the shell development. The CAD modelled file (STL) of the designed geometry is sliced and processed to develop the base of the part using a FDM machine. Having made the base of the part, the inner and outer boundaries (the shell) for the first layer of the part are developed using the FDM machine. The FDM machine is idling whilst finishing the rest of the layer work. Then, a layer of UV curable resins (SOMOS 10120 [8] for the experiments) is manually deposited inside the shell, approximately filling the entire shell. Thicker layer of material can be deposited with improved materials which resulting in faster process [9, 10]. Thereafter the deposited resin layer is exposed to a UV light source (by manually feeding the externally arranged lamp into the build chamber of the FDM machine) in order to cure the resin layer. This process is repeated until the complete part is built (Figure 2). Then, the built part with the shell is removed from the building base (platform) and immersed in a special water bath (for the experiments FDM Waterworks was used) in order to remove the shell. If the part is built with the shell, only the base of the part with support structure material is dissolved. The finished part is then taken out from the tank once all the unnecessary material is dissolved.

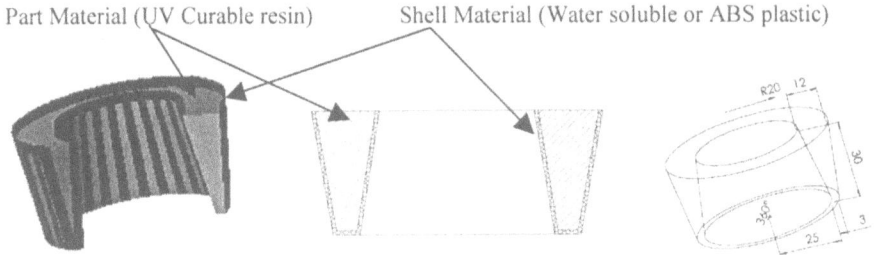

Figure 1: Key elements of a work-in-progress part of SALM detailing the dimensions

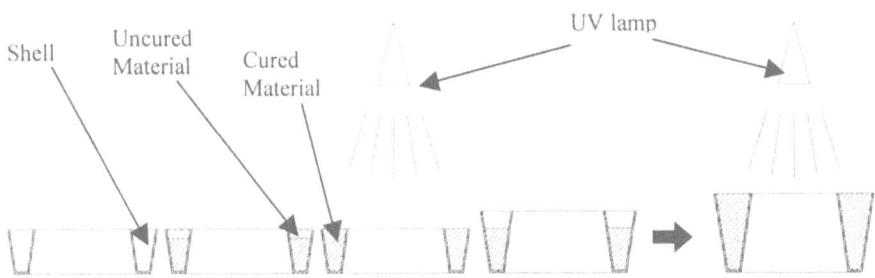

Figure 2: Production steps of a part using SALM

3. Results and Outcomes

The fundamental experiments have been carried out. Results show the SALM process is feasible for developing improved productivity and quality in rapid prototyping parts. For the basic concept realisation experiments, geometries without overhangs and undercuts have been built. Figure 3, 4, 5 and 6 show some parts built using the technology.

(a)   (b)   (c)

Figure 3: CAD and Physical views of a built part

Figure 4: A simple cup made (with the shell) using the SALM method

Figure 5: A simple cup (right) made with the shell using the SALM method. A separate shell (left) built using FDM 3000 for testing

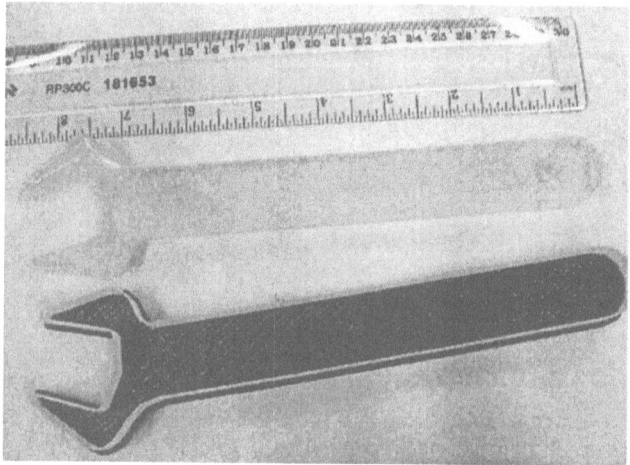

Figure 6: A simple spanner (upper) made (without the shell) using the SALM method. A shell (below) for the spanner is separately built using FDM 3000.

3.1 SALM process defects

At present, only parts with simple geometries (no overhangs and undercuts) have been made. These help to understand the feasibility of producing parts using the SALM process. Since the SALM is novel process, it is important to identify all the possible defects of the process at this level of investigation. The early experiments show achieving perfectly flat top facing surfaces is not feasible. This arises due to the surface tension and capillary action (when building thin webs etc.) of the viscous resins. The resin layer close to the shell (edge) is rising along the shell.

Hence, the top facing surface close to the edge becomes curved (Figure 7). As a solution to overcome this issue, a milling cutter is to be introduced that mills the non-flat area of the surface in order to flatten the surface. This is the primarily identified alternative for the issue. Further investigations have to be undertaken to identify the right solution.

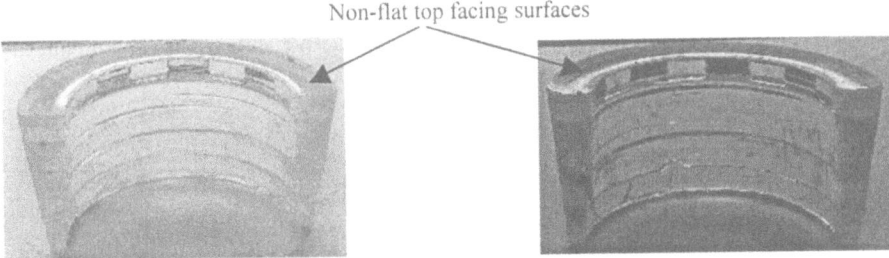

Figure 7: Sectioned views of parts created for the investigation of the flatness of top facing surfaces

The other defect of SALM products is change in bond strength between successive layers. There are a few possible causes which have been identified that affect the bond strength, presence of foreign particles (debris from FDM process which is use to build the shell, presence of air bubbles (Figure 8) and the surface finish of the cured layer prior to putting the resin on top to make the next layer. Basic studies have been carried out on these issues. Further investigations are to be done to find the right solution to eliminate or minimise the causes affecting bond strength.

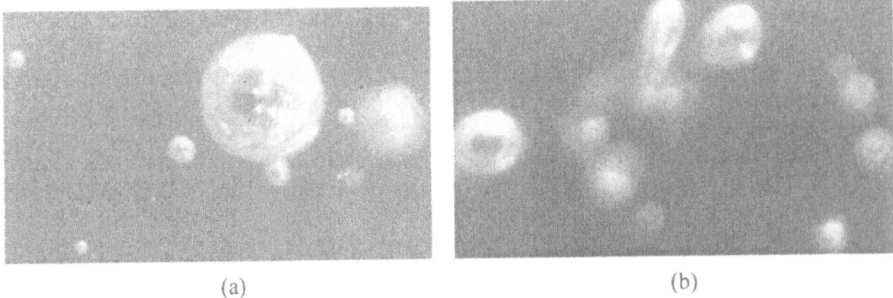

Figure 8: Microscopic view of a defected surface with some air bubbles 200 magnifications

4. Conclusions

The fundamental experiments have been carried out for simple part geometries. Parts have been made with two basic options (with shell and without shell) prescribed. Results show that the SALM process is feasible for developing improved productivity and quality in parts. The basic experiments were carried out with some WaterClear 10120 (SL material developed for solid state laser curing) photopolymer for the part material. Since, the SALM process is based on UV curing; the material used for the experiments does not produce identical results as expected in the real process. Hence, with ideal materials, the results would be expected to be better than those obtained. The experiments were carried out with the existing slicing technology embedded in the FDM 3000 system. A better performance can be achieved by introducing appropriate techniques of adaptive slicing [11, 12]. An easily removable shell other than P 400 SR (Soluble Release from Stratasys Inc.) will make the process more attractive and reduce the amount of additional work involved with the shell removal.

References

[1] http://www.envisiontec.de/02hperfa. htm [accessed 23/10/03]
[2] Forth M, The technology breakthrough that the industry has been waiting for!, TCT News-current issue articles, http://www.time-compression.com/x/guide.htm [accessed 19/03/2004]
[3] US patent no 6,658,314 Gothait; Hanan (Rehovot, IL), System and method for three dimensional model printing, December 2, 2003 and 6,259,962 Gothait; Hanan (Rehovot, IL), Apparatus and method for three dimensional model printing, July 10, 2001
[4] http://www.objet.co.il/home.asp [accessed 25/07/03]
[5] Cohen D, A New Era of Rapid Prototyping, TCT News - current issue articles, http://www.time-compression.com/x/guideArticle.htm?id=6614 [accessed 19/03/04]
[6] P400 SR, Material Safety Data Sheet, August 22, 2000, Stratasys Inc., Minneapolis, MN, USA
[7] P400(ABS), Material Safety Data Sheet, November 04, 1994, Stratasys Inc., Minneapolis, MN, USA
[8] WaterClear^{TM} 10120 Product Data Sheet, DSM Somos, New Castle, DE 19720, USA
[9] Carter W and Lamb K, Cationic UV Curing of Thick Sections, The Dow Chemical Company, USA
[10] Lee J H, Prud'homme R K and Aksay I A, Cure depth in photopolymerization: Experiments and theory, Journal of Material Research, Volume 16, Number 12, December 2001
[11] Pandey P M, Reddy N V and Dhande S G, Real time adaptive slicing for fused deposition modelling, International Journal of Machine Tools & Manufacture 43 (2003) 61–71
[12] Han W, Jafari M A and Seyed K, Process speeding up via deposition planning in fused deposition-based layered manufacturing processes, Rapid Prototyping Journal, 1 April 2003, vol. 9, no. 4, pp. 212-218(7) MCB University Press
[13] Dickens P M, Rapid Manufacturing, Proc. TCT'99 Conference, East Midland Conference Centre, Nottingham, UK, 12-13 Oct 1999, pp. 119-123
[14] Dickens P M, Update on Rapid Manufacturing, Proc. TCT'00 Conference, Cardiff, UK, 10-11 Oct 2000, pp. 87-90
[15] Wohlers T, Wohlers Report 2000, pp.212- 216, Wohlers Associates, Inc. Fort Collins, Colorado 80525 USA

Laser Welding of Low-Porosity Aerospace Aluminium Alloy

Ing. Geert Verhaeghe[1] and Dr Paul Hilton[2]

[1]TWI Technology Centre (Yorkshire) Ltd, Sheffield, UK; geert.verhaeghe@twi.co.uk
[2]TWI Ltd, Cambridge, UK; paul.hilton@twi.co.uk

Abstract: Aluminium is currently the preferred material and riveting the preferred joining method for the manufacture of thin-gauge airframe structures. Although the potential of laser welding as a low-distortion alternative for such applications is recognised, questions are still being raised about the weld quality, and in particular the porosity levels, that can be achieved in aluminium. This paper focuses on the cleaning of parent material and filler material prior to welding, the use of a twin-spot energy profile in the laser beam focus and the use of a low-moisture shielding gas and shielding gas delivery, and their individual and combined influence on the presence of weld metal porosity for Nd:YAG laser welds in 3.2mm thickness 2024 aluminium alloy. The paper describes how, through careful selection of processing conditions and aforementioned factors, fully penetrating, square-edge butt welds were achievable with levels of weld metal porosity lower than those specified in the stringent weld quality class of standards relevant to the aerospace industry, including the European BS EN ISO 13919-2:2001 and the American AWS D17.1.

1. Introduction

Currently, the preferred manufacturing route for aircraft fuselage structures is riveting and the principal material for these structures is aluminium. Recent analyses, however, have indicated that a move from riveted to welded airframe structures could lead to manufacturing cost savings in the region of 30% [1]. Laser welding is one of the processes currently being considered for this, because of the high processing speeds, low heat input, low distortion, good weld quality and the overall flexibility that the process offers [1-3]. Although possible, the laser welding of aluminium is generally perceived to be difficult because of the initial high surface reflectivity and the high thermal conductivity of aluminium, both of which contribute to the risk of weld imperfections such as lack of penetration or cracking, in certain alloys. Weld metal porosity is also frequently associated with the laser welding of aluminium [4-8] and this weld imperfection is the subject of the work described in this paper.

Porosity is generally categorised as either fine or coarse, typically differentiated at an average pore diameter of 0.5mm. Fine porosity appears as a distribution of spherical pores and is generally understood to originate from hydrogen or from the rejection of dissolved shielding gases on solidification. Coarse porosity, which can have a detrimental effect on a welded joint's mechanical performance [6], is characterised by larger, more irregularly shaped voids, randomly distributed throughout the weld bead. These are generally considered to be the result of low boiling point constituents causing keyhole instabilities and are typically present in partially penetrating welds [1,7,8].

The work described in this paper, carried out as part of a programme named CEMWAM (Cost Effective Manufacture: Welding of Aerospace Materials), initiated by a number of leading UK industrial companies, Research and Technology Organisations (RTOs) and Universities, details Nd:YAG laser welding trials aimed at reducing both the fine and coarse porosity in laser welded 3.2mm thickness 2024 aerospace aluminium alloy. Particular efforts were on reducing the fine porosity, and especially fine porosity resulting from hydrogen-entrapment. Hydrogen dissolves very rapidly into the aluminium weld pool [4] but has a very low solubility in solid aluminium. With a high-speed process such as laser welding, the time available for diffusion is sufficiently low that a certain amount of hydrogen can become entrapped in the solidifying weld pool [7]. Hydrogen can originate from the parent material, the filler wire or the shielding gas and these sources were all investigated. In addition, the performance of twin-spot Nd:YAG laser welding was assessed to establish whether this technique resulted in a reduced level of porosity, because of the resultant elongated weld pool created when using this technique [9].

Square-edge butt welds were produced using a 3kW flashlamp-pumped continuous wave (CW) Nd:YAG laser focused into a single 0.45mm diameter spot, on 150mm wide and 300mm long samples. The samples were subjected to a *primary cleaning operation*, whereby the long edges were cold band sawn and dry machined to ensure a good joint fit-up and the samples degreased with acetone immediately after the dry machining operation. Industrial grade helium (purity grade 5.0 or 99.999% pure) conforming to BS EN 439:1994 [10], typically containing around 5ppm moisture, was used for shielding both top and bottom of the weld pool. A 1.2mm diameter 2319 filler wire was introduced into the leading edge of the weld pool [11].

Process parameters such as welding speed, laser focus position (in relation to the material surface), filler wire position (in relation to the laser-workpiece impingement point) and shielding gas flow rates, were varied to obtain visually acceptable, in accordance with BS EN ISO 13919-2:2001 [12], fully penetrating, square-edge butt welds. The absence of cracks in the weld metal and heat-affected zone (HAZ) was confirmed through radiographic and macro-metallurgical examination. The presence of micro-cracks in the weld metal or HAZ was not examined. Subsequent welds were produced using the process parameters established earlier, but with different conditions for parent material cleanliness, filler wire cleanliness and condition and delivery of process shielding gas. These factors, widely considered to be the main causes of hydrogen-induced weld metal porosity, were varied in a controlled fashion, as detailed further. In addition, the effect of a twin-spot, i.e. two spots of 0.45mm diameter separated in the beam focus, versus a single-spot laser energy distribution on the level of weld metal porosity was also investigated.

A selection of the welds was radiographed and a pore count carried out over a 100mm longitudinal section of the weld, representative of the entire weld length, to quantify the levels of weld metal porosity. These levels and the equivalent pore length/area per given weld length/area calculated from these levels, were compared with the acceptance levels of the *stringent* weld quality class defined in two international standards, BS EN ISO 13919-2:2001 [12] and AWS D17.1:2001 [13], and one company *internal* standard ABP 2-4102 [14]. The European standard was selected because it is specific to laser welding of aluminium, whereas the American standard and the internal standard were chosen because they are specific to fusion welding for aerospace applications. The acceptance criteria that needed to be fulfilled included the diameter of the largest pore, the minimum distance between adjacent pores and the equivalent projected pore area (for the European standard) or pore length (for the American and internal standard) per given length of weld. In this work, the second criterion, i.e. the minimum distance between adjacent pores, was not considered as it was fulfilled for all welds.

2. Results and Discussion

2.1 Parent material cleanliness versus porosity

In addition to the initial dry machining of the sample edges after cutting, they were subjected, just prior to welding, to a *secondary cleaning* operation. The secondary cleaning methods examined comprised linishing, abrading, scraping, machining or chemical etching (or chemi-etching), each followed by an acetone degrease with a lint-free cloth to remove residual dirt, moisture or lubricant. The acetone degreasing was also investigated on its own as a secondary cleaning operation. For the chemi-etch cleaning, the samples were soaked, at least six times, for two minutes, in a nitric and hydrofluoric acid solution, followed by a water rinsing. All cleaning operations, except machining and chemi-etching, were carried out manually and, with the exception of acetone degreasing, performed to remove the porous oxide layer, a potential source of contaminants and moisture, and thus porosity [15].

Whether the parent material was cleaned or not prior to welding made little difference visually, but produced an obvious distinction in the resulting level of weld metal porosity, as could be seen by comparing radiographs. Notwithstanding some variability, little difference was apparent in the total level of weld metal porosity, i.e. pore size and distribution (or equivalent pore length/area per given weld length/area) between the different secondary cleaning methods. This is demonstrated by the linished and chemi-etched samples shown on the left of Figure 1. If anything, the chemi-etched samples exhibited less porosity than those linished, presumably because of the uniformity and non-manual operation of the chemi-etching. For the machining and chemi-etch sample preparation, low levels of porosity could only be achieved when cleaning was carried out less than 24 hours before welding, and with the samples kept in a dessicating unit. The spherical nature of the coarse porosity found in some of the welds and the fact that the welds were fully penetrating would indicate that this coarse porosity originated from large volumes of entrapped gas rather than the result of keyhole instabilities [2].

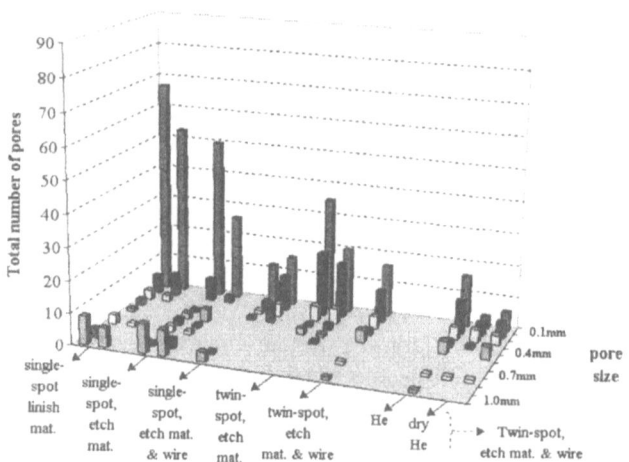

Figure 1 Weld metal porosity for cleaned (parent material) samples, cleaned filler wire and twin-spot energy distribution

2.2 Filler wire cleaning versus porosity

To assess the individual effects of parent material and filler material cleaning on weld metal porosity, welding was carried out on samples that were chemi-etched less than four hours prior to welding, with as-received filler wire as well as with filler wire that was chemi-etched immediately before welding. A clear drop in the total level of weld metal porosity, i.e. pore size and distribution, or equivalent pore length/area per given weld length/area, was noticeable for the cleaned filler wire, as demonstrated in Figure 1. The achieved reduction in porosity meant that all welds produced with the chemi-etched filler wire passed both investigated criteria of the European and internal company standard. The equivalent pore length per given weld length of some of the samples however, still remained too high to pass the third criterion (total pore length for given weld length) of the American standard AWS D17.1.

2.3 Twin spot energy distribution versus porosity

Additional trials were carried out to examine if an elongated weld pool created by a twin-spot laser energy profile would delay the rapid solidification of the weld pool, thus allowing more time for hydrogen to escape [9]. Based on earlier work [16,17], welds were produced using a twin-spot energy profile with a 0.27mm separation and a 50/50 energy distribution between the two 0.45mm diameter spots.

Figure 1 shows that the levels of coarse porosity for welds produced with a twin-spot energy profile were lower when compared to those produced earlier with a single-spot energy profile. However, a consequence of the twin-spot technique was that welding was carried out at lower travel speeds, i.e. between 0.8 and 1.0m/min, compared with the single-spot speed techniques where the welding speed was 1.75m/min. With the twin-spot technique, the total laser power is shared between the two spots, creating a lower power density in each of the spots, necessitating a lower travel speed compared with the single-spot technique, to produce fully penetrating welds in a given material thickness. A similar, but smaller influence of the twin-spot technique and the inherent lower travel speed (compared with the single-spot technique) on reducing the level of coarse weld metal porosity was observed for laser welds in 3.2mm thickness 6056 aluminium alloy. The presence of coarse porosity in this alloy, however, was less pronounced, indicating that besides energy beam profile, welding speed and spot size, material grade or alloy composition can also have an effect on the level of coarse porosity in laser welds

2.4 Shielding gas moisture content versus porosity

To examine how moisture in the shielding gas contributes to the presence of fine weld metal porosity, welds were also produced using a high purity, low dew-point *research grade* helium shielding gas and a 'modified' shielding gas delivery system. The latter comprised the shortest, i.e. maximum 2m, length of polyurethane tubing possible between gas cylinder and coaxial shielding nozzle, and, due to the hygroscopic nature of polyurethane, tubing that was 'acclimatised' several hours prior to welding, purged for several minutes at the onset of the welding trials and for at least two minutes between subsequent welds. The welds were carried out on chemi-etched samples, with chemi-etched filler wire and using a twin-spot energy profile, based on the thus far accumulated experience.

Welds produced using this 'modified' delivery system and a *research grade* (low-moisture content) helium shielding gas demonstrated a considerable reduction in pore numbers compared with welds made under exactly the same conditions but using industrial grade helium, as can be seen in Figure 1. The resulting equivalent pore length per given weld length of these welds were considerably lower than any of those achieved before. In fact, from all welds produced in this work, these were the only welds that passed all three porosity criteria for each of the standards, even the third criterion (total pore length per given weld length) of the most stringent standard AWS D17.1.

3. Conclusions

A welding procedure was developed for producing fully penetrating, square-edge butt welds in 3.2mm thickness 2024 aluminium using 3kW CW flashlamp-pumped Nd:YAG laser power. By controlling the process conditions, it was possible to achieve a level of weld metal porosity lower than that defined for the *stringent* quality class in BS EN 13919-1:1997, a typical aerospace industry standard, and even the most rigorous of standards considered, i.e. standard AWS D17.1:2001:

- A focus position on or 1mm below the material surface helps achieve full penetration welds with weld profiles that conform to BS EN ISO 13919-2:2001.
- A high-purity, low dew-point *research grade* helium shielding gas, delivered through a moisture and condensation-free shielding gas delivery system, produces less weld metal porosity compared with industrial grade helium gas.

- Removing the porous oxide layer prior to welding contributes to reducing the weld metal porosity in laser welded 2024 aluminium alloy. Linishing, scraping, machining or chemical etching can be used, but the elapsed time between material preparation and subsequent welding needs to be as short as possible (less than 24 hours recommended) to avoid atmospheric moisture pick-up.
- A further reduction in weld metal porosity can be achieved by cleaning the filler wire, for instance with a chemical etching cleaning operation.
- The use of a twin-spot laser energy profile with a 0.27mm spot separation and a 50/50 energy distribution between two 0.45mm diameter spots helps eliminate coarse porosity in 3.2mm thick 2024 aluminium, but has less of an effect on pores smaller than 0.4-0.5mm diameter.

References

[1] Dunkerton S.B., *Welding for the 21st century*. Materials World, vol.10, no.3, March 2002, pp.18-19.

[2] Ion J.C., *Laser beam welding of wrought aluminium alloys*. Science and Technology of Welding and Joining, Vol.5, No.5, pp.265-275.

[3] Clarke J.A., *Laser welding aluminium alloys*. Industrial Laser Solutions, vol.15, no.6. June, pp.17-18.

[4] Gingell A.B.D. and Gooch T.G., *Review of factors influencing porosity in aluminium arc welds*. TWI report 625, Oct 1997.

[5] Mathers E., *Arc welding aluminium – a guide to best practice*. JoinIT section of TWI website www.twi.co.uk, Sept 2002.

[6] Katayama S, Matsunawa A, Kojima K, Kuroda S: 'CO$_2$ laser weldability of aluminium alloys. Report 4: Effect of welding defects on mechanical properties, deformation and fracture of laser welds'. Welding International, Vol.14, No.1, Jan 2000, pp.12-18.

[7] Sekhar N.C., *Review of laser welding of aerospace aluminium alloys*. CEMWAM report, CEMWAM II/ PT1B/TWI/038, July.

[8] Seto N., Katayama S., Matsunawa S., *Porosity formation mechanism and suppression procedure in laser welding of aluminium alloys*. Welding Int. 2001 vol.15, pp.191-202.

[9] Iwase T., Shibata K., Sakamoto H: 'Real Time X-Ray Observation of Dual Focus Beam Welding of Aluminium Alloys'. Proceedings of ICALEO 2000, Section C, pp.26-34.

[10] BS EN 439:1994, *Welding consumables - shielding gases for arc welding and cutting: Welding consumables - shielding gases for arc welding and cutting*.

[11] Gittos M.F. and Scott M.H., *Selection of filler materials for arc welding aluminium alloys*. The Welding Institute Research Bulletin, vol.29, Feb 1988.

[12] BS EN ISO 13919-2:2001, *Welding – Electron and laser beam welded joints – Guidance on quality levels for imperfections*.

[13] AWS D17.1:2001, *Specification for fusion welding for aerospace applications*.

[14] ABP 2-4102 (Appendix C), *Acceptance standards for aluminium and aluminium alloys*. Internal British Aerospace Plc standard, issue 3, Jan 1996.

[15] *Welding Aluminium: Theory and Practice*. The Aluminum Association, Third edition, Nov 1997, ISBN 89-080539.

[16] Lugan A., *Twin-spot laser welding of 2024 aluminium alloy*. TWI report 765, Mar 2003.

[17] Hohenberger B., Chang C.L., Schinzel C.,et al, *Laser welding with Nd:YAG- multi-beam technique*. Institut fur Strahlwerkzeuge (IFSW), University of Stuttgart, Proceedings of ICALEO 1999 conference, Section D, pp.167-176.

Ultrafast Pulse Laser Interference for Sub-micromachining

Ho Shook Foong[1] and Bryan Ngoi Kok Ann[2]

[1] Precision Engineering & Nanotechnology Centre, Nanyang Technological University, N3-B4c-03, 50 Nanyang Avenue, Singapore 639798. HoShookFoong@pmai!.ntu.edu.sg
[2] Precision Engineering & Nanotechnology Centre, Nanyang Technological University, N3-B4c-03, 50 Nanyang Avenue, Singapore 639798. MBNgoi@ntu.edu.sg

Abstract: Submicromachining with ultrafast pulse laser has been demonstrated utilising the phenomenon of laser light interference. This technique has been proven to overcome the feature size limit of conventional laser micromachining. By first interfering the laser light, and then using the central bright fringe of the interfered beam to machine, it has been proven that the effective ablation spot size can be reduced and subsequently, reducing the size of the ablated features. Preliminary results show a 300% reduction in feature size by machining with the interfered laser beam compared to the conventional direct laser beam. 300 nm holes were successfully ablated on a 100 nm thick gold film on a fused silica substrate using the interfered laser beam compared to 1 μm holes ablated using the conventional direct laser beam at the same laser energy and machining parameters.

1. Introduction

Ultrashort pulse laser machining is one of the forerunners of today's precision micromachining technology. With this technology, features down to micron ranges of high precision and exceptional machining quality can be achieved. And with the advent of femtosecond pulse lasers, ablation in the sub-focal spot is now possible, deposing the previous belief that the minimum achievable machining spot size is dependent and limited by the smallest focal spot size given by the laser wavelength. Features as small as 1/10th of the laser focal spot size has been reported, the smallest being an amazing 300 nm in diameter or approximately 10% of the laser focal spot size of 3 μm [1,2].

1.1 Mechanism of sub-spot size ablation with ultrafast lasers

Sub-spot size micromachining using ultrafast pulse lasers was first demonstrated by Pronko P.P. et al. in 1995, overcoming previous notions that the minimum achievable machining spot size is limited by the smallest focal spot size of the laser wavelength [1,2,3].

One of the amazing characteristics of ultrashort pulse laser ablation is that the threshold value can be precisely determined, in contrast to the broad threshold bandwidth of long pulse laser ablation. This enables ultrafast pulse lasers to create features substantially below that of the central wavelength of the laser pulse itself.

The ultrafast laser pulse has a Gaussian beam profile with peak intensity in the centre of the beam which smoothly decreases radially outward from the centre. By properly choosing the incident laser fluence, it is possible to control only a fraction (the central peak region) of the focused Gaussian beam within which the fluence is greater than the threshold of the material, to machine. Therefore, the ablated region can be restricted to a very small area, much smaller than the diameter of the focused spot size [1]. This limited area of ablation can be as small as 1/10th of the focus spot size itself. The principle of this technique is illustrated in Figure 1(*a*).

However, the sub-spot size ablation mechanism has a limit as to how small a feature it can produce, and it has reached its lowest limit when it hit the 1/10th of the laser focal spot size range.

To overcome this hurdle, a novel concept of first interfering the laser light, and then using the central bright fringe of the interfered beam to machine, thus effectively reducing the effective ablation spot size and subsequently, reducing the size of the ablated features, has been contrived.

1.2 Mechanism of sub-micromachining with interfered ultrafast laser beam

As mentioned, there is a limit to how small the features can go down to. Up till now, the smallest feature size reported using laser micromachining is 300 nm, or approximately 10% of the original laser spot size of 3 μm [1,2]. There has not been significant achievement in the reduction of ablation features beyond this limit. Therefore, a novel concept of first, interfering the laser light, and then using the central fringe of the interfered beam to machine, thus effectively reducing the effective machining spot size and subsequently, the size of the ablated features, has been contrived.

The principle of this technique is illustrated in Figure 1(*b*).

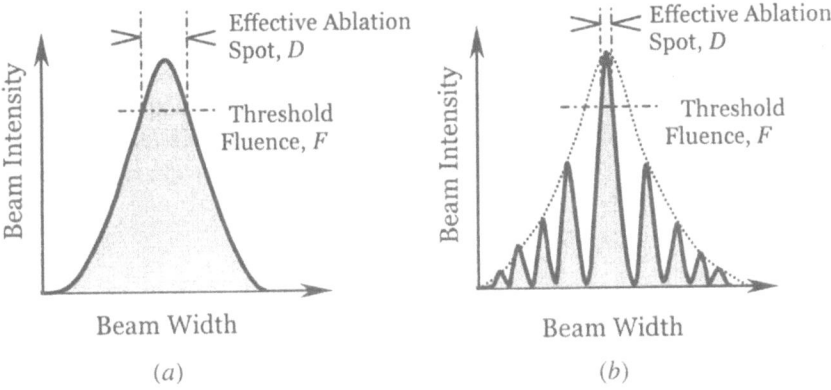

Figure 1: (*a*) 2D profile of the direct non-interfered Gaussian laser beam
(*b*) 2D profile of the interfered Gaussian-like laser beam

By comparing the intensity profile of the interfered beam and the conventional non-interfered beam in Figure 1(*a*) and Figure 1(*b*), it can be seen that by the careful control of the threshold fluence of the laser, it is possible to manipulate the interfered beam that only a portion of the central fringe will machine. Therefore, the effective ablation spot can be further reduced as compared to machining with the direct laser beam.

2. Experimental Setup

A commercial femtosecond chirped pulse amplification (CPA) based Ti:sapphire pulse laser system (BMI Alpha-1000S) is used. The laser system comprises of a Ti:sapphire oscillator, a pulse stretcher, a regenerative amplifier pumped by a Nd:YLF laser, and a pulse compressor. The system provides laser pulses at 800 nm with a variable pulse output energy of up to of 800 mW or 800 µJ, and a pulse width of 150 fs at the pulse repetition rate of 1 kHz.

The beam wavelength of the laser system falls between 790 nm and 810 nm and is frequency centred at 800 nm in the red light spectrum region. However, most important manufacturing materials such as metals, ceramics and silicon are poor absorbers of red wavelength light as compared to the ultraviolet (UV) wavelength range [2]. Therefore, for the optimisation of machining efficiency, the fundamental beam from the laser system ($\lambda = 800$ nm) is frequency doubled by transmitting the beam through a 0.5 mm thick non-linear SHG BBO crystal (Second Harmonic Generator) to generate pulses of wavelength 400 nm in the visible UV.

Also, the machining feature size is dependent upon the wavelength of the laser beam used [2,3]. The smaller the wavelength of the laser light used, the smaller the machining feature size obtained. Therefore, it is desirable that wavelength from the lower range of the spectrum of the laser light be used for machining.

A Fabry-Perot etalon interferometer is used to generate interference fringes on the femtosecond pulse laser. The schematic of the experimental setup is shown in Figure 2 below.

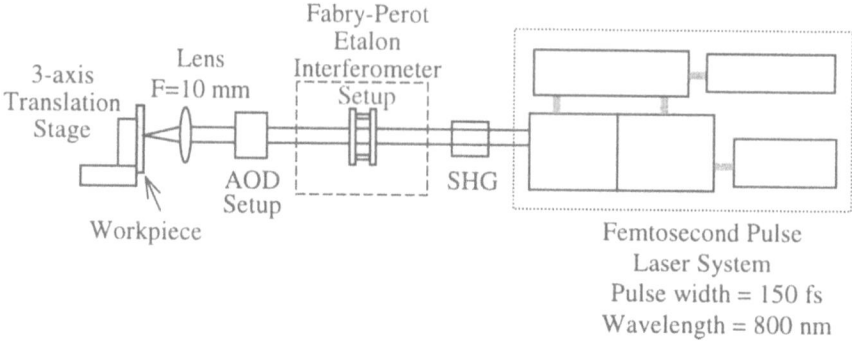

Figure 2: Schematic of the experimental setup

The 150 fs laser pulses at wavelength of 800 nm is transmitted through the SHG to frequency double the wavelength to 400 nm. The laser beam is then passed through the Fabry-Perot etalon to obtain circular interference fringes. Finally, the interfered laser beam is scanned across the stationary workpiece using an acousto-optic deflector (AOD) and focussing lens setup. The workpiece specimen used in this investigation is a 100 nm thick pure gold film sputtered on a 2" × 2" × 3 mm fused silica substrate.

3. Results and Discussion

The interfered laser beam from the setup in Figure 2 is scanned across the gold thin film workpiece. The Fabry-Perot etalon setup is then removed and the direct laser beam is used to machine using the same beam energy and machining parameters. The machined features obtained with both the interfered beam and the direct beam are then analysed and compared using an atomic force microscope (AFM).

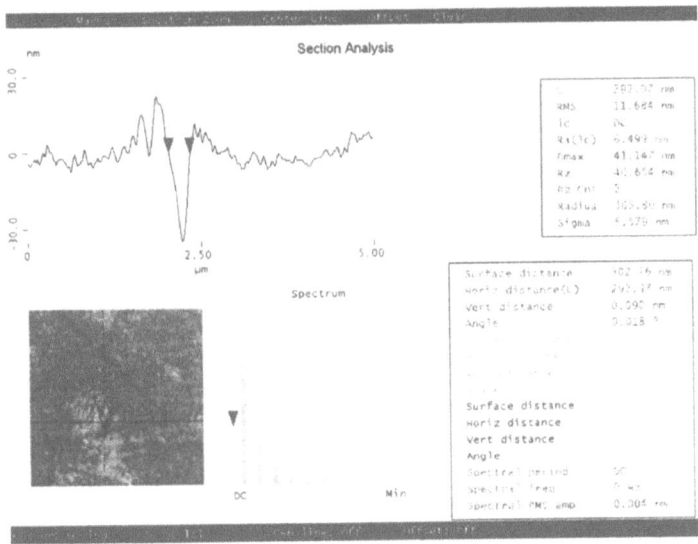

Figure 3(a): The AFM analysis of the feature machined using the interfered laser beam. The diameter of the feature obtained is 292.97 nm (≈300 nm).

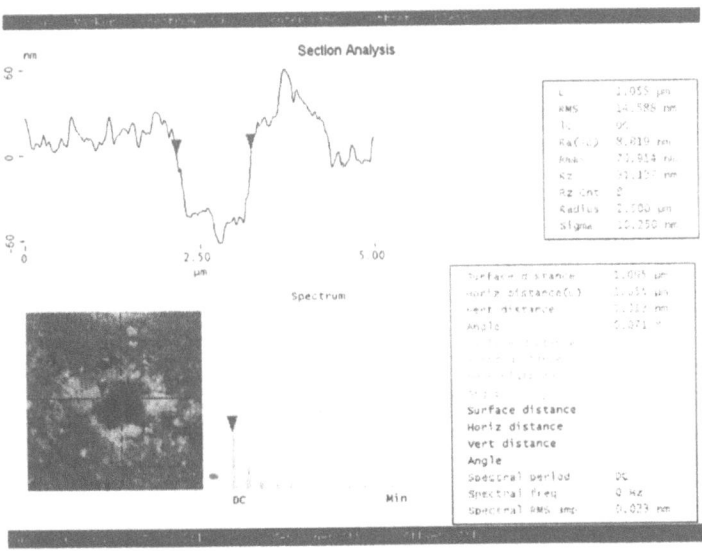

Figure 3(b): The AFM analysis of the feature machined using the direct laser beam. The diameter of the feature obtained is 1.055 μm (≈1 μm).

Figure 3(*a*) is the AFM analysis of the feature obtained on the 100 nm thick gold film using the interfered laser beam while Figure 3(*b*) is the feature obtained using the direct non-interfered laser beam. The pulse energy used for both cases is 5 nJ and the machining parameters are kept constant throughout.

A 300 nm (0.3 μm) hole was machined using the interfered laser beam compared to the 1 μm hole machined using the direct laser beam at the same beam energy. This is equivalent to a 300% decrement in feature size.

4. Conclusion

300 nm holes were successfully ablated on a 100 nm thick gold film sputtered on a fused silica substrate using the interfered laser beam compared to the 1μm holes ablated using the conventional non-interfered laser beam at the same pulse energy and machining parameters. This is equivalent to a 300% reduction in feature size, a substantial improvement in feature size obtained.

5. Applications

Potential applications for this technology includes the drilling of microvias in printed circuit boards/printed wire boards (PCB/PWB) in the semiconductor industry, micro-drilling of microscale fluidic devices such as ink-jet printer nozzles and drug delivery systems, direct writing on mask for lithographic applications and many more.

References

[1] Pronko, P.P., Dutta, S.K., Squier, J., Rudd, J.V., Du, D., & Mourou, G., 1995, *Machining of sub-micron holes using a femtosecond laser at 800 nm.* Optics Communications, **114**, 106-110.
[2] Simon, P., & Ihlemann, J., 1997, *Ablation of submicron structures on metals and semiconductors by femtosecond UV-laser pulses.* Applied Surface Science, **109**, 25-29.
[3] Liu, X., Du, D., & Mourou, G., 1997, *Laser Ablation and Micromachining with Ultrashort Laser Pulses.* IEEE Journal of Quantum Electronics, **33**, 1706-1716.

CO$_2$ and Diode Laser Welding of 5182 Sheet Aluminum Alloy

Jinhong Zhu[1], Lin Li[2], Zhu Liu[3]

[1]School of Materials Engineering, Henan University of Science and Technology, Luoyang 471003, China   Zhu.john@163.com
[2]Laser Processing Research Center, Dept. of Mech., Aero. & Manu Eng., UMIST, Manchester M60 1QD, UK   Lin.li@umist.ac.uk
[3]Corrosion and Protection Center, UMIST, Manchester M60 1QD, UK Zhu.Liu@umist.ac.uk

Abstract: A 2 kW CO$_2$ laser and a 1.5 kW diode laser were used in an experimental study to investigate the basic phenomena in welding 5182 aluminum alloy sheets. The work has shown that both keyhole and conduction limited welds can be achieved with the CO$_2$ laser and only conduction limited welding can be achieved the diode laser. The effects of laser power density, welding speed, joint fit-up conditions, shielding gases and surface preparation on the weld quality are reported.

1. Introduction

Lightweight alloys, such as aluminum alloys, are increasingly used in automotive and aerospace industries. Joining technology plays an important role in the successful application of aluminum alloys. There are several methods adopted for aluminum alloy joining, e.g., TIG or MIG arc welding, riveting, resistance spot welding and laser welding. Some problems have been experienced in the past for welding of aluminum alloys using CO$_2$ and Nd:YAG lasers. These problems include poor mechanical strength of the joints and material elongation (up to 80% of that of the parent material) [1-3]. Also keyhole instability and welding defects often occur if the welding condition is not well controlled. Porosity and poor weld geometries are often observed. Another challenge is the excessive electrical energy consumption of the conventional high power lasers as CO$_2$ lasers have 10% and YAG lasers have 1-3% electrical to optical energy efficiencies. High power diode lasers can have more than 30% energy efficiency. However, due to poor coherence and multiple beams used, diode lasers cannot be focused to comparative spot size as other high power lasers, and therefore, are mostly limited to conduction welding applications [4,5].

In this paper, a comparison between a 2 kW CO$_2$ laser and a 1.5 kW diode laser welding of 5182 aluminum alloys is reported. An experimental study has been conducted in order to investigate a practical method for laser welding of the aluminum alloy with minimum defects.

2. Experimental Procedure

Table 1 shows the composition of the 5182 aluminum alloy used in this investigation. The experimental specimen consisted of workpieces of 1.2 mm thickness cut into 25x40 mm². Bead-on-plate welding was carried out initially to find a suitable parameter range. Butt welds were then carried out. In order to compare the effect of surface conditions, some samples were welded as received and some were sandblasted. The butt weld edges were prepared as sheared or polished.

Table 1 — Chemical Composition of 5182 Aluminum Alloy

Material	Mg	Mn	Si	Zn
5182	4.44	0.35	0.20	0.07

A 2 kW Rofin-Sinar DC-020 RF excited slab CO_2 laser with a TEM_{00} mode was used in the experiment. Focusing was performed using a parabolic mirror having a 150 mm focal length, giving a minimum spot size of 160 um. A 1.5 kW Laserline LPL160 diode laser with 50%:50% mixture of 808+940 nm wavelengths was also used in the investigation. The focused spot size was 1.8 x 3.2 mm2. The two lasers are shown in Figure 1.

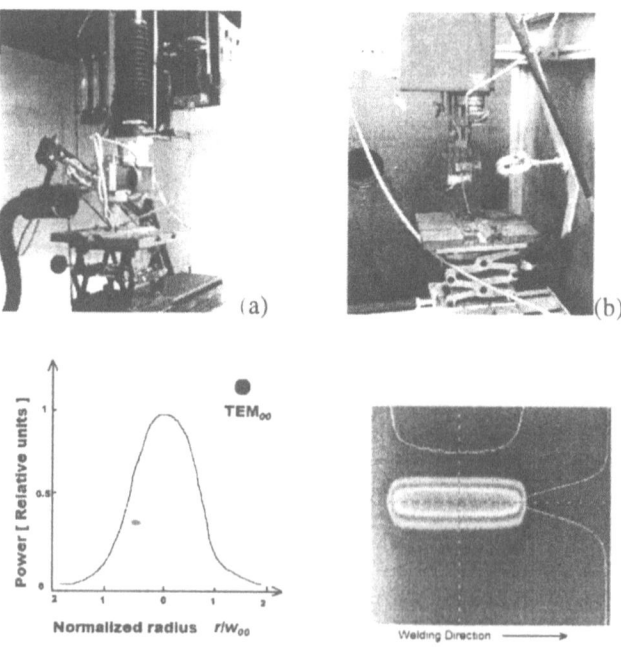

Fig.1. CO_2 and diode lasers: (a) 2 kW CO_2 laser and beam profile, (b) 1.5 kW diode laser and beam profile.

During the welding process, the CO_2 and the diode laser beams were kept stationary and the workpieces were traversed with CNC x-y tables. Argon and Helium gas were used respectively as shielding gases. In some of the welding experiments, the shielding gas was fed from a pipe from above the workpieces. The gas feeding had a 30 degree angle to the horizon plane with the bottom part of the workpieces being left open. Also some experiments were carried out with workpieces placed in a gas box having a slot opening on the top. These are illustrated in Figure 2. After the welding, the weld quality was initially evaluated by visual inspection. Then the samples were sectioned, ground and polished. The mounted specimens were etched using 10% sodium hydroxide solution, and the morphology and microstructures were examined using optical microscopy.

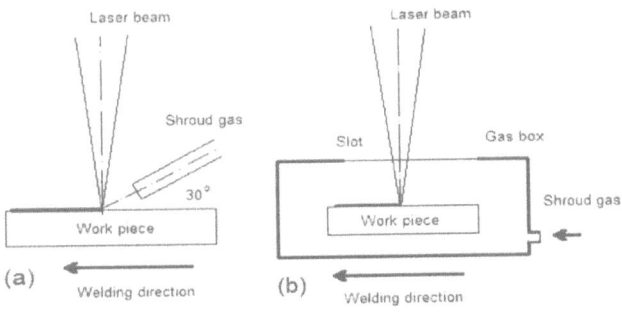

Fig.2. Welding configuration (a) with pipe gas flow, (b) in gas box

3. Results and Discussions

3.1 Welding process and optimization

It was found that, with the CO_2 laser welding, using focused laser beam, keyhole welding can be achieved whilst a defocused beam can result in conduction limited welding. The diode laser can only achieve conduction welding, as shown in Figure 3.

3.1.1 Keyhole welding

The optimized parameters for CO_2 keyhole welding were: laser power at 2 kW, and a welding speed of 150~180 mm/sec. In this study, the beam was focused on the surface. With the limitation in power, although keyhole welding could be achieved, the operating window was very narrow and instability and defects could easily occur. As shown in Figure 3 (a), some surface discontinuity still remained. Keyhole welding is desirable in many applications due to the small material distortion and heat-affected zones (HAZ). The keyhole process runs on a dynamic balance. Therefore, instability and welding defects occur easily if the welding condition is not well controlled. Power density (determined by laser power and spot size) and interaction time (determined by welding speed) are the basic factors influencing keyhole stability. Keyhole welding could only occur if the laser power density is above a threshold value. For this particular material, the threshold power density was found to be 107 W/cm^2. Welding speed could also have some influences on achieving a keyhole welding, but its effect was relatively small.

	Top	Bottom
a. CO_2 laser focusing		
b. CO_2 laser defocusing		
c. Diode laser		

Fig.3. Typical welding samples

For steel welding, the keyhole process is rather stable due to the good balance between viscosity and surface tension. In aluminum alloy welding, low viscosity of the material may cause molten pool resonant easily and hence irregularity. Alloying elements like magnesium and silicon can decrease the viscosity further. The size of interaction area may also affect the keyhole process balance. According to previous studies [6-8], defocusing of laser beam is useful to achieve a stable weld. For a higher power laser, e.g. more than 5 kW, it is easy to achieve the keyhole effect with a larger spot size. For a 2 kW laser as used in this study, the spot size must be kept to a minimum in order to achieve keyhole welding but the process is much unstable. Thus, a more powerful laser is needed to achieve better welds.

3.1.2 Conduction welding

Conduction welding can be easily obtained using the CO2 laser (by defocusing) and the diode laser at the focus, as shown in Figure 3 (b) and (c). In the CO2 laser welding, a series of phenomena were observed. When the laser power density was just below the keyholing threshold, there was a rapid change from the keyhole mode to the conduction mode. It was still at a relatively high speed and a small focus spot. A narrow, shallow and smooth bead on plate weld was obtained, but it failed to obtain a butt weld. As the laser was further defocused (spot size around 0.5 mm) and welding at a slower speed (25-30 mm/s), the weld beads became wider and penetration increased and a satisfactory butt weld was achieved. If the laser was further defocused, the heating area became too wide and the penetration depth started to decrease. With the diode laser welding, due to the large spot size, only conduction welding was possible and all welds were made at the focus point. The top surface of conduction weld was smooth, while lines can be observed at the bottom of the welds. The optimized parameters for the CO_2 and diode conduction welding were similar. For the CO_2 laser, at 2 kW and a spot size of around 0.5 mm the optimum welding speed was 25-30mm/s (1.5-1.8m/min). The diode laser with 1.5 kW required a welding speed of 20-25mm/s (1.2-1.5m/min). The diode laser had slightly higher energy efficiency due to better beam absorption.

3.2 Effect of joint fit-up and shield gases

3.2.1 Gap and edge condition

In the keyhole welding, it was found that the gap between the plates should be kept less than 10% of the material thickness. If the gap was larger, poor weld geometry may occur at either the root or the top as can be seen in Figure 4 (a) and (b). In the conduction butt-welding, a large gap between the plates can cause molten metal segregation. Sheared edge preparation was, however, acceptable. If not strictly aligned, e.g. with declined edges and thus a variable gap or a small angle the ends, some polished sample were still difficult to be welded.

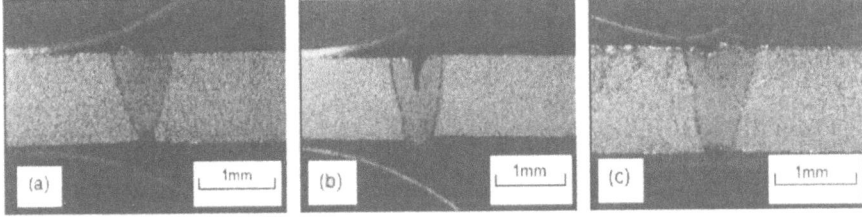

Fig.4. Cross sections of CO_2 keyhole welding samples (butt) showing: (a) a notch from the bottom, (b) a notch from the top and (c) a good weld

3.2.2 Surface preparation and the effect of shielding gases

Surface cleaning and sandblasting can increase laser absorption. It was found that 5182 alloy was not sensitive to hydrogen pores, but oxide was indeed very harmful. Some welds made with the oxide remaining on the surface were brittle and easy to break, some with root not fully molten as shown in Figure 5 (a).

Shielding gas was very important. A weak gas flow (<5 l/min) was not enough to protect welding pool, while a strong gas flow (>25 l/min) may stir the welding pool too much, or even cause defects. Figure 5 (b) shows the weld surface shape affected by gasflow. An important consideration was the gas flow configuration. If the shield gas was fed opposite to the welding direction an irregular weld bead and penetration would occur because the gas would blow the molten metal out of the weld pool. It was found in this study that Ar and He gases were similar for 5182 keyhole and conduction welding. If the conduction welding was carried out in a gas box, a reasonable weld could be achieved more easily at $1.2m^3/h$ (20 l/min). Figure 5 compares welds under different conditions. Further microstructures inspection proved that the crystals in welding zone are much refined.

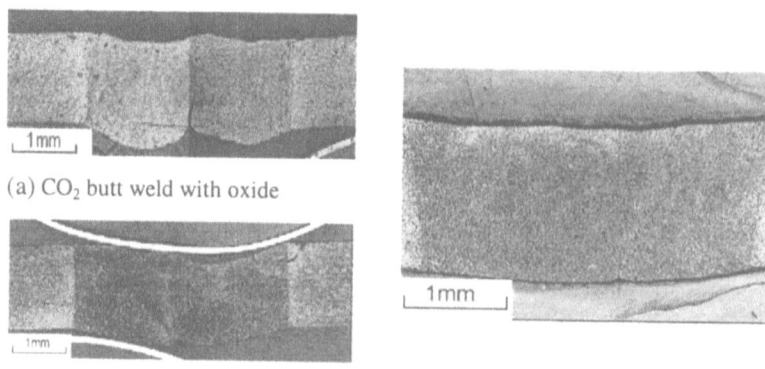

(a) CO$_2$ butt weld with oxide

(b) Diode butt weld with gas blow effect   (c) optimized conduction butt weld

Fig.5. Conduction butt welding

4. Conclusions

In the present study, a 2 kW CO$_2$ laser and a 1.5 kW diode laser have been used for 5182 aluminum alloy welding. It has been found that keyhole welding is possible with CO$_2$ laser, but the operating window is small and surface discontinuity remains as a problem.   By defocusing, the CO$_2$ laser can produce conduction welding with good quality.   Diode laser is similar but with a higher efficiency which is of importance today. The optimum operating parameters are nearly the same for the two lasers.   Fit-up gaps, edge conditions, surface preparation and gas flow could also affect the welding quality. They must be controlled carefully to produce solid and reproducible welds.

References

[1] K. Behler, J. Berkmanns. A. Ehrhardt etal. Laser beam welding of low weight materials and structures, Materials and Design, Vol 18, Issues 4-6, 1 Dec 1997, Pages 261-267

[2] S. Ramasamy and C. E. Albright. CO$_2$ and Nd:YAG laser beam welding of 5754-O aluminium alloy for automotive applications, Science and Technology of Welding and Joining 2001 Vol. 6 No. 3, 182-190

[3] John F. Ready, Dave F. Farson. LIA Handbook of Laser Materials Processing, Laser Institute of America, Magnolia Publishing, Inc. 2001

[4] Lin Li. The advances and characteristics of high-power diode laser materials processing, Optics and Lasers in Engineering 34 (2000) 231-253

[5] Friedrich Bachmann. Industrial applications of high power diode lasers in materials processing, Applied Surface Science, Volumes 208-209, 15 March 2003, Pages 125-136

[6] E. Schubert, M. Klassen, I. Zerner etal. Light-weight structures produced by laser beam joining for future applications in automobile and aerospace industry, Journal of Materials Processing Technology, Vol 115, Issue 1, 22 August 2001, Pages 2-8

[7] C. Mayer etal. Laser welding of Al-Mg alloys sheets process optimization and weld characterization, Material Science forum, Vol217-222,1679-1684,1996

[8] Pastor M, Zhao H, Martukanitz RP, etal. Porosity, underfill and magnesium loss during continuous wave Nd : YAG laser welding of thin plates of aluminum alloys 5182 and 5754, Welding Journal 78 (6): 207S-216S JUN 1999

Laser Surface Treatment of Laser Welded Duplex Stainless Steel

Edoardo Capello[1], Moreno Castelnuovo[1], Barbara Previtali[1], Maurizio Vedani[1]

[1] Politecnico di Milano Dipartimento di Meccanica, Via Bonardi 9, 20133 Milano, Italy, barbara.previtali@polimi.it

Abstract: It is well known that, when Duplex Stainless Steel are welded, the optimally balanced structure of 50α-50γ is lost, yielding to a ferrite rich bead.

In the paper the problem of recovering the laser welded microstructure of duplex stainless steel is investigated by means of a laser treatment approach. A procedure to determine the processing window is presented. The conditions to obtain the desired microstructure have been identified, based on evaluation of two critical thermal parameters, the peak temperature and the cooling rate by an analytical thermal model.

Experiments performed on a 2.11 mm thick plate of a 22Cr-5Ni-3Mo duplex stainless steel, allowed to successfully recover the welded beads by using the calcuated thermal conditions, showing that the proposed solution could be a helpful tool to reduce the number of iterations usually needed by the traditional experimental approach.

1. Introduction

Duplex Stainless Steels (DSSs) have found a widespread application during recent years owing to their excellent mechanical and corrosion behaviour achieved by a balanced phase distribution made up by approximately $50\%\alpha$-$50\%\gamma$. However, when welded by laser, DSSs lose their good properties due to the high cooling rates imposed by the process. The as welded structure becomes strongly unbalanced with an average content of austenite (γ phase) in the bead of about 20%. Therefore, a post-heat treatment is necessary to recover the desired proportion between α–ferrite and γ-austenite phases. Recently, the possibility to modify the weld bead by a Laser Heat Treatment (LHT) as an alternative to the conventional post-weld furnace treatment (widely reported in literature [1]) has been investigated. Several works have been published, dealing with the recovery of the expected balance between the γ and α phases in the top surface of the bead [2]. These works have positively demonstrated the feasibility of the LHT process but they also highlighted the intrinsic difficulty to define the operational range of the process parameters to effectively treat all the bead thickness and not only the top surface. Therefore, the following step towards the industrial exploitation of this unconventional process is the control and the optimisation of the process conditions, in order to obtain a completely treated bead with a well-balanced microstructure. To control the process conditions, in terms of treated thickness and of degree of structure transformation, an accurate model of the temperature field in the workpiece is needed. The present work is aimed at studying the thermal conditions allowing a complete recovery of the bead structure by a LHT process. First a thermal model, appositely developed by Woo and Cho [3], has been tested and applied to the LHT of a 2.11 mm thick sheet of DSS. Then, the model has been used to define the thermal treatment cycles and to validate the metallurgical conditions required to achieve the partial α→γ transformation. The operative technological range required to recovery of the original structure, when possible, has been outlined, based on the previous established conditions. Experiments performed on 2.11 mm thick plates of a DSS, according to the technological nomogram, allowed to successfully transform the welded beads.

2. Modelling of the Laser Heat Treatment Process

In the LHT process aimed at the modification of the microstructure in the solid phase, the correct choice of process parameters is undoubtedly a critical point. In the case of the recovery of the original 50%α-50%γ structure for DSSs, the transformation into an optimal structure requires to hold the material above about 1050°C for a well-defined time. In the LHT process, this time directly depends on the interaction time between the workpiece and the laser beam. After this phase, it is necessary to provide a rapid cooling of the heat treated zone. In the case of the LHT process, due to both the low heat input rate and the high thermal capacity of the cold surrounding material, the cooling of the treated zone occurs in a short period of time. Therefore, the thermal conditions that should be obtained in the treated area are: *i)* avoiding melting of the steel; *ii)* heating the zone to a sufficiently high temperature; *iii)* allowing the α−ferrite to partially transforms in γ−austenite; *iv)* cooling the weld bead in a short time down to room temperature. Experimentally, the control of the afore mentioned conditions, has revealed to be very difficult due to the rapid heating and cooling thermal cycles characterising a laser material processing. Therefore, in this work a thermal model, recently developed by Who and Choo [3], has been used to describe the temperature field in the workpiece. This model is a quite accurate three-dimensional transient temperature model, which seems to be the appropriate tool to predict the temperature field in the LHT process.

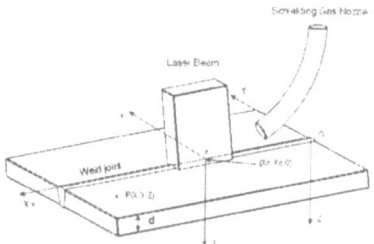

Figure 1 Laser beam configuration and coordinates systems.

The configuration considered by the model is reported in Figure 1. The LHT process is achieved by the movement of a rectangular spot beam, releasing a constant power density along the X direction. The main assumptions of this model are: *i)* the workpiece is a homogeneous plate with semi infinite X and Y dimensions, *ii)* the surface radiation is negligible, *iii)* the thermal properties do not vary with temperature, *iv)* prior to laser heating the material is at room temperature. The temperature distribution solution of the thermal problem in the moving coordinates *(x,y,z)* at time *t*, due to a rectangular distributed moving heat source with centre in *(X_c, Y_c, Z_c)*, is [3]:

$$
T(x,y,z,t)-T_0 = \int_0^t \frac{\alpha\eta P}{8kdRS}\sum_{n=0}^{\infty} A_n \exp[-\mu_n^2(t-\tau)] \times \left[\cos\left(\frac{\mu_n}{\sqrt{\alpha}}Z\right) + \frac{\beta_1\sqrt{\alpha}}{\mu_n}\sin\left(\frac{\mu_n}{\sqrt{\alpha}}Z\right)\right] \times \left\{\mathrm{erf}\left[\frac{R+X_c(\tau)-x-X_c(t)}{\sqrt{4\alpha(t-\tau)}}\right] + \right.
$$
$$
\left. +\mathrm{erf}\left[\frac{R-X_c(\tau)+x+X_c(t)}{\sqrt{4\alpha(t-\tau)}}\right]\right\} \times \left\{\mathrm{erf}\left[\frac{S+Y_c(\tau)-y-Y_c(t)}{\sqrt{4\alpha(t-\tau)}}\right] + \mathrm{erf}\left[\frac{S-Y_c(\tau)+y+Y_c(t)}{\sqrt{4\alpha(t-\tau)}}\right]\right\} d\tau
$$

(1)

where $\beta_1 = h_1/k$ and $\beta_2 = h_2/k$, while the constant μ_n and A_n are respectively:

$$
\tan\frac{\mu_n}{\sqrt{\alpha}}d = \frac{(\beta_1+\beta_2)\mu_n\sqrt{\alpha}}{\mu_n^2 - \beta_1\beta_2\alpha} \qquad A_n = \frac{\mu_n}{\mu_n^2 + \alpha\beta_1^2 + 2\alpha\beta_1/d}
$$

(2)

The meaning and the values of all the symbols and constants used in Equation 1 and Equation 2 are shown in Table 1 [3].

Table 1 Symbols and units.

Symbol	Meaning	Units	Value
R, S	X,Y Half-length spot	mm	8, 4
d	Workpiece thickness	mm	2.11
α	Thermal diffusivity	$m^2 s^{-1}$	6.4×10^{-6}
k	Thermal conductivity	$Wm^{-1}C^{-1}$	30
η	Absorption coefficient	%	0.81
h_1, h_2	Forced and natural convective coefficient	$Wm^{-2}C^{-1}$	200, 20

The model in Equation 1 and Equation 2, once calibrated, has been used in the study of the LHT process for two purposes:
1. To determine the thermal and metallurgical conditions required for a 50%α-50%γ structure recovery.
2. To identify the technological range of the laser parameters allowing a complete recovery of a laser welded bead.

3. Laser Heat Treatment Process Conditions

The material investigated was a 22Cr-5Ni-3Mo (UNS S32205) DSS in the form of plates of 2.11 mm thickness. Bead on plate welds have been obtained by laser CW welding. The welding process parameters (power: 5 kW, feed rate: 55 mm/s, focal height: 0 mm) were selected in order to achieve the best results in terms of full penetration, suitable bead shape ratio and absence of relevant welding defects. The average ferrite volume fraction in the weld metal, estimated by manual counting method according to ASTM E562-95 standard, was 20±5%. Post weld heat treatments were performed on the oxidised top surface of the welded bead. The laser used for the experimental phase was a 6 kW CO_2 CW with a multimode distribution equipped with an integrator mirror. A flow of N_2 shielding gas was used, due to the well-known austenite-stabilizing properties of N_2. Single pass treatments have been performed on the top surface of the workpiece along the previous welding line direction. Due to the presence of the integrator mirror, only two process parameters, power beam P and feed rate v, were selected to set the thermal field in the workpiece and, as a consequence, the recovery of austenite in the weld bead. A set of preliminary experiments was performed to assess, on a purely empirical basis, the process parameter range where the recovery of austenite could be obtained without exceeding the melting temperature T_F ($T_F \approx 1450°C$ for the UNS S32205). According to these preliminary results, the levels of the process parameters P and v listed in Table 2 were determined.

Table 2 Laser heat treatment parameters and resulting induced thermal cycle data.

# Trial		1	2	3	4	5	6	7	8
P	[W]	600	600	840	840	1000	1000	1250	1250
V	[mm/s]	1	1	2	2	3	3	5	5
H_{max}	[mm]	1.2	1.8	1.9	1.7	1.8	1.1	1	1.4
$\Delta t_{800-500}$	[s]	7	7.5	6	5.75	5	4.75	2.75	3.5
$T_{\alpha-\gamma}$	[°C]	850	855	940	925	975	990	820	880

In order to validate the above described thermal model, a comparison between the real and the modelled temperature field is needed. Therefore, a direct measurement of the temperature field during the LHT process was performed by a K-type uncoated micro thermocouple directly welded on the specimen bottom surface at the welding line. In the quasi-stationary region of each heat treated seam, a prismatic sample was sectioned and polished for metallographic observations. Measurements on macrographs of cross-sections allowed to estimate the maximum width H_{max} of the recovered layer of the laser treated weld bead. By the thermal model, the isothermal lines through the thickness were plotted, allowing to calculate at H_{max} the theoretical temperature at which a significant amount of α-ferrite transformed into γ-austenite ($T_{\alpha-\gamma}$), as shown in Figure 2. It should be noted that no condition in Table 2 led to melting at the top of the bead. In a following phase, by the time-temperature curves, the $\Delta t_{800-500}$ parameter, that is the time necessary to pass from 800 °C to 500 °C on cooling of the weld bead [4], was determined at the top of bead, where it assumes the minimum value. By this approach, the minimum temporal condition for an effective the LHT process was found. Based on the results in Table 2, it has been chosen as minimum $\Delta t_{800-500}$ for the heat treatment the value $\Delta t_{\alpha-\gamma}$=2.75 s. Greater values of $\Delta t_{800-500}$ seem to guarantee the α→γ transformation in accordance to the values reported in literature [5]. $T_{\alpha-\gamma}$=1028 °C was assumed as the minimum value for the α→γ transformation which corresponds to the upper limit of a 95% confidence interval based on the value in Table 2.

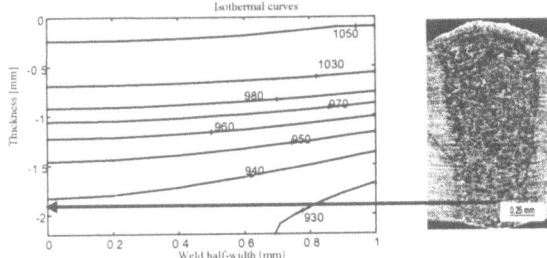

Figure 2 $T_{\alpha\text{-}\gamma}$ minimal transformation temperature determination (trial 3 in Table 2)

The thermal and temporal conditions necessary to the LHT process are summarized in Equation 3:

$$\begin{cases} T_F = 1450°C > T(z) > T_{\alpha\text{-}\gamma} = 1028°C & per \quad 0 < z < H_{max} \\ \Delta t_{800\text{-}500}(z) > \Delta t_{\alpha\text{-}\gamma} = 2.75s & per \quad z = 0 \end{cases} \tag{3}$$

4. Optimisation of the LHT Process

According to Equation 3 when H_{max} is equal to the workpiece thickness d (d=2.11mm), the weld bead results completely recovered.

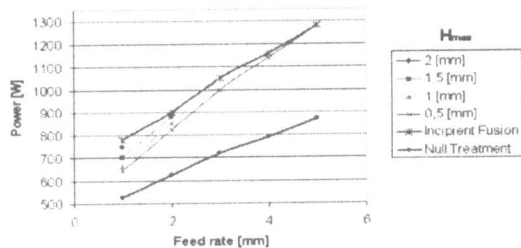

Figure 3 Technological nomogram

As observed in the nomogram given in Figure 3, obtained by plotting the iso-treatment lines at different H_{max} as a function of the process parameters P and v, only an experimental point satisfies the condition of full-thickness treatment (P=750 [W] v=1 [mm/s]). Moreover, Figure 3 shows that the technological area where the LHT process can be performed is extremely narrow, since it is limited at the lower boundary by the no-treatment condition ($T < T_{\alpha\text{-}\gamma}$) and at the upper boundary by the onset of melting ($T=T_F$). A second set of investigations aimed at reproducing the full treatment of the bead was performed. The experimental results showed that the 2.11 mm thick DSS plate could be successfully recovered in the selected condition, confirming the validity of the process parameter selection.

Figure 4 reports both a macrograph of a fully heat treated weld bead and a micrograph of the fusion line region of the heat treated weld (on the right side the heat affected zone is visible).

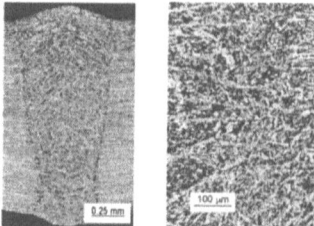

Figure 4 Macroscopic and macroscopic aspects of the fully heat treated weld bead.

As can be observed, the weld bead is effectively treated along the full thickness. Moreover, from quantitative measurements, performed by manual counting method according to ASTM E562-95 standard, it could be stated that the austenite content in the bead in average increased up to 57%±4.5%, as shown in Figure 5.

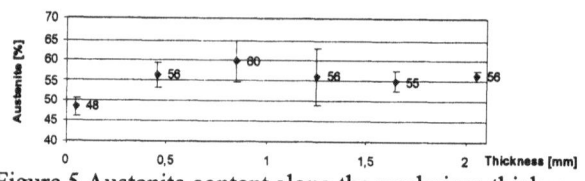

Figure 5 Austenite content along the workpiece thickness

5. Conclusions

In the present paper the problem of obtaining a recovery of the laser welded microstructure of a DSS by means of a post weld laser heat treatment was investigated. The temperature distribution in a workpiece treated by a laser beam was firstly determined by an analytical thermal model. The thermal conditions required to obtain the desired microstructure were identified. These conditions, together with the thermal model, were used to identify the suitable technological range of the process parameters. It was found that the process window for the LHT process is quite narrow and therefore its determination imposes an accurate parameters selection.

References

[1] Bonollo F., Tiziani A., Penasa M. et al., 1994, *An experimental approach to CO$_2$ laser welding of duplex stainless steels (UNS S 31803)*, Eurojoin 2, Italy.
[2] Capello E., Chiarello P., Previtali B., Vedani M., 2003, *Laser welding and surface treatment of 22Cr-5Ni-3Mo duplex stainless steel*, Mat. Science & Eng., A351, 334-343.
[3] Woo H.G., Cho H.S., 1999, *Three-dimensional temperature distribution in laser surface hardening processes*, Surface and Coatings Technology, 373, 695-711.
[4] Liou H., Hsieh R., Tsai, 2002, *Microstructure and stress corrosion cracking in simulated heat-affected zones of duplex stainless steels*, Corrosion Science, 44, 2841-2856.
[5] Ramirez A., Lippold J., Brandi S., 2003, *The Relationship between Chromium Nitride and Secondary Austenite Precipitation in Duplex Stainless Steels*, Metallurgical and Materials Transactions A, Volume 34A, 1583-1597.

Laser Assisted Jet Electrochemical Machining of Hard to Cut Alloys

P T Pajak[1], A K M De Silva[1], D K Harrison[1], J A McGeough[2]

[1] Glasgow Caledonian University, School of Engineering Science and Design, City Campus, Cowcaddens Road, Glasgow G4 0BA, U.K.
[2] The University of Edinburgh, School of Engineering and Electronics, King's Buildings, Edinburgh E9 3JL, U.K.

Abstract: Laser Assisted Jet Electrochemical Machining (LAJECM) is a hybrid process which combines electrolyte jet and low power laser beam (375mW) to facilitate material removal. The main purpose of the laser is to enhance dissolution from the workpiece in the specific area giving faster dissolution in axial rather in lateral direction and in result higher dimension precision. Experimental analysis using stainless steel, titanium alloy and high carbon alloy has proved that laser assistance can yield up to 33% higher volumetric rate and up to 65% better dimensional accuracy. It has been also proved that laser enhances current efficiency during the machining process.

1. Introduction

Electrochemical Machining (ECM) is a well known and established method for producing a wide branch of parts for aerospace, automotive, defence and recently also for medical applications (e.g. implants) [1, 2]. ECM is usually used where the machined surface has to characterise of minimal mechanical and thermal stress, hence the highest reliability. These machining advantages can be granted to the nature of the material removal mechanism (electrochemical dissolution) in ECM. However, the complexity of the ECM dissolution process makes it difficult to theoretically predict, control and successfully utilise all the ECM benefits. Mainly the machining precision is adversely affected by unwanted stray machining, which limits the dimensional accuracy. Due to this stray effect investigations aimed at localising and hence gaining better control of ECM dissolution [1].

The ECM process improvement investigations involve such techniques as the use of an insulated electrode, the use of smaller inter electrode gaps, the use of low concentration electrolytes, the use of pulsed current, etc. There have been also developments aiming at precision and micro-ECM applications and machining of highly passivating and non-conductive materials [2, 3]. Another major ECM development is its combination with other machining processes (hybrid processes) [4]. There have been several attempts to hybridise ECM with different physical or chemical mechanisms in order to achieve better process localisation [5, 6]. One of these hybrid methods is the parallel combination of an electrolyte jet with a laser beam, which forms the basis for Laser Assisted Jet-Electrochemical Machining (LAJECM).

2. Machining Localisation in LAJECM

Laser Assisted Jet Electrochemical Machining (LAJECM) is a hybrid method. It combines a relatively low power (375mW) laser beam with electrolyte jet giving non-contact tool-electrode. The predominant mechanism of material removal is electrochemical dissolution supported by laser thermal energy transmitted to the workpiece. The use of a focused laser beam to improve electrochemical dissolution by enhancement of the kinetics of the electrochemical reactions to a specific area (also called *localised zone*) has been found by Datta [6]. The major laser influence is attributed to the increase in temperature in that specific area. Therefore, electrolyte temperature rises, causing higher electrolyte conductivity and higher current density. The reaction products transportation processes by diffusion are also improved. Secondly, higher temperature in the localised zone leads to lower reaction's activation energy, thus dissolution reactions are initiated easier. Figure 1a and 1b gives an outline of Jet-ECM and Laser Assisted Jet-ECM.

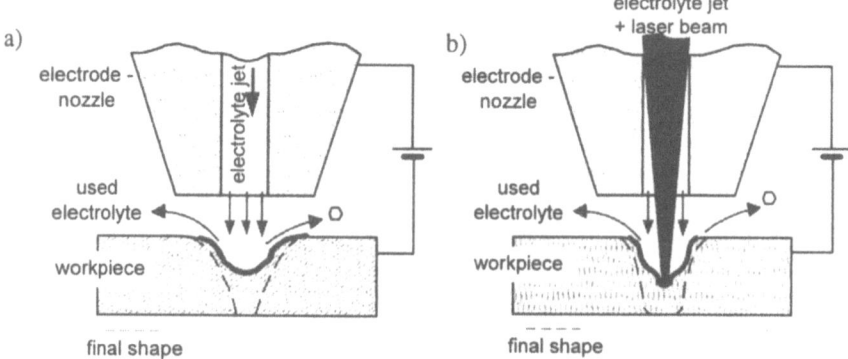

Figure 1. Outline of a) Jet-ECM, b) Laser Assisted Jet-ECM

Generally the machining localisation mechanisms lead to differentiation in dissolution speed. As shown in Figure 1b, in LAJECM dissolution is more concentrated in axial (deeper cavity) rather then in lateral direction, but in JECM (Figure 1a) dissolution proceeds more equally in every direction. This illustrates that the laser localisation mechanism enables not only an increase in dissolution velocity (removal rates), but also a reduction of stray machining, hence better machining precision can be achieved.

3. Instrumentation and Methodology

The main parts of the research apparatus are schematically shown in Figure 2. An appropriate electrolyte flow and hence velocity is maintained by varying electrolyte pressure. The jet of electrolyte is formed by the 1mm diameter nozzle placed in the bottom of a jetcell. The distance between electrodes (workpiece-anode and nozzle-cathode) can be varied. The Nd:YAG laser of wavelength 532nm and power density

47.5 W mm^{-2} is coaxial to the electrolyte jet. A machining chamber houses the jetcell and a specimen stage.

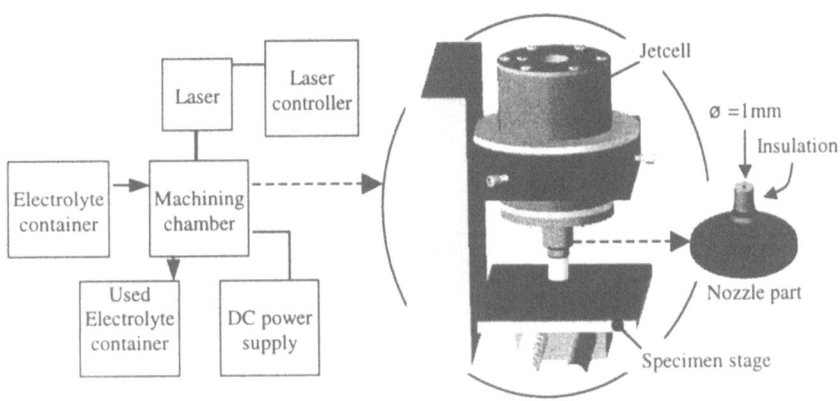

Figure 2. Diagram of Laser Assisted Jet Electrochemical Machining apparatus

Three relatively hard to cut materials were taken for this investigation – stainless steel (type Cr13Ni8), titanium alloy (type Ti90Al6V4) and high carbon steel (~1.1%C). For different machining variables holes were machined for JECM and LAJECM. In order to evaluate material receptivity for JECM and LAJECM, some of their properties are introduced in Table 1.

Table 1. Characteristic material properties for JECM and LAJECM

	Density [g cm^{-3}]	Thermal conductivity [W m^{-1} K^{-1}]	Electrical conductivity [$\mu\Omega^{-1}$ mm^{-1}]
Stainless Steel	7.81	16.3	0.12
Titanium Alloy	4.42	6.0	0.06
High C. Steel	7.20	50.0	0.55

For a single ECM process electrical conductivity has the priority, because the higher electrical conductivity, the higher current can be achieved. For LAJECM both – thermal and electrical conductivity have to be taken into account. The higher thermal conductivity, the sooner heat from the localised machining area escapes, lower temperature in that area is achieved and hence the laser cannot improve machining to such a degree as it does for lower thermal conductivity materials. However, heat and electric conductivity usually increase or decrease together as indicated in Table 1.

4. Experimental Results

The experimental analysis aims to show that machining with laser assistance can improve machining removal rates and also dimensional precision. Current densities are measured to show a direct reason for higher machining rates.

4.1 Volumetric Removal Rate

Material Volumetric Removal Rates (VRR) were investigated for the given materials with and without laser assistance using different process variables. The average values results for LAJECM are presented in Figure 3a and 3b.

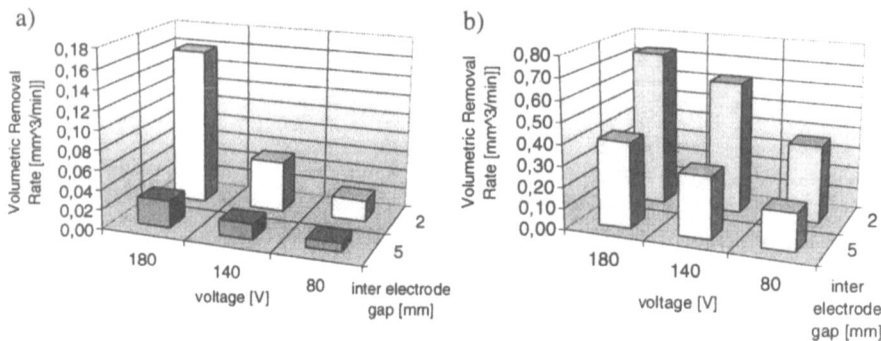

Figure 3. Volumetric Removal rates in relation to Inter Electrode Gaps (IEG) and voltage for Laser Assisted Jet-ECM: a) stainless steel, b) titanium alloy

Generally, according to ECM fundamentals, VRR increases with voltage and adversely with IEG. The VRR for titanium alloy is significantly higher for LAJECM. As it was presumed, titanium alloy was easier to machine due to its poorer thermal conductivity compared to stainless steel regardless of its poor electrical conductivity. Measurements of current during machining time have revealed that for the same variables set, both materials had similar average current levels (e.g. for U=140V, IEG=2mm, the order of 0.5A), hence the material property of electrical conductivity is not a key factor in LAJECM.

In order to ascertain the laser influence on VRR, its values for LAJECM and JECM were compared. Figure 4 shows relative VRR increase in LAJECM over JECM.

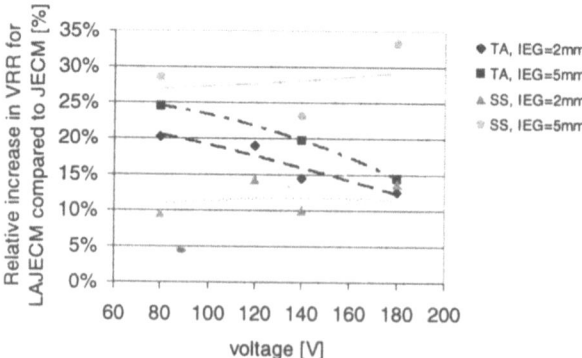

Figure 4. Relative increase of Volumetric Removal Rate for LAJECM compared to JECM

Usually laser assistance facilitates machining more for lower voltages and higher IEGs. For these machining variables setups, the process time to produce a hole is longer giving the laser more time to transfer heat energy to the workpiece. Therefore, machining is more effectively localised in a specific area compared to smaller IEGs when laser assistance is less effective due to shorter machining time and higher current which dominates the process.

4.2 Dimensional precision

In LAJECM significantly higher material removal appears locally in the laser localised zone. Material is removed faster in the axial direction as stated in section 2. This phenomenon significantly improves dimensional precision of produced holes. Figure 5a shows percentage hole taper reduction in LAJECM for investigated materials.

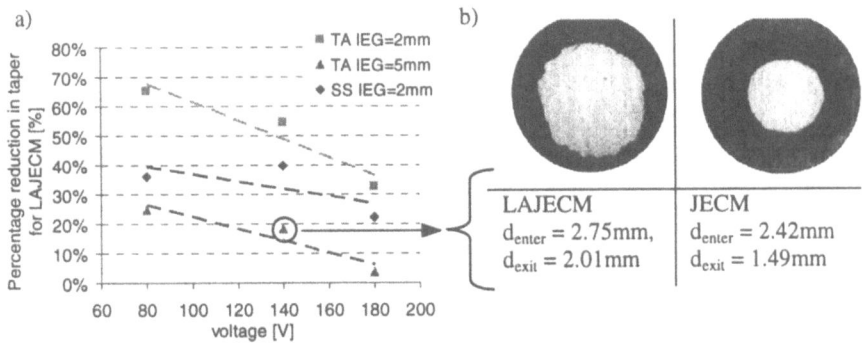

Figure 5. a) Percentage reduction in hole taper in LAJECM, b) examples of holes produced in titanium alloy by LAJECM and JECM

Consequently, the taper decreases most for machining conditions of the highest VRR. The highest taper reduction was achieved for titanium alloy. Figure 5b shows examples of photomicrographs of LAJECM and JECM of titanium alloy. Generally it was noticed that taper reduction depends on the mechanism of material dissolution. Material which dissolves linearly with time, such as stainless steel, usually does not yield so high a reduction in taper as materials of non-linear dissolution characteristics such as titanium alloy. This observation may be a subject of further chemical analysis.

4.3 Current densities and current efficiencies

Measurements of current density on a rod-shaped specimen (high carbon steel) have shown that for LAJECM higher current densities and hence higher material removals are achieved. The rod-shaped specimen was used in order to take precise measurements of current in the laser localised zone from a known area of diameter 1mm. Figure 6a shows the results of volumetric removal rate versus current density for LAJECM and JECM. This indicates a relationship "reason – result" (higher current density – higher material removal). Figure 6b shows current efficiency for the same current densities as in Figure 6a.

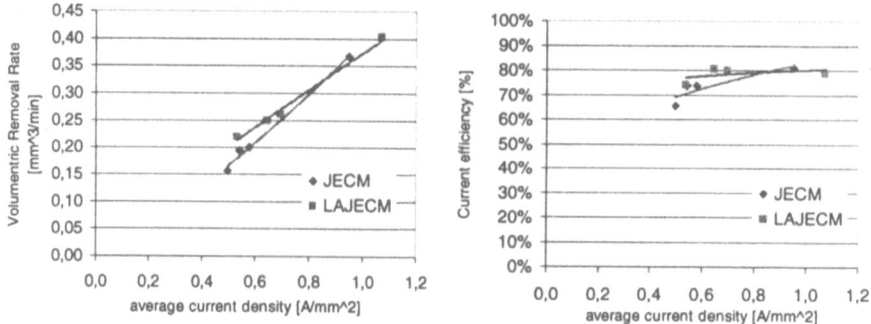

Figure 6. High carbon steel, starting IEG = 2mm, a) Volumetric Removal Rate versus current density. b) Current efficiency versus current density

It is characteristic that differences between VRR values are bigger for smaller current densities than for larger current densities in the case of LAJECM (e.g. for the first pair of LAJECM and JECM points the difference in VRR is about 0.06 mm^3/min and the difference in current density is only 0.04 A/mm^2, but for the last pair of LAJECM and JECM points the difference in VRR is 0.04 mm^3/min and the difference in current density is 0.12 A/mm^2). Thus, the machining improvement is better for smaller current densities, which are for smaller voltages and higher IEGs. For these machining conditions electrochemical dissolution is less dominant, hence the laser can better enhance dissolution. This stays consistent with the observation in the differences of volumetric rates for LAJECM and JECM using different machining variables (Figure 4).

Current efficiency for machining of high carbon steel oscillates around 70 – 80%. Again for smaller current densities current efficiency is higher for LAJECM and exactly for these current efficiency values significantly higher VRR are achieved as shown in Figure 6a. This suggests that current efficiency is also subject to laser influence. Generally, as stated in section 2, laser energy increases electrolyte conductivity hence current density and material removal giving in the end higher current efficiency values compared to JECM.

5. Summary

Experiments have shown that Laser Assisted Jet Electrochemical Machining improves material removal rates and machining precision. The highest absolute values of volumetric removal rate were achieved for titanium alloy (up to 0.73 mm^3/min), and LAJECM has yielded material removal up to 20.3% compared to single Jet-ECM. On the other hand, machining precision (expressed by taper reduction) improved in the order of 65%. The opposite situation has taken place for stainless steel. Absolute volumetric removal rates were much lower than for titanium, but for stainless steel LAJECM has yielded material removal up to 33% and machining precision up to 38% compared to Jet-ECM.

Current density and material removal analysis was carried out on high carbon steel. It has shown that laser assistance always yields material removal, however the most efficient is for smaller current densities (the order of 9.8%). Experimental analysis has also revealed that LAJECM has improved machining of high carbon steel in the lowest degree compared to other materials. As discussed this can be incorporated into the material properties such as electrical and heat conductivity factors.

References

[1]   Rajurkar K.P., D. Zhu, J.A. McGeough, J. Kozak, A.K.M. De Silva, 1999, *New Developments in Electro-Chemical Machining*, Annals of the CIRP, Vol.48/2

[2]   McGeough J.A., Pajak P.T., De Silva A.K.M., Harrison D.K., 2003, *Recent Research and Developments in Electrochemical Machining*, International Journal of Electric Machining, No.8

[3]   De Silva A.K.M., Altena H.S.J., McGeough J.A., 2003, *Influence of Electrolyte Concentration on Copying Accuracy of Precision-ECM*, Annals of the CIRP, Vol.52/1

[4]   Kozak J., Rajurkar K.P., 2000, *Selected Problems of Hybrid Machining Processes*, Advances in Manufacturing Science and Technology, Vol.24, No.2

[5]   Kozak J., Rajurkar K.P., 2001, *Laser Assisted Electrochemical Machining*, Transactions of the North American Manufacturing Research Institution of SME, Vol.XXIX

[6]   Datta M., Romankiw L.T., Vigliotti D.R. et al., 1989, *Jet and Laser-Jet Electrochemical Micromachining of Nickel and Steel*, Journal of Electrochemical Society, Vol.136, No.8

[7]   Pajak P.T., De Silva A.K.M., McGeough J.A., Harrison D.K., 2003, *Investigations in Laser Assisted Electrochemical Machining*, Proceedings of the 3rd International Conference on MMSS, Krakow, Poland

[8]   Pajak P.T., De Silva A.K.M., Harrison D.K., McGeough J.A., 2004, *Modelling the aspects of precision and efficiency in Laser Assisted Jet Electrochemical Machining (LAJECM)*, the ISEM XIV Proceedings (to be published in April 2004)

[9]   De Silva A.K.M., Pajak P.T., Harrison D.K., McGeough J.A., 2004, Modelling and Experimental Investigation of Laser Assisted Jet Electrochemical Machining, Annals of the CIRP, Vol.53/1 (to be published in July 2004)

Modelling The Effects Of Laser Beam Geometry On Laser Surface Treatment Of Metallic Materials

Shakeel Safdar, Lin Li and M.A.Sheikh

LPRC, UMIST, Department of Mechanical, Aerospace and Manufacturing Engineering, Manchester, M60 1QD. s.safdar@postgrad.umist.ac.uk

Abstract: Laser material processing has been carried out in the past using circular or rectangular beams. This paper presents an investigation into the effects of different beam geometries on heating/cooling rates and temperature gradients. The beam geometries presently investigated are circular, rectangular and triangular. Finite element modelling technique has been used to simulate the steady state and transient effects of a moving beam for laser surface treatment (e.g. transformation hardening,) on mild steel and stainless steel. The temperature distributions, cooling rates and thermal gradients have been compared.

1. Introduction

Laser processing of materials has progressed over the past years as a major industrial tool. Change in laser power or scanning speed affects the heating/cooling rate and temperature gradients on the material surface thereby modifying the material microstructure and properties. However variation of these parameters might adversely affect the overall material processing such as heating depth and coverage rate. Another possible method of varying the heating /cooling rate and thermal gradient (without changing input power or scanning speed) is by modifying the geometry of laser beams.

Beam patterns and geometries are important in heat transfer as they affect local interaction times and heat transfer rates. However the effect of the laser beam shape on processing condition and parameters has received very little attention. The majority of laser surface processing is performed using the standard circular or rectangular beam shapes. Other beam geometries may offer advantages by altering the temperature distribution and cooling rates across and along the beam. The beam geometry can be varied by the use of various optical systems and diffraction optics [2].

Chen et al. [1] conducted some theoretical work on beam geometries; it was found that a square shape led to lower cooling rates when compared to line and rectangular beam shapes. The later shapes having the longer dimensions in the direction of motion. The reduction in cooling rate was thought to be due to longer interaction time. *Sandven* [2] investigated the effects of circular and square beam on laser surface transformation hardening of steels. He found out a relationship between the lateral heat losses and beam spot. *Kock* [3] suggested that the shape of the hardened zone in heat-treating varies with the shape of the spot. A circular spot produces a hemispherical or meniscus-shaped hardened zone while a square spot produces a relatively flat based or even hardened zone. This is due to the energy absorbed per unit area across the surface perpendicular to the direction of travel of laser beam

Analytical investigation of the effect of beam geometries in laser surface processing is a demanding task, mainly due to the complex geometries, boundary conditions and material properties. The general expressions found in literature to determine the temperature distribution in the work piece usually consider semi-infinite body and a one-dimensional heat flow normal to the input heat flux. However in reality heat diffuses in all directions into the work piece from the moving laser spot and not only normal to the surface. Therefore assumptions necessary for analytical solution may prove to be too compromising for realistic results in this case. The evolution of numerical techniques and their use in high-speed computers has enabled complicated calculations to be performed in relatively short periods of times with acceptable accuracy. Finite element analysis has been used in the past to model various laser processes [4 - 9].

This paper presents an initial investigation of the effects of various beam geometries on cooling rates and other useful parameters. A finite element model has been constructed to analyse the effect of various beam shapes on laser processing. A commercial finite element package, ANSYS, has been used to simulate the process for each beam geometry.

2. Finite Element Model

The beam geometries analysed are shown in the Table.1. A total of five different beam geometries were simulated each with the same effective area (i.e. $10\ mm^2$). The basic shapes were circular, rectangular and triangular. However both rectangular and triangular beam were simulated for forward and reverse directions respectively. A uniform power distribution was assumed. The scanning speed of laser was set at $10\ mm/sec$. The power density was kept constant to allow comparison between different beam shapes at similar speeds. The power density was set at $2\ kW/cm^2$ to avoid melting of material. The processing of a single surface track length of $20\ mm$ was modelled.

Shape	Circle	Rectangle (i) & (ii)	Triangle (i) & (ii)
Geometry			
Dimension(mm)	d =3.56	a=2 b=5	c = 4.80 d = 4.16

Table1: Beam Geometries

The material modelled was stainless steel AISI 304. The size of the model was *50x50x10 (mm)*. The thermo- physical properties were assumed to be independent of temperature to reduce computational time. The material was assumed to be homogenous and isotropic. The ambient temperature was set at *20 °C*. Material properties of the AISI 304 are shown below.

Property	**Value** (at 300 K)
Density	7900 kg/m^3
Specific Heat (Cp)	477 J/kg K
Thermal Conductivity (k)	14.9 W/m K

Table2. Material properties [11]

The temperature field caused by the irradiation and travelling of laser beam is transient in nature. A three-dimensional transient heat transfer model was therefore formulated to simulate the process. The surface was only subjected to input laser heat flux. Convection and radiation boundary conditions were not used as the convective and radiative heat losses are very small [10]. Moreover they make the calculations highly non-linear thus causing convergence problems and increasing the computational time by a great amount. The governing equation for a transient thermal analysis in FEM is written in the matrix form as follows.

$$[C]\{\check{T}\} + [K]\{T\}=\{Q(t)\}$$

Where

[C]	Specific Heat Matrix
$\{\check{T}\}$	Time derivative of Temp
[K]	Thermal Conductivity Matrix
{T}	Temperature
{Q (t)}	Load (Heat) Matrix

The laser beam was modelled as a moving heat flux which traverses step wise in scanning direction, changing its spatial position in small time steps. Each area sector of the surface was subjected to a heat load for a time period equal to the interaction time dictated by the processing speed. Figure 1 shows schematically the simulation of the motion of the circular beam. At one time step a group of areas representing the shape of the modelled beam were subjected to heat load. After one time step a certain group of areas, depending on the position and motion direction were free of heat load where as the group of areas on the opposite side of beam shape in the direction of motion, which were free of heat load in previous time step, were subjected to heat load. This method was followed for all the time steps simulating the beam motion in one direction for the entire analysis time period.

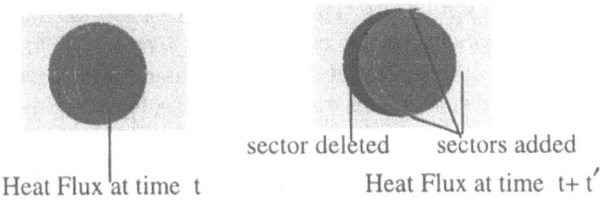

sector deleted   sectors added

Heat Flux at time t   Heat Flux at time t+ t$'$

Figure-1: Simulation of moving laser beam

3. Modelling Results and Discussion

As shown earlier a track of *20 mm* was simulated for different beam shapes. The global coordinate system was placed such that the laser scanning started at origin. i.e. *(0, 0, 0)* and traversed along the x-axis and finished at *(20, 0, 0)*. The beam geometry was equally weighted across the x-axis. To analyze the effect of various beam shapes on temperature distribution, a point in the middle of the track *x=10*, *y=0, z=0 (i.e. 10,0,0)* was selected. All the graphs and plots shown below are for the selected point.

Isotherms on the surface of the solid for the three beam shapes are shown in figure 2. The isotherms in front of the leading edge of the beam reflect the beam geometry, whereas the isotherms behind the trailing edge of the beam are almost similar in shape.

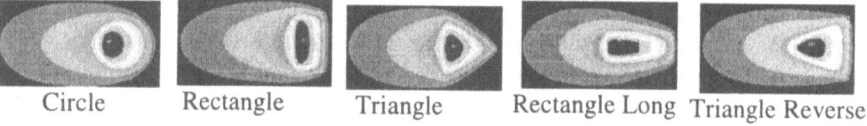

| Circle | Rectangle | Triangle | Rectangle Long | Triangle Reverse |

Figure 2. Isotherms on the top surface

Figure 3 and 4 similarly shows isotherms in the y-z plane and x-z plane respectively. Isotherms in the x-z plane (depth) are significantly different for the three beam shapes.

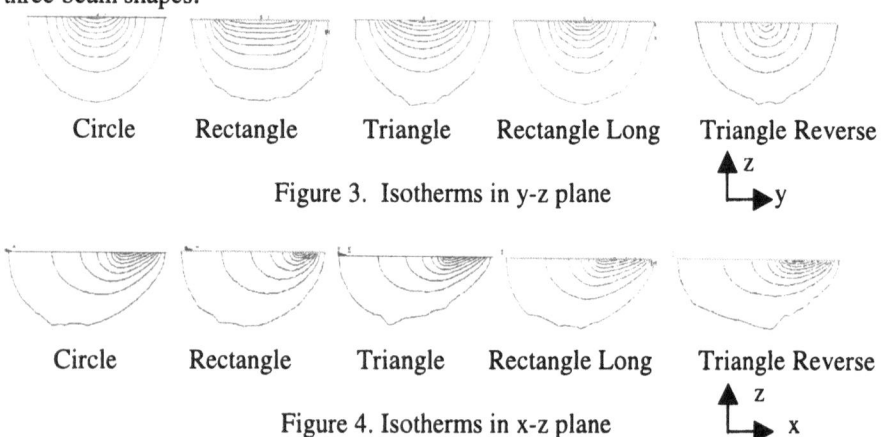

Circle   Rectangle   Triangle   Rectangle Long   Triangle Reverse

Figure 3. Isotherms in y-z plane

Circle   Rectangle   Triangle   Rectangle Long   Triangle Reverse

Figure 4. Isotherms in x-z plane

Figure 5 shows the time-temperature history for different beam shapes. The maximum temperature attained is for Rectangular Long beam *(1683.5 K)* followed by circular beam *(1680.8 K)*. The rectangular beam has the least temperature *(1461.3 K)*. It is shown that the beams having longer dimension in the scanning direction with respect to its lateral dimension, attains higher temperature in spite of the fact that power density was same for all the beams. However the triangular beam had a longer dimension in the scanning direction as compared to the circular beam, but the temperature attained is more for circular beam. This is due to the fact that the ratio of difference in dimensions in the scanning direction *(4.16/3.56)* is lesser than the ratio in difference in then lateral direction *(4.8/3.56)*.

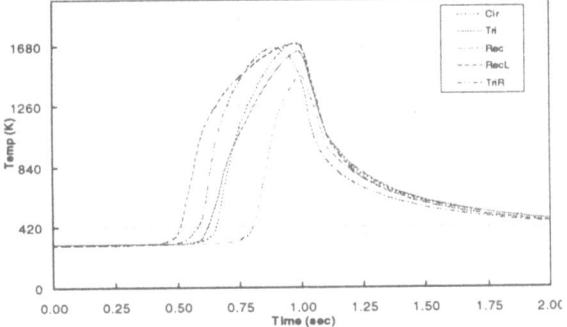

Figure 5. Temperature vs. time predicted by F.E model

Figure 6 shows the heating/cooling rate for the different beams. Triangular reverse beam has the maximum heating rate *(7639.31 K/sec)*. It is significant to note that though rectangular beam had the least maximum temperature but has the second maximum heating rate *(7495.96 K/sec)* followed by the circular beam *(7174.23 K/sec)*. The Rectangular Long has the maximum cooling rate *(-6077.97 K/sec)* followed by the circular beam *(6069.87 K/sec)*. From this graph it can be seen that the beam with shortest dimension in the scanning direction tends to have a high heating rate but has the lowest cooling rate. This is because of the fact that the leading edge of the beam is closer to the trailing edge thus the rise in temperature from leading edge of the beam to the trailing edge of the beam is very rapid however the same fact inhibits a higher cooling rate.

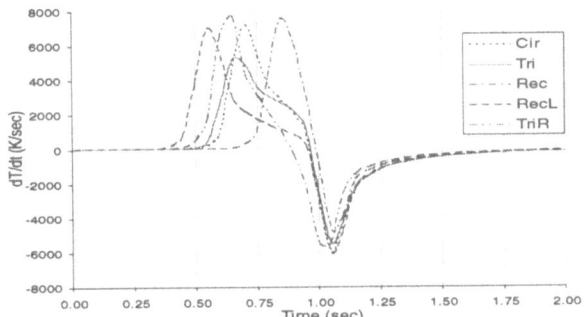

Figure 6. dT/dt vs. time

The above fact is further illustrated in figure 7. It shows the temperature distribution within the beam. i.e. from leading edge of the beam to the trailing edge. All the beams trailing edges are at the same point *(i.e. at (10,0,0))* from the start to allow a reasonable comparison. The absorption starts at leading edge of the beam and the maximum temperature is attained just before the trailing edge of the beam. This temperature remains almost steady till the trailing edge with the exception of triangular reverse beam. This is because the beam geometry is tapering down vary rapidly towards the trailing edge thus the area of heat flux is now lesser to sustain that temperature in the bulk material.

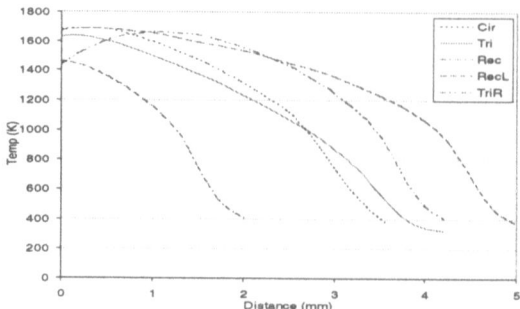

Figure 7. Temp distribution from L.E to T.E

Figure 8 shows the temperature distribution across the trailing edge of the beam. The distance is *8* units to the left and right from the middle of trailing edge (*. i.e. at point (10, 0, 0)*). The temperature distribution across the trailing edge is reflective of the beams spread at trailing edge. Rectangular beam has the maximum spread therefore it has almost a top hat temperature distribution whereas the circular beam with a point at trailing edge has a Gaussian temperature distribution. The temperature distribution in depth is shown in figure 9 (at the selected point).

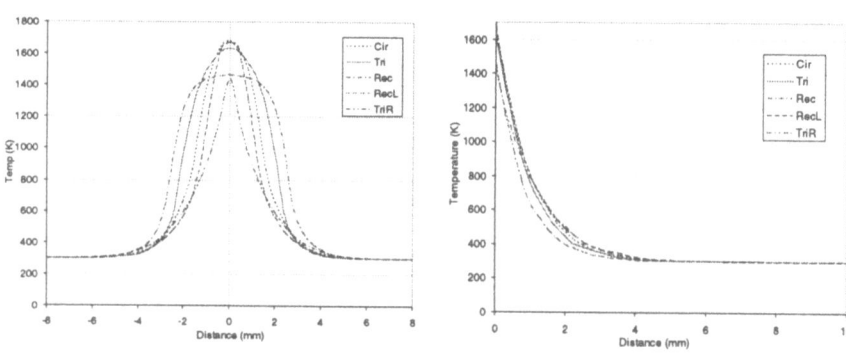

Figure 8. Temp dist across the beam   Figure 9. Temp distribution in Depth

The temperature distribution on the surface and in the depth affects the temperature gradient in different directions. figure 10,11,12 and13 shows the variation of temperature gradient at the considered point with respect to time in the x, y and z direction.

The thermal gradient can be validated by analytical equations. The heating/cooling rate is related to the thermal gradient in the direction of motion by

$$(\partial T / \partial t) = - V . (\partial T / \partial x)$$

Therefore

$$(\partial T / \partial x) = - (1/V) . (\partial T / \partial t)$$

We have already plotted heating/cooling rate vs. time in Figure 6. If we see Figure 10, we will see that the it is an inverted plot of heating/cooling rate due to the negative sign in above equation and the magnitude is reduced due to the division by scanning velocity of laser beam (which in our case is *10 mm/sec*).

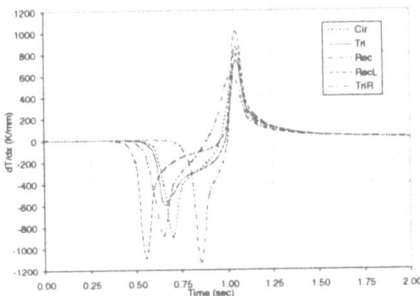

Figure 10. Therm Grad in *x* direction vs. Time

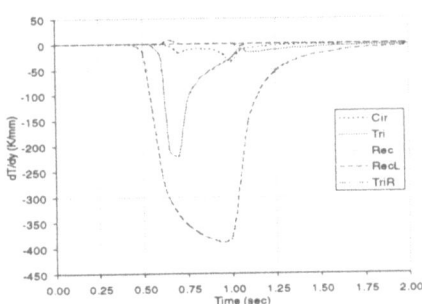

Figure 11. Therm Grad in y direction vs.. Time

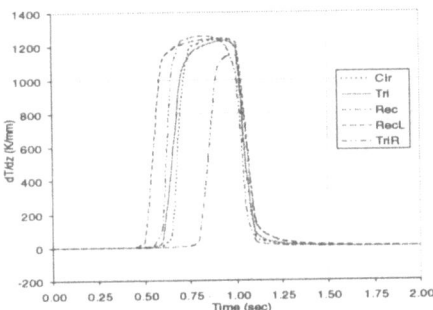

Figure 12. Therm Grad in *z* direction vs. Time

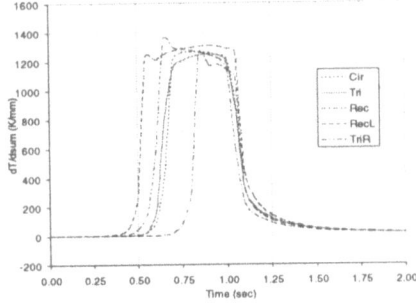

Figure 13. Overall Thermal Gradient vs.. Time

4. Conclusions

An initial investigation into the effects of laser beam geometry on temperature distribution (within and across beam geometry), heating/cooling rate and thermal gradients has been presented. The results have been calculated using Finite Element Analysis. A total of five beam geometries were considered. A point in the middle of the laser-scanned track was selected for calculating different parameters for subsequent comparison between different beam shapes. Rectangular Long beam had the maximum attainable temperature followed by the circular beam. Rectangular beam had the second maximum heating rate whereas circular beam had the maximum cooling rate. Thermal gradients were also calculated with respect to each axis for the selected point. An overall thermal gradient. i.e. the resultant of the vector sum of the thermal gradient in each axis was also calculated. Triangular reverse had the maximum resultant thermal gradient followed by rectangular long and circular beam. Thermal gradient in the scanning direction for the different beam shapes had similar plots but there was a significant difference in their magnitude. Rectangular beam had the maximum thermal gradient for the heating phase whereas rectangular long beam had the maximum thermal gradient for the cooling phase. From this initial study it is evident that modifying the laser beam geometry varies the heating/cooling rate and thermal gradient. This fact can be utilised for optimisation of different laser processes.

References

[1] Chen, Y.X., He, Y.Y., Jun, P.S., Nian, S.J., The Role of Beam shape in Convection and Heat Transfer in Laser Melted Pool. Proceedings of ICALEO '90, Nov. 1990, Boston, USA, Vol. 82: pp. 480-491.

[2] Sandven, O., Laser Surface Transformation Hardening. Metals Handbook, 9th Ed., Vol. 4: Surface Engineering, Pub. ASM, pp 507-517.

[3] De Kock., J., Lasers Offer Unique Heat Treating Capabilities. Industrial Heating: 2001 via URL www.industrialheating.com.

[4] Wei, P.S., Ho, C.Y., Shian, M.D., Hu, C.L., Three-Dimensional Analytical Temperature Field and its Application to Solidification Characteristics in High- and Low-Power-Density-Beam Welding. Int. J. Heat and Mass Transfer. Vol. 40: 1997, pp. 2283-2292

[5] Shuja, S.Z., Yilbas, B.S., 3-Dimensional Conjugate Laser Heating of a Moving Slab. App. Sur. Sci. Vol. 167: 2000, pp. 134-148

[6] Kar, A., Scott, J.E., Latham, W.P., Effects of Mode Structure on Three Dimensional Laser Heating Due to Single or Multiple Rectangular laser beams. J. Appl. Phys. Vol. 80: July, 1996, pp. 667-674

[7] Dai, K., Shaw, L., Thermal and Stress Modelling of Multi-Material Laser Processing. Acta Mater, Vol. 49: 2001, pp. 4171-4181

[8] Matsumoto, M., Shiomi, M., Osakada, K., Abe, F., Finite Element Analysis of Single Layer Forming on Metallic Powder Bed in Rapid Prototyping by Selective Laser Processing. Int. J. Mach. Tools Manuf. Vol. 452: 2002, pp. 61-67

[9] Shankar, V., Gnanamuthu, D., Computational Simulation of Heat Transfer in Laser Melted Material Flow. AIAA 24th Int. Aerospace Sci. Meeting, Reno, Nevada: 1986, pp. 1-10

[10] Yeung, K.S., Thornton, P.H., Transient Thermal Analysis of Spot Welding Electrodes. Supplement to the Welding Journal: January, 1999, pp. 1s-6s

[11] Incropera, Dewit, Fundamentals of Heat Transfer, 4th Ed., Pub. John Wiley and Sons, Inc.

MACHINE TOOL ACCURACY AND COMPONENT INSPECTION

A Novel 3D Laser Ball Bar for The Accuracy Assessment of Multi-axis Machine Tools

Kuang-Chao Fan[1], Hai Wang[2]

[1]Department of Mechanical Engineering National Taiwan University1, Sec. 4, Roosevelt Rd Taipei, Taiwan, ROC. fan@ntu.edu.tw
[2]Department of Mechanical Engineering Ming-Chi Institute of Technology 84, Gung-Juan Rd., Tai Shan Taipei, Taiwan, ROC. whai@ns1.mit.edu.tw

Abstract:

A novel design of 3D Laser Ball Bar (LBB) is proposed in this research for the three dimensional measurements of any moving object in real time phase. This system is based on the spherical coordinate principle containing only one precision Laser linear measurement device and two precision Laser rotary encoders in the gimbals base with a precision linear guide for radial sliding motion. Such a system can be dragged by any 3D moving target with a magnetic head and freely moved in the space. Three sensors simultaneously record the ball positions and transform into the Cartesian coordinate in real time. It aims to measure the tool position relative to the worktable at any working point no matter the inspected machine is of 3D Cartesian type or parallel type multi-axis machines. Practical applications are addressed with respect to different error measurements, such as linear errors, circular errors, volumetric errors, and to different types of machine tools, such as Cartesian type and parallel type. All can be done simply with the developed 3D LBB system.

1. Introduction

Techniques for performing accuracy testing of CNC machine tools can be found in many standards, such as the ISO 230 or ASME B5.54 [1]. Most of the existing linear measurement instruments are one dimensional, such as the laser interferometer or step gauge. For the circular test of 2D motion, as specified in the ISO 230-4 [2], some instruments have been developed, such as the double ball bar (DBB) [3], Contisure [4], and the latest laser ball bar (LBB) [5, 6]. Although these instruments are capable of 2-axis error measurements, they are still sensitive to one dimension only.

A novel design, which integrates the merits of LBB and laser tracking system (LTS) [7], is proposed in this research for the three dimensional measurements of moving object in real time phase. This system is based on the spherical coordinate principle containing only one precision Laser linear measurement device and two precision Laser rotary encoders in the gimbals base with a precision linear rail for radial motion. Such a system can be dragged by any 3D moving target with a magnetic head and freely moved in the space. Three sensors simultaneously record the ball positions and transform into the Cartesian coordinate in real time. Having calibrated by a HP Laser interferometer, the systematic accuracy can be compensated and enhanced to a higher degree. As this system is operated in a passive mode in 3D space, the cost is cheap. It is called the 3D Laser Ball Bar (3D-LBB).

2. Design Principle of 3DLBB

2.1 Structural design

Figure 1 illustrates the system configuration of the 3D-LBB. It is constructed in spherical coordinate of which the center of the gimbals mount is the origin (O_r). Mounted onto the gimbals center is a precision linear slide with effective traveling length of 480mm and straightness accuracy of 3µm made by Swiss Schneeberger Co. A standard steel ball mounted on the carriage and the ball's center to the origin (O_r) is perfectly in line with the slide center .

The movement of the 3D-LBB is generated by the precision steel ball, which can be dragged by a magnet socket carried by any moving object. The radial motion (R) of the ball is detected by a laser Doppler scale (LDS, model 109N, made by Optodyne Co.), whose beam passes through the origin (O_r) and towards to ball center, and the beam is reflected back by a reflector at the carriage's side plate. The LDS has wavelength stability to 0.1 ppm and system accuracy to 1.0 ppm.

The pitch (θ) and the yaw (ϕ) motions of the slide with respect to the gimbals base are detected by two precision Laser rotary encoders (model K-1, made by Canon Co.) individually. Each encoder has very fine scales of 81,000 ppr. With an additional 16-division interpolator board (model 16-2), the resolution can achieve to one arc-sec. An additional reference socket is located on the gimbals base mount for the reset point of this spherical coordinate system.

2.2 Coordinate transformation

To obtain the tool point Cartesian coordinate components, the equations of coordinate measurement can be easily derived as shown in Figure 2.

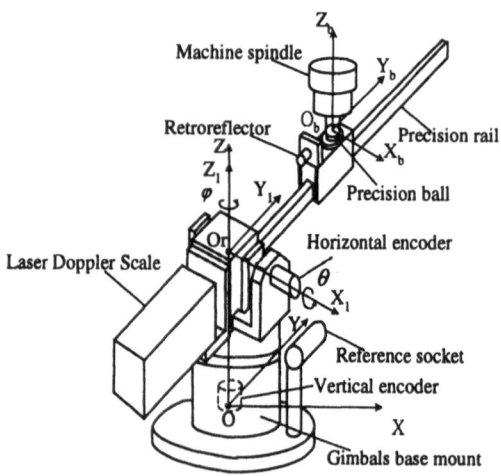

1.   The structure of 3D LBB.

$$
\left.
\begin{aligned}
X &= R \cos\theta \sin\varphi \\
Y &= R \cos\theta \cos\varphi \\
Z &= \sin\theta
\end{aligned}
\right\}
\qquad (1)
$$

where, $R = R_0 + R_1$, R_0 is the initial distance from the ball center to origin (O_r) when the steel ball is attached to the reference socket, and R_1 is the relative displacement from initial distance. Figure 3 shows the picture of the developed prototype 3D-LBB.

3. Accuracy Calibration

As an instrument for the spatial position measurement, the system must be more accurate than the inspected machine itself. Error sources, therefore, must be identified and properly calibrated to improve the instrument's system accuracy.

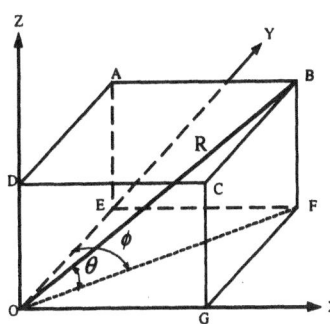

2. Spherical and Cartesian coordinates

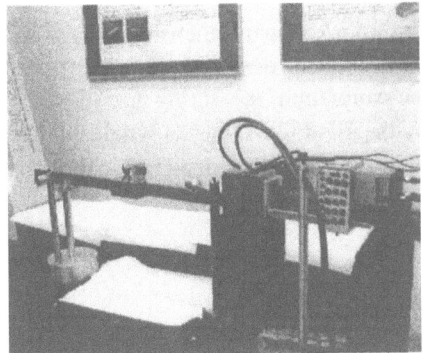

3. The prototype of 3D-LBB

3.1 Calibration of the R axis

As shown in Figure 4, a HP 5528A laser interferometer is used to compare with the reading of the LDS, and the 3D-LBB is mounted on the table of a CMM. Since the error has an obvious tendency and repeatability, the R errors can be compensated by the best fitting line. The modified R reading is expressed by Eq. (2). After compensation, positioning errors of the LDS can be maintained within ± 0.3 μm.

$$
R_{mod} = R - 2 \times 10^{-4} R + 0.00006 \qquad (2)
$$

where R_{mod} is the modified R value, which is the direct reading from LDS.

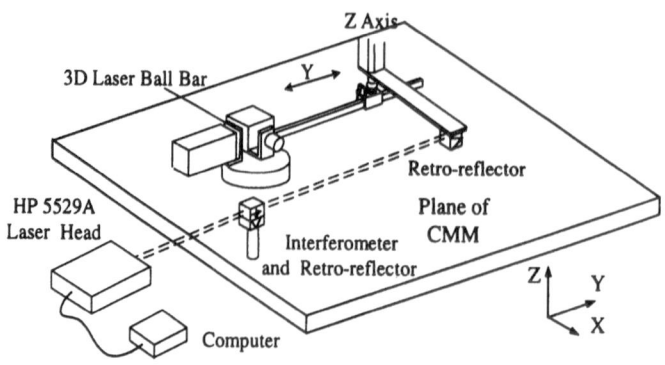

4. Set-up for R accuracy calibration.

3.2 Calibration of the φ axis

In this experiment, the rail-carriage assembly is carefully aligned to parallel with the X-Y plane and directed to the Y-axis of a CMM, as shown in Figure 5A. During the calibration, measuring arm moves in the X direction and positions are recorded by the HP 5528A, the φ angle changes in trigonometric relationship, as shown in Figure 5B, and the equation is expressed as

$$\varphi_i = \arccos\left(\frac{R_0^2 + R_i^2 - X_i^2}{2R_0 R_i}\right)$$

(3)

$$\Delta\varphi_i = \varphi_{ir} - \varphi_i$$

where φ_i is the nominal angle at the *ith* position, φ_{ir} is the actual readout from the vertical encoder. The calibrated φ_i errors can be expressed by Eq. (4). After correction, the φ errors can be maintained within around ± 1.5 arc-sec.

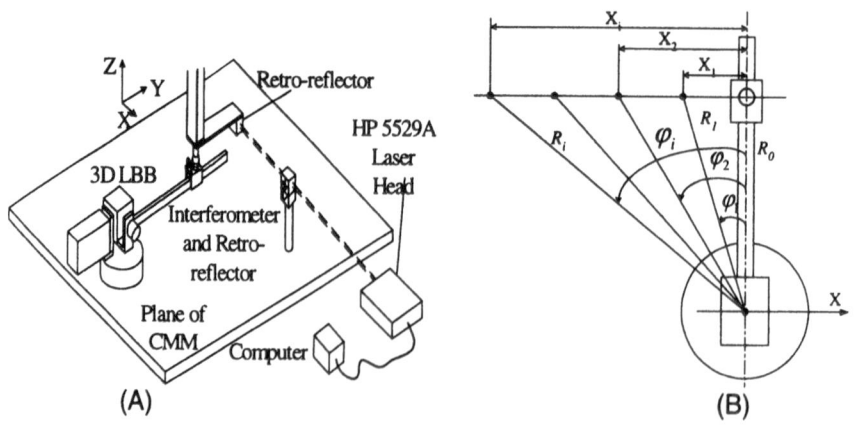

(A)   (B)

5. (A) Set-up for ϕ calibration; (B) principle of calibration

$$\varphi_{mod} = \varphi_r - \left(-1.17 * \varphi_r - 0.2\right) \qquad (4)$$

where φ_{mod} is defined as the modified value, and φ_r is the readout from the vertical encoder.

3.3 Calibration of the θ axis

The setup and the principle of calibration are similar to Figure 5 except the laser beam is bent to the spindle direction. The residual errors are about ±1.2 arc-sec.

4. Applications

4.1 Volumetric error measurement of a serial-parallel type machine tool

The machine tool is of a serial-parallel type consisting of a 3-dof parallel spindle platform with two angular orientations and one linear motion in Z-axis, and a conventional X-Y table, which carries the workpiece. The experimental setup is shown in Figure 6. Giving commands to change two angular orientations of the spindle platform and keep the tool tip spatial positions, the spatial position error could be measured by the 3D LBB. In this experiment, there are four different spatial paths to be measured and each path contains nine spindle poses of uniform spread. Experimental results of volumetric errors are plotted in Figure 7.

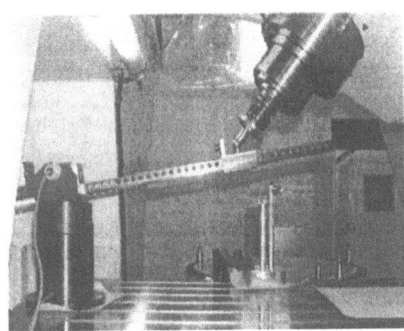
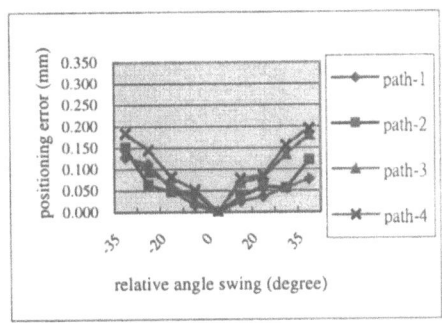

6. Set-up for accuracy check of   7. Parallel platform spatial positioning error
   a serial-parallel machine tool.

4.2 Calibration of a Cartesian type machine tool

Performance tests on a traditional vertical machining center, conforming to ISO-230 specifications, were carried out to measure the positioning errors and straightness errors of an axis using the 3D LBB. The experimental results are plotted in Figure 8. The volumetric error measurement can be directly obtained, as shown in Figure 9. Circular tests can also be easily obtained by this equipment with proper setup.

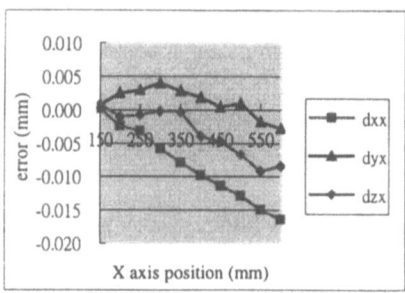

8. X axis positioning error

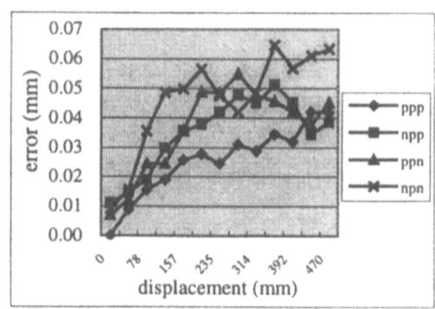

9. Machining center volumetric error

5. Conclusions

A versatile and novel 3DLBB system was developed for measuring the volumetric error of multi-axis machines. Its accuracies in $\delta_R = \pm 1.5 \mu m$, $\delta_\phi = \pm 1.6 arc \sec$ and $\delta_\theta = \pm 1.2 arc \sec$ were determined by identifying and compensating the possible error sources. In addition to the spatial position measurements, this instrument can also be used to check other kinds of errors in precision machine metrology, such as spindle thermal drift, robot-contouring accuracy etc.

Acknowledgements

The work reported forms part of a research program funded by the National Science Council of the Republic of China on the study of ultra-precision machine system.

References

[1]   ASME. B5. 54, 1993, Methods for Performance Evaluation of Computer Numerically Controlled Machining Centers. Version 1.0.

[2]   ISO 230-4, 1998, Acceptance Code for Machine Tools, Part 4 : Circular Measurements. International Standard.

[3]   Bryan, J. B., 1982, A Simple Method for Testing Measuring Machines and Machine Tools. *Precision Engineering*, Vol. 4, No2.

[4]   Burdekin, M., and W. Jywe, 1992, Optimising the Contouring Accuracy of CNC Machines using the Contisure System. *Proc. of the 29th Int. MATADOR Conf.*

[5]   Schmitz, T., and J. Ziegert, 2000, Dynamic Evaluation of Spatial CNC Contouring Accuracy. *Precision Engineering*, Vol. 24, pp. 99-118.

[6]   Ziegert, J.C., and C.D. Mize, 1994, The Laser Ball Bar: a New Instrument for Machine Tool Metrology. *Precision Engineering*, Vol. 16, No. 4, pp. 259-26

[7]   API Co., 2003, Laser Tracking System, http://www.apisensor.com.

Intelligent Tools: Novel Technology for In-process Measurement of Pressure Distribution on Optical Functional Surfaces

Dr. U. Klaeger[1], S. Gronwald[2] and Dr. C. Weber[3]

[1]Fraunhofer Institute for Factory Operation and Automation, Sandtorstrasse 22, 39106 Magdeburg, Germany, Uwe.Klaeger@iff.fraunhofer.de
[2]Fraunhofer Institute for Factory Operation and Automation, Sandtorstrasse 22, 39106 Magdeburg, Germany, Susan.Gronwald@iff.fraunhofer.de
[3]IGAM - Engineering Company for Applied Mechanics Ltd., Steinfeldstrasse 3, 39179 Barleben, Germany, weber@igam-mbh.de

Abstract: One of the most important prerequisites for providing high performance optical parts is creating high-precision surfaces with marginal variations of shape. Intelligent tools, which detect the smallest changes on tool and workpiece during the polishing process, represent a new generation of tools. The performance of current piezoceramic fibers and foils has reached a state that makes it possible to integrate adaptive structures in application-oriented solutions. However these are still linked with high costs especially in the early phrases of product development. By applying Rapid Prototyping processes that work generatively (in layers), in particular the advantageous features of Layer Laminate processes (LLM, LOM), intelligent tools can be manufactured not only quickly and accurate to dimension but also cost effectively and efficiently. This paper reports current research results using the example of the manufacture of ultraprecise functional surfaces on optical lenses and silicon wafers.

1. Introduction

At present, producing high-precision surfaces with small variations in shape (in part up to 50 nanometers) is a complicated process [1]. Both the temperature generated by friction and the dispersion of the polishing suspension lead to inaccuracies of similar magnitude. Thus optical lenses with high-precision requirements and wafers with low structural widths can be produced only with great effort.

2. Approach

An extremely promising solution for the problem described is the use of intelligent shape adaptive tools. The smallest changes on tool and workpiece are detected during the polishing process. Afterward the shape of the tool is adjusted by piezoelectric actuators in such a way that the surface is machined flawlessly. A deterministic polishing process for optical functional surfaces and semiconductor materials requires being able to regulate the control of material removal, which according to PRESTON can be specified for every point of the surface of the part as

$$\frac{dz}{dt} = K \cdot p \cdot v_r$$

p: pressure
v_r: relative speed
K: process constant

The approach using shape adaptive tools aims at controlling pressure distribution in the effective slit.

The dimensioning of shape adaptive tools requires modeling and simulation using FEM in order to analyze both the deformation behavior of the shape shell itself and the deformations caused by contact between optical functional surfaces and tools in the polishing process. Contact analysis in particular requires complex nonlinear mathematical models.

To arrive at a stable numerical solution for these contact problems, underlying analyses are required to guarantee the reliability of the numerical models for systematic simulation calculations (Figure 1).

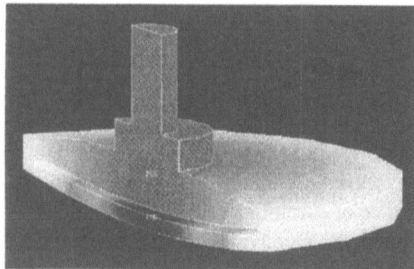 

Figure 1: FEM model and polishing tool derived from it

Knowledge of the pressure distribution in the polishing surface is extremely important in order to produce the greatest accuracy of shape. This can only be determined when workpiece (lens) and tool (polishing dish) are in direct contact. For this reason, it is technologically necessary to integrate the sensors directly in the tool, close to the area of contact.

3. Applications

3.1 Ultraprecision machining of wafers

The most important procedure for the production of ultraprecise surfaces in the semiconductor industry and microsystems technology is planarizing using chemical-mechanical polishing (CMP). Current technologies achieve accuracies in the range of 20-50 nm [2]. The edge of the wafer where the greatest deviations occur is especially critical.

As current structural widths of 90 nm further decrease to 60 nm, removal profile accuracies of 10 nm over the entire surface will be required in the future. Apart from enlarging a wafer from 200 mm (8") to 300 mm (12") and simultaneously better utilizing its surface, approximately 2.4 times as many chips can be accommodated on an 8 inch wafer.

Therefore, the tools with which influence can be systematically exerted on the removal profile in CMP are extremely interesting for high polishing process quality. Adaptive tools, with surfaces determinately adjustable by appropriate actuators when process parameters such as polishing pressure, temperature or slurry used have little influence, have proven to be an outstanding solution.

Figure 2: Carrier with electronic module

With actuator travel of 25µm, the piezostacks employed in the carrier (Figure 2) allow a resolution of approximately 1 nm (open loop). When a positioning control (closed loop) is employed, resolutions are achieved in the range of 0.3 nm. Although only three actuators each touch per ring, the results of calculations show a very uniform progression of deformation and pressure in the wafer's circumferential direction. Polishing tests were conducted to characterize the removal behavior on a 6'' variant of the system.

The pressure distribution of the actuated carriers measured with pressure measuring film between polishing pad and wafer before polishing is identical with the removal profile obtained (Figure 3). On the one hand the results confirm the Preston hypothesis and on the other hand they demonstrate that the pressure between the wafer and polishing cloth is suited in principle as a control variable to regulate the polishing process.

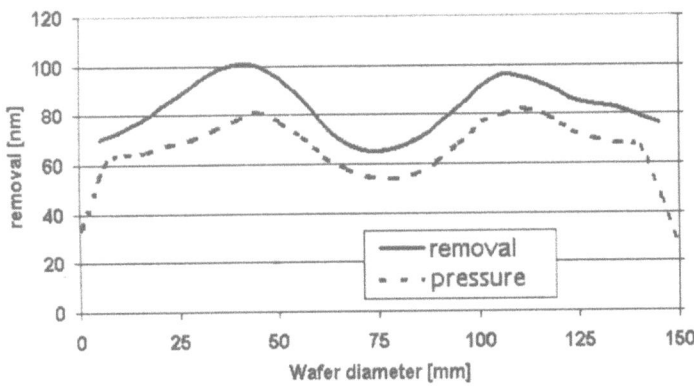

Figure 3: Measuring the removal and pressure profile

Although only three piezoactuators touch on every ring, the results show a very uniform, rotationally symmetrical distribution of the pressure and removal profile along the wafer's circumference (Figure 4).

By systematically controlling the rings, the uniformity of removal (within wafer non-uniformity - WIWNU) could be improved by 50% in a corrective step.

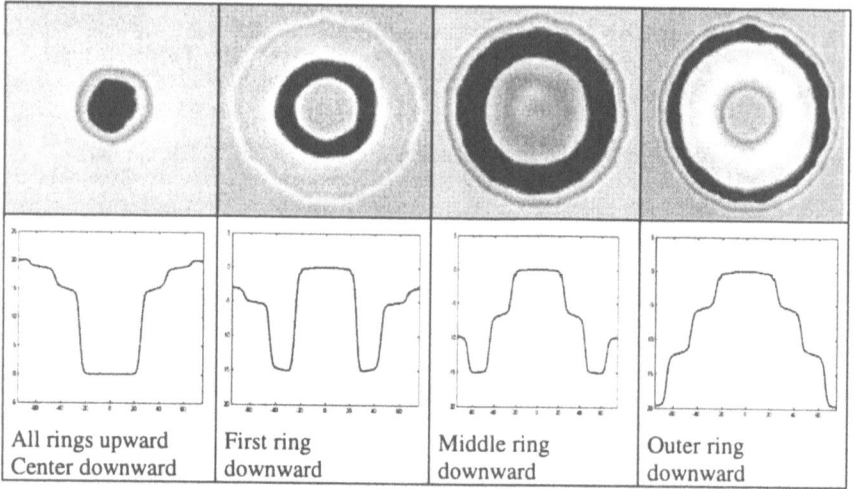

| All rings upward Center downward | First ring downward | Middle ring downward | Outer ring downward |

Figure 4: Measured pressure distribution and linescans

3.2 Ultraprecision machining for optical lenses

In view of the excellent experiences with the adaptive tools, tests are being conducted at present with the objective of improving the polishing process by using in-process measuring of pressure distribution on optical functional surfaces (lenses, mirrors). For want of alternatives for information transmission, current approaches are frequently based on radio solutions.

The insufficient interference resistance caused by obstructions and reflections in manufacturing facilities turn out to be disadvantageous.

Optical information transmission is also not recommendable because of the sensitivity to contamination (polishing suspension).

In the production process, functional surfaces are presently produced by grinding with diamond tools (macrogeometry). This is followed by CNC polishing with large-area forming tools, which achieves a transparent surface with an accuracy of approximately 1 μm. A corrective CNC process with small-area active zones (computerized polishing correction, ion beam machining, magneto-rheological polishing) is employed to achieve the final precision necessary. The initial situation is a massive tool with a fixed radius (Figure 5).

CNC polishing in the Synchrospeed method is based on transmitting the shape of the polishing dish to the lens. The achievable accuracy of shape of the polishing dish depends on the diameter of the lens to be machined.

Initial tests demonstrate that here too the CNC polishing can be improved substantially by adjusting the variable surface geometries using an adaptive tool. The conventional polishing process can be shortened considerably as a result.

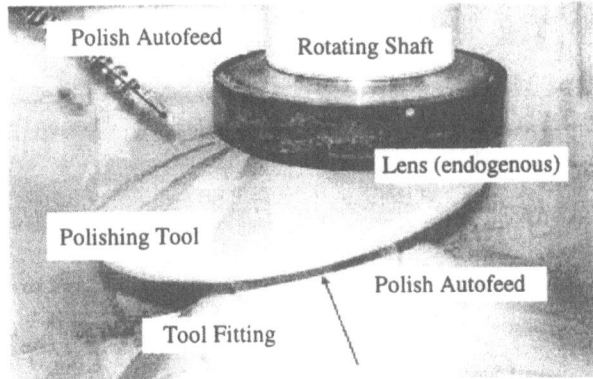

Figure 5: CNC polishing process for optical lenses

4. Generatively Produced Tools for Prototypes

Especially in the early phases of product development, the sizeable development risk frequently makes calculating costs not at all or barely possible. In order to minimize such risks, the authors developed a new concept for the applications described, which uses Rapid Prototyping processes that work generatively (in layers), to integrate sensory components.

Above all the advantageous features of layer laminate processes (LLM, LOM) promise fast and dimensionally accurate manufacturing of intelligent tools. A composite of various foil materials (metal, piezoceramic, paper) was used to develop a tool in shell construction, the outstanding features of which are its great rigidity and stable in-process acquisition of measuring data (Figure 6).

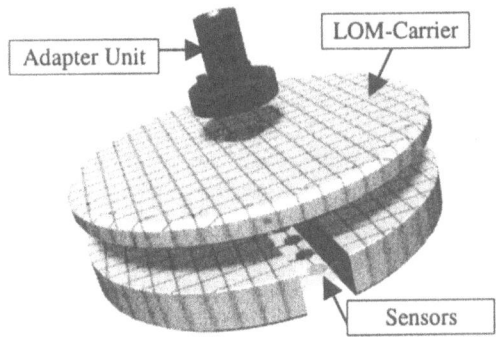

Figure 6: Prototype of the shape adaptive polishing tool produced generatively

At the same time a model of the carrier was constructed in the concept finding phase (Figure 7).

This makes it possible, already at the start of testing, to configure structural elements later integrated in mass produced tools. On the one hand defective designs can be spotted relatively early. On the other hand inexpensive models and prototype tools can be made available at an early stage.

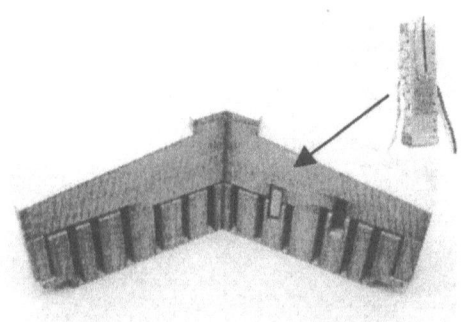

Figure 7: LOM prototype of the carrier with integrated actuator

5. Outlook

Using this new approach, the authors have managed to considerably raise the allowable limit values for the speed and acceleration of moving systems. Foregoing mechanical contact not only improved process stability but above all the security and dependability of energy transmission too. Downtimes and maintenance effort are reduced. At present, tests are being run to improve the complete system in the polishing process. The basic research activities can be outlined as follows:

- Continuous monitoring of the workpiece mold (pre-process);
- Monitoring of the tool mold (state of wear);
- Monitoring of the state of truing (pre-process);
- Infiltration of the base material to increase the strength of the part;
- Adaptation of the complete system (redesign of the feeding/ bonding system);
- Tool making/tool optimization.

References

[1] Herold, V., 2000, *Technical Process for Making Tools for Machining Surfaces.* Patent No. DE 198.34.559 A1, München.

[2] Oliver, M. R. (Ed.), 2004, *Chemical-Mechanical Planarization of Semiconductor Materials, Vol. 69, Springer Series in Materials Science.* Springer-Verlag Berlin.

A Simulated Investigation on Surface Texture/Topography Generation in Precision Turning Processes

Xichun Luo[1], Kai Cheng[2], Robert Ward[3] and Richard Holt[4]

School of Technology, Leeds Metropolitan University, Calverley Street, Leeds LS1 3HE, UK   [1]x.luo@leedsmet.ac.uk,   [2]k.cheng@leedsmet.ac.uk,   [3]r.ward@leedsmet.ac.uk, [4]r.holt@leedsmet.ac.uk

Abstract: The paper presents a modelling and simulation approach for investigating the generation of precision surfaces in relation to the affecting factors from the machine tool, cutting tool, workpiece materials and operation conditions in precision manufacturing qualitatively and quantitatively. The modelling approach integrates dynamic cutting force model, thermal model, vibration model, chatter model, tool wear model, machining system response model and surface topography model as a whole. Some nonlinearities, such as spindle motion error, generation of built-up-edge, etc. in the machining process are modelled in the light of their physical features. The cutting force and surface texture parameters are the outputs of the modelling approach. The whole modelling and simulation is implemented by MATLAB & Simulink programming. Precision turning trials are conducted to validate the modelling and simulations developed. The simulation results are quite consistent with experimental results to certain extent. They all show that surface texture/topography generation is highly affected by feed rate and some nonlinear factors in the precision diamond turning process.

1. Introduction

Precision manufacturing concerns the creation of products/components with high form and dimensional accuracy, and surface quality, which will affect the products/components' specified functionality. The achievable quality of the precision machined surfaces is shown to be affected by four main issues, i.e. the machining process, machine tool performance, tooling geometry including the cutting edge quality and the workpiece material. Therefore, a more scientific approach is needed for building up a theoretical basis for bridging the gap between the surface machined and the determining factors from the four main issues above, and to further explore that basis of which the machining of workpiece surfaces with respect to the desired surface integrity and its intended functional performance.

There have been tremendous endeavours at predicting cutting force and surface roughness by using numerical methods, such as Finite Element analysis [1], and empirical methods such as Response Surface Methodology [2] and Neural Network approach [3]. However, the generation of a machined surface is a dynamic process in which the machining dynamics and structural dynamics should be fully taken into account so as to ensure that modelling and simulation is precise and realistic. There are only few researchers studying the effects of machining process variables and tooling characteristic in the dynamic environment [4] [5].

The work presented in this paper attempts an analytical scientific approach for investigating the generation of high precision surfaces in relation to the dynamic factors from the machine tools, tooling, workpiece material and operation conditions and therefore to control the surface texture/topography generation as desired.

2. Modelling Approach and Simulation Implementation

2.1 Inputs modelling

The inputs to the modelling approach are listed in Table 1. They can be classified as linear inputs and nonlinear inputs. The linear inputs are constants or linear functions. The nonlinear inputs can be modelled in the light of their physical features. For example the usage of coolant will change the friction coefficient or friction angle between the tool rake face and chip, so the coolant can be modelled based on the variation of friction angle. The details of the nonlinear modelling can be found in the authors' other paper [5].

Table 1: The inputs to the modeling approach.

Source	Linear factors	Nonlinear factors
Machine tool	Modal parameters of machining loop Straightness of the slideway	Spindle rotational run-out Spindle axial run-out
Cutting tool	Rake angle, Clearance angle Tool nose and cutting edge radius	Tool wear
Operation condition	Depth of cut, Spindle speed Feed rate	Coolant Environmental vibrations
Material property	Initial shear stress Initial shear angle	Hard spots, Chip formation The variation of shear angle

2.2 Cutting mechanics and dynamics modelling

The cutting mechanics and dynamic modelling includes the modelling of dynamic cutting forces, vibrations, chatter, temperature, tool wear and machining system response. The dynamic cutting force model is deduced based on the elastoplastic mechanics. The random vibration, due to the environmental vibrations, hard spots in workpiece material, generation and removal of built-up edge (BUE) are modelled in the light of their physical features. The regenerative vibrations and shear localized chip formation are modelled based on their impacts on depth of cut and shear stress respectively. The temperature and tool flank wear are model based on empirical formulations [5].

The cutting system can be described using second-order spring-damper vibratory functions in the X, Y and Z directions, which is then transformed into frequency domain by Laplace transform to form the so-called machining system surface response model. The responses will be the dynamic displacements of the cutting tool and workpiece at the cutting point. The detail of the cutting mechanics and dynamics modelling can be found in the authors' papers in reference [6].

2.3 Surface texture/topography generation

Superposition of the dynamic displacements with the ideal tool positions will be the real tool positions onto the workpiece. Surface topography model is used to calculate the intersection points of the tool paths according to the real tool paths and cutting tool profile. The machined surface topography can be constructed by trimming the line above the intersection [6]. Currently the outputs of the modelling approach include cutting forces, dimensional error and surface texture/topography, bearing and lubricant functionality parameters, which are listed in Table 2. There will be more functionality parameters included in the outputs of the modelling approach in the future.

Table 2: Outputs of the modelling approach.

Surface topography	Surface texture	Surface functionality
S_q - RMS deviation S_{sk} - skeness of height distribution S_{ds} - density of summits	S_{tr} - surface texture aspect ratio S_{td} - texture direction	S_{bi} - surface bearing index S_{ci} - core fluid retention index

2.4 Implementation of the modelling approach

Figure 1 illustrates the implementation of the model approach in MATLAB Simulink environment. It is comprised of inputs module, machining mechanics and dynamics module and output module. The Constant blocks are some linear inputs. The nonlinear function blocks and small sub-system blocks stand for some nonlinear inputs. Those nonlinear inputs are connected with the machining mechanics and dynamics module by some Manual Switch blocks. So it is very convenient to add or remove nonlinear factors during the simulation process. The long block in the middle of Figure1 is the cutting force model. The vibration model, chatter model, thermal model and tool wear model are the big sub-system blocks connected with force model. The details of implementation of these models are all hidden in the respective subsystems within the whole model hierarchy. Machining system respond module is constructed using Transfer Function blocks in Simulink. The output consists of dynamic displacement of the workpiece and the tooling system and material removal rate module. The real tool path and 3D surfaces are visualized by MATLAB programming. The surface characterization parameters are calculated by MATLAB programming.

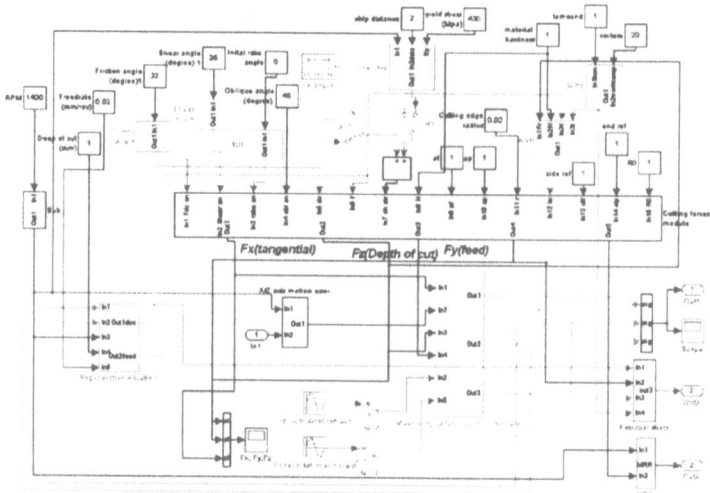

Figure 1 Implementation of the modelling approach.

3. Simulation and machining verification

Nearly 90 simulations on face turning and cylindrical turning of Al alloy and low carbon steel were run to study the effects of linear and nonlinear factors on surface texture/topography generation. A total number of 83 turning trials have been carried out on a lathe using carbide tool insert and high speed steel cutting tool to evaluate and validate the simulation results. The cutting forces and machined surfaces are measured by a Kistler dynamometer and Zygo optical surface profiler respectively. In most cases the simulated cutting forces are well agreed with the measured results (about 29.8% to 50.3% lower than the measured results). The difference between the simulated surface roughness and measured surface roughness are much high (around 40.7% to 86%) It may be caused by the estimated static machining structural parameters, but the simulation results are still in the reasonable deviation scale. The details of machined verification can be found in reference [6].

4. Discussions

4.1 Machining dynamics

In a face turning process (Spindle speed = 1400 rpm, depth of cut = 0.3969 mm, feed rate = 0.0265 mm/rev), the average cutting forces in the three directions are 24.83 N, 41.48 N and 43.24 N respectively. The main frequencies of cutting forces in the tangential and radial directions are 793 Hz and 235 Hz. They are 44 and 10 times of the spindle rotational frequency (1400 rpm / 60 = 23.3 Hz). Figure 2 illustrates the dynamic displacements between cutting tool and workpiece. The average amplitudes of vibration between cutting tool and workpiece in the tangential and feed directions are 0.18 μm, 0.52 μm and 0.3 μm respectively. The dominant power spectrum peaks of the vibrations in the radial direction are found to be at low frequencies in Figure 3. They are 24.2 Hz (f_1), 30.4 HZ (f_2), 19.8 HZ (f_3) and 2.3 Hz (f_4) respectively. f_1 is very close to the frequency of spindle rotation. It is also the frequency of regenerative vibration and rational spindle run-out. f_2 is close to the frequency of spindle axial run-out. f_3 is nearly the frequency of generation and

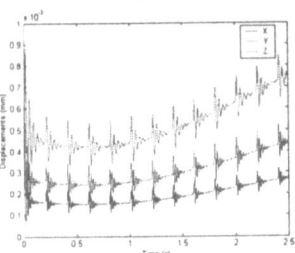 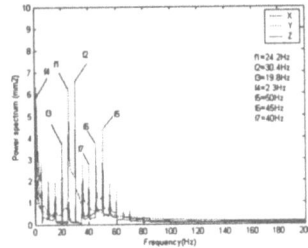

Figure 2 Simulated machining dynamics. **Figure 3** FFT of the machining dynamics.

Remarkable modes of vibrations in the tangential direction are found to be at frequencies f_5, f_6 and f_7 of 50, 45, 40 Hz respectively. The first mode of vibration at frequency f_5 is close to the frequency of hard spots in the workpiece materials. f_6 is shear oscillation due to the coupling of the cutting forces. f_7 may be the frequency of formation of shear-localized chip. The spectral analysis of vibrations between cutting tool and workpiece suggest that the machining system or process has the damping function to remove the effects of high frequency vibrations. The vibrations between cutting tool and workpiece are mainly cause by the spindle run-out, regenerative chatter, environmental vibration and generation and removal of BUE.

4.2 Machined surface spatial analysis

In Figure 4, the peak to valley height by simulation in a cylindrical turning process is about 4.2 μm, which is very close to the measured results of 5.176 μm. Both the simulated and measured power spectrum plots show the significant power spectral appear at 6.31 cycle/mm and 6.3 cycle/mm, which is just the spatial frequency of feed rate ($1/f$=1/0.1588=6.29 cycle/mm). In the simulated power spectral density plot, the spatial frequency of regenerative vibration and spindle run-out, which is around 19.7 cycle/mm can also be observed. In the measured power spectrum plot, some spatial frequency of 0.8, 1.3, 12.5 and 18.7 cycle/mm are also very obvious. They are the spatial frequencies of environmental vibration, BUE and harmonic of the spatial frequency of feed rate. This suggests that feed rate and spindle run-out and regenerative vibration has significant effects on machined surface topography.

300

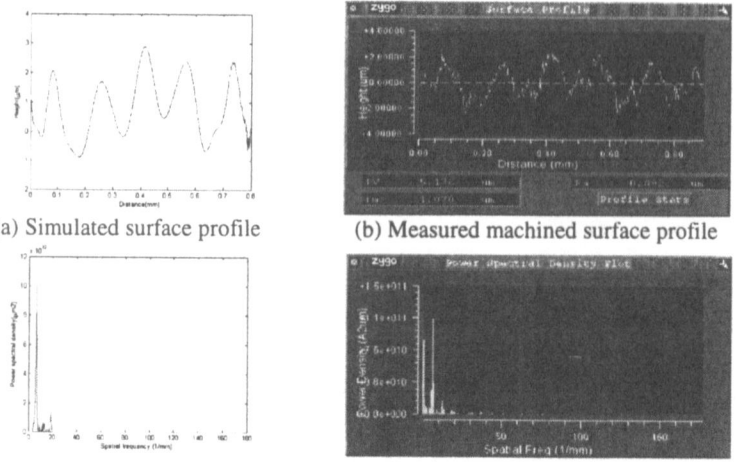

(a) Simulated surface profile   (b) Measured machined surface profile

(c) Power spectral density by simulation (d) Power spectral density by measurement

Figure 4 Surface profile and power spectral density by simulation and measurement.

(Spindle speed = 1400 rpm, depth of cut = 0.5 mm, feed rate = 0.1588 mm/rev)

5. Conclusions

An integrated model has been proposed to simulate the surface texture/topography generation in precision turning processes. The machined trials show the modelling and simulation can accurately present the dynamic cutting process and predict cutting forces and surface texture/topography to a certain extent. It is found that the machined surface topography is affected by low frequency factors. Feed rate plays the most significant role on the machined surfaces topography. The surface topography is also highly affected by nonlinear factors such as spindle run-out and regenerative vibration, BUE and environmental vibration.

References

[1] Moriwaki T., et al., 1993, Combined stress, material flow and analysis of orthogonal micromachining of Copper", Annals of the CIRP, 42(1), pp. 75-78.
[2] Baras J. S., et al., 1996, Designing response surface model-based run-by-run controllers: a worst case approach, IEEE Transactions on Components, Packaging & Manufacturing Technology, Part C Manufacturing, 19(2), pp. 98-104.
[3] Iain R. K., et al., 2000, Optimum selection of machining conditions in abrasive flow machining using neural network, Journal of Materials Processing Technology, 108 (1), pp. 62-67.
[4] Altintas Y, 2000, Manufacturing Automation: Metal Cutting Mechanics, Machine Tool Vibrations, and CNC Design, Cambridge University Press, Cambridge.
[5] Luo X., et al. 2003, Nonlinear effects in precision machining engineering materials, Proceedings of 18th ASPE annual meeting, pp. 489-493.
[6] Luo X., The effects of machining process variables and tooling characterisation on the surface generation: modelling, simulation and application promise, IJAMT (in press).

Taking Digitised Points Quality into Account in Geometrical Specification Measurement by Laser Sensor

Jean-François Fontaine [1], Jean-Pierre Gonnet[2], David Joannic[3],

Laboratoire de Recherche en Mécanique et Acoustique, Université de Bourgogne, site d'Auxerre, Route des Plaines de l'Yonne, 89 000 Auxerre, France. [1]fontaine@iut-dijon.u-bourgogne.fr

Abstract: Today, non-contact measurement techniques like laser sensors are used more and more in industrial applications. They allow to obtain many points in a short time. However these techniques have a less good accuracy than the classical processes using contact probe for metrology applications. The work presented in this paper, is part of a research project on the in-process product inspection. In the case of in-process products inspection, the throughput time must be the shortest with a good dimensional accuracy. So, it is necessary to increase the measurement accuracy of the points cloud for using the laser scanners. For this first approach, we suggest to build a measurement procedure based on the digitised points quality: the points having the required quality alone are kept for the numerical calculation. To do that, a quality cartography giving the measurement accuracy is developed. This one is based on the sensor dispersion modelling .We show, by a concrete example, how it is possible to increase the accuracy of the inspection results in introducing the local measurement uncertainty.

1. Introduction

Today, the non-contact measurement techniques are very used in industrial applications like reverse engineering. In this case, the accuracy is not a determining factor because generally the measured points are smoothed to obtain a CAD model. Theses techniques allow to measure skew surfaces. They are competitive in regard of the classical techniques because the rate of points acquisition is greater. However, the measurement accuracy is often in the same order than the tolerance range of the specification. So, it is necessary to control the measurement uncertainties for assuring the results. So, in the objective to build 3D-inspection device, we suggest to control the measurement uncertainties for assessing the results. This paper presents the first issues.

2. Non-contact Measurement Uncertainties

The non-contact measurement techniques are generally based on laser scanning or on structured light projection. We describe the methodology used with the first kind of device: auto-synchronised laser scanner Hyscan 45C. This working principle is the following: one face of turning mirror ensures a line laser beam projection on the measured object surface. The second face permits the reception of the reflected ray by a CCD sensor (Fig. 1). Then, the measured point co-ordinates can be calculated by the triangulation principle[1] . In order to obtain the sensor motion during the scanning, the dice is fixed on a CMM (Fig. 2).

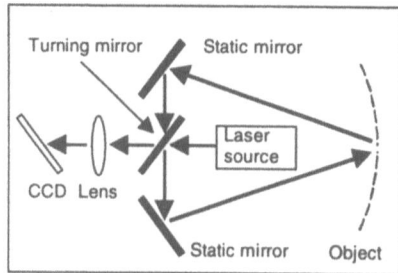

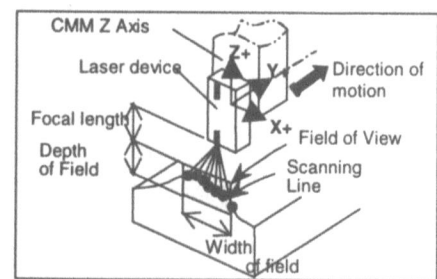

Fig. 1: Auto synchronised triangulation   Fig. 2. Measurement device on CMM

2.1 Intrinsic scanner uncertainties

The uncertainties come from several parameters and are depending of the used technology. In our case, we have for example the flatness defect of the laser beam, the defect of the CCD sensor, the bad focusing of the laser,... The intrinsic calibration step gives a correction for some parameters (i.e.: position and orientation between the CCD and the lens, ...). Some others are assessable with difficulty. One way consists to build a local uncertainty modelling in taking into account the influence of geometrical measuring parameters. This modelling can be compared to a global uncertainty modelling that consists to measure different parts whose geometry is assumed known.[2], [3].

2.2 Measurement uncertainties

These uncertainties depend on the measurement process parameters:
- alignment of scanner reference frame with CMM reference frame
- calculation of results: smoothing data, data treatment algorithm,...
- measured part properties: form defect, roughness, material reflectivity,...

Finally, due to the difficulty to build a complete uncertainty model, we limit this study to some influent parameters. We assume that systematic error can be corrected , then the uncertainties can be evaluated by repeatability measurement and statistical analysis.

2.3 Used uncertainty modelling

2.3.1 Measurement parameters uncertainty (u_{Mes})

We use the model obtained by Prieto [4] with an identical scanner (Hyscan3D). The measure of the "same point" has been repeated 128 times according to the distance between the scanner and the surface part (d) , the incidence angle in the laser plane

(α) and the incidence angle perpendicular to the laser plane (β) (Fig. 3). This model is local, it depends of the co-ordinates point. The variation range of the parameters are the following: $d \in [170mm, 230mm]$, $\alpha \in [-35°, +35°]$ and $\beta \in [-15°, +15°]$. The uncertainty model can be given from the variance of every parameter in the CMM referential frame (O,X,Y,Z):

$$u^2(d,\alpha,\beta) = \begin{pmatrix} u^2_{xx}(d) & u^2_{xz}(d) \\ u^2_{zx}(d) & u^2_{zz}(d) \end{pmatrix} + \begin{pmatrix} u^2_{xx}(\alpha) & u^2_{xz}(\alpha) \\ u^2_{zx}(\alpha) & u^2_{zz}(\alpha) \end{pmatrix} + \begin{pmatrix} u^2_{xx}(\beta) & u^2_{xz}(\beta) \\ u^2_{zx}(\beta) & u^2_{zz}(\beta) \end{pmatrix}$$

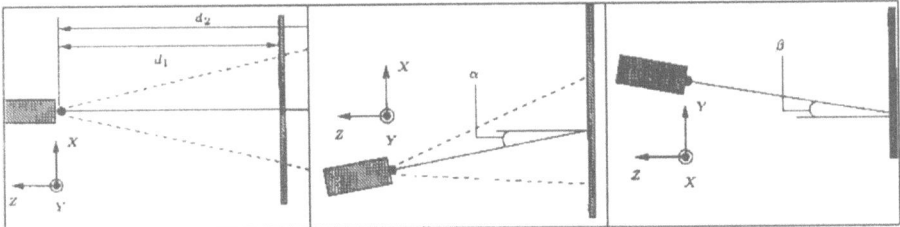

Fig. 3.a. Distance (d)   Fig. 3.b:Plane incidence (α) Fig. 3.c: Ortho. incidence (β)

The model gives the uncertainties according to d, α and β $u^x_{mes} = \sqrt{\frac{1}{2}(u^2_{xx} + u^2_{xz})}$ and $u^z_{mes} = \sqrt{\frac{1}{2}(u^2_{zx} + u^2_{zz})}$ respectively on the x and z directions . Figures 4.a and 4.b give the calculated value of u^z_{mes} according to d and α for $\beta=0°$ and $\beta=15°$. It is symmetric when α et β are negative

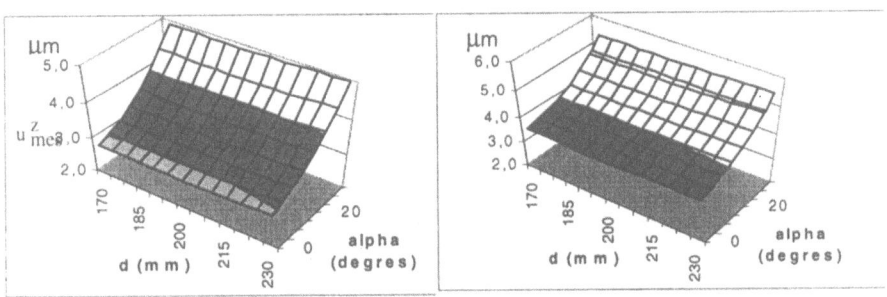

Fig 4.1: Uncertainty in Z-direction for $\beta=0$ Fig 4.2: Uncertainty in Z-direction for $\beta=15°$

2.3.2 Scanner- MMT Alignment uncertainty ($u_{Réf}$)

The alignment uncertainty has been established in repeating 30 times the alignment procedure after orientation changing. This procedure consists to determine the roll, pitch and yaw angles of the transformation between both reference frames and to scan a reference sphere. The uncertainty of the angles are less than 10^{-6} radian, so we will neglect it in the model. The position uncertainty evaluated by the sphere is equal to 4,5 µm, 5,4 µm and 4,3 µm respectively in the three directions X, Y and Z

2.3.3. MMT motion uncertainty (u_{MMT}).

The uncertainty of the motion has been estimated for an identical MMT to 0,8µm, 1,2 µm et 0,3 µm on the axis X,Y,Z [5]. We assume that it is negligible because it takes into account the contact sensor performance.

2.3.4 Surface finish quality uncertainty (u_{Mat}).

Our example is based on the measurement of a part in aluminium alloy. The material has bad properties of reflectivity. That imposes to apply a spray, so the generated uncertainty is close to 6 µm [6].

2.3.5 Global model of uncertainty

We assume that all these uncertainties are independent and are governed by a gaussian law. So we can applied the additive property . The composed uncertainty is the square root of the variances sum: $u_c = \sqrt{u_{Mes}^2 + u_{Réf}^2 + u_{MMT}^2 + u_{Mat}^2}$.

To simplify the modelling, we apply a majorant of the uncertainty components. Table 1 gives the uncertainty cartography according to the measurement parameters d and α

Table 1: Uncertainty maxi: (*) with scanner realignement, (**) with a fixed orientation

u_c (*) µm	0	10	20	35	u_c (**)µm	0	10	20	35
180	8,3	8,4	8,6	9,2	180	6,7	6,8	7,0	7,7
200	8,3	8,4	8,6	9,2	200	6,6	6,8	7,0	7,7
220	8,3	8,4	8,7	9,3	220	6,7	6,8	7,1	7,8

3. Proposed Process for Inspection

To warrant the measurement results, we propose to follow the inspection process based on the uncertainties control in respecting different steps:
1°) Determination of the measurement uncertainties admissible for all points (P_{ij})of the surface S_j in respect to the controlled geometrical specification
2°) Determination of parameters (d, α, β) for each point Pij such as $u_c(P_{ij}) < u_c (S_{j,})$
3°) Research of the optimal scanner trajectory in respect of previous conditions
4°) Measure of all points for a given scanner orientation
5°) Association of points cloud with nominal surfaces
6°) specifications Control in giving a confidence interval.

4. Example

We are considering the part represented on figure 5:

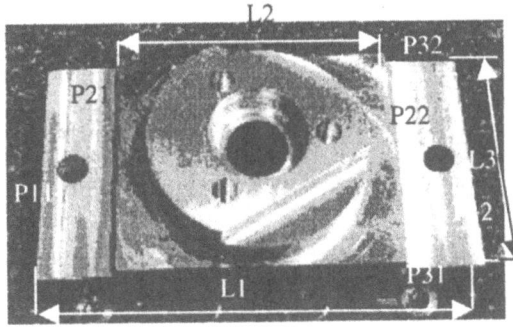

The considered specifications are the following lengths :

L1=150 mm +/- 0,2
(between the planes P11 et P12)

L2= 105 mm +/- 0,05
(between the planes P21 et P22)

L3= 75 mm +/- 0,1
(between the planes P31 et P32)

Fig 5 : Tested part

1°) The condition on the length measurement uncertainties between P_{ij} and P_{il} is given by the standard (i.e: NF 02-204):

$$\frac{u_c^i}{IT_{jl}} < \frac{1}{8k} \qquad (1)$$

where k is the confidence ratio of gaussian law (i.e.: k=2 confidence = 95,4%) and it is the tolerance interval of the length (L_i) between the planes P_{ij} and P_{il}.
The measurement uncertainty of the specification L_i is depending of the inspection method: to determine the length L_i we associate to the measured points M_{ij} of the real reference plane P_{ij}^{th} a plane P_{ij}^m with the least squares method , to calculate the set of the distances between the measured points M_{il} and the plane P_{ij}^m . The length u_c^i uncertainty is obtained by propagation of the plane P_{ij}^m uncertainty and the point M_{il} uncertainty.
2°) the uncertainty maxi for the specification is calculated from (1) (Table 2)

Table2: Measurement uncertainties and admissible measurement positions

specification	tolerance Interval	Admissible uncertainty	inspected planes	Measurement configuration			
.				d_{mini}	d_{maxi}	α_{maxi}	β_{maxi}
Length L1	0,4 mm	25 μm	P11 et P12	170	230	35	15
Length L2	0,1 mm	6,25μm	P21 et P22	non- controllable for k=2			
Length L3	0,2 mm	12,5 μm	P31 et P32	170	230	25	10

3°) The trajectories are defined in respect of uncertainties (see Fig. 6)
5°)- 6°) The specified lengths are calculated after determination of the planes P_{11}^m , P_{12}^m and P_{13}^m . The acceptation is made according to the confidence ratio (Table 3).

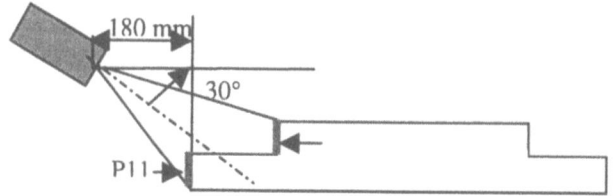

Fig: 6: Example of an admissible position to measure the planes P11 and P21

Table 3: Results of inspection with confidence level

	measured length maxi (mm)	measured length mini (mm)	Reduced specification IT red=$IT - 4ku_c$	Conformity	confidence level
L1	149.883	149.968	150 mm +/- 0,176	Yes	95%
L2	104.980	105.047	105 mm +/- 0,026	No	95%
L3	74.909	75.087	75 mm +/- 0,076	No	95%

5. Conclusion

We have introduced a 3D-inspection procedure which takes into account the measurement uncertainties. In the case of non-contact 3D-measurement, this approach is justified because the uncertainties are close to the tolerance of specifications. It is possible to build an uncertainty model based on statistical confidence. However, for using non-contact sensor in 3D inspection, it will be necessary to determine the systematic errors to have more realistic model.

References

[1] Rioux A., 1984, *Laser range finder based on synchronised scanner*, Applied Optics, November 1984, vol. 23, N21, p. 3837-3844
[2] Che C., Ni J., 2000, *A ball-target-based extrinsic calibration technique for high-accuracy 3D metrology using laser stripe sensors*, Precision Engineering, V 24, pp 210-219
[3] El-Hakim S.F.,Beraldin J.A., *Configuration design for sensor integration*, SPIE proceedings, vol 2598, Videometrics IV, Philadelphia, 1995, p. 224-285
[4] Prieto F., *Métrologie assistée par ordinateur: apport des capteurs 3D sans contact*, thèse de doctorat, INSA de Lyon, décembre 1999.
[5] Linares J.M., Bourdet P., Sprauel J.M., 2002, *Quality measurement on CMM* Integrated design and manufacturing in mechanical eng., Ed. Chedmail et all, pp. 219-226
[6] Contri A., *Qualité géométrique de la mesure de surfaces complexes par moyens optiques*, Thèse de doctorat, ENS de Cachan, 28 novembre 2002

Simulation of Precision Grinding Process for Predicting Surface Roughness

T. A. Nguyen[1], and D. L. Butler[2]

[1]Nanyang Technological University, 50 Nanyang Avenue, Singapore 639798,
nta@pmail.ntu.edu.sg
[2]Nanyang Technological University, 50 Nanyang Avenue, Singapore 639798,
mdlbutler@ntu.edu.sg,

Abstract: A numerical approach is proposed for the simulation of a precision grinding process. The approach takes into consideration the non-Gaussian distribution of grain protrusion heights by simulating the grinding wheel topography as a random field. Furthermore, an algorithm is proposed to identify the active grains on the simulated grinding wheel topography. The cutting, ploughing or rubbing of the grains is determined by estimating the attack angle of the active grains. Then, the workpiece topography is generated by mapping the active grain topography into the workpiece surface, using the kinematics relationship between the grinding wheel and the workpiece. The workpiece surface roughness, predicted by the simulation is compared with the experimental result to justify the proposed approach.

1. Introduction

Simulation of the grinding process has been the subject of intensive research for the past twenty years, yielding a vast amount of valuable information. There exist various types of simulation based on either the theory of abrasive wear, or the empirical models of the process. The focus of this paper, however, is the kinematic simulation of the grinding process, in which the kinematic relationship between the grinding wheel and the workpiece is used to generate the workpiece surface [1-8]. For the realistic simulation of the grinding process, any kinematic simulation scheme has to consider two issues: the generation of the grinding wheel and the interaction between the workpiece surface and the abrasive grains. With regard to the generation of the grinding wheel surface, one popular approach adopted by most researchers is meshing the grinding wheel with simple shape abrasive grains such as spheres and cones [1-5]. In reality, the shapes of abrasive grains are much more complex. An alternative to the above-mentioned approach is to use the measured wheel topography [6-8]. However, a realistic simulation of the grinding process requires a vast amount of wheel measurement, thus, hindering the broad use of the approach.

The other issue to be addressed in the simulation of the grinding process is the interaction of the abrasive grains with the workpiece surface. It is well known that grinding involves cutting, ploughing and rubbing of the abrasive grains at the workpiece surface. However, until now most of the suggested schemes only considered the cutting action of the grains [1-8].

In this paper, these two above-mentioned issues are addressed. The three dimensional topography of the grinding wheel is treated as a random field. As the distribution of surface heights of the grinding wheel is frequently non-Gaussian, a methodology is proposed for generating a non-Gaussian random field. Then the 3D topographical data of the grinding wheel is mapped to the 3D surface texture of the workpiece. An algorithm is proposed to identify the active abrasive grains and their attack angles, based on which the interaction modes of the abrasive grains and the workpiece are determined.

2. Generation of the Grinding Wheel Topography

From the mathematical point of view, the topography of the grinding wheel can be considered as a two-dimensional spatial random field $X(r)$, defined as a family of random variables $\{x\}$ at points with coordinates $\{r\}=\{(r_1,r_2,...,r_n)\}$ in a nth-dimensional "parameter space". Thus, the methods of digitally generating sample functions of random fields would be applicable for the grinding wheel topography. Unfortunately, most of the methods are only capable of simulating Gaussian fields [9] while grinding wheel surfaces often show non-Gaussian characteristics.

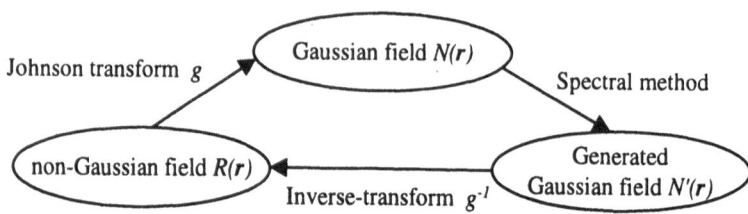

Figure 1: A methodology for simulating non-Gaussian grinding wheel topography.

For effectively simulating a large amount of the grinding wheel topography used in the simulation, a methodology, which is the combination of the spectral representation and the Johnson transformation, is proposed (Fig. 1). The spectral representation method is an effective algorithm, but limited for generating non-Gaussian random fields [10]. To overcome its limitation, it is suggested that a non-Gaussian random field $R(r)$, having the covariance function K_R is transformed to Gaussian one $N(r)$ having the covariance function K_N. Then the spectral representation method can be applied to generate a sample field $N'(r)$ with the same covariance K_N. The generated field then is inverse- transformed to a non-Gaussian field $R'(r)$. It can be shown that the generated $R'(r)$ and sampled $R(r)$ random fields will have the same marginal distribution function and covariance structure if the transformation g and its inverse g^{-1} are monotonic, and their Jacobians are non-zero, which is satisfied by Johnson transformation [11].

3. Interaction of the Abrasive Grains with the Workpiece Surface

A typical surface of the grinding wheel contains a large number of abrasive grains. However, only a small fraction of these abrasive grains will probably contact the workpiece surface. These active abrasive grains can be identified from the topographical array by slicing the topography at the specified depth of cut (Fig. 2a). The clusters of the cutting points remaining on the grinding wheel topography can be classified as the active abrasive grains. Strictly speaking, these abrasive grains are only static active, as in the grinding process they may or may not contact the workpiece surface.

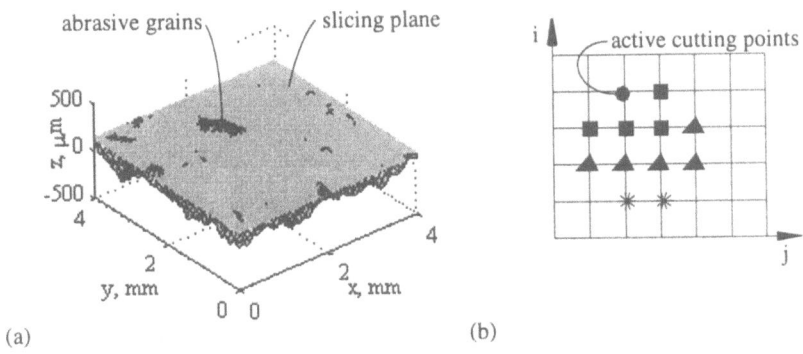

(a)   (b)

Figure 2: Active abrasive grains on the grinding wheel topography.

In order to identify the active abrasive grains from the topographical array, a search algorithm is proposed. The search process starts with the first point of the abrasive grain (circle point in Fig. 2b). By searching the neighbourhood points of the circle point, the next search front is identified (all the rectangular points). The searching of all the neighbourhood points is repeated for the rectangular points for identifying the successive search front (all the triangular points). The process continues until all the adjacent points are identified. All the found points are considered as belonging to the same abrasive grains. Strictly speaking, the two closely-positioned clusters of the cutting points can possibly belong to the same abrasive grain. This situation can be remedied by considering the distance between two clusters. If the distance is smaller than the average grain diameter, they can be considered as belonging to the same grain. However, in this simulation, two non-adjacent clusters are considered as two separated grains.

In grinding metals three distinct phases can be distinguished at the interface of the abrasive grain and the workpiece: rubbing, ploughing and cutting [12]. In this paper, it is suggested that the interaction of the abrasive grain with the workpiece depends on the attack of the abrasive grain. It means there are certain critical attack angles α_p at which the abrasive grain transits from cutting to ploughing, and α_r at which the abrasive grain transits from ploughing to rubbing. If the abrasive grain tip is approximated as a sphere, the attack angle α of the abrasive grain can be estimated as follows (Fig. 3a)

$$\alpha = arccos\left(\frac{R-r}{R}\right) \tag{1}$$

where R is the radius of a spherical tip, and r is the grain depth of cut. The radius of the abrasive grain tip can be found by inversion of the curvature of the grain. Since the sum of the curvatures of a surface at a point along any two orthogonal directions is equal to the sum of the principal curvatures [13]. The curvature of the abrasive grain can be defined as the arithmetic mean summit curvature of all the topographical points forming the abrasive grain

$$\kappa = -\frac{1}{2n}\sum_{k=1}^{n}\left(\frac{h_{i+1,j}+h_{i-1,j}-2h_{ij}}{\Delta x_s^2}+\frac{h_{i,j+1}+h_{ii,j-1}-2h_{ij}}{\Delta y_s^2}\right) \tag{2}$$

where n is the number of the topographical points forming the abrasive grain, and h_{ij} is the height of these points.

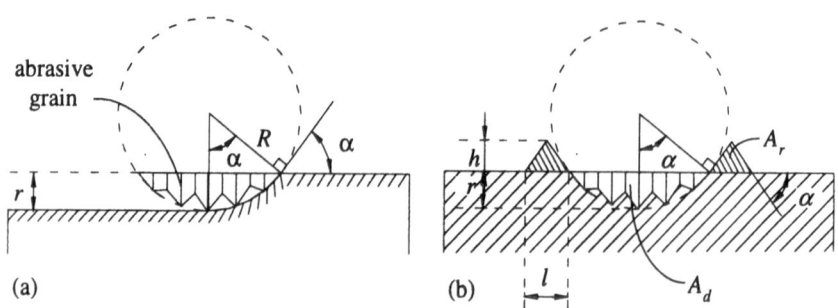

Figure 3: The interaction between the abrasive grain and the workpiece surface.

A portion of the displaced material, when the grains cut or plough, will remain on the workpiece surface and form side ridges along the groove. For the simulation, the cross section of the side ridges can be modelled as an isosceles triangle with base angle α equivalent to the attack angle of the abrasive grain (Fig. 3b). The perpendicular bisector h and the base side l of the side ridge can be estimated from the area A_r of the side ridge. The area A_r is assumed to be proportional to the area of the groove A_d by some cutting efficiency ratio. A similar approach was used by Chen and Rowe [5], except that a ridge was approximated as a parabola.

4. Model Verification

For the purpose of model verification, a grinding experiment was conducted with a wheel speed of 1500 rpm and a table speed of 0.1 m/sec on an Okamoto surface grinding machine. The workpiece material was mild steel, and the grinding wheel was WA80J8V. The wheel was trued with 30 µm depth of cut, 4 passes and a crossfeed of 0.2 mm/rev. Then it was dressed with 15 µm depth of cut, 2 passes and a crossfeed of 0.1mm/rev. After dressing the wheel surface was replicated at 4 locations, and its topography was taken on Talyscan measuring system. The sampling space was 40 µm. It was crucial to decide the sample spacing as the grinding action of one grain in the model was determined based on the grain curvature. The optimum sample spacing required to measure active grain was based on the equation recommended by Blunt and Ebdon [14]. The critical attack angles α_p and α_r for the abrasive-workpiece interaction were chosen based on Xie and William [15].

Table 1: Selected 3D parameters [13] for the experimental and simulated surface.

Parameter, unit	Sq, µm	Ssk	Sku	Sal, mm	Std	SΔq	Sds, mm^{-2}	Ssc, µm^{-1}	Sbi
Experiment	0.78	-0.10	2.60	0.04	0	0.03	51.34	8e^{-4}	0.63
Simulation	1.05	-0.19	2.7	0.04	0	0.04	14.97	10e^{-4}	0.64

Selected averaged 3D surface characterization of the ground and the simulated surface is given in Table 1. The workpiece was measured on the same Talyscan system. The simulation was performed on the PC Pentium II-500MHz. A total of 4 simulations were conducted. Computer time for each simulation was approximately 10 min. A comparison between the two data sets shows that amplitude (*Sq, Ssk* and *Sku*), autocorrelation (*Sal, Std* and *SΔq*) and functional (*Sbi*) parameters and are in good agreement. Other spatial and hybrid parameters (*Sds* and *Ssc*) have large variation. The variations can be explained by the fact that the simulated surfaces have less high frequency components compared to the real surface and thus appear to be smoother. The topography of the experimental and simulated ground surfaces is shown in Fig. 4.

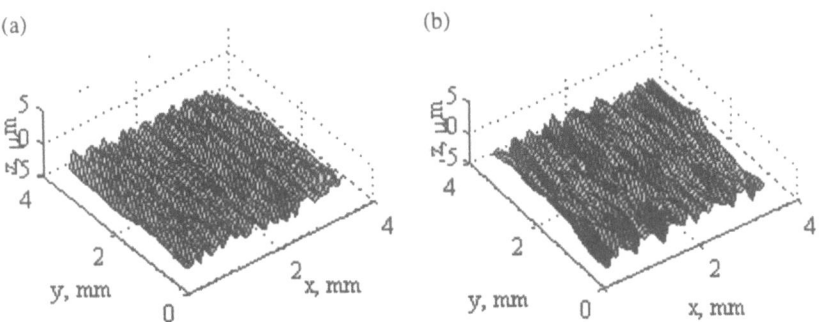

Figure 4: The topography of the experimental and simulated surfaces.

5. Conclusion

A numerical approach for simulating a precision grinding process is developed in the paper. It takes into consideration the non-Gaussian characteristics of the grinding wheel surface and the complex behaviour of the abrasive-workpiece interaction, which were often ignored in previous studies. The good agreement of the simulated surface with the experimental result indicates the validation of the approach.

References

[1] Koshy P., Jain V.K., and Lal G.K., 1997, Stochastic Simulation Approach to Modeling Diamond Wheel Topography. *International Journal of Machine Tools & Manufacture*, Vol. 37, No. 6, pp.751-761.

[2] Cooper W., and Lavine A.S., 2000, Grinding Process Size Effect and Kinematics Numerical Analysis. *Journal of Manufacturing Science and Engineering*, Vol. 122, No. 1, pp. 59-69.

[3] Warnecke G., and Zitt U., 1998, Kinematic Simulation for Analyzing and Predicting High-performance Grinding Processes. *Annals of CIRP*, Vol. 47/1, pp. 265-270.

[4] Gong Y.D, Wang B., and Wang W.S., 2002, The Simulation of Grinding Wheels and Ground Surface Roughness Based on Virtual Reality Technology. *Journal of Materials Processing Technology*, Vol. 129, No. 1-3, pp.123-126.

[5] Chen X., and Rowe W.B., 1996, Analysis and Simulation of the Grinding Process, *International Journal. of Machine Tools and Manufacture*, Vol. 36, No. 8, pp. 871-896.

[6] Inasaki, I., 1996, Grinding Process Simualtion Based on the Wheel Topography Measurement. *Annals of CIRP*, Vol. 54, No. 1, pp. 347-350.

[7] Wang, Y., and Moon, K.S., 1997, A Methodology for the Multi-Resolution Simulation of Grinding Wheel Surface. *Wear*, Vol. 211, No. 2, pp. 218-225.

[8] Salisbury E. J., Domala K. V., Moon K.S., Miller M.H., Sutherland J.W., 2001, A Three-Dimensional Model for the Surface Texture in Surface Grinding. *Journal of Manufacturing Science and Engineering*, Vol.123, No. 4, pp.576-590.

[9] Cressie N., 1993, *Statistics for Spatial Data*. John Wiley & Sons, New York.

[10] Shinozuka M., and Deodatis G., 1996, Simulation of Multi-Dimensional Gaussian Stochastic Fields by Spectral Representation. *Applied Mechanics Reviews*, Vol. 49, No. 1, pp. 29-53.

[11] Johnson N.L., 1949, Systems of Frequency Curves Generated by Methods of Translation. *Biometrika*, Vol. 36, No.1/2, pp. 149-176.

[12] Hahn R.S., and Lindsay R.P., 1982, Principles of grinding, In: Bhateja C. and Lindsay R. (ed.) *Grinding Theory: Techniques and Troubleshooting*. SME, Michigan, pp.3-10.

[13] Stout K.J., Sullivan P.J., Dong W.P., Mainsah E., Luo N., Mathia T., and Zahouani H., 1993, *The Development of Methods for the Characterisation of Roughness in Three Dimensions*. Commission of European Communities, Luxembourg.

[14] Blunt L., and Ebdon S., 1996, The Application of Three-Dimensional Surface Measurement Techniques to Characterising Grinding Wheel Topography. *International Journal of Machine Tools and Manufacture*, Vol. 36, No. 11, pp. 1207-226.

[15] Xie Y., and Williams J.A., 1996, The Prediction of Friction and Wear When a Soft Surface Slides Against a Harder Rough Surface. *Wear*, Vol. 196, No. 1, pp. 21-34.

Co-operative Engineering Approach and Parametric Tolerancing

Salim Boukebbab[1], Idriss Amara[1], Jean Marc Linares[2] and Jean Michel Sprauel [2]

[1] Laboratoire de Mécanique, Faculte des Sciences de l'Ingénieur, Université de Mentouri-Constantine, Campus Châab Ersas, 25000 Constantine, Algérie, email: boukebabb@yahoo.fr
[2] Laboratoire EA (MS)[2], IUT d'Aix-en-Provence, Avenue Gaston Berger, F 13625 Aix-en-Provence, France, email: linares@iut.univ-aix.fr

Abstract: Suggesting a concept, which arises from the systemic approach allows a better description of the functionality of parts during tolerancing and provides a wider scope for the other functions taking place after design. An adaptive tolerancing method is suggested. It takes account of the properties of interfaces (degrees of freedom or freedom space). A control process is created to validate the tolerancing according to the previously suggested methodology. Mastering the transfers of the degrees of freedom, allows thus to strongly decrease the number of scraps while respecting the expected function.

1. Introduction

The world economy first translates into an increase in profits and a need for competitivity for all mechanical industries [1]. Within the past twenty years, efforts were focused on individualized processes. New words appeared, as well as new definitions for each craft (SPC, SMED...). This situation partly accounts for the failure of simultaneous engineering at its beginning. The systemic approach of industrialization problems brings now answers at a wide scale for use. A new tool suggestion for design must absolutely account for the simplifications induced on the whole industrialization line.

Any company present in a sector open to competition must satisfy the customer's requirements about the technical characteristics and the price of the product. Functional tolerancing rises from such constrains at the design level. Tolerancing is described according to the tools suggested by the current standards. What concerns us is the realization of the global function of the product. First, the design engineer gives volume to the kinematics block diagram through a design study. Then the tolerancing phase is activated. After this step of the industrialization process, the biggest part of the product final cost is potentially defined. Mastering costs thus greatly depends on the foresight with which these implications will be understood and managed. With this research we suggest extending the use of tolerancing requirements (maximum material requirement, least material requirement envelope requirement) by adaptive tolerancing in the industrialization process.

2. Fundamental Tolerancing Principles

At the tolerancing phase, the designer has tools at his disposal (straightness, flatness, perpendicularity, position, etc.) and principles (maximum material requirement, least material requirement, envelope requirement). According to the selected items, he can then define in advance the work of the control function of the so toleranced parts. When the independence principle is used, the defects are assumed to be independent, thus leading to classify them. In Fact, from the micro-geometrical conditions of surfaces to the geometrical ones, the following different classes which permit to codify the elementary defects of real surfaces are defined by actual standards: roughness, undulation, form, orientation and dimensions (Fig.1).

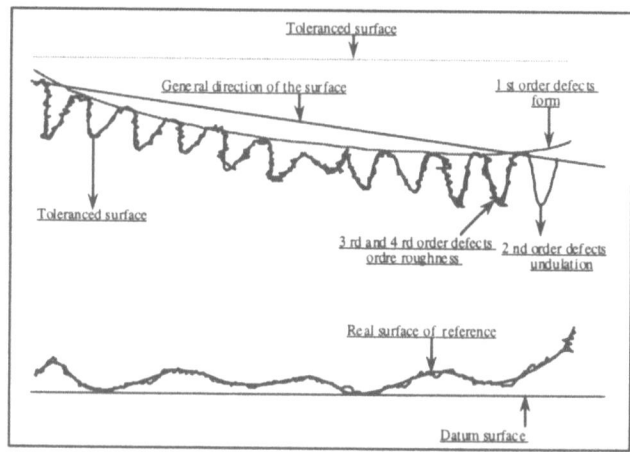

Figure 1: Classification of defects

Tolerancing following this principle directs the control function towards the metrology process [2]. For this aspect of the industrialization process, the control function uses metrology as a defect extraction tool (Surface sampling, Surface Association, Verification, etc.). Inspection is thus performed comparing each associated surface with the related nominal one [3]. It can nevertheless be noticed that the control of a global functional condition may require more than one inspection process. This methodology is based in most cases, on a sampling of the real surfaces and integrates the limits of the filtering methods. Such degraded view of the analyzed surfaces will therefore lead to uncertainties. Inspection includes several steps:

    a) Measurement of the real geometrical elements
        a.1) Measurement equipment

a.2) Mechanical or electronical smoothing of undesired defects

a.3) Acquisition method

b) Processing of the measured data

b.1) Filtering of defects by mathematical modeling

b.2) Association of perfect Geometrical Features to the data

b.3) Calculation of uncertainties

c) Comparison between the design model and the associated surfaces.

The designer can also use the maximum material, the least material and the envelope requirements. In this case, defect globalization is achieved. This tolerancing mode results in a new aspect of the industrialization process. In that case, contrary to the previous one, the inspection performs the control of a local functional condition. It materializes the expected function by a gauge.

The fitting of the real surface, defined by its envelope, into the gauge marks the success of the control procedure. Such tolerancing mode is closer to the needs of the functional flows. It defines more efficiently the functionality of the surfaces.

The envelope requirement applies in the many cases for which an assembly function is needed. The assembly (fitting) function is indeed transcribed into a maximum material requirement condition. In that case, writing the tolerance mechanism into a set of equations leads to create two groups (Fig 2):

Type I: Tolerance of a given feature towards one single datum surface, both elements being at the maximum material condition.

Type II : Tolerances of a given feature towards a multiple datum reference system, all the elements being at the maximum material condition.

The control of such requirements is based on the use of gauges. The gauge reconstructs the functional environment present in the assembled mechanism. However, the costs induced by the manufacturing of material gauges now result in building virtual (computer) ones [4]. Inspection is a binary process. We can however notice some limits concerning the use of the requirement principles. In the second case (type II multiple datum reference systems), there is no formulation at the level of the design and control functions that would allow to master the tolerance and manage the transfers from the clearances to the variations in dimensions of the specified surface.

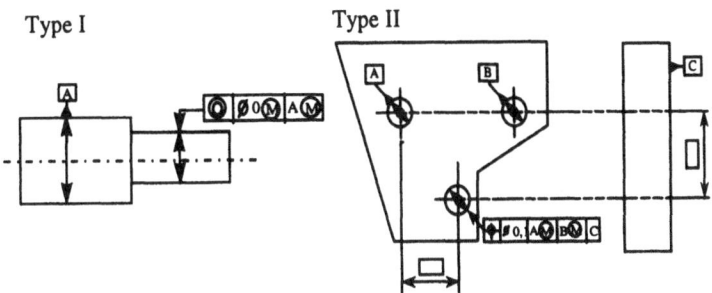

Figure 2 : Group I and II

316

Moreover, in the most recent releases of software's for coordinate measurement machines, only the cases of such transfer in type I tolerancing are implemented [5]. The interfaces between two groups of surfaces for assembly often show freedom potentialities (clearance, diameter variation). Neglecting the clearances in these interfaces results in increasing some constraints of functional tolerancing. To conclude this paragraph we can show the existing relations between design and control according to the use of requirements (Fig. 3).

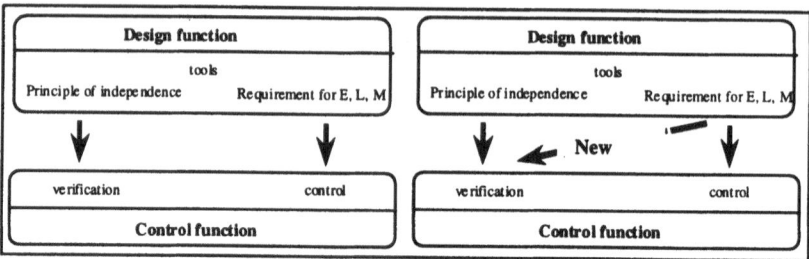

Figure 3: 3rd path

This illustration brings to the fore the failures in industry when gauges are replaced by a coordinate measurement machine which basically performs rather verification measurement then control in the present state of software's. The introduction of computer virtual gauges or any other method using the data obtained by a coordinate measurement machine will not reach the perfection of material gauges, since it is based on a sampling of the analyzed features which only gives a partial and degraded view of the real surfaces.

However the ease of use, flexibility and the high level of automation of coordinate measuring machines are major advantages for industry because it leads to greatly reduce the costs of verification and control. Now in parallel with the virtual gauge concept, a third way may be defined to improve the actual procedures. For both case (type I and II tolerancing) it is possible to supply another alternative for inspection, based on the optimization of the acquired points and managing the transfers between clearances and variations in dimensions. In the following part of the study, we shall suggest a concurrent engineering method (design, control) for an adaptive tolerancing based on the requirement principles.

3. Toward an Adaptive Tolerancing

Our tolerancing methodology is based on the notion of Functional Group (FG). A functional group is a set of entities that participate together in the realization of the function. A tolerancing methodology based on the notion of Functional Group brings to the fore the duality which exists between the internal tolerancing and the clearances [6]. The freedom spaces are thus composed of clearance and internal tolerancing. This liberty permits to release the tolerancing constraints. This notion of liberty space requires the settlement of a new description of the tolerance area which isn't currently define in the ISO standards.

The spaces of freedom are of complex form and are only defined by implicit equations. To facilitate their description, in order to use them in design, we suggest replacing these implicit equations by vectorial function approximations: one sheet Hyperboloid with an elliptical basis (fig.4)

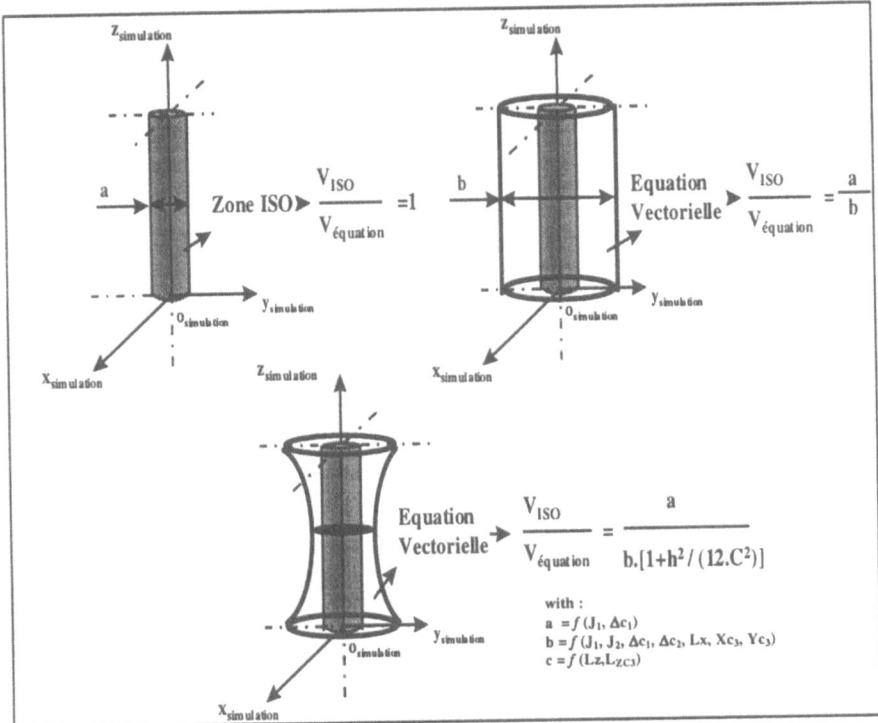

Figure 4 : Comparison freedoms spaces boundaries with tolerance zone ISO

4. Design Function

For design, we suggest to model these spaces of freedom. This feature opens into complex parametrical equations. A comparison between different possible modelling ways (rotation matrix, approximation for development of second order and approximation for development of first order) has been achieved to test the robustness of the different simulations. By the use of statistics, we intend to simplify these parametrical equations using easily usable functions (fig.5).

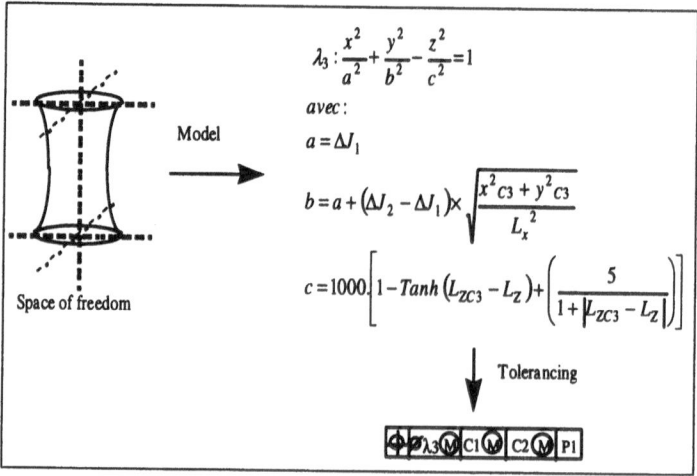

Figure 5 : Model

A reliable proposal for an adaptative tolerancing of the functional needs for fitting is getting possible by using the interfaces potentialities.

5. Control Function

To comply with the philosophy of « concurrent engineering », we now have to suggest a methodology for the control (fig. 6). Two possibilities have been explored:

- Control incorporated transfer flows for Type I and II tolerancing
- Control or pairing with associated surfaces (level1)

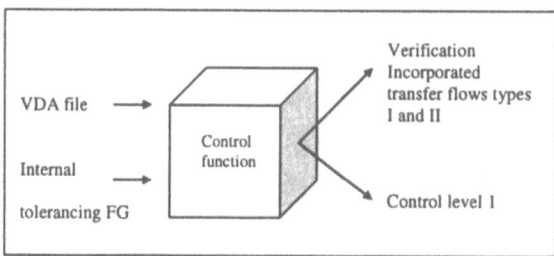

Figure 6 : Methodology for the control

The proposed control is called of level 1 since it is based on a sampling of the analyzed surfaced and therefore integrates the same kind of uncertainties which exist in verification procedures using coordinate measuring machines. The necessary inputs for the control are files containing all the points M_i of the measured surfaces defined in the measurement reference system [7]. The data of the assembled Functional Groups are also required.

6. Conclusion

In this paper a concurrent engineering approach of tolerancing and control has been proposed to improve the design and manufacturing of mechanical systems [8]. This approach is based on a new concept: the Functional Group which allows defining the functional needs of a given part in the mechanical assembly. From these constrains it leads to tolerances based on the requirement principles (maximum material requirement, least material requirement, envelope requirement). Such tolerances direct the inspection towards the verification by gauges which reconstruct the functional environment present in the assembled mechanism. Such procedure requires however to manage the transfers between clearances and dimension variations. The suggestion for simplified transfer equations presents new possibilities for adaptive tolerancing in the required functional state without constraining the other functions of the elementary industrialization process.

References

[1]   C. MARTY, 1995, *Concurrent Engineering and economic effects of design decisions*, Colloque International INRIA, Grenoble, France, pp 07-12.

[2]   J. WEN-YUH, L. CHIEN-HONG, C. CHA'O-KUANG, 1999, *The min–max problem for evaluating the form error of a circle*, Measurement, **26**, pp 273–282.

[3]   ASME Y 14.5 M-1994, *Dimensioning and Tolerancing*, The American Society of Mechanical Engineers, New York, 1994.

[4]   D. M. ROBINSON, 1997, *Geometric tolerancing for assembly with maximum material parts*, in: CIRP Seminar on Computer Aided Tolerancing, Toronto, Canada.

[5]   J. M. LINARES, S. BOUKEBBAB, JM. SPRAUEL, 1999, *Parametric tolerancing*, 6th CIRP International seminar on Computer Aided Tolerancing, University of Twente, Enschede, The Netherlands, pp 167-176.

[6]   J. M. LINARES, S. BOUKEBBAB, JM. SPRAUEL, 1998, *Co-operative engineering approach : tolerancing, control*, CIPR Seminar, Production Technology Centre, Berlin, Germany, pp 145-156.

[7]   P. E. PAIREL, 1997 ,*The "Gauge model": a new approach for coordinate measurement*, Proc. of the XIV IMEKO World Congress, Tampere, Finland, pp. 278-283.

[8]   ISO TS 17450-1, *Geometrical product specification*, General Concepts, Part 1: Model for geometric specification.

Comparative Study of the Volumetric Positioning Accuracy of CNC Machining Centres Using the Latest Laser Measurement Technology

Mr. Ondrej Svoboda

Research Center for Manufacturing Technologies,
Czech Technical University in Prague,
Horska 3, 128 00 Praha, Czech Republic. o.svoboda@rcmt.cvut.cz

Abstract: Machine tool manufacturers are striving to achieve maximum accuracy of their products. Geometric accuracy together with high speed has become the most important parameter of all CNC machine tools. Traditionally only the axial positioning errors have been checked and compensated (if necessary). Nowadays it is not sufficient to control only this type of error. Straightness, squareness and angular errors have to be also taken into account. Furthermore in connection with the increasing capabilities of modern control systems more complex compensations can be performed resulting into an outstanding volumetric accuracy.

Reported here are the results of geometric measurements done by means of the latest laser measurement technology. Recently a set of mid-sized CNC machining centers was tested in order to compare the significance of the above-mentioned types of geometric errors. All measurements were done according to the international standards ISO 230-2, ISO 230-6 and by the laser vector method. Finally the effect of a volumetric compensation based on the output of the laser vector method is discussed as a possible action in order to improve the machine tool accuracy.

1. Introduction

According to the ISO 230 standard geometric deviations of each axis are divided into three groups. The first group is represented by the deviation in the direction of movement – linear errors $Dx(x)$, $Dy(y)$ and $Dz(z)$, where subscript is the error direction and the position coordinate is inside the parenthesis. The second group contains the horizontal and vertical straightness deviations - $Dz(x)$, $Dz(y)$, $Dy(z)$, $Dy(x)$, $Dx(y)$ and $Dx(z)$. The last group covers the angular deviations – pitch $Ay(x)$, $Ax(y)$, $Ax(z)$, yaw $Az(x)$, $Az(y)$, $Ay(z)$ and roll $Ax(x)$, $Ay(y)$, $Az(z)$. The accuracy between the pairs of axes is additionally described by the squareness errors – Bxy, Byz and Bzx. Therefore a standard three axes machining center is generally influenced by 21 different types of geometric errors (with the rigid body assumption).

2. Measurement Data

The following text shows the results of positioning accuracy measurements done by different methods on a set of 10 machining centers from various manufacturers.

2.1 Types of performed measurements

All the performed measurements were realised by three methods: firstly according to the ISO 230-2 (positioning accuracy and repeatability of each axis separately), secondly according to the ISO 230-6 (diagonal positioning accuracy) and finally by the laser vector method. The laser vector method is an extension of the diagonal displacement test where the laser beam is pointing in the body diagonal direction of the working volume of a machine tool. However instead of moving the x, y and z-axis together this technique uses a sequential step diagonal path –first move x only, stop and collect data, then move y only, stop and collect data, then move z only, stop and collect data and so on until the opposite corner of the diagonal is reached. After a mathematical analysis of the obtained data the vector method gives all the linear errors, a complete set of the vertical and horizontal straightness errors and additionally the three squareness errors as noted in the introduction.

Figure 1: Measurement by the laser vector method on a vertical machining center

2.2 Overview of the tested machine tools

The total number of tested machine tools was ten. Eight were made by the German manufacturer Deckel Maho Gildemeister (DMG), one by the UK`s Bridgeport and one by the Czech company Kovosvit MAS. The DMG machines are for better illustration inscribed with a number behind each type description (e.g. DMU80T-2). A brief description of the machines can be found in Table 1.

Table 1: Measured machines` parameters

Machine No.	1	2	3	4	5
Machine id.	DMC60H-1	DMC60H-2	DMC65V-1	DMC65V-2	DMU80T-1
Manufacturer	DMG	DMG	DMG	DMG	DMG
Type	horizontal	horizontal	vertical	vertical	vertical
Axis stroke (X/Y/Z) mm	600 / 560 / 560	600 / 560 / 560	650 / 500 / 500	650 / 500 / 500	880 / 630 / 630
Control sys.	Sinumerik 840D	Sinumerik 840D	Sinumerik 840D	Sinumerik 840D	Heidenhein iTNC530
Meas. system	direct	direct	direct	direct	direct
Spindle char.	15000rev / 20kW	15000rev / 20kW	18000rev / 15kW	18000rev / 15kW	10000rev / 30kW
Service hours	2589	1655	3550	3338	2847

Machine No.	6	7	8	9	10
Machine id.	DMU80T-2	DMU80T-3	DMU80T-4	VMC500	MCV1000
Manufacturer	DMG	DMG	DMG	Bridgeport	MAS
Type	vertical	vertical	vertical	vertical	vertical
Axis stroke (X/Y/Z) mm	880 / 630 / 630	880 / 630 / 630	880 / 630 / 630	650 / 500 / 500	1016 / 610 / 720
Control sys.	Heidenhein TNC430	Heidenhein iTNC530	Heidenhein TNC430	Heidenhein TNC410	Heidenhein iTNC530
Meas. system	direct	direct	direct	indirect	direct
Spindle char.	10000rev / 30kW	10000rev / 30kW	10000rev / 30kW	7000rev / 8.5kW	12000rev / 35kW
Service hours	4081	1672	3723	892	437

2.3 Comparison of the results for individual machine tools

A compete set of results is shown in Table 2. Measurements according to ISO 230-2 were performed along the three edges of the machine working volume. These are identified by the marks I, II and III. The angular errors are derived from the linear positioning by respecting the Abbe offsets and utilising the equations (3-2) stated in [1] or equations (23), (24), (25) stated in [2]. The diagonal positioning accuracy is described by the parameter Ed (diagonal systematic deviation of positioning) according to ISO 230-6. The remaining geometric errors were evaluated from the laser vector method: Dx(x), Dy(y), Dz(z), Dz(x), Dz(y), Dy(z), Dy(x), Dx(y), Dx(z), Bxy, Byz, Bzx and can be found in the lower part of Table 2.

Table 2: Measurement results

Meas. techn.	Error type	Pos.	Machine No.									
			1	2	3	4	5	6	7	8	9	10
			Maximal deviation [μm], resp. [μm/m]									
ISO 230-2	Dx(x)	I	9.5	5.3	16.5	24.0	35.8	23.5	10.7	20.5	7.6	12.3
		II	7.2	7.3	31.1	22.5	47.7	24.1	12.0	54.3	X	X
		III	X	X	19.2	19.0	51.6	28.4	X	29.4	X	X
Calc	Ay(x)		X	X	53.0	4.0	-12.0	-7.0	X	14.0	X	X
	Az(x)		X	X	15.0	26.0	-36.0	5.0	X	84.0	X	X
ISO 230-2	Dy(y)	I	15.8	7.8	15.3	18.4	20.3	14.3	16.2	5.5	13.6	15.7
		II	12.0	8.7	4.9	20.4	18.3	19.2	17.1	6.3	X	X
		III	X	X	13.2	24.9	22.9	21.6	11.2	12.1	X	X
Calc	Ax(y)		X	X	60.0	-12.0	11.0	-3.0	7.0	-8.0	X	X
	Az(y)		X	X	38.0	6.0	2.0	-3.0	-9.0	28.0	X	X
ISO 230-2	Dz(z)	I	X	36.3	10.9	10.6	14.0	16.5	6.8	6.6	23.3	14.1
		II	14.3	17.8	14.9	7.1	15.5	19.2	8.4	8.7	X	X
		III	25.2	21.1	10.1	7.7	18.0	15.2	7.7	15.7	X	X
Calc	Ax(z)		X	-99.0	-2.0	5.0	-5.0	-8.0	-3.0	-15.0	X	X
	Ay(z)		X	-72.0	5.0	-14.0	7.0	15.0	-7.0	-6.0	X	X
ISO 230-6	Ed		15.9	33.4	34.4	38.3	45.4	31.8	15.8	41.5	33.2	26.9
Laser vector method	Dx(x)		2.7	8.4	20.2	7.8	18.6	11.8	1.7	16.8	12.8	6.9
	Dy(x)		2.9	2.9	7.5	2.9	3.9	5.6	6.2	2.8	7.1	15.6
	Dz(x)		2.4	3.4	9.2	4.1	2.5	3.1	2.4	1.9	8.5	6.6
	Dy(y)		2.2	8.2	15.2	8.3	14.0	8.9	1.5	12.6	8.2	9.4
	Dz(y)		2.3	2.8	2.3	1.2	2.0	4.0	7.3	3.3	2.3	3.5
	Dx(y)		2.4	8.8	6.7	11.9	5.2	4.5	10.3	4.0	18.4	7.9
	Dz(z)		2.6	9.7	10.8	4.2	10.8	6.8	2.7	9.7	15.3	7.8
	Dy(z)		6.1	13.1	5.3	23.3	25.1	23.9	5.2	9.3	27.5	21.3
	Dx(z)		15.9	28.2	7.2	5.2	5.2	2.1	8.5	15.8	25.6	6.4
	Bxy		-1	15	-18	-8	5	3	15	-8	56	11
	Bxz		41	-52	-31	-7	7	4	-18	-39	64	-37
	Byz		-18	-18	-8	-67	-53	-48	-16	-27	73	-7

3. Discussion

Respecting the measured data from Table 2 the following observations can be made:

Horizontal machining centers No.1 and No.2 show significant deviations measured according to ISO 230-2 in the Z-axis, furthermore the pitch and yaw errors of this axis represent the major piece of the total error budget. The diagonal displacement accuracy (ISO 230-6) varies eminently between the two machines whereas in case of machine No.2 the value 33.4µm indicates a relatively poor volumetric performance. From the laser vector method it can be found that the worst accuracy shows the Z-axis (corresponds with ISO 230-2 results), respective the straightness of this axis. In addition the squareness between the X and Z-axis seems to be of major importance in the laser vector method analysis of both machines.

Vertical machining centers type DMC65V (No.3, 4) have the main ISO 230-2 deviations in the X-axis and in some positions also in the Y-axis. This fact indicates extensive pitch and yaw angular errors in these axes. The diagonal displacement test has shown a significant error exceeding 34µm on both machines. Results of the laser vector method vary for both machines. In case of machine No.3 the most important is the $Dx(x)$ error, then the Bxz squareness and the $Dy(y)$ linear error. The ranking of errors of No.4 is as follows, 1st Byz, 2nd $Dy(z)$, 3rd $Dx(y)$.

Geometric errors of the DMU80T (No.5, 6, 7, 8) have a relation with the number of service hours. Machine No.7 is the newest one, thus shows the best accuracy, the remaining machines (over 2500 service hours) achieved considerably worse results particularly in the X-axis, which has a longer stroke than the Y and Z-axis. Furthermore in case of machines No.5 and No.8 the X-axis contains a large angular error $Az(x)$. All the older machines also have a minimally double diagonal positioning error compared with No.7, this fact indicates the suitability of using the diagonal displacement tests for time trend monitoring. Basically the error budget for the DMU80T type is: 1st $Dx(x)$ and the angular errors of the X-axis, 2nd Byz squareness, 3rd Ed. The possibility of improving the volumetric accuracy was tested on machine No.8 by applying a modified part program for the diagonal displacement test. Corrections obtained in this part program were generated based on the results of the laser vector method. A comparison of the original state and the measurement with the corrected part program can be seen in Figure 2. The parameter Ed was reduced from 24.5µm without correction to 6.7µm with the corrected part program.

Machine No.9 (VMC500) was measured in a reduced manner though it can be stated that the Z-axis has the worst linear error according to ISO 230-2. The diagonal positioning error exceeds 33µm, which represents a significant value in the total error budget. The laser vector method results signalize extensive squareness errors in all three coordinate planes, in addition both straightness of the Z-axis seem to play an important role.

Results of a reduced set of measurements on machine No.10 (MCV1000) are well balanced between 12 and 16µm according to ISO 230-2 in all axes. The diagonal displacement error is just bellow 27µm in this case. The laser vector method shows the main errors in Bxz squareness and $Dy(z)$ straightness.

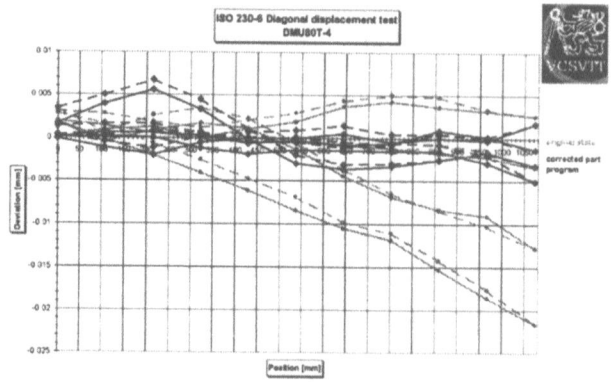

Figure 2: Comparison of the diagonal displacement accuracy with and without compensation

4. Conclusions

It is clear that the volumetric positioning accuracy of the tested machine tools depends on the major errors found in the individual axes. The diagonal displacement test performed according to ISO 230-6 was found to be a good indicator of the volumetric performance. For an overall improvement of the machine accuracy it is highly recommended to compensate for both the linear and the straightness errors as shown from the application on machine No.8.

References

[1]   Weck M., 1992, *Werkzeugmaschinen Fertigungssysteme Band 4: Messtechnische Untersuchung und Beurteilung*, VDI Verlag Dusseldorf, Germany.

[2]   Wang C., Liotto G., 2003, *A Theoretical Analysis of 4 Body Diagonal Displacement Measurement and Sequential Step Diagonal Measurement*, Proceedings of the Lamdamap 2003 Conference, Huddersfield, England.

[3]   ISO 230-2: 1997, *Test Code for Machine Tools-Part 2: Determination of accuracy and repeatability of positioning numerically controlled axes*, an International Standard, ISO, Geneva, Switzerland.

[4]   ISO 230-6: 2002, *Test Code for Machine Tools-Part 6: Determination of Positioning Accuracy on Body and Face Diagonals (Diagonal Displacement Tests)*, an International Standard, ISO, Geneva, Switzerland.

This research has been supported by the Czech Ministry of Education under the grant LN00B128.

MANUFACTURING PROCESS MODELLING

A Numerical Study of the Aluminum Pipe Extrusion Process

Mohammad Movahhedy, Ali Rezaei

Department of Mechanical Engineering, Sharif University of Technology, Azadi Ave., Tehran, IRAN, movahhed@sharif.edu

Abstract: The pipe extrusion process is simulated using the arbitrary Lagrangian-Eulerian finite element method. 2D and 3D models are presented and it is shown that the 2D models can be used effectively for qualitative study of the effects of the process parameters and optimizing the die geometry. The models are used for study of the effects of die web angles on the process

1. Introduction

Extrusion is one of the most common manufacturing processes for production of various profiles. The design of extrusion dies is of outmost importance in reduction of energy required for forming. The dies are traditionally designed based on experience, but in recent years, analytical and numerical methods such as slip line field, upper bound [1], finite elements and finite volume [2] have been applied extensively to simulate material flow in extrusion and optimize process parameters. However, the studies have mostly focused on solid sections. For tubular sections, due to the presence of legs and portholes, the material flow in the die is not axisymmetric and a more demanding modelling is required. In hot extrusion, the material deformation is also dependent on strain rate and temperature and a viscoplastic material flow model is used. In this article, the hot extrusion of aluminum pipes is studies using the Arbitrary Lagrangian Eulerian (ALE) finite element method. The process is first modelled as a 3D process, and it is shown that the 2D plane strain models may also be used for qualitative study of the die geometry. The effects of die angles in material flow are studied.

2. Process Modelling

Porthole dies are normally used for hot extrusion of aluminum pipes. The die consists of webs, portholes, welding chamber and bearing area, as shown in figure 1 [3]. When the material flows into the die, the webs, which hold the mandrel in the middle, divide the material flow into channels. The material past the webs is welded again in the welding chamber and is extruded through the bearing which forms it to the shape of the final product. The behaviour of the hot aluminum during extrusion may be modelled as viscoplastic flow. The modified Sellars-Tegart equation may be used to model the viscoplastic behaviour, including small elastic deformations [4].

$$\sigma_f(\dot{\epsilon}, T) = s_m \ arcsin \ h((\frac{\dot{\epsilon} + \dot{\epsilon}_o(T)}{A} exp(\frac{Q}{RT}))^{\frac{1}{m}})$$

In this equation, s_m is the parameter dependent on material in MPa, $\dot{\epsilon}$ is the strain rate in 1/s, Q is the activation energy in J/mol, T is temperature in Kelvin, and R is the universal constant of gases. $\dot{\epsilon}_0$ is a temperature dependent parameter used for elastic recovery of material, usually considered between 0.01 and 0.001.

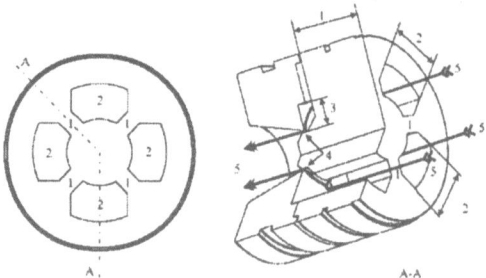

Fig 1: Portholes die for the extrusion of round tubes 1-Legs 2-Portholes 3-Welding Chamber 4-Bearing 5-Aluminum flow [3]

During the deformation, the heat is generated as a result of large plastic deformation and friction between the part and the walls of the die. The frictional heat is much smaller than the heat due to plastic work, because the velocity of the material flow is not very high. The heat due to plastic strain is calculated from the following equation:

$$r^{Pl} = \eta \sigma : \dot{\epsilon}^{Pl}$$

where r^{Pl} is the heat generated, σ is the stress tensor, $\dot{\epsilon}^{Pl}$ is the plastic strain rate, and η is a factor the percentage of plastic work converted to heat. This factor is considered to be 90% in this work. The friction between the die and workpiece material is modelled using the Coulomb friction law, and a coefficient of friction equal to 0.1 is considered. In applying the heat boundary conditions, the heat transfer between the workpiece and die is neglected and it is assumed that the part and the die are both initially in 450C [5]. During the deformation, the part temperature increases, but the die will remain in the same temperature. This assumption is not expected to create significant error, because the die and part temperature remain close during the deformation. From numerical point of view, two different approaches are usually used in numerical modelling of forming processes; Lagrangian and Eulerian. In the Lagrangian approach, the motion of material during deformation is followed, because the nodal points are attached to material points. In this way, it is easy to follow the history of material deformation. However, the mesh may undergo distortion due to the large deformation, resulting in premature ending of the numerical process. The Eulerian approach, on the other hand, focuses its attention to a particular point in a space and the deformation of the material points passing this point at given times is studied. While in this approach, it is harder to follow the material deformation history, the mesh is fixed in space and is not distorted. However, the boundary of the deformation region should be known a priori, because they may not be updated during the deformation easily. An alternative approach, which combines the advantages of both formulations, is the arbitrary Lagrangian-Eulerian (ALE) method. In this approach, the mesh can have a motion independent of material deformation. Therefore, the motion of the mesh can be designed in accordance with the nature of deformation, and thus, avoid mesh distortion on the one hand and update the boundaries on the other. The mesh motion can also be limited to a certain region, and thus reduce the computation time [6, 7].

In this article, ALE finite element approach is used for modelling the Aluminum pipe extrusion process. For a simplified plane strain model of this process, as explained later, the ALE boundary conditions are given in figure 2 and table 1.

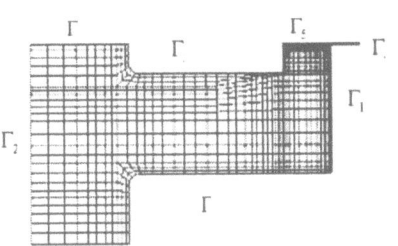

Fig 2: Boundary condition for plane strain problem

Table 1: boundary conditions for plane strain problem

$\Gamma1$	Die walls	Contact with friction	Lagrangian boundary
$\Gamma2$	Material entrance	Ram velocity V_1	Eulerian Boundary
$\Gamma3$	Web	Contact with friction	Lagrangian boundary
$\Gamma4$	Material exit		Eulerian boundary
$\Gamma5$	Symmetric boundary	symmetry	symmetry

3. 3D Modelling and Results

Because the pipe extrusion is not symmetric due to the presence of webs, in the first stage, a 3D model is created. If four webs are used in the die, modelling a quarter of the die area is sufficient. Around 15000 brick elements are used as shown in figure 3. A denser mesh is used around the bearing area, and the pipe thickness is 2mm, billet diameter is 200mm, the height of welding area is 30mm, web length is 80mm and it thickness is 30mm. The ram velocity is considered to be 5 mm/sec.

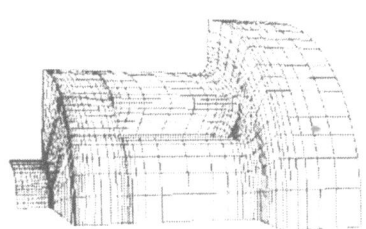

Fig 3: 3d model with brick element

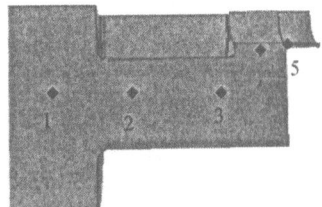

Fig 4: Selected points for identify steady state

The simulation is started by material entering the die area and continues until steady state condition is reached. In order to identify this state, points in different positions in the die are considered as shown in figure 4 and the history of velocity and stress is recorded at these points. Figure 5 shows the material velocity at these points. It is seen that after around 0.75s, the velocity of the material reaches an almost constant value, and this time may be considered as the time sufficient for reaching the steady state. Figure 6 shows the contours of Von Mises stress for the 3D model. The inset figure at right shows the magnified region around the bearing area where the maximum stress occurs. Figure 7 shows the pressure in the die after steady state is reached. The largest pressure is observed in the entrance region and its intensity is reduced toward the bearing area. In fact, in the bearing area, shear stresses have the largest contribution to the Von Mises stress. Another result of this simulation concerns the velocity of material flow, depicted in figure 8. It is observed that the material velocity increases steadily toward the bearing area. In the bearing area the gradient of velocity is very high, reaching 0.8 m/s. In the vicinity of the walls of the portholes and web, the velocity is much slower, showing sticking type of friction as the material approaches the bearing. Figure 8 also shows a big dead metal zone formed at the corner of the welding zone. A study of the material flow around the web also shows that dead zones are alslo formed at both ends of the web, as shown in figure 9. The presence of dead zones represents an increase in shear stresses in the material, with the result that a large portion of the energy is spent for redundant work. Furthermore, the dead zone at the end of the web will practically mean that the height of the welding chamber is reduced and material welding will not be done efficiently. Figure 10 shows the contour of temperature distribution in the material. The temperature reaches a high value of 530c in the bearing area due because the bulk of the plastic work is performed in this region. Frictional work also results in higher temperature around the walls.

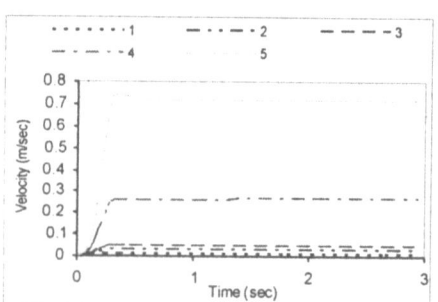

Fig 5: History of material velocity at selected point in Fig 4

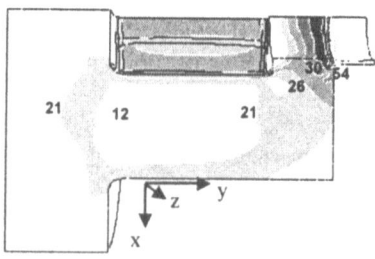

Fig 6: contour of Von Mises stress for the 3D model (MPa)

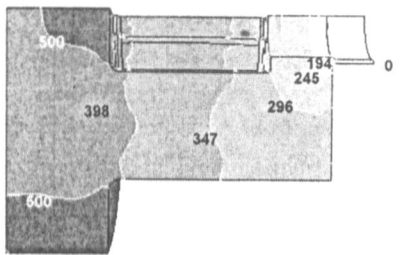

Fig 7: contour of pressure (MPa)

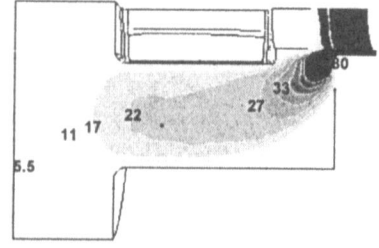

Fig 8: contour of velocity (mm/s)

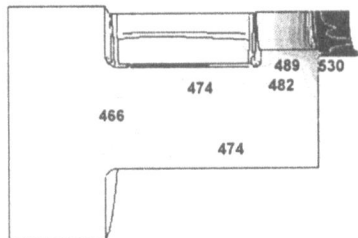

Fig 9: coutors of velocity around the web (mm/sec)   Fig 10: counter of temperature (°C)

4. 2D Modelling and Results

3D modelling of the pipe extrusion is the most accurate approach, in practice it is not computationally efficient. In particular, the use of a 3D model in optimization of process parameters will be very time-consuming and expensive. For this reason, it would be more efficient if approximate 2D models of the process can be created which are representative, albeit with lower accuracy and perhaps qualitatively, of the actual process. If these models can be validated, it will save enormous energy and time. Two type of 2D models may be considered; axisymmetric model and plane strain model. The plane strain model is generated by sectioning the die along its length and parallel to its webs, as shown in figure 11. Since a section as above will not include the bearing area, this area is added to the lower part of the model as shown in figure 12. The underlying assumption here is that the material flow in the middle of a porthole approaches plane strain conditions. This model is useful in study of the effects of web geometry and the welding chamber on the process.

Fig 11: Die section for plane strain state   Fig 12: Die and material flow in plane strain state

The first step in the 2D study is to compare the results of the 2D models with corresponding 3D results, and thus validating the approximate models. The 2D results are not obviously expected to be as accurate, but a qualitative agreement with 3D results will validate the use of 2D models in parametric studies. Figure 13 show the pressure distribution for the plane strain model respectively. Figure 14 show the velocity profile. Comparing these results with those given in previous sections shows that there is good agreement qualitative agreement between the 2D and 3D models. This allows us to use the 2D models in the study of the effects of the process parameters.

334

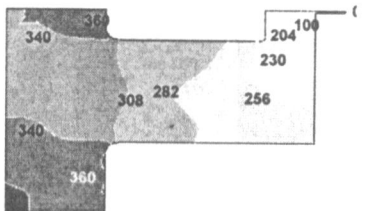

Fig 13: Pressure distribution for the plane strain state (MPa)

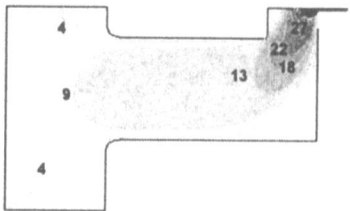

fig 14: Velocity distribution for the plane strain state (mm/sec)

5. The Effects of Web Geometry on Pipe Extrusion Process

One of the objectives of this study is to obtain the best web geometry for efficient pipe extrusion. The 2D plane strain model is used for this parametric study. The feed angle of the web, shown in figure 15, is changed in a range between 0 to 16 degrees. This study shows that increasing this angle reduces the maximum hydrostatic pressure only slightly. Figure 16 shows the change of pressure along the symmetric boundary of the welding chamber which also shows a slight increase of pressure with an increase in the angle. Figure 17 show the velocity profile for different angles along the symmetric boundary of the welding chamber. It is seen that raising the angle will result in larger material velocity and smaller dead zone. The height of dead zone will decrease from 5mm to 1.8mm when the angle increases from 0 to 16 degrees. Figure 18 shows the total work vs. time for various web angles. It is observed that at 16 deg. the work is reduced. It may be concluded that the increase of web angle will shrink the dead zone and thus reduces the redundant energy. However, it does not considerably affect the temperature and pressure distribution. These two parameters are very important in welding of the material in the welding chamber. The other important factor in welding is the height of the dead zone; when the dead zone is larger, the contact area between the incoming material is smaller and the welding of the material will be less efficient. Finally comparision of the above results with that of reference [4, 5] shows a very good agreement.

Fig 15: feed angle of the web

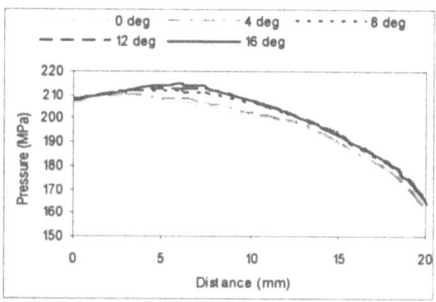

Fig 16: change of pressure along the symmetric boundary of the welding chamber

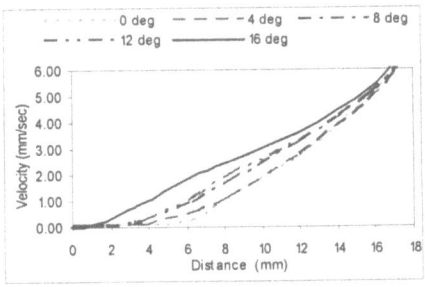
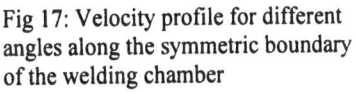
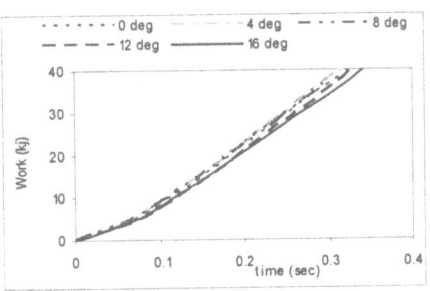

Fig 17: Velocity profile for different angles along the symmetric boundary of the welding chamber

Fig 18: External work vs. time for various web angle

6. Conclusions

The ALE finite element method is used for modelling of the aluminum pipe extrusion problem. This method is very useful because it avoids mesh distortion. 2D models of the pipe extrusion problem are presented and validated through comparison with a corresponding 3D model. It is shown that the 2D models can qualitatively represent the actual process, and thus can be used in parametric study of the process and save computational time. The 2D model is used for study of the effects of the web angle on the process, showing a decrease of the consumed energy with an increase in the web angle.

7. References

[1]   William F. Hosford, Robert M. Caddel, "Metal Forming Mechanics and Metallurgy," Prentice –Hall, 1983, PP. 168-201
[2]   Taylan Altan, Soo-lk Oh, Harold L. Gegel, "Metal Forming Fundamentals and Application," American Society for Metals, 1983, pp.196-197, 329-334
[3]   T. H. J. Vaneker, "Development of an Integrated Dasign Tool for Aluminum Extrusion Dies," Print Partner Ipskamp, Enschede, Netherlands, 2001, pp. 19-33
[4]   J.Lof, J.Huetink "FEM Simulation of the Material Flow in the Bearing Area of the Aluminum Extrusion Process," Proceeding of the Seventh International Aluminum Extrusion Technology Seminar, May 16-19, 2000, Chicago 111inois, pp. 211-220
[5]   H.G.Mooi, P.T.G.Koenis, J.Huetink, "An Effective Split of Flow and Die Deformation Calculations of Aluminum Extrusion," Journal of Materials Processing Technology, 88 (1999)
[6]   Ted Belytschko, Wing Kam Liu, Nrian Moran, "Nonlinear Finite Element for Continua and Structures," John Wiley & Sons, 2000, pp. 393-443
[7]   Klaus-Jurgen Bathe, "Finite Element Procedures," Prentice –Hall, 1996, pp. 671-675

Integration of LS-DYNA in the Process Chain of an Automotive Manufacturer

Jens Buchert[1], David K Harrison[2], Anjali K M De Silva[3], Herbert Bauer[4]

1 Glasgow Caledonian University, Glasgow, United Kingdom; Jens.Buchert@bmw.de
2 Glasgow Caledonian University, Glasgow, United Kingdom; D.K.Harrison@gcal.ac.uk
3 Glasgow Caledonian University, Glasgow, United Kingdom; A.DeSilva@gcal.ac.uk
4 University of Applied Sciences, Aalen, Germany; Herbert.Bauer@fh-aalen.de

Abstract: The introduction and evaluation of the simulation tool LS-DYNA in manufacturing process simulation is the basis for a new view in the virtual development chain. The explicit LS-DYNA forming simulation can map complex processes very exactly. One intention is to achieve more accurate computer models with the transfer of the part properties caused by production for subsequent crash and strength calculations. Using case studies a view of the transferable data is analysed to highlight their effects on the part.

1. Introduction

Today nearly all software tools required for virtual product development are in use in the automobile industry. Simulations in the area of crash, strength and manufacturing are very much state-of-the-art. In the process chain the tools work separately from each other. However, the basis of the individual simulations is the CAD data. The computer models described for strength and crash simulation are more exact than ever and boundary conditions are still improving in recent years. In spite of these developments there is still a considerable difference between the virtual results and reality. The gap between simulation and reality depends on the manufacturing process in which the part geometry and the material characteristics are transformed. In order to be able to consider the manufacturing process and the changes in material and geometry, data transfer between the individual simulation steps must take place. This step will close the virtual development chain and assure the complete process. In this investigation, the focus is on the hydroforming process to evaluate the influences caused by production processes. With the simulation tool LS-DYNA, the manufacturing steps are virtually processed and compared with the results of real parts. The results are examined and evaluated thereafter regarding their influence on the following simulations steps.

338

2. Virtual Development in the Automobile Industry

In order to involve a closed process chain, the results of the forming simulation must be available for the crash and strength simulation. This demand guarantees the process chain illustrated in Figure 1. The "results" arrows contain the possible data transfer of thickness, stretch and residual stress distribution to the downstream simulation.

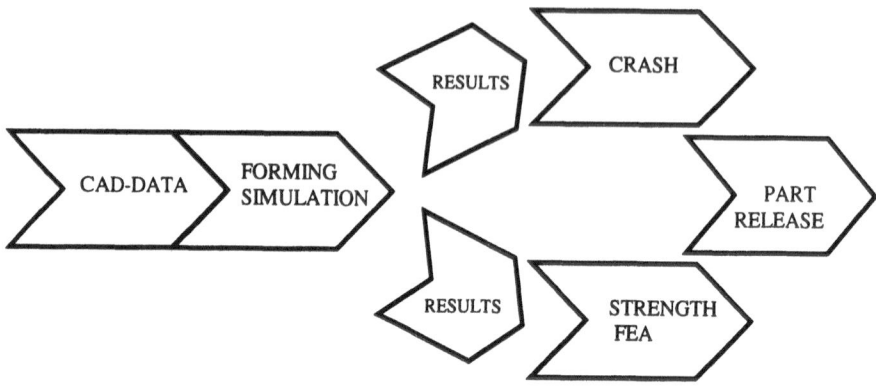

Figure 1: Virtual chain with integrated data transfer

2.1 Integration of LS-DYNA in the forming simulation

The hydroforming process represents a very complex range of the forming techniques; the evaluation of LS-DYNA takes place with the simulation of this process. An execution in other technologies (e.g. deep drawing, stretch-forming) is guaranteed after the positive evaluation of the hydroforming simulation with the software. There are many different forming steps integrated in the process e.g. bending followed by pre-forming is the basis for the best forming result after the high-pressure step. Thereby, the tasks for the software are highly complex and multifunctional, bending processes with rotary tools are standard as are also pre-forming steps in tools with slide valves. In Figure 2 the hydroforming chain is presented.

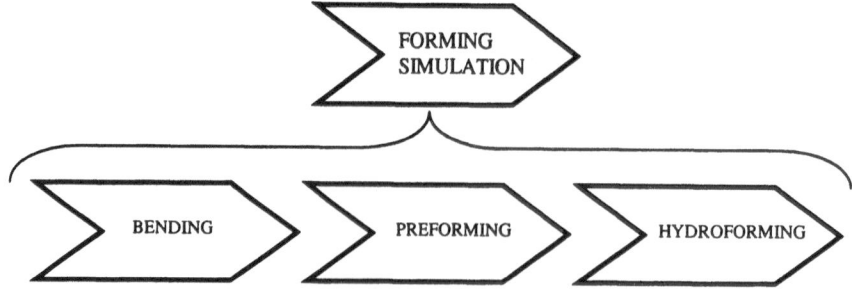

Figure 2: Process steps in hydroforming

The mesh generation for the LS-DYNA simulation is made with the pre processor ANSA, this step is based on the CAD data. All necessary tools must also exist as a CAD model. With full parametric systems such as CATIA V5, this tool generation is very readily achieved, e.g. a bending tool which automatically transfers the changes of the pipe diameter into the CAD data. For each process step prefabricated input files are prepared. Only the part's ID as well as the material, must be adjusted for the new simulation. The analysis of the results is carried out via LSPOST, the standard LS-DYNA post processor.

3. Validation of LS-DYNA Models via a Hydroforming Simulation Example

The accuracy of the final result always depends on the results of the previous simulation. A thickness and stretch distribution caused by the bending process can influence the preforming and hydroforming steps and thereby affect the feasibility of the part. Figure 3 shows the process chain of hydroforming via an exemplary component. After the bending operation in the first step, the curved pipe is pressed in a preforming tool to a flat, oval shape. The last step is a calibration process with a high pressure of up to 1500 bar.

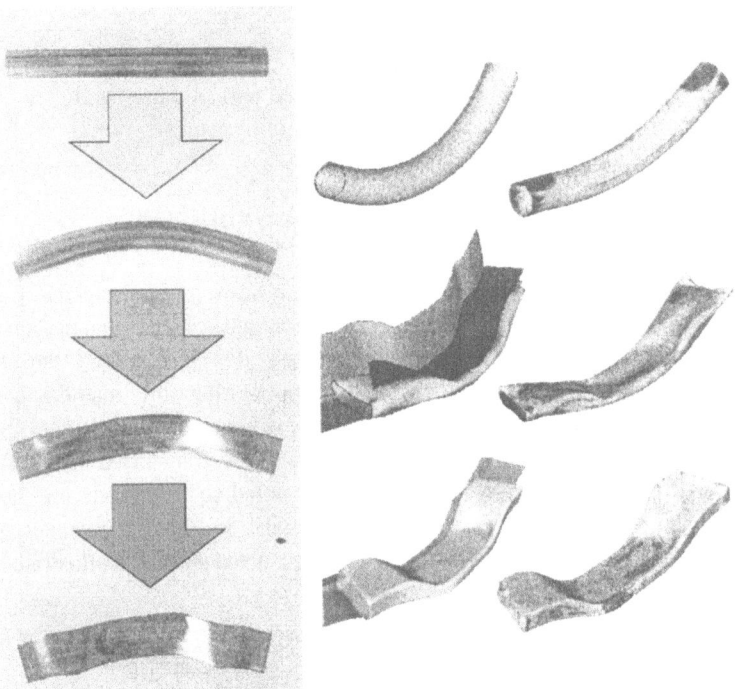

Figure 3: An example of the process chain in tube hydroforming

In order to evaluate the results of the simulation, a complex part is selected with a 3D bending geometry and a preforming operation. For the individual forming steps the thickness and stretch distribution were transferred into the following simulation. The comparison of the simulation results with the accomplished measurement of a finished part shows a perfect match of the thickness distribution [1]. Figure 4 shows the measured part sections, the measuring points in each section are radial on the tube. The measuring sections 1 and 2 are in areas with bending influence and the large cross section expansion has a high influence on the wall thickness distribution.

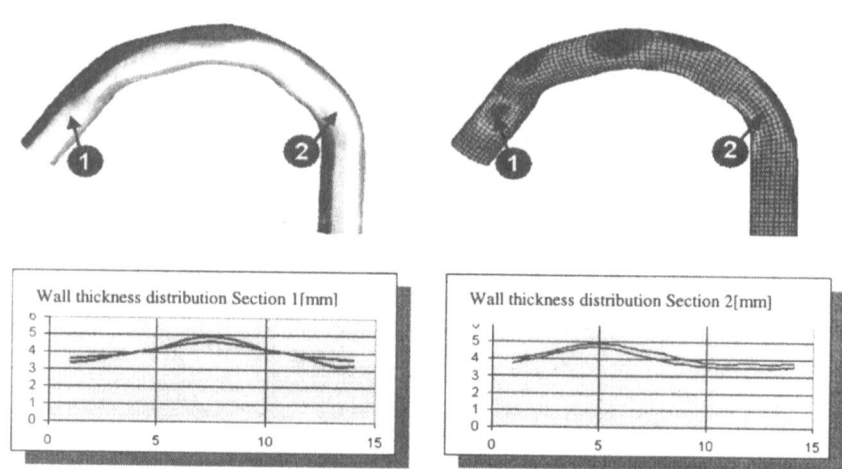

Figure 4: Wall thickness distribution on a real part (red line) and simulated part (blue line); (x-axis = number of measuring points)

4. Data transfer to Crash and Strength Simulation

The positive results of the forming simulation are the basis for the transfer of the data in the crash and strength simulation. As a result of the manufacturing simulation the thickness, stretch and residual stress distribution and the new geometry is available. Geometry change can happen after the manufacturing process caused by springback or wrinkles, thus the new geometry will deviate from the CAD model. These changes in geometry can also be transferred by surface reconstruction of the formed mesh [2]. The reconstructed surface forms the basis for the new mesh used in crash and strength calculations. In Figure 5 a scatter-plot from the manufacturing simulation and the new reconstructed surface is illustrated.

Figure 5: Scatter-plot out of the forming simulation and the reconstructed surface

4.1 Transfer of thickness, strain and residual stress distribution

Substantial parameters for the strength of a part are thickness distribution and the residual stress of the part produced, these two parameters have a direct influence on the strength calculation. The assumption of the thickness distribution is done by the mapping of the forming results on the new mesh geometry for crash and strength calculation. Usually the mesh of the forming simulation is finer than in other simulations, for the mapping process the average of a few elements is transferred on the new element. In areas with large thickness jumps (e.g. radii, beads, edges or tailored blank) the mesh geometry in crash and strength simulations must be refined to gain good results after the mapping. In section 3 the perfect agreement of simulation and reality is pointed out and this secures the transfer of thickness distribution.

At the moment, the results of the residual stresses are not compared with reality. In the future, an analysis of this area will be required to validate the results and give a conclusion over the quality of the residual stress output.

The strain distribution in a part is an indicator for the work hardening of the material. In the flow curve, the plastic behaviour for each material is shown. After the point where the elastic area ends, the strain which is necessary to form the material grows. A part with a constant strain distribution needs a higher stress than the yield strength before a new plastic deformation takes place. This affects the safety factor of the forming process which increases and the part will be passively laid out. In order to be able to convert the goals of weight reduction, the implementation of this work hardening effect is a further advantage and positively affects the safety factor. The determination of the new safety factor will take place in the post processing stage: for each element the new yield strength is defined over the strain condition and the flow curve. In Figure 6 this work hardening effect is shown on the flow curve of steel.

5. Outlook

The transfer of characteristics caused by the production process closes the virtual product development chain. The closed simulation chain allows the prediction of material behaviour that arises in test components. The view of the total process opens new potential concerning lightweight construction, costs and operating safety. In further investigations the influence of the residual stress on the part behaviour will be examined. The basis of the research is again a hydroforming part, because of the considerable knowledge of the simulation and the process already accumulated. Also, an important point is the investigation of the time behaviour of the material, which shows a possible destabilization, and a decay of yield strength.

Flow curve

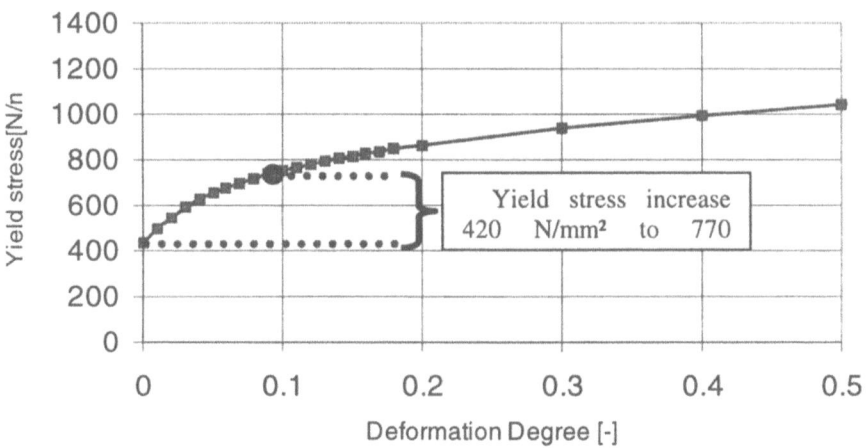

Figure 6: Flow curve of a typical high strength steel used in the automobile industry

References

[1] J. Spörer, T. Delker, A. Zisler & M. Delker (2003) "Hydroforming at BMW", 3rd Hydroforming of Tubes, pp. 91-109, Extrusions and Sheet Metals, Stuttgart/Germany, ISBN 3-88355-321-2

[2] P. Gantner, D. K. Harrison, A. K. M. De Silva & H. Bauer, (2003) "More Realistic Virtual Prototypes by means of Process Chain Optimisation", pp. 245-254, 4th European LS-DYNA Users Conference 2003, Ulm/Germany

Technical Assessment of Finite Element Software for Modelling Manufacturing Processes

Rushabh J Vora[a], Mohammed A Sheikh[b]

[a] Wolfson School of Mechanical & Manufacturing Engineering, Loughborough University, Loughborough, Leicestershire, LE11 3TU, UK. Email: - r.vora@lboro.ac.uk
[b] Department of Mechanical, Aerospace & Manufacturing Engineering, UMIST, P.O. Box 88, Manchester M60 1QD, UK

Abstract: Finite element analysis is a technique where a complex region defining a continuum is discretised into finite elements. The behavior of each element is predicted by mathematical equation whose summation approximately simulates the actual response of the Part. Finite element method has steadily increased its importance in simulation of manufacturing processes as the benefits of computationally determining the effects of various process parameters has decreased the shop floor trials. The objective of the paper is to carry out a technical assessment of finite element software like DEFORM for modelling manufacturing processes. DEFORM is a simulation system whose application ranges from various forming and heat treatment processes used in metal forming industry. Factors like software's capability in handling object geometries, range of materials available in the database, its control over process parameters and simulation were investigated. The assessment was made on the basis of the efficacy of the software for a particular process. Accuracy of the software was checked directly by comparing the results with those obtained from shop floor trials.

1. Introduction

In the late 1970s and early 1980s the use of computer-aided techniques (Computer aided engineering, design and manufacturing) in metal forming industry increased considerably [1]. However, accurate determination of various process parameters became possible only when the finite element method was developed [2].

1.1 Finite element formulation for deformation analysis of metal forming processes

Discretization of a finite element problem consists of the following steps: -

Description of the finite element: The geometry of an element, in general, is uniquely defined by a finite number of nodal points. The shape and the order of shape functions characterize the element to produce an element strain-rate matrix and an element stiffness equation.
A set of nodal point velocities in vector form is represented as:

$$v^T = \{v_1, v_2, \cdots, v_n\} \qquad (1)$$

where n = total number of freedoms in the model.

The shape functions for the element defines an admissible velocity field locally in terms of velocities of associated nodes. For example, for a two-dimensional 4-noded rectangular element, admissible velocity fields can be defined uniquely over the element by the shape functions (N_α) and the nodal velocity components as:

$$u_x(\xi,\eta) = \sum_\alpha q_\alpha(\xi,\eta)u_x^{(\alpha)} \tag{2}$$

$$u_y(\xi,\eta) = \sum_\alpha q_\alpha(\xi,\eta)u_y^{(\alpha)} \tag{3}$$

Setting up of a global system of equations: *The element equations can be assembled to give:*

$$K\Delta v = f \tag{4}$$

where K is the stiffness matrix; Δv represents the nodal velocities corrections; and f is the residual nodal point force vector [3].

Applying the contact boundary conditions: The total surface S is given by:

$$S = S_u + S_f + S_c \tag{5}$$

where S_u and S_f define the parts of surface where velocities and tractions are prescribed. S_c is the contact surface between the tool (master) and the work-piece (slave).

Friction conditions at die metal interface greatly influence metal flow, formation of surface and internal defects, stresses acting on dies, and load and energy requirements. In order to evaluate the performance of various lubricants and to predict forming pressures, it is necessary to express this interface friction in terms of a factor or a coefficient. The friction shear stress, f_s, is expressed by Coulomb law as: $f_s = \mu p$, where μ is the friction coefficient, and p = compressive normal stress at the interface (or die pressure). Friction can also be expressed as $f_s = mk$, where m is the friction factor ($0 \le m \le 1$), and k is the shear strength of the deforming material.

For numerical calculations, the frictional stress (f_s) and the relative sliding velocity (u_s) are modelled by:

$$f_s = -mk\frac{u_s}{|u_s|} \cong -mk\left(\frac{2}{\pi}\tan^{-1}\left[\frac{u_s}{u_0}\right]\right) \tag{6}$$

where u_0 is the initial velocity.

It is assumed that the relative sliding velocity u_s can be approximated in the terms of nodal values $V_{s\alpha}$ by using shape functions as:

$$u_s = \sum_\alpha q_\alpha V_{s\alpha} \tag{7}$$

Solution of the global system of equations: From a variation formulation, and arbitrariness of δv_I

$$\frac{\partial \pi}{\partial v_I} = \sum_j \left(\frac{\partial \pi}{\partial v_I}\right)_{(j)} = 0 \tag{8}$$

v, δv_I are the nodal velocities and their variations respectively; (j): the jth element.

The above stiffness equation is generally nonlinear and the solution is obtained by employing an iterative procedure such as the Newton-Raphson method. Here, on linearization by Taylor expansion

$$\left[\frac{\partial \pi}{\partial v_I}\right]_{\underline{v}=\underline{v}_0} + \left[\frac{\partial^2 \pi}{\partial v_I \partial v_J}\right]_{\underline{v}=\underline{v}_0} \Delta v_J = 0$$

$$\text{or } K\Delta v = f \tag{9}$$

where v_0 is the assumed velocity (updated according to $v_0 + \alpha \Delta v$); K: Stiffness matrix; f: residual of the nodal force vector.

Time increment and geometry updating: The deformed geometry of the workpiece in the case of two dimensions is obtained by updating the co-ordinates of the nodes (Lagrangian mesh system) by:

$$\left.\begin{array}{l} x_i(t_0 + \Delta t) = x_i(t_0) + u_x^i + \Delta t \\ y_i(t_0 + \Delta t) = y_i(t_0) + u_y^i + \Delta t \end{array}\right\} \tag{10}$$

where, (x_i, y_i): Co-ordinates of node i,

t_0 = Time at current configuration and

Δt = Time increment.

The strains are updated in a similar manner from the strain-rate solution [2].

2. Software - DEFORM

DEFORM (**D**esign **E**nvironment for **For**ming) is an engineering software that enables designers to analyse metal forming processes [3]. It is an implicit software code and follows a Lagrangian approach for updating the algorithm.

DEFORM 3D (Version 4.0) is used for three-dimensional simulations; DEFORM 2D models axi-symmetric and plain strain problems; DEFORM PC-PRO and DEFORM PC are variants for simulations on personal computers; DEFORM HT provides heat treatment process simulation capability; and DEFORM TOOLS adds to the overall presentation capability of the DEFORM system.

DEFORM 2D and 3D are available on all popular UNIX platforms (HP, SGI, SUN, DEC and IBM), as well as on PCs running Windows NT.

3. Applications

3.1 Non-isothermal spike forging

A benchmark problem of non-isothermal spike forging is analysed for determining the stresses in the dies. It is selected here to explore the capability of DEFORM-3D in forging and die stress analysis [3].

The height of the billet is taken as 2.25 inch and the top die velocity is set at 2 in/sec. The top die and the bottom pad are meshed and imported from IDEAS. The billet, on the other hand, has been meshed in DEFORM itself.

Dies are made up of H-13 whilst the material of the billet is AISI–1025. The temperature of the billet is 2000^0F and the temperatures of top and bottom dies are 300^0F and 400^0F respectively.

After importing object geometries, meshing, and defining the boundary condition, the inter-object relationships are defined as per Table 1.

Object	Relation	Shear-friction	Heat-transfer Coefficient
Billet-top die	Slave-Master	0.3	0.004
Billet Bottom die	Slave-Master	0.3	0.004

Table 1. Inter-object interface

The initial step for the spike forging problem is shown in Figure 1.

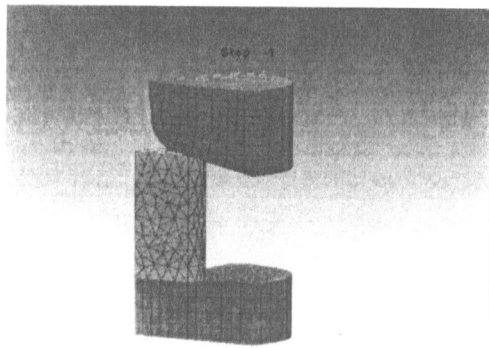

Figure 1. The initial step

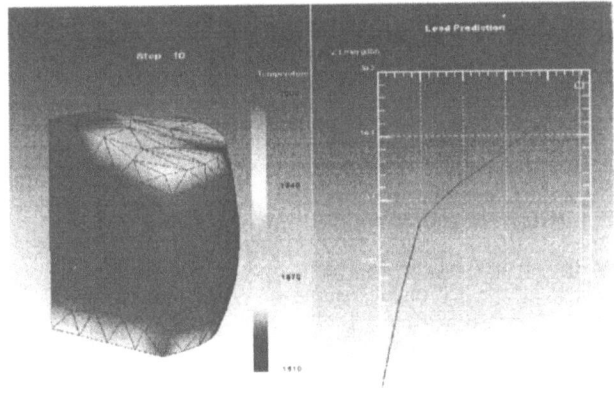

Figure 2. Temperature Profile   Figure 3. Load vs. Stroke

The results are obtained for various state variables such as strain, strain rate, effective stress and temperature. From the load/ stroke curve shown in Figure 3, the total load required at the top die to deform the billet was observed. The temperature profile at the end of 10 steps (defined for the simulation) is shown in Figure 2.

Here, the maximum and minimum temperatures are 2000^0F and 1510^0F respectively. The load/stroke graph of Figure 3 shows that the load gradually rises with the maximum (19.4 klb) at the end of the last step. This represents the maximum force required for the deformation. From a sensitivity study of various other parameters, it was found that the main factor, which affects the load of the press is the coefficient of friction between the billet and the bottom die. A reduction in the value of the friction coefficient from 0.3 to 0.1 would significantly change the maximum load.

On examining the temperature profile in Figure 2, greater chilling is seen at the contacts between the dies and the billet. This effect should be minimized in order to lower the forging loads.

The effective stress distribution for the billet is shown in Figure 4. It can be seen that the maximum stress of 37.1 ksi occurs in region 'A' which is in direct contact with the top die. For further examination of effective stress in this region, the billet is also sliced in a plane normal to the billet and oriented to view the cross-section of the billet.

3.2 Die stress analysis

The stresses which are developed in the dies at the end of the above forging process are now analysed for the integrity assessment of the dies. The work-piece is removed and the forces exerted on the dies by the work-piece are interpolated.

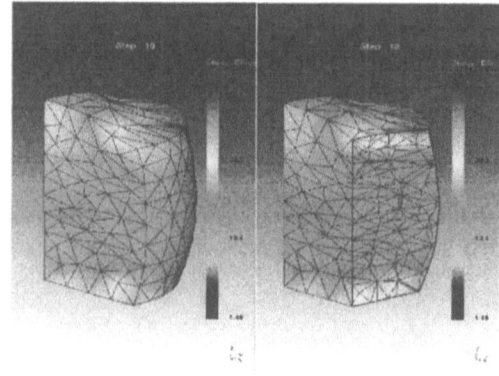

Figure 4. Effective stress

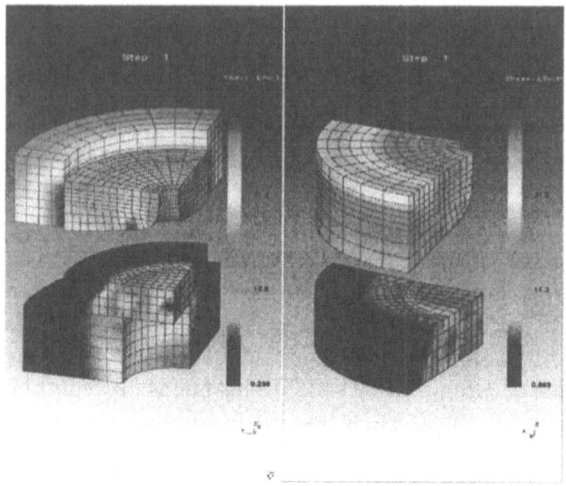

Figure 5. Die stress analysis

The effective stress distribution within the dies is shown in Figure [5], where the regions of high stress have been marked. The maximum stress in the die is 31.8 ksi (220 MPa). This is much lower than the yield stress value of 372 MPa for H-13 (die material) and is thus acceptable.

4. Discussion`

DEFORM is a reliable software in metal forming industry. The forging and die stress analysis was performed effectively, where DEFORM was able to estimate the forging loads and the stresses in dies at the end of the simulation.

Application ranges from forging, extrusion, machining, die stress analysis, cogging, glass pressing, shape rolling, drilling to predicting phase transformation, ductile fracture, micro structural evolution, machining distortion & chip *morphology*. DEFORM has separate templates for extrusion, cogging (DEFORM

3D) and hammer, machining, and rolling (DEFORM 2D), due to which the processes can be simulated very accurately and in less pre-processing time.

Advantages of the software are in its wide range of application and its user friendly graphic user interface. Important features include its extensive material database, capability to create user defined material data input, good geometry handling capability, and good control over process parameters. DEFORM is capable to produce accurate results which rages from stresses, strains, temperatures, Load-stroke information, point tracking as well die strain, grain flow, material flow, die fill, defect formation and ductile fracture.

The software is still being developed in the areas of rotary forming and extrusion processes. As DEFORM uses solid elements, it is hard to simulate thin surfaces made up of shell elements. It is a straight forward application in the metal forming industry but is not suitable for solving structural and dynamic problem.

5. Conclusions

The areas of application of the Finite Element Method to model manufacturing processes are potentially broad. It can be used for the simulation of many complex processes where theoretical analysis of the process parameters would be difficult. DEFORM is an example of some effective finite element programs developed for metal forming, ring rolling and roll forming processes respectively. It can be inferred from the paper that DEFORM is capable of simulating metal forming processes effectively.

References

[1]   Kalpakjian, S. Manufacturing processes for engineering materials, Addison-Wesley, NY, 1991.
[2]   Kobayashi. S, Soo-Ik Oh, and Altan, T. Metal forming and finite element method, Oxford University Press, NY, 1989.
[3]   Scientific Forming Technologies Corporation, DEFORM-3D, V 4.0, Columbia, Ohio,1993.
[4]   SHAPE-RR User Manual V 1.3, SHAPE Co. Ltd (Korea).
[5]   www.imsteel.com/h13.htm, Sept. 2003.
[6]   Zienkiewicz, O C. The Finite Element Method in Engineering science, McGraw-Hill, Maidenhead, 1971.

Investigation of Asymmetrical Plate Rolling by Arbitrary Lagrangian Eulerian Finite Element Technique

F. Farhat-Nia[1] , M. Salimi[2,*] and M.R. Movahhedy[3]

[1]Mech. Eng. Dept., Khomainishar Azad University, IRAN zh_farhat@yahoo.com
[2,*]Mech. Eng. Dept., Isfahan University of Technology, IRAN, salimi@cc.iut.ac.ir
[3]Mech. Eng. Dept., Sharif Univ. of Tech., Tehran, IRAN, movahhed@sharif.edu

Abstract: In this paper asymmetric plane strain rolling is analyzed by using an elastic-plastic ALE finite element method. Results of the ALE finite element investigation of curvature development due to inequality in work rolls/plate surface finish (interface friction) and speed mismatch are presented. Reasonable agreements were found between the numerical method and experimental results. The ALE technique is found to be a convenient method for simulation of processes such as asymmetrical rolling where the material is deformed in an unexpected shape.

1. Introduction

In recent years the asymmetrical rolling process has become more important due to realization of improved properties of the product surface and lower forming energy. Theoretical and experimental studies have been carried out to investigate the deformation mechanics of asymmetrical plane strain rolling. Johnson and Needham [1] experimentally investigated the rolling parameters on lead specimens. Their experiments showed that the speed ratio or speed mismatch, the diameter ratio, and the surface roughness of the work rolls affect the rolling force, the rolling torque and strip curvature. Collins and Dewhurst [2-3] presented a slip line field solution covering a range of geometries of asymmetric hot rolling of strip and compared it with experimental data. Pospiech [4] investigated the effect of thickness reductions on curvature developments for the asymmetrical rolling conditions. Salimi and Sassani [5] based on the slab method of analysis developed an analytical model for the general case of asymmetrical plane strain rolling due to unequal roll diameter, unequal surface speed of the rolls and different contact friction. Pietrzyk et al [6] simulated the steady state asymmetrical plate rolling by the finite element method. Hamuyu et al [7] studied the asymmetrical rolling process in unequal surface speed conditions using the rigid perfectly plastic model. Shivpuri et al [8] used an explicit integration FE method. Coa et al [9] studied the curvature development due to speed mismatch, Richelsen [10] investigated the influence of the degree of deformation and initial thickness on bending using the FE method. Knight et al [11] used the FE technique to analyze the effect of asymmetrical factors. Some papers are not essentially able to predict the developed curvature. In this paper asymmetrical plane strain rolling is analyzed using an elastic-plastic ALE finite element method.

In the development of the numerical analysis, the following assumptions were made:

A. The material being rolled is considered elastic-plastic and work hardening material.
B. The analysis is undertaken in steady state conditions.
C. The deformation in the lateral direction is negligible, i.e. the plate undergoes plane strain deformation.
D. The frictional stress between the sheet and the rolls is considered as follows:

$$\tau_f = \begin{cases} \mu p & for \quad \mu p < \bar{\sigma}/\sqrt{3} \\ \bar{\sigma}/\sqrt{3} & for \quad \mu p \geq \bar{\sigma}/\sqrt{3} \end{cases} \tag{1}$$

where μ is the friction coefficient and p is the pressure at the interface. (In this paper friction factor m and friction coefficient μ are chosen in such a way that they represent a unique frictional behaviour). The neutral planes are the positions at which the frictional stresses change sign.

E. The rolls are assumed to be rigid and the material being rolled is assumed to obey the Von- Mises yield criterion.

2. The Finite Element Model

In general, most of the steady state processes use Eulerian finite elements in which the elements are fixed in space and material flows through the elements. Thus, Eulerian elements undergo no distortion due to material motion and cannot properly predict the outgoing material shape, as the treatment of the moving boundary and interface is difficult with this method. To follow the material movement and to estimate the plate curvature Lagrangian meshes are more convenient. Unfortunately for simulation of processes such as asymmetrical rolling where the material is severely deformed and the outgoing plate may be deformed in an unexpected shape, the Lagrangian method may fail due to element distortion. Therefore, the ALE technique, which combines the advantages of the Eulerian and Lagrangian methods, has been employed.

In the ALE description of motion, the velocity of material itself (v) and the mesh velocity ($\bar{v}$) are two different velocities, which are defined in the ALE formulation. Although these velocities are independent, there is a one to one mapping between the material domain and the grid domain and the two domains have identical boundaries. Hence:

$$(\bar{v} - v)n = 0 \tag{2}$$

where n is the unit vector normal to the boundary surface. The physical interpretation of this equation is that the surface particles remain on the surface and no normal convective velocity occurs across the boundary. The relationship between the material time derivative $f^{\cdot}$ and the referential time derivative f' is given by [13]:

$$f' = f^{\cdot} + \frac{\partial f}{\partial x}(\bar{v} - v) \tag{3}$$

where x is the material coordinate. Equation (3) makes it possible to track the material deformation history in the ALE technique.

Transforming the principle of virtual work to the computational reference domain at time t and taking the time derivative with the computational nodal point held constant, the ALE equation is written as follows [12]:

$$\int_{^t v} \frac{\partial \delta u_i}{\partial x_j} ({}^t\sigma_{ij} - {}^t\sigma_{ik} \frac{\partial' \bar{v}_j}{\partial' x_k} + {}^t \sigma_{ij} \frac{\partial' \bar{v}_k}{\partial' x_k} + ({}^t\bar{v}_k - {}^t v_k) \frac{\partial' \sigma_{ij}}{\partial' x_k}) d'V$$

$$- \int_{^t v} \delta u_i ({}^t f_i^B \frac{\partial' \bar{v}_k}{\partial' x_k} + ({}^t\bar{v}_k - {}^t v_k) \frac{\partial' f_{ij}^B}{\partial' x_k}) d'V - \int_{^t s} \delta u_i [{}^t f_i^s \frac{\partial' \bar{v}_k}{\partial' x_k} - {}^t \sigma_{ik} {}^t n_j \frac{\partial' \bar{v}_j}{\partial' x_k} \qquad (4)$$

$$+ ({}^t\bar{v}_k - {}^t v_k){}^t n_j \frac{\partial' \sigma_{ij}}{\partial' x_k}] d's = \int_{^t v} \delta u_i {}^t f_i^B d'v + \int_{^t s} \delta u_i {}^t f_i^s d's$$

where the left superscript t indicates the time at which the quantity occurs, f^B and f^s are body and traction forces, σ_{ij} is the Cartesian component of Cauchy stress, f_i is the material rate of change of the force f_i, n_j is the unit vector normal to the boundary and δu_i is the virtual displacement component within the domain. For a given model with N degrees of freedom, the discretization of the above equation leads to a nonsymmetrical stiffness matrix with 2N unknowns while the number of equations is only half of the number of unknowns. The relations between the material velocities and the mesh velocities at the nodal points may supply the supplementary equations, i.e. an explicit mesh motion scheme in ALE. This method has the potential of producing a higher quality mesh. In the ALE code that was used in this work, the transfinite mapping method [14] is used. The general form of relation between the mesh and material displacements in ALE may be set as;
$\bar{v}_i = a_i + B_{(i)} v_i$ such that the motion of nodes in different parts of the mesh can be controlled by the choice of the mesh motion parameters a_i and $B_{(i)}$ with no summation on i, for each degree of freedom i as follows:
- if $a_i = 0$ and $B_{(i)} = 1$ then $\bar{v}_i = v_i$ and the degree of freedom is Lagrangian.
- if $a_i = 0$ and $B_{(i)} = 0$ then $\bar{v}_i = 0$ and the degree of freedom is Eulerian.
- if $a_i \neq 0$ then the degree of freedom is an ALE one. In this case a_i is given by:
$a_i = {}^{t+\Delta t} x_i^g - {}^t x_i^g$, where x_i^g is the grid coordinate.
This paper presents the results of the ALE finite element investigation of curvature development due to inequality in work rolls/plate surface finish (interface friction) and speed mismatch. A general 2 dimensional ALE finite element code, which is developed by Gadala et al [12], is employed. Due to asymmetry, the whole specimen is discretized. A total of 800, 4-node quadrilateral elements are used in the finite element model. The mesh nodes within the roll gap are taken as Eulerian points. To track the material boundary at exit, the mesh nodes in all other boundaries are treated as Lagrangian points in both horizontal and vertical directions. The motion of internal nodes is specified by the mesh motion scheme. Material strip with an initial thickness $h_i = 6$ mm, Young's modulus $E = 68$ GPa, Poisson's ratio $v = 0.3$ and initial yield stress of $\sigma_{yo} = 50.3$ MPa is considered. The material has isotropic strain hardening with the following relationship:

$$\bar{\sigma} = \sigma_{yo} (1 + \frac{\bar{\varepsilon}^P}{0.05})^{0.26} \qquad (5)$$

where $\bar{\sigma}$ and $\bar{\varepsilon}^P$ are the effective stress and effective plastic strain respectively.

The work roll radii are considered equal ($R = 105$ mm for the case studied).

3. Results and Discussion

Figures 1 and 2 illustrate the effect of roll speed ratio v_2/v_1 upon rolling force and rolling torque for different thickness reductions, respectively (Quantities at the upper roll are denoted by subscript 1 and at the lower roll by subscript 2). As expected, both the rolling force and rolling torque decrease with increasing the roll speed ratio whereas they increase with increasing the thickness reduction.

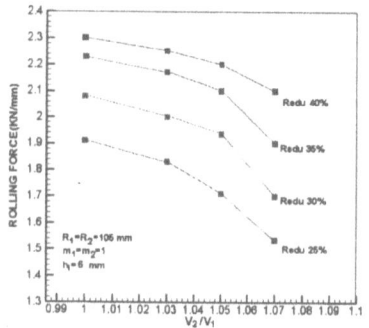 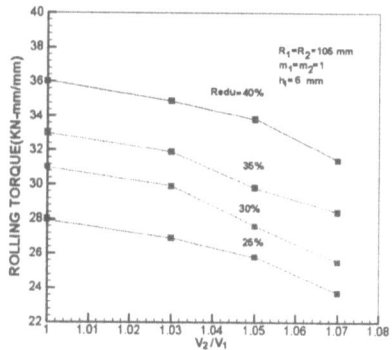

Figure1 Variation of roll force with speed ratio   Fig .2 Variation of torque with speed ratio

Figure 3 Shows the deformed shape and the contours of equivalent plastic strain for 40% reduction in thickness, equal roll speed and equal friction coefficient at upper and lower interfaces (i.e. symmetrical rolling condition). It is seen that the rolled material leaves the roll gap in a quite straight shape and yielding is indicated to occur before the point of contact at the boundaries while at the centre it is closer to this point. The deformed shape and the distribution of effective plastic strain for 40% reduction in thickness, equal work rolls speed ratio and a friction factor ratio of $m_1/m_2=2.5$ is shown in Figure4. As Richelsen [10] reported in this condition the plate is bent toward the roll with the greatest friction. It can be demonstrated by this method that increasing the reduction in thickness will increase the curvature. In Figure5 the distribution of equivalent plastic strain for different friction factors at the interface and for 25% reduction in thickness is shown. In this figure the roll speed ratio is set, so that the outgoing plate leaves the plastic region in an unbent shape.

Figure6 represents the effect of roll speed ratio upon the strip curvature index for $m_1/m_2= 2.5$ and reduction of 25%. The curvature index is defined as the ratio of slab upper surface length by slab lower surface length [5]. As indicated for $v_2/v_1=1$ the strip is bent towards the roll with higher friction factor (K>1). By increasing the lower roll speed, the strip curvature reduces. At about $v_2/v_1=1.05$ the rolled plate remains straight, and for $v_2/v_1>1.05$ the plate is bent towards the lower roll. Slab method values from Ref. [5] are also shown in the same figure.

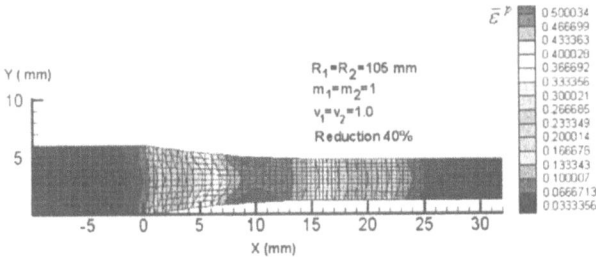

Figure3 The effective plastic strain distribution for symmetrical rolling condition.

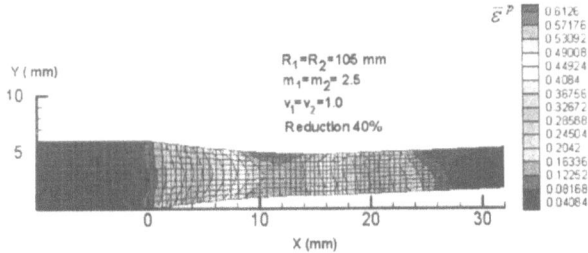

Figure 4 Deformed shape and effective plastic strain distribution for rolling with different interface frictions.

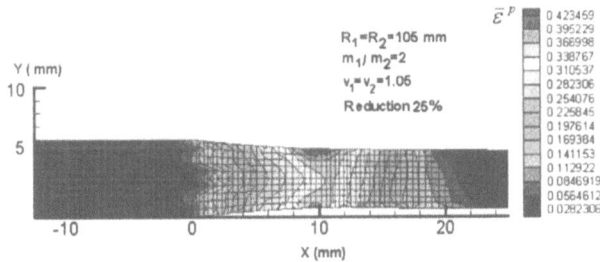

Figure 5 Effective plastic distribution for rolling with different interface friction and surface speed.

4. Conclusion

In this paper the asymmetrical rolling process is simulated using a general 2D ALE finite element code. It is shown, by this method, that when differential interface friction is applied to the plate, the predicted curvature is always towards the roll with greatest friction. The magnitude of curvature is shown to increase with increasing reduction. Roll speed mismatch is shown to have a marked effect on curvature and can be used as a means of controlling strip curvature. The results are compared with some available references, which are shown to be in good agreement. This verifies the validity as well as the suitability of the ALE technique in comparison with the conventional pure Lagrangian and Eulerian method.

356

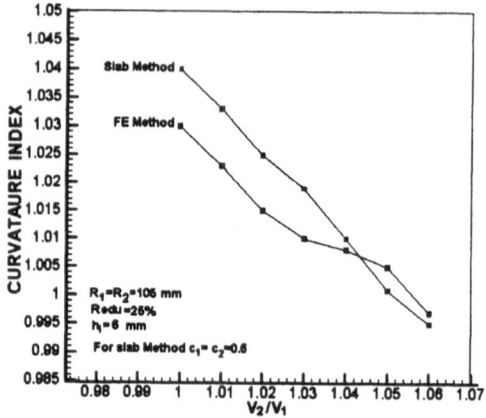

Figure 6 Effect of speed ratio on curvature index.

References

[1] Johnson W., G., Needham, 1972, *Further Experiment in Asymmetrical Rolling*. Int. J. Mech. Sci. 8443-445.

[2] Dewhurst P., Collins I.F., Johnson W., 1975, A *Theoretical and Experimental Hot Rolling*. Int. J. Mech. Sci. 17, 643-651.

[3] Collins I. F., Dewhurst P., 1975, A *Slip Line Field Analysis of Asymmetrical Hot Rolling*. Int. J.Mech. Sci.17, 643-651.

[4] Pospiech J., 1987, *A Note on the influence of Some Factors Affecting Curvature on the Flat Rolling of Strip*. J. Mech. Work. Tech., 15, 64.

[5] Salimi, M., Sassani F., 2002, *Modified Slab Analysis of Asymmetrical Plate Rolling*. Int. J. Mech. Sci., 44, pp. 1999-2023

[6] Pietrzyk, M., Wilk, K. and Kusiak, H., 1993, *Steady State FEM Simulation of the Strip bending in Asymmetrical Rolling Process*. In Proc. Metal Forming 93, Krynica, pp.50-55.

[7] Hamuzu S., Yamada K., Kawanmi T., et al, 1987, *Rigid plastic Finite Element Analysis of asymmetrical rolling, In Computational plasticity*. eds. D.R.J. Owen, E. Hinton and E. Onate, Pineridge Press, Swansea, pp. 1087-1096.

[8] Shivpuri R., Chou P.C., Lau C.W., 1982 *Finite Element Investigations of Curling in Non-symmetric rolling of flat Stock*. Int. J. Mech. Sci., 30: 625-635

[9] Coa G. R., Hall F. R., Hartley P., et al, 1993, *Elastic Plastic Finite Element Analysis of Asymmetric Hot Rolling*. In Proc. 1st Int. Conf. Modelling, of Metal Rolling Processes, The Institute of Material, London, pp.542-552.

[10] Richelsen A.B., 1997, *Elastic-Plastic Analysis of the Stress and Strain Distributions is Asymmetric Rolling*. Int. J. Mech. Sci., Vol. 39, No.11, pp 1199- 1211.

[11] Knight C. W., Hardy S. J., Lees A.W., et al, 2003, *Investigations into the Influence of Asymmertric Factors and Rolling Parameters on Strip Curvature During Hot Rolling*. J. Mater. Process. Technol. 134:180-189

[12] [12]Wang J. and Gadala M.S., 1997, *Formulation and Survey of ALE Method in Nonlinear Solid Mechanic. Finite Elements in Analysis and Design*, 24: 33- 55.

[13] [13]Hughes T.J.R., Liu W.K., Zimmermann T.K., *Lagrangian Eulerian Finite Element Formulation for Incompressible Viscous Flows*. Computer Methods in Applied Mechanics and Engineering, 29: 329-349.

[14] [14] Habor R., Shepard M.S., Abel J.F. et al, 1981, *A General Two Dimensional, Graphical Finite Element Preprocessor Utilizing Discrete Transfinite Mappings*. International Journal for Numerical Methods in Engineering 17: 1015-1044.

Cold Extrusion Modelling of Aluminium with Experimental Verification

P. Tiernan, M.T. Hillery, B. Draganescu

Dept. of Manufacturing & Operations Engineering, University of Limerick, Ireland,
peter.tiernan@ul.ie

Abstract: This paper describes experimental and finite element analysis (FEA) of the cold extrusion of high-grade (AA1100) aluminium. The influence of die half angle, reduction ratio and die land on the extrusion process was investigated. A forward extrusion die was designed and manufactured for the purpose of the experimental research. A finite element model of the cold extrusion process was developed in parallel with the experimental programme. Using this model, a thermo-mechanical simulation of the process was carried out using ELFEN, FEA software, specifically produced for metal forming simulation. Data obtained from the FE model included die-workpiece contact pressure, effective stress and strain and material deformation velocity. The correlation between the experimental and FEA data obtained in this research is presented and discussed.

1. Introduction

Extrusion, though one of the most important manufacturing processes today, is a relatively young metalworking process. In cold extrusion, which is used for the manufacture of special sections and hollow articles, the material is generally made to flow in the cold condition by the application of high pressure. The high pressures force the material through a cavity enclosed between a punch and a die. Cold extrusion can be used with any material that possesses adequate cold workability - e.g., tin, zinc, copper and its alloys, aluminium and its alloys. Indeed it is for these metals that the process is more widely adopted. Low-carbon soft annealed steel can also be cold extruded. If the product cannot be fully shaped in a single operation, the extrusion process may be performed in several stages [1]. The initial stock from which cold extrusions are produced consists of round blanks, lengths cut from bars, or specially preformed blanks. Previous research has shown that the extrusion die geometry, frictional conditions at the die-billet interface and thermal gradients within the billet greatly influence metal flow in cold extrusion [2]. Many researchers have attempted to investigate the effects of various lubricants at the die-billet interface. These investigations have resulted in a number of standard friction tests, e.g. ring and bucket test [3-5]. Geometrical characteristics of the extrusion die influence both the extrusion process and the mechanical properties of the extruded material. Experimental investigations have been made to determine the effects of die reduction ratio, die angle and loading rate on the quality of cold extruded parts, extrusion pressures and flow patterns for both lead and aluminium

[6]. The geometrical features of the die land are a critical feature in obtaining defect free cold extruded parts. As the die land length directly influences the amount of friction at the die-billet interface, extrusion die designers use this geometrical parameter to control the metal flow from the die. Appropriate die land geometrical features will allow uniform distribution of residual stresses in the extruded part as it emerges from the die. The occurrence of central bursts in cold extruded parts has been investigated as a function of die geometry [7]. These researchers concluded that the central bursts could be considered as the process of ductile fracture resulting from the coalescence and growth of microvoids in the materials. The basic idea of many ductile fracture criteria is that fracture occurs when the value of a damage parameter reaches a critical value [8-11].

2. Experimental Programme

An extensive experimental programme of forward cold extrusion was undertaken in the present investigation. The aims were to analyse the effect of three geometrical variables, namely; reduction ratio, die angle and die land height on the extrusion force. Experiments were conducted using two different lubricants; zinc stearate and an oil-based lubricant that contained lead and copper additives. The billets were cut from bar of diameter 38mm in 20mm lengths. A force-sensing component, which was integrated into the die, was designed for a nominal load rating of 2MN. This rating was calculated using a modified upper bound equation below.

$$F = 2k_f \left[4\mu \cdot \left(\frac{H}{D} + \frac{h}{d} \right) + \left(\frac{\mu}{\sin \alpha} + 1 \right) \cdot \ln \frac{D^2}{d^2} \right] \frac{\pi \cdot D^2}{4} \qquad \text{........ (1)}$$

The experimental program undertaken is outlined in Table 1. The three parameters varied during the experimental work were the die exit diameter, d, die land height, h and die angle, γ.

Table 1: Experimental programme

Exp. No	1	2	3	4	5	6	7	8	9	10	11	12	13	14	15
d (mm)	9	21	9	21	9	21	9	21	5	25	15	15	15	15	15
h (mm)	2.8	2.8	5.2	5.2	2.8	2.8	5.2	5.2	4	4	2	6	4	4	4

3. Finite Element Analysis

A finite element analysis of the extrusion process was undertaken using the flow formulation approach using rigid plastic material elements. Nodal velocities defined in the Eulerian mesh were the unknowns in the flow formulation and were related to the material strain rate by standard kinematical expressions. A mechanical simulation of the extrusion process was performed using the finite element software. This was achieved by constructing an accurate 2D CAD model of the process. The model was meshed with appropriate elements and material properties and boundary conditions were added. As the meshed model became distorted during the simulation, a remeshing facility made the analysis of large deformations and strains possible.

4. Results

On completion of the FE simulations billet-die interface contact pressure, effective stress and strain and material deformation velocities were output to file and viewed. The magnitude of the extrusion force obtained from the FE simulation was compared to values acquired by both experiment and calculation. This comparison is presented in Figures 1 and 2. The calculated values were obtained by using Equation 1.

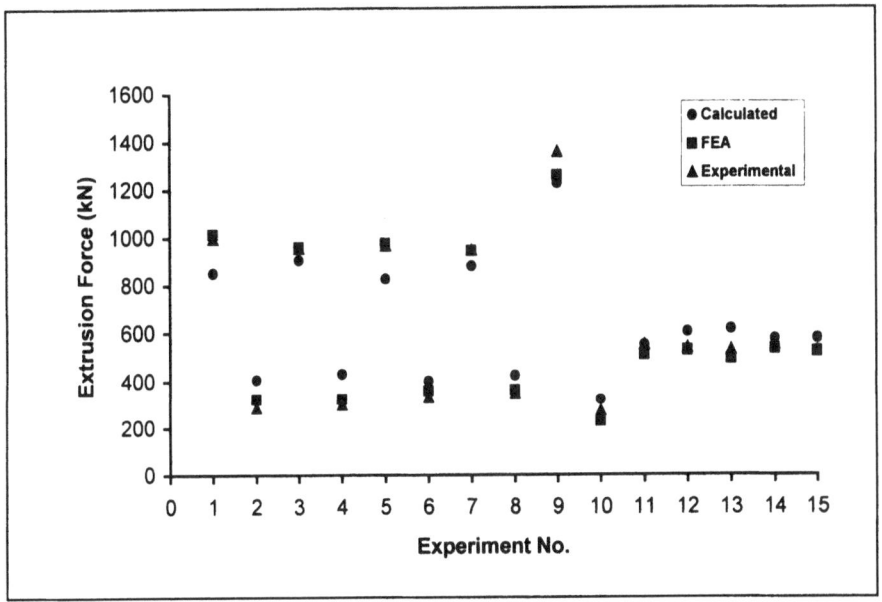

Figure 1 Extrusion force using Zn stearate lubricant

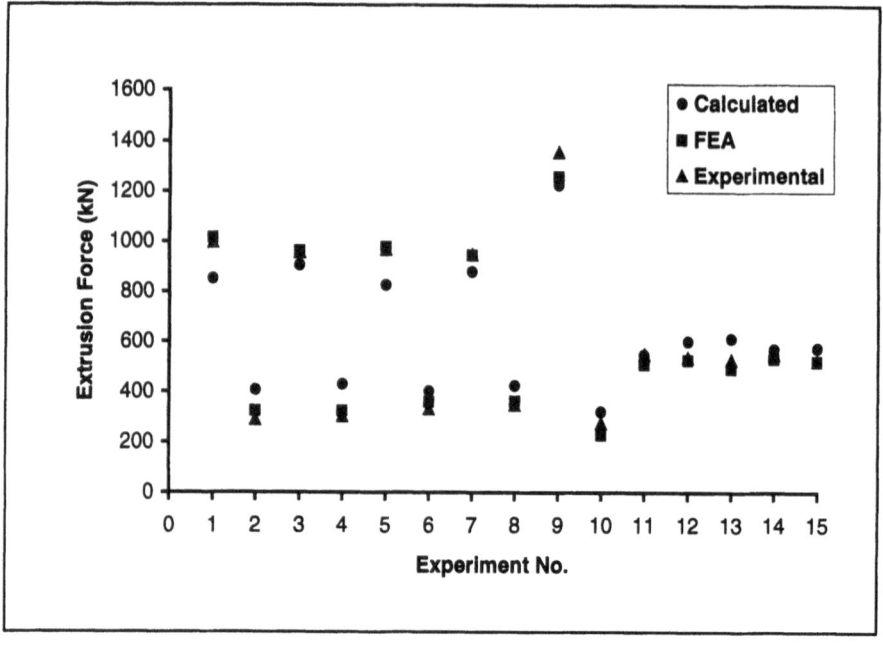

Figure 2 Extrusion force using Cu-Pb lubricant

5. Discussion

Fifteen manufactured die sets were successfully used in a purpose built extrusion tool to investigate the effect of die geometry on the extrusion force. It was possible to vary the reduction ratio between 60% and 98.4% and vary the die angle from 120° to 180°. Additionally the die land was varied from 2mm to 6mm. Two lubricants were used during the extrusion experiments. A finite element analysis of the extrusion process was carried out in parallel to the experimental programme. The high-grade aluminium billet was modelled as a perfect plastic material for the non-linear analysis during the finite element simulation. Output from the finite element program included contact pressure, effective stress and strain and material deformation velocities. The values for the magnitude of the extrusion forces obtained experimentally were compared to both the finite element results and data calculated from Equation 1. This comparison of results is illustrated in Figures 1 and 2 and show excellent correlation between the FE, experimental and calculated values.

6. Conclusions

A successful experimental programme of cold extrusion of high-grade (AA1100) aluminium was carried out with purpose-built tooling. Extrusion forces were successfully determined by incorporation of a load cell in the tooling set-up.

Experiments were conducted using two different lubricants; zinc stearate and an oil-based lubricant that contained high pressure lead and copper additives. There was no remarkable difference in the extrusion forces required for the different lubricants. The finite element results show good correlation with results obtained from the experimentation, thus confirming the accuracy of the finite element model. Furthermore extrusion force data obtained by calculation show excellent agreement with data obtained from both calculation and FE work. The largest extrusion force obtained was 1350kN. This force was measured when extruding the aluminium billet using a die with exit diameter, die angle, and land height of 5mm, 150° and 4mm respectively. This represents a reduction in area of the billet of 98.2%. Further research is planned to quantify the effect the extrusion process has on the hardness and surface roughness of the extruded component.

References

[1]   G.W. Rowe, Principle of Principles of Industrial Metalworking Processes, Arnold, London, 1977.
[2]   Altan, T., Oh, S.I., Gegel, H., Metal Forming; Fundamentals and Applications, American Society for Metals, Metals Park, Ohio, 1983.
[3]   Lee, C.H., Altan, T., Influence of Flow Stress and Friction Upon Metal Flow in Upset Forging of Rings and Cylinders, J. Eng. Ind., 775, 1972.
[4]   Shen, G., Vedhanayagam, A., Kropp, E., Altan, T., A Method for Evaluating Friction Using a Backward Extrusion-Type Forging, Journal of Materials Processing Technology, 33, pp. 109, Elsevier, 1992.
[5]   Lazzarotto, L., Dubar, L., Dubois, A., Ravassard, P., Oudin, J., Three selection criteria for the cold metal forming lubricating oils containing extreme pressure agents, Journal of Materials Processing Technology, Vol. 80-81, pp. 245-250, Elsevier 1998.
[6]   Onuh, S.O., Ekoja, M., Adeyemi, M.B., Effects of die geometry and extrusion speed on the cold extrusion of aluminium and lead alloys, J. Mat. Proc. Tech. Vol. 132, pp. 274-285, Elsevier, 1998.
[7]   Ko, Dae-Cheol, Kim, Byung-Min, The Prediction of Central Burst Defects in Extrusion and Wire Drawing, J. Mat. Proc. Technol, Vol. 102, pp. 19-24, Elsevier, 2000.
[8]   McClintock, F.A., A Criterion for ductile Criterion by the Growth of Holes, J. Appl. Mech., Vol. 35,1968, pp. 363-371.
[9]   Oyane, M., Sato, T., Okimoto K. and Shima, S., Criteria of ductile fracture and their application, J. Mech. Work. Technol., Vol. 4, 1980, pp. 65-81.
[10]   Cockcroft, M.G., and Latham, D.J., Ductility and the workability of metals, J. Inst. Met. 96 (1968), pp. 33-39.
[11]   Osakada, , K., Mori, K., Prediction of ductile fracture in cold forging. Ann. CIRP 27 1 (1978), pp. 135-139.

MANUFACTURING SYSTEMS

Market Dynamism, Manufacturing Flexibility and Type of Automation in Indian Industries: A Survey

P. B. Sharma[1], Suresh. Garg[2] and P. C. Gupta[3]

[1]Professor and Principal, Delhi College of Engineering, Delhi - 110042, India.
pbsharma48@yahoo.co.in
[2]Assistant Professor, Delhi College of Engineering, Delhi - 110042, India
suresh_kumar_garg@hotmail.com, Phone +91-11- 2549 6347,
[3]Scientist C, D,R.D.O., Lucknow Road, New Delhi, India

Abstract: The requirement of flexibility in manufacturing systems depends upon the uncertainties in the environment and the service level required. This flexibility is generally achieved by incorporating automation components like CNC machines, automated material handling systems, Robots etc. To get the best results from a system, it is essential that these three aspects, namely manufacturing flexibility, market uncertainty and use of automation and advance manufacturing technology should be considered together. This paper reviews the various issues linking these aspects and presents an empirical study to reveal the linkages, if any, in these aspects.

1. Introduction

With globalization of trade and liberalization of Indian industries, the scenario is changing from seller market to buyer market. The buyer is having a choice not only from Indian manufacturers, but global manufacturers also, who are providing their latest products. The half-life of the products and technology is reducing at very fast rate. New products are being introduced and the era of one model selling for years has gone. Information technology, e-business and the virtual factory concept are facilitating this phenomenon. To be competitive in this dynamic environment, organizations must be able to quickly respond to changes. They need to build flexibility into their manufacturing system. Automation components are incorporated in the manufacturing processes to achieve flexibility. But the incorporation of automation components is associated with high expenditure and increased complexity. Thus, there is a need to link the incorporation of automation components to the need of flexibility, both type and extent, which in turn depend upon the market dynamism. The theoretical model is presented in Figure 1.

This paper examines the inter-relationship between three attributes namely market dynamism, manufacturing flexibility and type of automation components with the help of the literature. Then the paper proposes a theoretical construct linking these attributes and also presents the results of an empirical study of Indian organizations to reveal the linkages, if any. The following hypothesis are examined in the paper:

1. All organizations are operating under the same market dynamism conditions.
2. All flexibility types are equally important to all organizations.
3. Type of flexibility desired is independent to the market dynamism.

4. All the automation components are equally incorporated in manufacturing systems.
5. Type of automation component incorporated is independent to the flexibility desired.

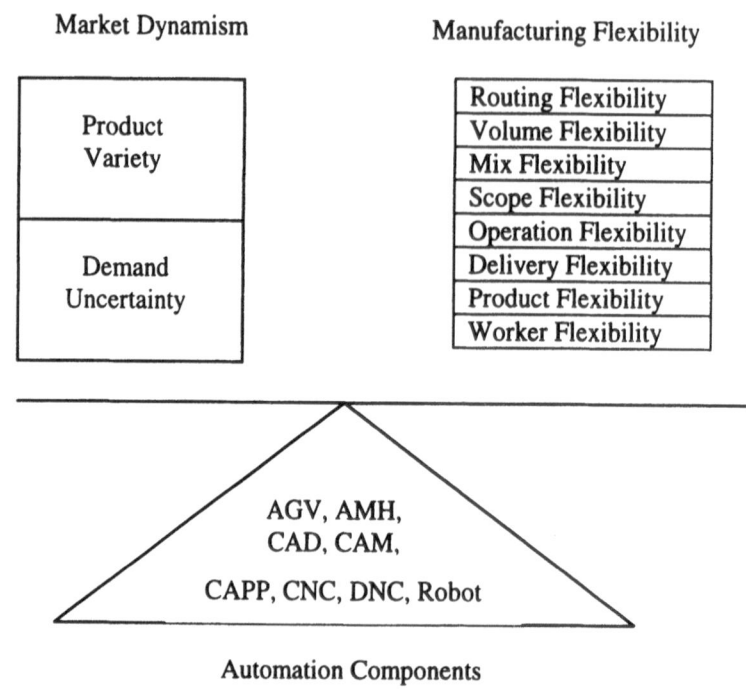

Figure 1: Theoretical Model

2. Market Dynamism, Flexibility and Automation

Market dynamism is characterized by:
- High rate of new product introduction and thereby high variety,
- Changes in demand of different models,
- Changes in total demand,
- Changes in loyalty of customers
- Fast changing technology etc.

These characteristics result in high product variety and high demand uncertainty, which are considered as the parameters of market dynamism.

Manufacturing flexibility is an important element of a firm's manufacturing strategy, providing the capability to respond quickly to shifts in market requirements [1, 2] show, through their empirical study, that manufacturing flexibility could be used to relieve problems caused by an uncertain and dynamic environment. Companies, which have high flexibility, are able to make the changes called for at low cost and in a short time period. A company with high product flexibility can introduce a new product quickly and cheaply, because it has developed rapid design routines, flexible tooling, standard production methods etc.

Companies, which are less flexible, will find that making changes is slow and costly. A Company with low delivery flexibility will find that changes in delivery dates causes expensive rescheduling problems such as overtime working, stock shortages and delays and other expenses caused by disruption. A more flexible company would have the resources to work around these changes.

In order to meet the conflicting requirements of high productivity and flexibility, flexible-manufacturing systems came into existence. It has been pointed out by [3] that the focus of competition in the global market place is shifting from quality and service towards flexibility. New systems are being made more and more flexible. For example with the advent of optical fiber, telephone, Internet and cable connection can be availed with one compact line. The new system will be flexible and cheaper. Excellent review of the literature on flexibility and definitions of flexibility is provided by [4 - 7].

Japan is one of the leading countries in adopting flexibility in manufacturing. It produced 16,555 CNC lathe machines in 1984, however its need was only for 10,551 CNC lathes, where as India produced 15 CNC lathes and its requirement was for 101 CNC lathes. Since 1970, Japan has been using flexible manufacturing systems. It is reported in manufacturing competitiveness frontiers that Japan has around 138 flexible manufacturing system installations and they took a major leap by adopting the concept of flexifactory.

Germany is one of the leading countries in the area of technology developments. There are 62 flexible manufacturing systems being used. Germany is the second country after Japan in manufacturing CNC lathes. India has been importing CNC machine tools mostly from Germany and Japan.

Aerospace, automotive, machine tools and power industries sectors widely use flexible manufacturing system in the USA. More than one hundred (112) flexible manufacturing systems have been installed in USA and the flexifactory concept has also been adopted in U.S.A. by Honda group.

U.K., France and Russia are also using flexible manufacturing systems in automotive, aerospace, machine tools and defence industries. These are also known as leading countries for the manufacturing of flexible manufacturing systems.

Automation is a technology concerned with the application of the mechanical, electronic, and computer based systems to operate and control production [8]. This technology includes:

- Automatic machine tools to process parts
- Automatic assembly machines
- Automatic material handling and storage systems
- Automatic inspection systems for quality control
- Computer system for planning, data collection and decision making to support manufacturing activities
- Industrial robots

Automated production systems can be best classified into three basic types:

1. Fixed automation
2. Programmable automation
3. Flexible automation

The development of advanced manufacturing technologies/automation in production has been, and will continue to be, the major technical driving force in facilitating manufacturing change. The electronics and information technologies provide new possibilities to produce a wide variety of products in efficient way. Modern manufacturing is beginning to adopt the flexibility in their conventional job shop production systems with productivity level that is far higher than before.

Indeed, productivity in flexible manufacturing system can now compete with mass production methods. Also it seems evident that advanced manufacturing technologies can help to produce goods of superior quality. Yet the new production systems seem to function efficiently on a much smaller scale (machine and personnel) than the older mass and large-scale production system. Advanced manufacturing technology is only one way of delivery flexibility [9].

3. Empirical Study of Indian Industry

This work aims to establish the importance level of various types of manufacturing flexibility and various sets of automation components required in achieving this flexibility in the Indian context, at present and in the future. In this study, three concepts of manufacturing namely Market Dynamism, Manufacturing Flexibility and Types of Automation are studied. Since this study is exploratory in nature, a survey methodology is used. The method of gathering relevant information from Indian industries was through a structured questionnaire. The questionnaire was based on a literature review and discussions held with practitioners and faculty. To assess the content validity a 'dry run' was made and few questionnaires were administered to leading practitioners and academicians. Based on their feedback the present questionnaire has been prepared, responses obtained, complied and analyzed.

Out of the 39 organizations who responded to the questionnaire, 24 are from the automobile sector and 15 from the machine tool sector. The annual turnover of the organizations varies from Rs. 7 millions to Rs. 13650 millions and the numbers of employees varies from 16 to 16000.

3.1 Market dynamism in Indian industry

In the questionnaire, market dynamism of an organization is measured via two parameters:
❑ Variety of products,
❑ Demand variation of the products
Respondents were asked to measure these parameters as high or low. Low product variety means less than 10 models/ product type, whereas low demand variation means range less than ±10% of mean demand.

The responses on market dynamism show that 30 organizations have high variety of products whereas remaining 9 have variety less than ten types. The demand variation of 13 organizations is more than ±10% whereas the remaining 26 organizations have fairly stable demands.

3.2 Manufacturing flexibility and market dynamism

Due to the variety of products and demand variations, organizations need to built flexibility into their manufacturing systems. The flexibility is required to meet the customer service level under dynamic and competitive market environment. In this study, the respondents were asked to rate the importance of various types of flexibility presently being used in their organizations. They were also asked to express their proposed level of requirement of flexibility for their manufacturing systems. Over 70 terms (type and measure) of flexibility have been defined in the literature [10] and often several terms refer to the same flexibility type and terms, which sometimes are identical, have quite different meanings. Therefore it is essential when studying flexibility to develop an *a priori* definition of the particular flexibility type which is to be examined. Multifunctional workers, as a flexibility type, is included in the study because the flexibility of plant depends much more on people than on any technical factor [11].

The respondents were asked to evaluate the relative level of importance of various types of flexibility from **1** (very low importance level) to **5** (very high importance level) in the part A of questionnaire. In part B of the questionnaire, the response on the present and proposed application level of various sets of automation components, to achieve the required types of flexibility, have been collected. The responses were studied using t-test and results are presented in Table 1. On a 5-point scale, the mean scores of the different flexibility types are also given in this table. Scope flexibility has the highest score of 3.769 and the next is mix flexibility with a score of 3.744. The need of these flexibility types is high, because the respondent organizations have indicated high product variety, but low variability of product demand. Routing flexibility and operations flexibility have mean score less than 3.0. The results of the significance test (t-test) show that mix flexibility, scope flexibility, delivery flexibility, product mix and worker flexibility have a level of importance which is more than moderate at a 5% level of significance. On the other hand, the level of importance of routing flexibility and volume flexibility, is below the moderate level.

Table 1: Importance Level of Flexibility Types

Sno.	Type of Flexibility	Mean	Std Dev.	T-test U=3	Significance at 5% level
1.	Routing Flexibility	2.744	1.39	-1.135	#
2	Volume Flexibility	3.368	0.997	2.275	#
3	Mix Flexibility	3.744	0.88	5.211	*
4	Scope Flexibility	3.769	0.931	5.091	*
5	Operation Flexibility	2.282	1.276	-3.468	# #
6	Delivery Flexibility	3.538	0.854	3.883	*
7	Product Flexibility	3.462	1.189	2.395	*
8	Worker Flexibility	3.692	0.922	4.626	*

$t_{0.05} = 2.37$ *Significantly above moderate
Difference is not significance   # # Significantly below moderate

The relationship between the need of flexibility type and market dynamism is studied using correlation analysis. For this purpose, the low product variety is given a score of 1 and high product variety is given a score of 5. Similarly, the low demand variability is scored as 1 and high as 5. The analysis shows that product variety has high correlation with scope flexibility. The correlation shows that the need / application of mix flexibility, scope flexibility, demand flexibility and product flexibility increases with increased variety of product. To meet the demand variation in the market place volume flexibility, mix flexibility and product flexibility play an important role.

3.3 Automation components and manufacturing flexibility

There are various automation components of advanced manufacturing technology, but the following principal automation components are included for the study:
1. Automatic Guided Vehicle System (AGVs)
2. Automatic material handling (AMH)
3. Computer aided design (CAD)
4. Computer aided manufacturing (CAM)
5. Computer aided process planing(CAPP)
6. Computer numerical control (CNC)
7. Direct Numerical Control (DNC)
8. Industrial Robots

The application level of automation components to achieve different flexibility types is given in Table 2. In each box two values are given. The upper value is the present level and the lower value is the desired/ proposed level of application.

Table 2: Application of Automation Components for different Flexibility Types

Automation Component	Flexibility Type						
	Routing	Volume	Mix	Scope	Operation	Delivery	Product
AGV	0.256*	0.231	0.103	0.026	0.179	0.231	0.000
	1.974#	0.359	0.256	0.153	0.282	0.333	0.077
AMH	0.368	0.256	0.333	0.256	0.231	0.487	0.126
	0.872	0.692	1.000	0.615	0.538	1.103	0.128
CAD	0.692	1.154	2.128	3.128	0.205	0.692	2.769
	0.974	1.102	2.974	3.718	0.307	0.948	3.512
CAM	0.718	0.538	0.821	0.769	0.385	0.718	1.205
	0.846	0.769	1.026	1.230	0.435	0.974	1.077
CAPP	0.795	1.615	1.641	1.487	0.360	2.179	1.205
	1.384	2.384	2.051	2.384	0.692	2.923	1.948
CNC	2.256	2.293	3.077	2.589	2.447	2.948	2.359
	2.307	3.794	3.480	3.611	3.153	3.487	2.897
DNC	0.205	0.512	0.000	0.153	0.128	0.256	0.384
	0.282	1.205	1.230	0.359	1.051	1.435	0.769
Robot	0.103	0.102	0.000	0.000	0.026	0.205	0.000
	0.512	0.487	0.394	0.000	1.103	0.236	0.000

* Present application Level   # Desired application level in future

Indian industries are using computer numerical control (CNC) machine tools at various application levels. They are used to achieve the various types of flexibility such as volume flexibility, mix flexibility, operation flexibility, and delivery flexibility, as these machine tools have extremely high flexibility with reasonable initial and running costs. In other words, CNC machine tools are one of the most important automation components to achieve manufacturing flexibility in Indian industries. It is also clear that the application level of CNC machine tools will further increase in future.

4. Conclusions

This paper has focused on finding the status and relationships between market dynamism, manufacturing flexibility and types of automation in Indian industries. The study shows that 76.9% of Indian industries have a high product variety i.e. there are more than ten models of their products. It means that manufacturers have to devote too much emphasis to the increasingly wider range of products. From the study it seems that Scope Flexibility and Mix Flexibility are the most important flexibility types, which are required to meet the increasingly product customization and increasingly wider range of products. Advanced manufacturing technology is a way to achieve manufacturing flexibility. At present, some companies have started implementing Computer Numerical Control, Computer Aided Design, Computer Aided Process Planning. Other automation components such as Automated Material Handling systems, Automated Guided Vehicle systems, Direct Numerical Control machine tools, etc. which are not widely used at presently, will be adopted by many of the Indian industries in the future.

References

[1] Boyer, K. K. and Leong, G. K., Manufacturing Flexibility at the Plant Level, Management Science, 24(5), 495-510, 1996.

[2] Swamidas, P. M. and Newwll, W. T., , Manufacturing Strategy, Envieonment Uncertainty and Performance: A Path Analytic Model, Management Science, 33(4), 509-524, 1987.

[3] Dixon J. R. Measuring manufacturing flexibility: An empirical investigation. European Journal of Operational Research. pages 131-143, 1992.

[4] Browne J. Dubois D. Ratmill K. Sethi S.P. and Stecke K.E. Classification of FMS. FMS magazine, page 114-117. 1984.

[5] Gerwin D. An agenda for research on the flexibility of manufacturing processes. International Journal of Operations and Production Management pages 38-49, 1986.

[6] Sethi A. K. and Sethi S.P. Flexibility in manufacturing: A survey. International Journal of Flexible Manufacturing Systems, 2(4): 280-328, 1990.

[7] Slack N. Flexibility as a manufacturing objective. International Journal of Operations and Production Management, 3:4-13, 1983.

[8] Groover, M. P. and Zimmer, E. W., CAD/ CAM, PHI, India, 1984.

[9] Gerwin, D., 1993, Manufacturing Flexibility: A Strategic Perspective, Management Science, 39(4), 395-410.

[10] Shewchunk, J. P. and Moodies, C. L., A Framework for Classifying Flexibility Types in Manufacturing, Computers in Industry, 33(2-3), 261-270, 1997.

[11] Upton, D. M., What Really Makes Factories Flexible?, Harvard Business Review, pp 74-84, July-August, 1995.

A Methodology for Evaluating the Efficiency of PLM Tools in Product Development Processes

Matteo Benassi[1], Monica Bordegoni[1] and Gaetano Cascini[2]

[1] Dip. di Meccanica – Politecnico di Milano, Via La Masa 34 – 20258 Milano Italy, [matteo.benassi; monica.bordegoni]@polimi.it
[2] Dip. di Meccanica e Tecnologie Industriali – Università degli Studi di Firenze, Via Santa Marta, 3 – 50139 Firenze - Italy, gaetano.cascini@unifi.it

Abstract: In order to be competitive in modern and growing markets, companies have to improve their production processes (shorter *time to market* and minor *product costs*) and reach increasingly higher quality standards. Within Product Lifecycle Management (PLM), Knowledge Management (KM) has proved to be key vector for business and product quality improvement, reducing lifecycle costs and time. Knowledge is often not explicitly expressed and often exists as personal and tacit know-how of people, that's why KM is considered a critical aspect for companies and it is strategic adopting KM methods and tools. Two issues have to be considered for the adoption of new KM methods and tools within the product lifecycle: 1) how to select appropriate and effective methods and tools; and 2) how to estimate benefits and impacts before adopting and/or integrating those methods and tools. The paper describes a methodology we have developed for defining and validating an integrated environment for studying and evaluating the adoption of knowledge and innovation management tools within product design process.

1. Introduction

Product development processes include several critical aspects. Among the others, today costs and time are more and more constraining for companies' competitiveness. Recent Product Lifecycle Management (PLM) solutions propose methodologies and tools aiming at improving the product development process and competitive engineering. They support a more closely integrated management of engineering activities of product lifecycle with process planning and manufacturing aspects. Within this broader view of product development and integration of its various aspects, Knowledge Management (KM) has proved to be a key enabler to reducing lifecycle costs and time, improving quality and helping to ensure safe products [1]. Knowledge-related issues are considered critical aspects for companies, since Knowledge is recognized to be one of the major assets of companies, but it is often not explicitly expressed and often exists as personal and tacit know-how of people. In current collaborative environments, Knowledge is managed not only within individual companies, but also across supply-chain relationships. Therefore, it is even more strategic adopting methods and tools supporting KM (EKM – Engineering Knowledge Management systems), where Knowledge is also distributed in collaborative environments.

Two issues have to be considered for the adoption of new KM methods and tools within the product lifecycle: 1) how to select appropriate and effective methods and tools; and 2) how to estimate benefits and impacts before adopting and/or integrating those methods and tools.

The authors have just concluded a research project proposing a solution to these two issues. The aim of the project was to identify and validate an integrated environment for studying and evaluating the adoption of knowledge and innovation management tools within product lifecycle (www.kaemart.it/ike). The main outcome of the project is a methodology that consists of guidelines for the adoption and integration of new technologies within the product development process. The paper describes the methodology developed, and one of the implemented study cases that apply and test the developed methodology.

2. Methodology

The methodology consists in a sequence of activities synthesized into a Roadmap that can be used for the selection and the evaluation of the efficiency and

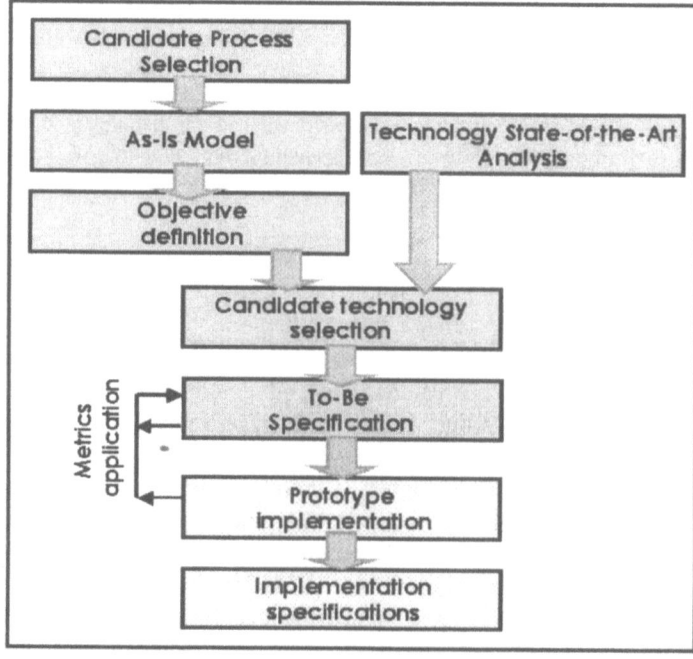

effectiveness of PLM/EKM tools in any product development process [2]. The Roadmap (figure 1) plans a set of steps, described in the following sections.

2.1 Process modelling and state of the art analysis

This activity consists in an initial selection and modelling of the product development sub-process as it is implemented in a company by interviewing company's experts, collecting data and structuring data (*Candidate Process Selection* and *As-Is Model* activities). Process data are organized and structured using the IDEF0 technique [3]. The outcome of this activity consists of IDEF0 diagrams of the selected process, providing a well-structured view of the process, including activities, actors, roles, resources, technologies, information flow, etc. Process models are subsequently analysed to identify the most critical aspects and those aspects that may be improved (*Objective definition activity*), in particular extracting information on knowledge: types of knowledge, use of knowledge within the process lifecycle, how knowledge is stored, and its impact on the process, etc. This allows us to explicitly defining activities related to knowledge in the process. This activity is conducted in collaboration with companies' experts (managers, designers, technicians) and process analysts, who help in structuring, rationalizing, and formalizing the process.

In parallel, technology experts update the state of the art of emerging tools and systems aimed to knowledge and innovation management, also evaluating the level of implementation of each technology functionality and their integrability (*Technology state-of-the-art analysis* activity). This activity provides a useful overview of possible solutions for satisfying specific process requirements.

2.3 Candidate Technology selection

Previous analysis information is collected into two matrixes, according to the QFD (Quality Functional Deployment) method adopted [4]: *Matrix 1* that reports the importance of generic knowledge management activities in respect to the process task of the selected product development cycle; *Matrix 3* that links PLM and EKM technologies with a set of functionalities related to knowledge management. An intermediate matrix has been defined, between Matrix 1 and Matrix 3 for correlating EKM activities and K functionalities. The matrix, named *Matrix 2,* is filled in by process experts together with technology experts. The matrix reports EKM activities along the rows and technology functionalities along the columns.

The column "*valuation*" reports the relative importance values for each EKM activity as resulting from Matrix 1. The importance of K-functionality related to each EKM activity is set by the process and technology experts on the basis of how the process performances have to be improved ($B_{i,j}$ values of Matrix 2). For each K-functionality we sum up the values measuring its relevance in respect to all EKM activities (weighted in respect to a specific process), and therefore, we obtain a ranking of the technology functionalities. These values are considered as weights in the "valuation" column of Matrix 3. Fig. 2 shows the matrixes and the candidate technology selection process.

The final selection of the candidate technology is done on the basis of the results

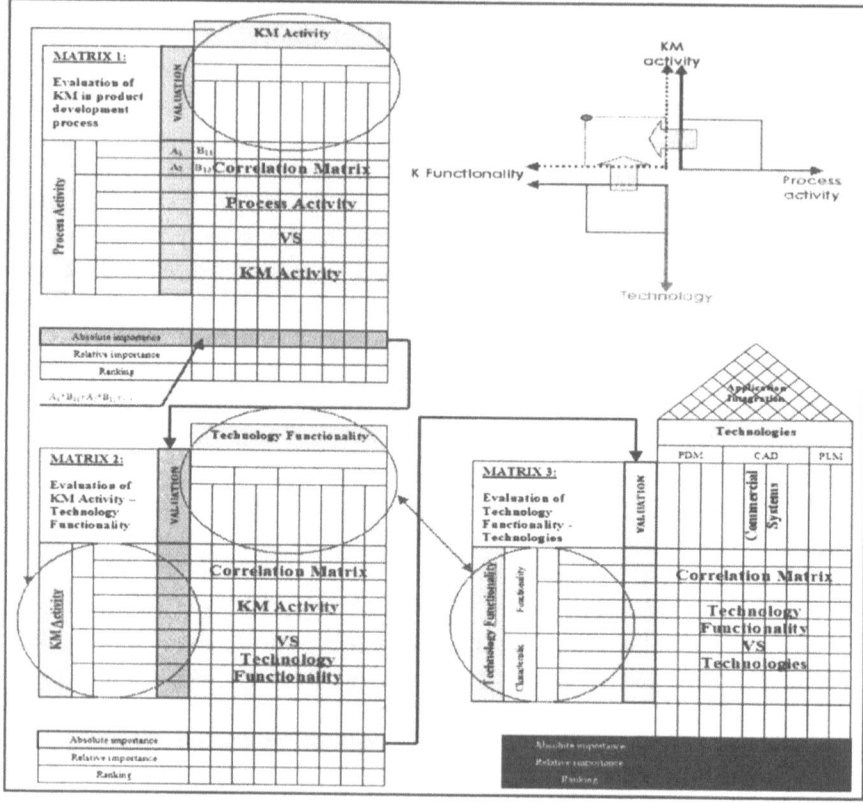

Figure 2 – Candidate technology selection method

obtained in Matrix 3. The decision is taken considering how a specific technology satisfies the product development process requirements (*Candidate technology selection* activity).

2.4 To-Be modelling and metrics

To evaluate the impact derived by introduction new selected technologies in the product development process, a modelling of the To-Be process including the selected technology and a simulation of the new process are performed (*To-Be specification* activity). Metrics parameters are defined in order to compare As-Is and To-Be processes (*Metrics application* activity). At first, metrics permits us to refine the To-Be process, by evaluating performances of To-Be simulated process versus As-Is process; once the new technological solutions have been integrated within the selected candidate process (*Prototype implementation* activity), the metrics is applied to evaluate the process performance enhancements, if any.

The final step of the methodology consists of defining specifications for new technology implementations within the real product development process on the basis of the performed activities and results (*Implementation specification* activity).

3. Study Case

The study case presented in this paper aims at evaluating the developed methodology, and has been developed in collaboration with JOBS, a company that designs and produces 3/5-axes operating machines and automated milling systems for aerospace, car industry, general mechanical and energy sectors.

The candidate process selected concerns the design of the tool-changer of a machine tool. As-Is modelling has been carried out by interviewing the designers involved in the machine development, and collecting data using the IDEF0 technique. The subsequent As-Is analysis underlines critical aspects and requirements within the current design processes, related to issues concerning Knowledge Management. Some of them are: simplify the reuse of existing projects; transfer singular experiences to the whole work-group; better manage design knowledge (better availability and quality of information); limit errors and loops, especially during preliminary design phases where they can play a more critical role; introduce a better support for novice designers.

The method for identifying the candidate technology has been applied, and has suggested that the RuleStream system (a KBE application capable of capturing and sharing design practices) would have been the most appropriate solution to satisfy the process requirements. Considering the identified technology, the To-Be process

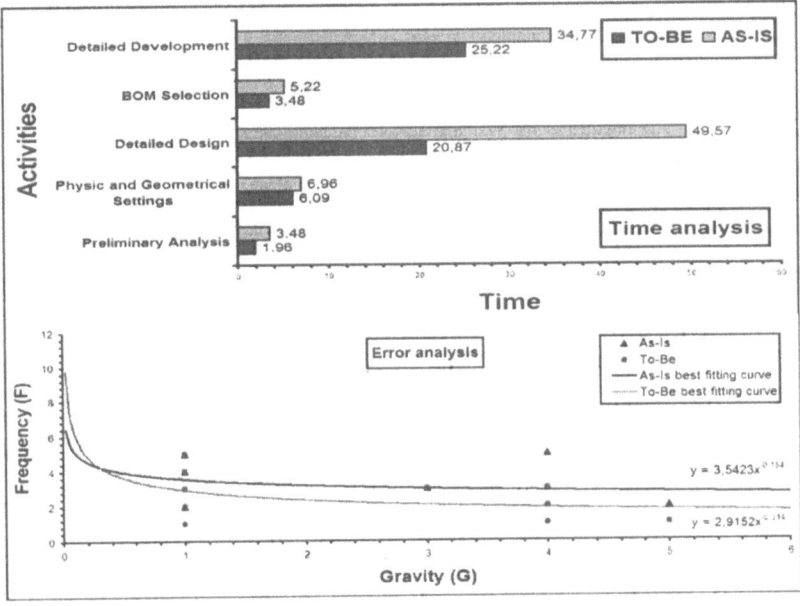

Figure 3 - Time and risk analysis graph

(modelled with IDEF0 technique) underlines a simplification in the overall product configuration process. In RuleStream, these users play different roles in respect to KM activities: senior users are delegated to introduce project properties, rules and procedures that will drive junior designers in the final configuration process. The application of the metrics points out better performances of the defined To-Be process: the analysis graphs (figure 3) show a strong decrement both in the time values (i.e.: Detailed Design, Preliminary Analysis and Detailed Development, where decreasing values reach 50% in some cases), and risk value (represented as area under best fitting curves of frequency and gravity errors) in many activities.

4. Conclusions

The paper presents the results of a research project aimed at studying a methodology for evaluating the adoption of PLM tools in product development processes. The research work has developed a roadmap that plans a sequence of activities. A candidate product development process is initially selected and analyzed. The analysis points out issues related to Knowledge Management that might require improvements. A method, based on the Quality Function Deployment method, has been set up for the identification of technological solutions that may improve the performances of the process. The effectiveness of the identified solution is proved through a simulation of the To-Be process, where product development processes are re-defined introducing the use of new EKM technologies. A metrics is used for measuring the performance of the To-Be processes, compared to As-Is ones. The roadmap has been evaluated through its application to selected case studies that underlines methodology .

4. Acknowledgements

The authors would like to thank MIUR (Italian Ministery for University and Research for supporting the research, and D. Pugliese, M. Pulli and M. Ugolotti from Politecnico di Milano, M. Cambi and D. Russo from Univ. Firenze, S. Filippi from Univ. Udine and F. Mandorli from Univ. Politecnica delle Marche for their contributions to the research. Special thanks also to Ing. Schiavi and Mr. Caminati from JOBS for their support on the test cases implementation.

References

[1] Cugini U. and Wozny N. Eds., (2002), From knowledge intensive CAD to knowledge intensive engineering, Kluwer Academic Publishers.
[2] Bordegoni M., Cascini G., Filippi S. Mandorli F. (2003), "A methodology for evaluating the adoption of Knowledge and Innovation Management tools in a product development process", ASME International Design Engineering Technical Conferences & Computers and Information in Engineering Conference, Chicago (IL), (to appear).
[3] IDEF0, Integration definition of functional modelling, Deaft Federal Information Processing Standard Publication 183 (FIPAPUB183). FIPA, USA, 1993.
[4] QFD, URL: www.qualisoft.com.

Six Sigma: Results in an Industrial Application

F. Aggogeri[1], E. Gentili[2]

[1] University of Brescia, via Branze 38, 25100 Brescia Italy, aggogeri@yahoo.it
[2] University of Brescia, via Branze 38, 25100 Brescia Italy, gentili@ing.unibs.it

Abstract: Six Sigma is a disciplined and rigorous approach to improve business performance as defined by customer satisfaction. Its goal is to increase profits by eliminating variability, defects and waste. The methodology is an implementation of a measurement system to collect data, analyse results and integrate the information into industrial processes. Six Sigma assures a strong reduction of variability, defined as 3.4 defects per million opportunities [1]. In fact, it utilises statistical tools and techniques that measure the process capability and the results of productive activities. Basically the Six Sigma problem solving method has five main phases: define, measure, analyse, improve and control (DMAIC). Six Sigma could be used as a new design for products and services (DFSS) tool. The Six Sigma project, outlined in this paper, underlines the different shades of improvement. The project company is a leader in the pharmaceutical sector. It knows and applies a particular version of the Six Sigma: the Lean Sigma, that merges the most important Six Sigma principles with the features and benefits of a lean organisation. This Six Sigma project focuses on the revision of a process to realise a product, the JNS. This process review takes into account the increased quality standards for the introduction of the product on the Japanese market.

1. Introduction

"Six Sigma is the application of the scientific method to design and operations management systems and business processes, which enable employees to deliver the greatest value to customers and owners" [2]. The methodology leaves a large field to interpret its adaptability in the different contexts; it delivers a clear structure that could be used and decomposed in many solutions. Its flexibility, if applied correctly, enables a significant improvement and saving also in service companies. Thus "Six Sigma is a management philosophy that attempts to improve customer satisfaction to near perfection" [3]. The performance increase is possible because of the reduction of the COPQ (Cost of Poor Quality); the COPQ are internal failures (scraps, rework), external failures (field failures, complaints, returned material), appraisal (inspection, audit) and prevention (quality planning, process control) [4]. Often, these costs absorb a significant share of the sales. In the Six Sigma language, we can identify these principles: a) the identification of the VOC (Voice Of The Customer), i.e., the customer requirements and needs, b) the mapping of the process to know the VOP (Voice Of The Process), it is the process features and information, and c) the translation of the VOP in the CTQ (Critical To Quality). They are all the features and parts of the process (or the product) that could be critical for the business performance. The CTQs are defined by the assessment of the VOC too. In this Six Sigma project the real problem was not to increase the system productivity, but to eliminate those process features that the Japanese market considers as defects. The problem solving method applied was the DMAIC (Define, Measure, Analyse, Improve, Control), but in some project aspects there are a few references to DFSS (Design for Six Sigma). The first step was the determination of the CTQ and the identification of all variables that could be external to the JNS process (for example supplier processes). "The activities, that cause the customer's critical-to-quality issues and create the longest time delay in any process, offer the greatest opportunity for improvement in cost, quality capital and lead time" [5]. In the second phase, the Six Sigma team identified the potential causes of the different unwanted effects. Thus it was possible to deploy many improvement actions, both in the short and medium term. Some of these actions were applied in the short term with immediate results. A quantitatively assessment was possible with a total visual inspection (100%) allocated in the final process of the JNS system. This activity took into account the process performance before and after the Six Sigma implementation. The inspection data quantified the saving and suggested the main aspects was to apply the DMAIC (recursive loop) a second time. This paper shows only some of the fundamental points of the project that are characterised by many small improvement aspects, necessary for the success.

2. Define

The first step of this phase is to define the scope and the duration of the project, identify the customer needs (VOC) and connect the requirements with process or system features (VOP). In this way, the information and the data are the inputs (CTQ) for the other Six Sigma steps. In the project, the Japanese market imposed on the company higher quality standards than in the other countries where the JNS was already commercialised. Thus a review of the JNS production stream was necessary. For a clearer understanding, main components of the JNS (device) will be explained in turn:
- the actuator: delivers the grip of the device and enables the spray exit;
- the solution: is contained in a glass vial with a small plastic stopper that is held by a needle (internal to the actuator) when the JNS is used;
- the vial: is covered by a cup;
- the label: shows the expiry date, bar code and general information relating to the product.

The preparation of the solution and the assembly process are undertaken in a classified area. Finally, the JNS is sold in a (single or double) blister. In this project, it was very important to identify the customer's real requirements. Thus, the first input was to identify the CTC (Critical to Customer) and to assess the complaints from the other markets, where the product had already been commercialised. A further assessment was the study of the complaints about similar products that had been sold on the Japanese market. The integration between the results obtained in this first step and the actual marketing experience suggested the publication of a document containing a more accurate definition of the terminology (TTS *Technical Terms of Supply*) and the different categories. In the report there was the Japanese AQL (*Accepted Quality Level*) that defined the level of defects in a batch beyond which there was the reject. The defects were divided into four classes, considering the AQL. For example, the category with AQL = 0 %, contained these defects typology. Other defect categories were: scrap on the devices, unreadable data, black spot on the blister or the device. The project was executed by a team, consisting of different personnel figures: the black belt, the project manager, the manager responsible for Japanese marketing, the manager of relationship for marketing, the person responsible for the JNS product unit, the maintenance operator and the internal consultant for improvement management. The second step of the "define phase" was to map the process. In this way, it was possible to identify the data already available in the company. The JNS value stream was divided into six macro - processes: the solution preparation, the filling, the autoclaving, the automatic inspection, the assembling and the blistering. In Table 1, as follows, we collected all the data to assess the performance.

Table 1: The JNS process performances.

	Process yield	Total yield	Working pieces	Batch Scraps
Preparation	0.998	0.998	80,838	162
Filling	0.980	0.978	79,221	1,617
Autoclaving	1.000	0.978	79,221	0
Inspection	0.990	0.968	78,429	792
Assembling	0.990	0.959	77,645	784
Total		0.959	77,645	3,355

In order to identify the CTQs, the team applied an important Six Sigma and Quality Management tool: QFD (Quality Function Deployment). This tool enables us to report the Critical to Customer information and compare it with the Process and the Product. We identified blistering (CTQ) as critical area on which the team should focus for immediate improvement.

3. Measure

"The measure phase is mainly concerned with identifying the key customers, determining what their critical needs are, and what are the measurable CTQs necessary for a successful designed product" [4]. To measure the performance level of the JNS processes, a visual inspection was implemented at the end of the value stream (after blistering), classifying the different defect categories and following the AQL. A data stratification share is shown in Table 2, whilst the total is reported for all defect classes. The inspection phase defined that the yield was 80%. Integrating this result with the other process yields we can obtain the system performance. So if we consider a Japanese product cycle of 81,000 pieces we can expect 18,884 defects.

Table 2: Visual inspection data.

Defect	Total	% Defects on total	% Defects class
Blister contamination	10,900	8.58	42.33
Device dirt	3,706	2.92	14.39
Scraps on device	3,410	2.68	13.24
Blister black spots r> 500μm	1,807	1.42	7.02
Blister damage	1,250	0.98	4.85
Total pieces inspected	127,005		
Total defects	25,750	20.27	100

It was possible to deploy a Pareto diagram to identify which defect category had a strong impact on the result. Then we compared the QFD quality results with inspection quantity results. We concluded that the total yield of the JNS system was 76.7% (we considered all the production process yields) and the critical area was the blistering. It was important to verify the effectiveness of the inspectors with repeatability and reproducibility of the measurement, so we implemented a MSA (Measurement System Analysis). We prepared a sample of 16 devices, with 50% bad and 50% good, and the operators measured them. We calculated the Effectiveness index (E), the Probability of False Reject (P(FR)), the Probability of False Acceptance (P(FA)) and Bias (P(FR)/P(FA)). The results (see Figure 1) show that the measurement system was not satisfactory and therefore training for operators was necessary to define the real defects.

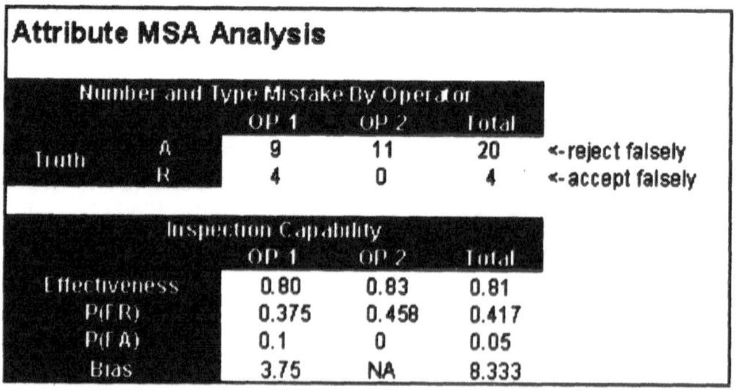

Attribute MSA Analysis

Number and Type Mistake By Operator

		OP 1	OP 2	Total	
Truth	A	9	11	20	←-reject falsely
	R	4	0	4	←-accept falsely

Inspection Capability

	OP 1	OP 2	Total
Effectiveness	0.80	0.83	0.81
P(FR)	0.375	0.458	0.417
P(FA)	0.1	0	0.05
Bias	3.75	NA	8.333

Figure 1: The MSA results.

The last step of this phase was an economic analysis to assess the COPQ. However the JNS is commercialised in a double blister while the data collected regarded a single blister, so the total scrap level was 64%. The reduction of the defects enables a significant saving as COPQ.

4. Analyse

The analysis phase has the scope of identifying the causes of an unsatisfactory performance. The team deployed a Kaizen Blitz (a technique that focuses attention on ongoing improvement and involves everyone) on the JNS line that enabled us to find the defect sources. For example, the Kaizen Blitz highlighted the defect sources on Machine 1. This machine assembles the actuator, the cup (within the vial) and finally the labels. This analysis tool enabled us to identify the possible scraps zones that could determine the Blister contaminations (as dust on conveyor belt or fibres of elevator belt). The Six Sigma team analysed all possible solutions to eliminate the rejects: e.g. on Machine 1 we assessed a different method to assemble the label on device (defect: not centred label). The team assessed if the photoelectric cell (that enables the reading of the distance between two label notches) could gauge the distance between two-label code bars. We considered applying Ishikawa Diagrams to identify the different causes, dividing them into six classes (6M). The main causes for the device dirt were the dump dust on the line, the maintenance, the label glue and label ink. We noted that in the measure phase the inspection system, as defined, was not reliable. There was a process review of the main suppliers (PET, label, actuator) that delivered their internal data and information (capability indexes, scraps, yields). The team studied all analysis results that suggested possible solutions.

5. Improve and Control

The first step of the improvement phase was to deploy a new training regime for the inspection operators. In this training, the team told operators which were the real defects and gave them a sample, containing the different defects, which were to be compared during the control. We deployed a new MSA with three operators (two of them were the old MSA operators) and thirty-four samples (50% good and 50% bad). We had an optimal result. In fact, the inspection yield increased with a good effectiveness, 90% vs. 81% of MSA 1. The second step was to improve the working line, dividing the actions in classes (Man, Method, Material, Machinery). For example, in the Man class, we suggested the wearing a new overall, reducing the blister contaminations (fibres, hairs). In the Method class the Six Sigma team imposed line cleaning before the production of pieces for the Japanese market, reducing blister contamination and device dirt defects. We changed the maintenance method working on all sources on device defects causing the scrap (i.e. maintenance of clamps that move the device). The team reviewed the supplier processes (Material class) and imposed high standards measures, suggesting a strict control on PET to reduce the black spots on the blister. "The best Six Sigma projects begin not inside the business but outside it" [6]. These actions enabled us to increase the process yield and to abate the defects, having a significant saving. In implementing the Six Sigma methodology, we can note a reduction of 26% of defects and the new system yield was 90% (see Figure 2). The team deployed the control phase for each defect class.

384

The use of Six Sigma shows also a strong decrease of COPQ and an assurance to customer satisfaction [7].

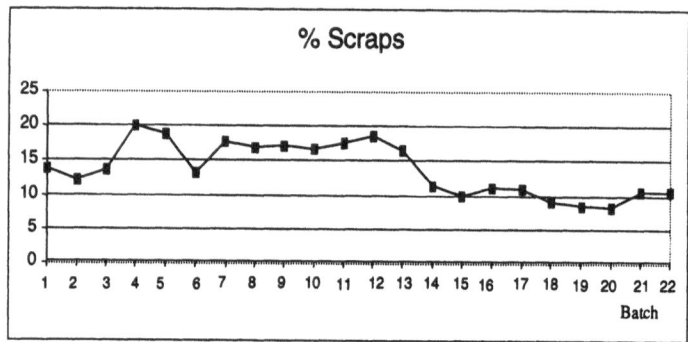

Figure 2: The system yield after the implementation of Six Sigma.

6. Conclusion

This project underlines the Six Sigma effectiveness and flexibility. The methodology assures an increased customer loyalty, more revenues, higher returns and increased earnings. It should be remembered that Six Sigma is more than a process improvement. It is part of a planned and monitored business strategy steered toward success. Thus the major key to obtain its successful implementation is the alignment of the organisation's visions, values and systems [8]. It is also fundamental to identify the sources of resistance to Six Sigma and to plan a strategy to overcome that resistance [9]. Therefore the application of Six Sigma confirms that this approach is essential to satisfy the customers need requirements reducing costs.

References

[1]   Chowdhury S., 2001. *The Power of Six Sigma*. Dearborn Trade Publishing Chicago
[2]   Pyzdek T., 2003, *The Six Sigma Handbook*. McGraw-Hill New York
[3]   Eckes G., 2001, *Making Six Sigma Last*. John Wiley & Sons, Inc. New York
[4]   De Feo A., Barnard W., 2004, *Juran Institute's Six Sigma*. McGraw-Hill New York
[5]   George M., 2002, *Lean Six Sigma*. McGraw-Hill New York
[6]   Pande P., Neuman R., Cavanagh R., 2000, *The Six Sigma Way: How GE, Motorola and other Top Companies are Honing Their Performance*. McGraw-Hill New York
[7]   Federico M., Beaty R., 2003, *Six Sigma Team Pocket Guide*. McGraw-Hill New York
[8]   Adams C., Gupta P., Wilson C., 2003, *Six Sigma Deployment*. Butterworth Heinemann Amsterdam
[9]   Eckes G., 2001. *The Six Sigma revolution*. John Wiley & Sons, Inc. New York

The International Material Data System: Global Data Collection for the End-of-life Vehicle Management

I. Pollok[1], B. K. Temple[2], D.A. Edgar[3], D.K. Harrison[4], S.C. Kinzler[5]

[1] EDS Operations Services GmbH, Ruesselsheim/Germany, ilona.pollok@eds.com
[2] Glasgow Caledonian University, Glasgow/UK, bkte@gcal.ac.uk
[3] Glasgow Caledonian University, Glasgow/UK, d.a.edgar@gcal.ac.uk
[4] Glasgow Caledonian University, Glasgow/UK, dha2@gcal.ac.uk
[5] FH Aalen, Aalen/Germany, susanne.kinzler@fh-aalen.de

Abstract: The International Material Data System (IMDS) is an internet-based application for the global collection of material data within the automobile supply chain. As outlined in the EU Directive 2000/53/EC concerning End-of-life vehicle management the data for future recycling has to be gathered before 2015. Eight automobile manufacturers together with the IT service provider Electronic Data Systems (EDS) developed and implemented the system in 2000. Subsequently, other car manufacturers have also joined and the participation of more companies is being negotiated.

The application was initiated by the Original Equipment Manufacturers (OEMs) but relies on a bottom-up approach to data entry. Thus the lowest tier in the supply chain (tier n) must be the first to provide the material data on their products and pass them to the next tier supplier (n-1) who would have ordered the product. They, in turn provide information on the product they have made from the tier n supplier. As the whole automobile supply chain is involved, the success of IMDS is based on three critical issues: firstly, every supplier must have access to good IT equipment; secondly, there must be a good inter-departmental co-operation and thirdly, there must be a good working relationship between the different tiers in order to glean accurate information. Assuming that the three critical factors are a precondition for a successful e-business solution and that the response to IMDS is an indicator of the ability to pass on the data through the electronic network, the paper concludes that the electronic measures for efficiency improvement have been less successfully applied and offers reasons why this may be so.

1. Introduction

The International Material Data System (IMDS) is an internet-based application for the global collection of material data within the automobile supply chain. As outlined in the EU Directive 2000/53/EC concerning End-of-life vehicle management the data for future recycling has to be gathered before 2015 in order to recycle 95% of the mass of a car by then. Germany's automobile industry voluntarily committed to earlier completion initiated by a consortium of eight automobile manufacturers - Audi, BMW, Daimler-Chrysler, Ford, Opel, Porsche, Volkswagen and Volvo - in association with the IT service provider, Electronic Data Systems (EDS). The system was implemented in 2000. Subsequently, eight other car manufacturers have also joined and the participation of more companies is being negotiated.

The application was initiated by the OEMs but relies on a bottom-up approach to data entry. Thus the lowest tier in the supply chain (tier 'n') must be the first to provide the material data on their products and pass them to the next tier supplier (n-1) who would have ordered the product. They, in turn provide information on the product they have made from the tier 'n' supplier. As the information passes up the supply chain every OEM ultimately receives a complete account of the environmental characteristics of their car.

As the whole automobile supply chain is involved, the success of IMDS is based on three critical issues: firstly, every supplier must have access to good IT equipment; secondly, there must be a good working relationship between the different tiers and thirdly, there must be a good inter-departmental co-operation in order to glean accurate information. This work assumes that the three critical factors are a precondition for a successful e-business solution and that the response to IMDS is an indicator of the ability to pass on the data through the electronic network.

The automobile industry is characterized by large fixed costs and overcapacity [1]. Their supply chain is a good example where collaboration can create higher value [2]. In addition, electronic processes can enable greater efficiencies [3]. To better understand the progress that has been made in the use of e-business techniques, IMDS was chosen for analysis as it is an example for an e-business project covering the entire automobile supply chain.

Although the automobile industry usually is very innovative and trend-setting, the findings support that e-business strategy development is not fully effective if it is to be judged by these critical factors. For example, there are considerable differences between the different company sizes concerning the importance to e-business in general as well as the IT infrastructure used. Equally, the level of "computerisation" within companies was far from complete, especially in the smaller companies on whose information the whole chain depends. While the automobile industry's supply chain is well established, the paper concludes that the electronic measures for efficiency improvement have been less successfully applied and offers reasons why this may be so.

2. Methodology

With a time delay of a year, two online questionnaires were designed. The first part of the two questionnaires dealt with IMDS application features like design, performance, services etc. The second part of both questionnaires contained general questions about the company size but also specific questions on the intra-company infrastructure and opinions on e-business importance.

The first survey was put online from the 04.12.2001 until 09.01.2002 and delivered 981 valid entries. The percentage of missing questions varies from about 6% to 7%. This leaves an average number of about 920 entries for analysis, which can be taken as representative population of about 17,000 users. The second survey was put online from the 04.02.2002 until 05.03.2002 and delivered 2,418 responses. The percentage of missing questions varies from about 9% to 10%. This leaves an average number of 2,180 entries for analysis, which is taken as representative population of about 24,000 users.

The assumption is that the above mentioned three preconditions have to be met by the companies in order to realise fast data delivery to the next tier of the chain. This paper considered the following question from the first questionnaire for testing the existence of electronic equipment and hence the IT possibilities within the companies: *Question 1 What is the percentage of PCs/work place in your company?*

As the second precondition is good inter-departmental cooperation the second question from the first questionnaire aimed at the commitment to integrating company systems and thus to improve inter-departmental information flow: *Question 2 Does your company use Enterprise Resource Planning software, e.g. SAP, BAAN?*

Another indicator for well-developed inter-departmental cooperation in a company was taken from the second questionnaire: *Question 3 How important is participation in e-business for your company? Question 4 How important is participation in e-business for your department?*

The two questions aimed at finding out whether e-business for employees is seen as equally important for their department as well as for the entire company they are working with. If a company is really characterized by "detached departments" [4] the company is considered being responsible – not the department the employees are actually working in.

For findings on the third precondition – a good working relationship between the different supplier tiers – the answers to the following two questions from the second questionnaire were analysed. The focus of both questions is the data flow through the entire supply chain: *Question 5 Which of the following words would best describe the relationship to your automobile industry customers? Question 6 According to your opinion, how are suppliers for new products selected in your company?*

3. Findings

The findings for each question are grouped according to the size definitions of the Chamber of Commerce in Heilbronn/Germany. Most of the companies in the survey have either 100-500 or more than 1,000 employees. This picture largely corresponds to the findings of Dahlhoff et al. [5]. In order to better compare the responses to the three sets of questions, the company sizes will be set to 100% in each question for every company size category.

Table 1 Percentage of valid responses compared with Dahlhoff et al.

Number of employees	Present work (Questionnaire 1)	Present work (Questionnaire 2)	Dahlhoff et al
less than 100	17%	20%	8%
101 to 500	39%	41%	26%
501 to 1000	12%	12%	12%
more than 1000	32%	27%	54%

3.1 IT infrastructure

For analysing the first precondition for electronic data provision the number of PCs in a company might be taken as an indicator of the extent to which the company is committed to electronic equipment. Figure 1 shows that the supply of PC workstations increases with company size. This is not unexpected, in so far as small companies are generally on the lower tiers whose contact with the OEM is indirect. Thus they are under less pressure to equip and indeed, have less means for regular investment into IT equipment and networks.

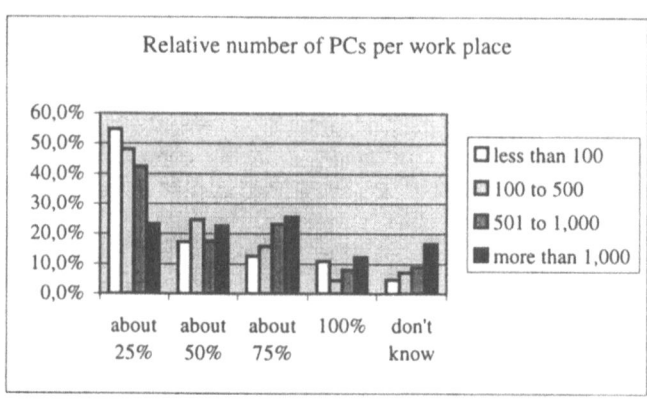

Figure 1 Relative number of PCs per workplace

3.2 Interdepartmental cooperation

According to Chopra and Meindl [6], Enterprise Resource Planning (ERP) employs an operational IT system, which is used company-wide in order to collect data and monitor processes in real-time. As ERP systems are to be considered "a key starting point for leveraging new Internet technologies" [7] the question was analysed whether ERP was used in the company. The responses are graphed in Figure 2 where they are shown as percentage responses in each category of company.

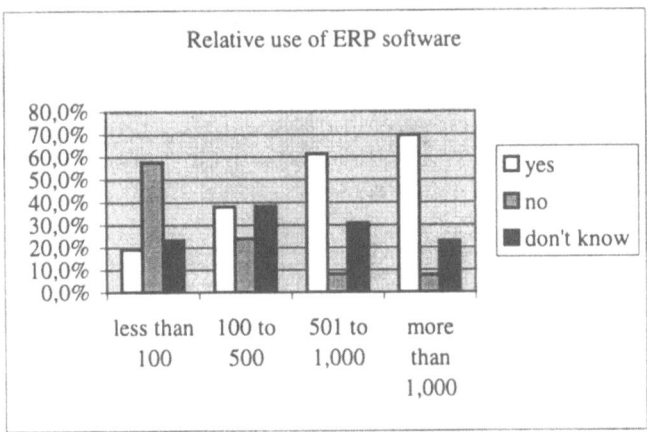

Figure 2 Relative use of ERP software

About half of the companies report the use of ERP software. Figure 2 clearly shows that the use of ERP software increases with the size of the company. One might have expected the level of "don't know" responses to increase with company size as found for IT equipment. In the case of ERP, most employees would be expected to use the system every day of their working lives and should be aware of ERP. Only in companies with between 100 to 500 employees was there major uncertainty. In small companies the clear "no" (58%) indicates that a high cost ERP solution is not implemented in this company category. Overall, there was a large number of responses in the "don't know" category, but this might be due to the fact that only two ERP systems (SAP and Baan) were given as examples and that some employees are not aware of working with an ERP system.

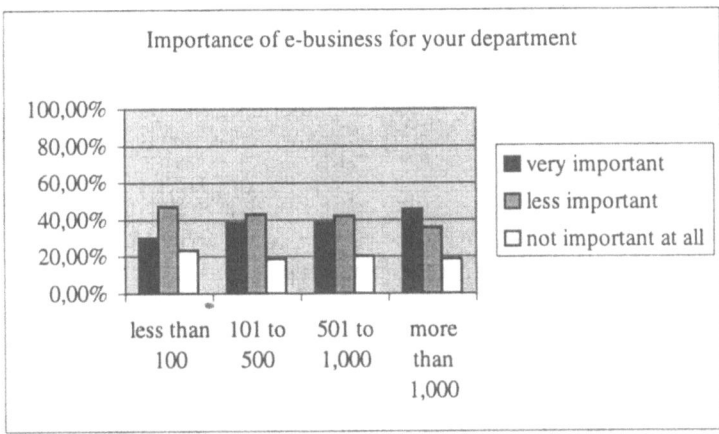

Figure 3 Importance of e-business for the department

The second question was on the importance of e-business for the department. The responses are shown in Figure 3. In order to compare the departments' involvement in e-business, the importance of e-business for the entire company was questioned, too. The responses are categorised in Figure 4.

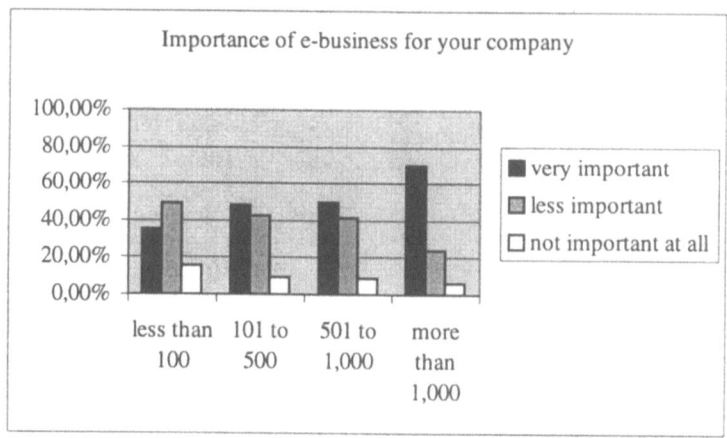

Figure 4 Importance of e-business for the company

There seems to be a clear shift from the smaller companies to the largest. The high percentage of smaller companies responding "less important" or "not important at all" seem to prove a lack of interest in new technology and/or low budgets. If the responses to the two questions are compared the different perspectives become obvious: asked about the importance of e-business to their departments the differences between the company sizes become less dramatic than in the previous question.

In fact, only within the smaller company size categories there seems to be no "yes it is important but not for me"-approach to the general involvement in e-business. In the large company categories there were very much lower percentages of importance for the department in contrast to the importance for the company. It can be concluded that there is a much larger level of involvement within smaller companies - if the importance of e-business has been acknowledged. In contrast to large companies, where increasing size often leads to an increased departmentalisation and hence to a certain ignorance of the developments in other parts of the company, this is not the case for smaller companies. Therefore, the smaller companies could have considerable advantages by exploiting the higher employee commitment.

3.3 Working relationship between the different tiers

The question about the relationship to the customers was answered differently by the companies of the different tiers:

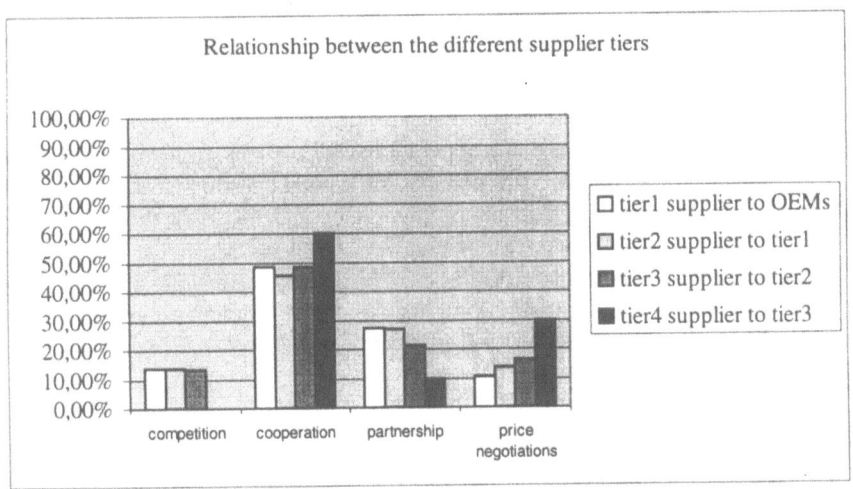

Figure 5 Relative relationship to the customers of different tiers

Only the larger tiers seem to characterise their relationship to their customers as cooperation- and partnership-driven. This is not surprising as they work closely with the OEMs and therefore have a better position in negotiations. Although cooperation was also very important for the smaller company categories, price negotiation was ranked the second term after cooperation. Two factors influence this position: Firstly, cooperation is ranked highest because smaller companies do have a closer, personal contact and interest in satisfying their customers' needs. Secondly, the lower tiers are closer to the raw materials industry where prices for commodities are easier to compare e.g. for steel.

However, all answers might have been biased to a certain extent by the fact that the questionnaire was positioned in an OEM application where the suppliers should answer as it might be expected.

4. Conclusions

It was a legal requirement that was the driving force for a common IT solution for recycling data collection in the automobile industry, but both the information and the data exchange necessary for the IMDS project has been shown to differ in companies of different sizes.

The automobile industry is dominated by small and medium-sized companies. Those two categories represented in both questionnaires about 60% of the industry. Concluding from the findings those two company categories are not well-equipped regarding the IT infrastructure. Especially at the lower tiers representing the smaller company categories the price is the most important argument. Hence the small companies are more likely to compete by price and they also chose their suppliers by price only. On the other hand, smaller companies seem to have the advantage of better information flow (e.g. the smallest part of "don't know responses" is found in this company category) but they are unwilling to invest in cost-intense solutions like ERP software that could improve supply chain performance.

The larger company categories are better equipped and better positioned towards their customers. The major obstacle in this company category seems to be the lack of inter-departmental cooperation which stems from the traditional view of detached departments. Additionally, the second questionnaire has revealed a low level of awareness of e-business importance and this is in the automobile industry, which usually is very innovative and trend-setting in the use of new technologies. We suggest that this lack of awareness offers an obstacle to a company's efficient operation and by implication, to the efficient use of IMDS.

In terms of the general involvement in electronic business, the findings suggest that there is an incomplete supply chain in so far as many of the smaller companies do not seem to possess either hardware or software compatible with ERP requirements. Without complete inter-operability, an integrated supply chain cannot function. To a high extent, the cooperation between the different tiers is lacking its preconditions to be met.

If the lack of awareness of e-business importance could be overcome and the relative high level of commitment of the employees would be used by the small and medium-sized companies, a very large part of the automobile industry could improve its performance and speed up the information flow by effectively using information technology.

References

[1]   PricewaterhouseCoopers, 2001, *Supplier Survival. Survival in the Modern Automotive Supply Chain*, p. 1
      http://www.pwcglobal.com/Extweb/pwcpublications.nsf/4bd5f76b48e282738525662
      b00739e22/4efba3d2e36509cc85256bde006bec7b/$FILE/supplier_survival_for_web.
      pdf

[2]   Sahay, B., 2003, *Supply Chain Collaboration: the key to value creation*, Work Study, Vol. 52, Number 2, ISSN 0043-8022, pp. 76-83

[3]   Prasad, S. and Sounderpandian, J., 2003, *Factors influencing global supply chain efficiency: implications for information systems* in: Supply Chain Management: An International Journal, Vol. 8, Number 3, ISSN 1359-8546, pp. 241-250

[4]   Tyndall, G. et al., 1998, *Supercharging Supply Chains: New Ways to Increase Value Through Global Operational Excellence*, WILEY & SONS, ISBN 0-471-25437-1, p. 18

[5]   Dahlhoff, H., Dudenhöffer, F. and Densing, C., 2000, *Marketing und E-Commerce – Ein neues Bild der Automobilzulieferer entsteht (Marketing and E-Commerce – A new picture of automobile suppliers emerges)*, VDA and Centre of Automotive Research, Druckerei Henrich GmbH, ISSN D946-0179, p. 3

[6]   Chopra, S. and Meindl, P., 2001, *Supply Chain Management - Strategy, Planning, and Operation*, Prentice-Hall, New Jersey, ISBN 0-13-026465-2, p. 343

[7]   PA Consulting Group (2002), *Ready to profit from the Potential of e-business? A study of e-maturity in the European automotive industry 2001-2002*, PA Knowledge Limited, p. 11

Scheduling with Selective Rerouting in Manufacturing Systems

Amol Singh[1], N.K. Mehta [2] and P. K. Jain[3]

[1] I. I. T. Roorkee, Research Scholar Department of Mechanical & Industrial Engineering, amolsdme@iitr.ernet.in
[2] I. I. T. Roorkee, Professor Department of Mechanical & Industrial Engineering, mehtafme@iitr.ernet.in
[3] I. I. T. Roorkee, Associate Professor Department of Mechanical & Industrial Engineering, pjainfme@iitr.ernet.in

Abstract: This work presents a selective rerouting methodology for part scheduling in the presence of shop floor disruption. Mean tardiness maximum tardiness and number of tardy jobs are the performance measures considered, to evaluate the effectiveness of the proposed methodology. In the proposed methodology, the parts in the queue of the failed machine are rearranged on the basis of earliest due date priority heuristics and the completion time of each rearranged part is computed on the failed machine as well as an alternative machine. If the completion time of a rearranged job at the alternative machine is less than the completion time at the failed machine then alternative machine is selected for processing otherwise the part remains in the queue of the failed machine.

1. Introduction

Present day manufacturing industries consist of complex systems and apply various approaches to make best use of their available resources. In this regard, production planning and control activities play an important role and ensure effective conversion of raw material into finished goods so that high productivity goals are achieved. Scheduling is the most important activity of production planning and control determining the effective flow of inventory from raw material stage to end product. Its importance and relevance to industry prompted researchers to study it from different perspectives over the past three decades. Reported literature on scheduling covers deterministic and stochastic problems, single machine and multi-machine problems; and, static and dynamic problems.

Nowadays, most manufacturing systems face uncertainties related to machine breakdowns, order canceling, rush orders etc. In this regard, generation of modified schedules or real time schedules are essential for better utilization of resources while minimizing the impact of system uncertainties. Machine breakdown is one of the most common problem faced by shop management during scheduling. An effective schedule often needs to be evaluated against many of the scheduling parameters such as processing times, inter-arrival times, resource availability etc. When a machine breakdown occurs, a decision has to be taken whether the new incoming jobs and the jobs in the queue of the failed machine should remain in the queue of the failed machine or be transferred to an alternative machine by using routing flexibility. In the present work jobs are selectively rerouted to the alternative machine to cope with the impact of breakdowns.

2. Literature Review

Most of the research in the area of production scheduling deals with predictive scheduling problems. However, reactive scheduling or schedule revision is also important for successful implementation of a scheduling system. In most of the practical environments, scheduling is an ongoing reactive process, which involves continuously changing environment and revision of pre-established plans. Scheduling research has traditionally ignored this "process view" of the problem, and focuses on optimization of performance under idealized assumptions of environmental stability.

Kutanoglu and Sabuncuoglu (2001) tested four different rerouting policies, namely all re-routing, queue re-routing, arrival re-routing and no re-routing to cope with the problem of machine failure. All these approaches rigidly follow a selected option without due consideration of the changing environment. Jain and Elmarghy (1997) considered machine breakdown, rush order arrival, increased order priority and order cancellations with shop floor uncertainties and proposed rescheduling algorithms for the above types of uncertainties. Mehta and Uzsoy (1998) proposed a predictable scheduling approach in which additional idle time is inserted in the schedule to absorb the impact of breakdowns. The amount and location of additional idle time is determined from the breakdown and repair time distributions as well the structure of the predictive schedule. In the present work, jobs in the queue of the failed machine and the new arrivals are re-routed selectively to improve system performance. An algorithm for selective re-routing has been developed for scheduling the jobs under machine breakdowns. Several performance measures based on tardiness are taken into consideration to evaluate the effectiveness of the re-routing of a selected job. Job is re-routed only when the outcome is positive. The details of the algorithm are presented in the subsequent sections.

3. Selective Rerouting Methodology

The following steps are performed for selective rerouting.

Step 1. Prioritize the queue of the parts at the failed machine (say machine k) on the basis of earliest due date (or any other measure) and select the first part for rerouting (say job j). The queue of parts is prioritized only to decide the sequence of the jobs for rerouting. It may be noted that the sequence of parts in the queue at the failed machine is not changed physically due to the above prioritization.

Step 2. Calculate the completion time for the selected job at the failed machine by summing up the processing time of the selected job j, the processing times of the jobs ahead of job j and the remaining down time of failed machine k as given below.

$$ct_{jk} = \sum_{i=1}^{i=n} p_{ik} + d_k \qquad (1)$$

Where,

n = the position of the selected job in the queue of the failed machine k.

$\sum_{i=1}^{i=n} p_{ik}$ = Sum of processing times of the jobs from position 1 to n (i.e. j^{th} job also)

on failed machine k.

d_k = Remaining down time of the failed machine.

Step 3. Identify the alternative machines (for job j) by taking routing flexibility into consideration and determine the new position (say m) of the part j in the queues of each of the alternative machines by using the earliest due date criterion. In this manner, provision of lateral entry in a suitable position in the queue at the alternative machine is made for all parts with tight due dates. Processing time of the selected job j on each of the alternative machines may change due to the difference in capability of the alternative machines. Calculate the completion time on each of the alternative machines for the selected job j by using the relationship:

$$ct_{ja} = p_{ja} + \sum_{i=1}^{i=m-1} p_{ia} + p_{ra} \tag{2}$$

Where,

p_{ja} = Processing time of job j on the alternative machine (say machine a).

p_{ra} = Remaining processing time of any under processing job on the machine a.

m = the position of the rerouted job in the queue of the alternative machine a.

$\sum_{i=1}^{i=m-1} p_{ia}$ = Sum of processing times of all the jobs ahead of job j from position 1 to $(m-1)$.

Step 4. Rerouting of job j will increase the completion time of each job after position m on the alternative machine and reduce the completion time of each job after position n at the failed machine. This net global effect is calculated for each of the alternative machines and summed up as below:

$$\text{Net global effect} = (p_{ja} \times q(a)) - (p_{jk} \times q(k)) \tag{3}$$

Where,

$q(a)$ = Number of jobs after m^{th} position at the alternative machine

$q(k)$ = Number of jobs after n^{th} position at the failed machine

p_{jk} = Processing time of job j on the failed machine k

p_{ja} = Processing time of job j on the alternative machine a due to the different capability of the alternative machine.

Step 5. Calculate the effective completion time ($ct_{ja(e)}$) of the rerouted job on each of the alternative machines and select the alternative machine with minimum ($ct_{ja(e)}$) among all the alternative machines. $ct_{ja(e)}$ is calculated as follows:

$$ct_{ja(e)} = ct_{ja} + \text{Net global effect} \tag{4}$$

If $(ct_{ja(e)})_{min} < ct_{jk}$, then reroute the job j to the alternative machine a otherwise let the job j wait in the queue of the failed machine.

Step 6. Repeat step 1 to 5 for all the jobs in the queue of the failed-machine.

4. An Example

A simulation model has been developed to evaluate the effectiveness of the selective rerouting methodology by considering mean tardiness, maximum tardiness and number of tardy jobs as the performance criteria. The simulation model consists of four machines and three part types. Each part type has four alternative routes. It has been assumed that all the processing times are known deterministically. Processing time along with operation sequence of each part type are given in Table 1.

Table 1: Processing sequence data

Part Type	Processing sequence	
1.	$M_1(20)$-$M_3(25)$-$M_2(20)$-$M_4(10)$,	$M_1(20)$-$M_3(25)$-$M_1(20)$-$M_4(10)$,
	$M_2(20)$-$M_3(25)$-$M_2(20)$-$M_4(10)$,	$M_2(20)$-$M_3(25)$-$M1(20)$-$M_4(10)$
2.	$M_2(20)$-$M_1(15)$-$M_3(15)$-$M_4(30)$,	$M_2(20)$-$M_1(15)$-$M_2(15)$-$M_4(30)$,
	$M_3(20)$-$M_1(15)$-$M_3(15)$-$M_4(30)$,	$M_3(20)$-$M_1(15)$-$M_2(15)$-$M_4(30)$
3.	$M_1(20)$-$M_4(15)$-$M_3(10)$-$M_2(10)$,	$M_1(20)$-$M_4(15)$-$M_4(10)$-$M_2(10)$,
	$M_1(20)$-$M_3(15)$-$M_3(10)$-$M_2(10)$,	$M_1(20)$-$M_3(15)$-$M_4(10)$-$M_2(10)$

Job arrival in the system is assumed to follow exponential distribution. Mean inter-arrival time is decided on the basis of shop utilization. If $\bar{P}$ represents the mean processing time of a job and U_s represents the shop utilization then mean inter-arrival time can be decided by the relationship given as below:

$$\lambda \quad = \bar{P}/M \times U_s \tag{5}$$

It is assumed that machine fails randomly while performing an operation on the job. Hence busy time approach has been considered for measuring the inter breakdown time. Gamma distribution has been chosen for generating the inter breakdown time and repair time (Law and Kelton 2000). The busy time between two successive failures (which is inter-breakdown time) is assumed to follow Gamma distribution ($\alpha = 1.4$, $\beta = MTTR/1.4$) and the duration of each breakdown (which is also known as repair time) is assumed to follow Gamma distribution ($\alpha = 0.7$, $\beta = MTTR \times e/(1-e) \times 0.7$).

Where, e = MTBF/(MTBF + MTTR)

Machine breakdown level is defined by the ratio of MTTR over the sum of MTTR and MTBF as,

$$BL_s = MTTR/MTTR + MTBF \tag{6}$$

Where BL_s represents the percentage of downtime of a machine.

5. Results and Discussion

The performance of selective rerouting policy with respect to mean tardiness, maximum tardiness and percentage of tardy jobs has been analyzed by developing a model of the manufacturing system with machine breakdowns and routing flexibility. In this analysis the percentage of time the machine is down is assumed to be (2.5%, 5%, 7.5, 10%) at MTTR = 5 $\bar{P}$. The effect of mean time to repair has been evaluated by considering the ($\bar{P}$, 2.5 $\bar{P}$, 5 $\bar{P}$, 7.5 $\bar{P}$, 10 $\bar{P}$) values of mean time to repair at 5% breakdown level. Due date tightness factor has been chosen as 3 and 6 for tight and loose due date respectively. Comparative results of selective rerouting and no-rerouting are summarized in Table 2 based on data of 2000 completed jobs with respect to mean tardiness, maximum tardiness and percentage of tardy jobs at various breakdown levels and mean time to repair.

Table 2: Mean tardiness (T), Maximum tardiness(Tmax) & Percentage of tardy jobs (Nt) at several breakdown levels (BLs) and Mean time to repair (MTTR) for no-rerouting (NR) and selective rerouting (SR)

Parameters		Tight due date			Loose due date		
		T	Tmax	%Nt	T	Tmax	%Nt
BLs=2.5%	NR	1475	4078	91.75	1424	3857	80.2
	SR	884	3067	85.70	832	2837	69.05
BLs=5%	NR	3087	7668	98.90	2933	7437	96.55
	SR	1840	5760	95.50	1730	5539	89.00
BLs=7.5%	NR	5184	11990	99.55	5017	11759	98.70
	SR	3040	10106	98.85	2890	9881	92.00
BLs=10%	NR	7485	16850	99.99	7285	16623	99.80
	SR	5113	16796	99.55	4944	16570	98.50
MTTR= $\bar{P}$	NR	1806	4281	98.25	1677	4059	92.85
	SR	1416	5521	94.75	1338	5295	83.05
MTTR=2.5 $\bar{P}$	NR	2791	5932	99.05	2646	5710	96.40
	SR	1772	5952	97.15	1644	5725	91.65
MTTR=5 $\bar{P}$	NR	3087	7668	98.90	2933	7437	96.55
	SR	1840	5760	95.50	1730	5539	89.00
MTTR=7.5 $\bar{P}$	NR	3583	8397	98.35	4347	8165	96.10
	SR	2327	7733	96.65	2265	7513	92.10
MTTR=10 $\bar{P}$	NR	4079	9736	96.05	3991	9512	92.25
	SR	1968	7053	93.7	1907	6822	85.55

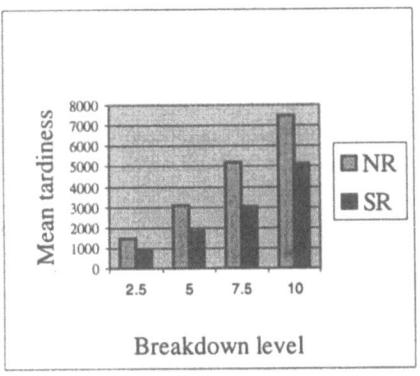 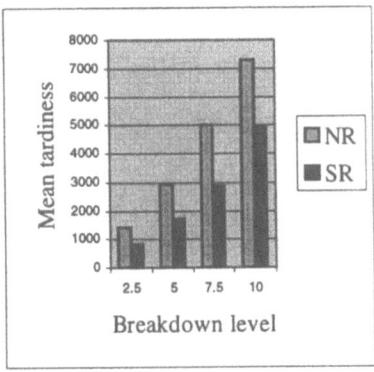

NR = No-rerouting ,   SR = Selective rerouting

Figure 1: Effect of rerouting at tight due date   Figure 2 : Effect of rerouting at loose due date

At tight as well as loose due date conditions, the selective rerouting approach provides better results than no rerouting with respect to mean tardiness, maximum tardiness and percentage of tardy jobs. Selective rerouting improves the system performance with respect to mean tardiness criteria as shown in figure 1 and figure 2. This approach improves the tardiness and decreases the waiting time of rerouted jobs. A similar comparative analysis has carried out in respect of queue rerouting arrival rerouting and all rerouting also and the results will be presented in a future work.

6. Conclusion

This paper presents an algorithm of real-time scheduling in case of machine breakdown in a manufacturing system using routing flexibility. This algorithm has been compared with no-rerouting under identical experimental conditions. It has been concluded that selective rerouting algorithm improves the system performance with respect to mean tardiness, maximum tardiness and percentage of tardy jobs and provides better results than no rerouting. Simulation results indicate that routing flexibility has a significant impact on the performance of the system in case of machine breakdowns.

References

[1] Abumaizar, R. J. & Svestka, J. A. 1997, Rescheduling job shops under random disruptions, *International Journal of Production Research*, Vol. 35, No. 7, pp. 2065-2082.

[2] Jain, A. K. & Elmaraghy, H. A. 1997, Production scheduling / rescheduling in flexible manufacturing, *International Journal of Production Research*, Vol. 35, No. 1, pp.281-309.

[3] Kutanoglu, E. & Sabuncuoglu, I. 2001, Routing-based reactive scheduling policies for machine failures in dynamic job shops, *International Journal of Production Research*, Vol. 39, No. 14, pp. 3141-3158.

[4] Mehta, S. V. & Uzsoy, R. M. 1998, Predictable Scheduling of a job Shop Subject to breakdowns, *IEEE Transaction on Robotics and Automation*, Vol. 14, No. 3, pp. 365-378.

[5] Sabuncuoglu, I. & Bayz, M. 2000, Analysis of reactive scheduling problems in a job shop environment, *European Journal of operational Research*, Vol. 126, No. 3, pp. 567-586.

[6] Yamamto, M. & Nof, Y. 1985, Scheduling/rescheduling in the manufacturing operating system environment, *International journal of production research*, Vol.23, No.4, pp. 705-722.

[7] Akturk, S. and Gorgulu, E., 1998, Match-up scheduling under a machine breakdown, *European Journal of Operation Research*, Vol. 112, No. 1, pp 81-97.

[8] Law, A. M. & Kelton, W. D. 2000, *Simulation Modeling and Analysis*, Third Edition, McGraw-Hill, New York.

[9] Singh, A., Mehta, N. K. and Jain, P. K., Selective approach for rerouting in the presence of machine breakdowns, The 14th International DAAAM Symposium "Intelligent manufacturing & automation: Focus on reconstruction and development" 22-25th October 2003, Sarajevo, pp. 417-418.

Manufacturing Cell Formation using Heuristic Rule Based Logic (HERBAL)

[1]I. Marghalany, [1*]V.I. Vitanov and [2]G. Sapundgiev

[1]Cranfield University, SIMS, Cranfield, MK42 0AL, UK, v.vitanov@cranfield.ac.uk
[2]Faculty of Automation, Technical University of Sofia, Sofia, Bulgaria

Abstract: This paper presents decision support methodology to facilitate the implementation of cellular manufacturing systems. The methodology is based on a new developed cell formation algorithm called heuristic rules-based logic (HERBAL). It has the facility to incorporate relevant production data such as production volumes and operation runtimes. It sorts weighted input machine-component matrices to obtain solution matrices with minimum actual intercellular operation workloads. Computational results show that the proposed methodology can provide better cell formation solutions that reflect real manufacturing situations.

1. Introduction

Cellular manufacturing (CM) is a concept that converts a traditional job shop production system into several manufacturing cells. Each cell contains all facilities needed to produce specific families of products that are similar in their design or manufacturing features. Cell formation (CF) aims to identify component families and machine cells where each cell of machines processes its corresponding family of components with minimum inter-cellular flows. This design process has been recognized by researchers as a complex problem, so it often proceeds in stages.

In the last three decades numerous methods based on different approaches have been developed to identify component families and their associated machine cells. Most of them employ binary machine-component incidence (MCI) matrix model that is based solely on the process routings information. The basic input data are machines m_i (row headings) where $i = 1, 2, ...M$, and components p_j (column headings) where $j = 1, 2, ...P$. The binary form of the MCI matrix limits its elements value a_{ij} to be 1 when component j needs to visit machine i, otherwise $a_{ij} = 0$. Cell formation methods sort matrix columns and rows to obtain diagonal blocks of ones that indicate component families (block columns) and machine cells (block rows). This form of matrix is often referred to as block diagonal matrix (BDM). The basic objective is to obtain the maximum number of 1's inside the diagonal blocks and the minimum number of 1's outside the blocks (often called exceptional elements).

2. Cell Formation Algorithms

In recent years many algorithms that incorporate operation sequences and production volumes of components have been published in the literature such as Gupta and Seffoddini [1], Nair and Narendran [2], and De Guio and Barth [3]. These algorithms provide optimal or near-optimal solutions with respect to the quantitative goodness measures, but most of them provide poor qualitative solutions such as the formation of irrational cell sizes especially in the case of ill-structured input matrices.

One of the new developed cell formation methods called Heuristic Rules-Based Logic (HERBAL) has been tested on binary input matrices and has shown promising and efficient results that can lead to feasible and near-optimal manufacturing cell formation [4]. It can guarantee rational cell formation solutions for any MCI matrix as far as there is one exists. Figure 1(a) shows a simple example of a binary input MCI matrix, where its non-binary form that involves workloads information is shown in Figure 1(b).

	P1	P2	P3	P4	P5	P6	P7
M1		1			1	1	1
M2	1		1				
M3		1			1	1	1
M4	1	1			1	1	
M5			1	1		1	

Figure 1 (a) Binary input MCI matrix

In order to evaluate the quantitative goodness of the CF solution matrix that is obtained from a non-binary input MCI matrix, the Quality Index (QI) measure (5) is employed that can be obtained using the following expression:

$$QI = 1 - (ICW / PW)$$

where ICW is the total intercellular workload and PW is the total workload of all operations.

	P1	P2	P3	P4	P5	P6	P7
M1		2175			33800	150	5850
M2	4550		250				
M3		5850			2175	250	6552
M4	4550	2175			325	150	
M5			33800	4550		250	

Figure 1(b). Non-binary MCI matrix (operation workloads entries)

3. Description of the HERBAL Method

The proposed method (HERBAL) implies heuristic procedure and rules that are used to identify component families and their corresponding machine groups based on columns and rows permutations.

The CF solution matrix shown in Figure 2(a) is obtained by implementing the proposed HERBAL algorithm on the binary MCI matrix example of Figure 1(a). There are 2 exceptional elements in the Figure 2(a). These are a_{41} and a_{56}. The corresponding workload values of these exceptional elements are obtained from the non-binary form of the input MCI matrix shown in Figure 1(b), which equal to $a_{41} =$

4550 and $a_{56} = 250$. The sum of these exceptional workloads (4800) provides the real weight of the total units of intercell flows.

	P2	P5	P6	P7	P1	P3	P4
M1	2175	33800	150	5850			
M3	4550	2175	250	6552			
M2					4550	250	
M4	2175√	325√	150√		4550		
M5			250√			33800	4550

Figure 2(a) CF solution matrix

	P2	P5	P6	P7	P1	P3	P4
M1	1	1	1	1			
M3	1	1	1	1			
M4	1	1	1		1√		
M2					1	1	
M5			1√			1	1

Figure 2(b) Non-binary solution matrix

Thus in order to find out the effect of employing the non-binary input MCI matrix instead of the binary form on the final CF solution matrix, the same cell formation algorithm (HERBAL) is implemented on the non-binary matrix shown in Figure 1(b). The obtained CF solution matrix is shown in Figure 2(b). There are 4 exceptional elements in the non-binary solution matrix. However, it is found that they correspond to a sum of 2900 exceptional workloads, which provide less total units of intercell flows.

	1 2 3 4 5 6 7 8 9 10 11 12 13 14 15 16 17 18 19 20 21 22 23 24 25 26 27 28 29 30 31 32 33 34 35 36 37 38 39 40 41 42 43
1	1 1
2	1 1 1 1 1 1 1 1
3	1 1 1 1 1
4	1 1 1 1 1 1 1
5	1 1 1 1 1 1 1 1 1 1 1 1 1
6	1 1 1 1 1 1 1 1 1 1 1 1 1 1 1 1 1 1 1
7	1 1 1
8	1 1 1 1 1 1 1 1 1 1 1 1 1 1 1 1 1 1 1 1
9	1 1 1 1 1 1 1 1 1 1
10	1 1 1 1 1 1 1
11	1 1 1 1 1 1
12	1 1 1 1 1
13	1 1
14	1 1 1 1
15	1 1 1 1 1 1 1
16	1 1 1 1 1 1 1 1

Figure 3. (16 x 43) Machine-component binary input matrix

A well-known practical engineering problem composed of 16 machines and 43 components has been selected from the literature [6] to illustrate the proposed method's capabilities. The binary input matrix of the problem is shown in Figure 3. Other production data such as routing sequences, components volume and operations runtimes are taking from reference [1].

Operation's workload of each entry w_{ij} in the machine-component matrix is given by demand quantity D_j of component j multiplied by operation's runtime R_{ij} on machine i per one component. Weighted matrices employ workload entries w_{ij} instead of previously binary entries a_{ij}. This type of matrices reflects the real production system in a more accurate way. In most industrial cases, operation workloads often have great diversity values among their entries w_{ij} of its represented input matrix. Therefore it is more convenience to convert workload values from their original unit of measurement to the standard z score unit that is computed using the mean and standard deviation of matrix entries population. For example, the weighted matrix showed in Figure 4 displays standard z scores for entries of the weighted MCI matrix shown in Figure 1(b).

	P1	P2	P3	P4	P5	P6	P7
M1		0.18		0.41	3.20		0.53
M2	0.41		0.01				
M3		0.53			0.18	0.60	0.01
M4	0.01	0.18			0.60	0.60	
M5			3.20	0.41		0.01	

Figure 4. Weighted MCI matrix (standard z scores)

Table 1 displays the cell formation result for the selected problem using the HERBAL method. A software package called HERBAL-easy has been employed to randomize the initial configuration of the input matrix more than 100 times and obtain the best result based on QI measure. The cell formation result based on weighted similarity coefficient [1] is shown in Table 2.

Table 1 Cell formation solution using HERBAL method

Cell /	Machine No's	2,1,6,9,16,14
Family No 1	Component No's	4,38,2,34,18,7,32,42,28,40,10,6
Cell /	Machine No's	12,11
Family No 2	Component No's	30,27
Cell /	Machine No's	3
Family No 3	Component No's	36,17,35
Cell /	Machine No's	8,4,13,15,5
Family No 4	Component No's	22,8,3,11,33,15,41,16,21,29,1,23,19,12,20,24,43,5,9
Cell /	Machine No's	10, 7
Family No 5	Component No's	26, 39, 13, 25, 31

Table 2. Cell formation solution using weighted similarity coefficient

Cell /	Machine No's	2,9,16,1
Family No 1	Component No's	42,40,38,37,28,18,10,4,2,32
Cell /	Machine No's	3,14,6,10,15
Family No 2	Component No's	39,31,26,19,43,17,14,13,36,7,6,35,34
Cell /	Machine No's	4,11,5,8
Family No 3	Component No's	20,30,29,41,16,15,27,33,12,11,9,8,24,23,5,3,21,1
Cell /	Machine No's	7
Family No 4	Component No's	25
Cell /	Machine No's	12,13
Family No 5	Component No's	22

Table 3 shows cell formation measures of four different solutions that have been obtained using McAuley's similarity coefficient (7), Gupta and Seffoddini's similarity coefficient, and the proposed HERBAL method. There are two solutions obtained using HERBAL method. One is based on binary input MCI matrix and the other is based on the non-binary form of the same input MCI matrix. It can be seen from Table 3 that the quality index has improved by 5% in the case that based on weighted (non-binary) input MCI matrix. It has been noticed that the solution obtained using McAuley's similarity coefficient and HERBAL (binary-based) have formed blocks of nonzero entries with less exceptional elements. This is true because their main objective was to obtain block diagonal matrix with minimum number of ones. Whereas cell formation methods that employ weighted similarity coefficient or HERBAL (weighted-based) algorithm have aimed to obtain block diagonal matrix with minimum amount of intercellular operation workloads.

Table 3. Summary of results for four solutions

Measure	McAuley's Binary Similarity Coefficient	Gupta & Seffoddini Weighted Similarity Coefficient	HERBAL Binary	HERBAL Weighted
QI	0.765	0.829	0.826	0.865
Exceptional Elements	28	41	29	32

5. Conclusions

The HERBAL method has outperformed other methods that use McAuley and Gupta and Seffoddini similarity coefficients, with respect to QI measure. Also, by comparing cellular configurations shown in Tables 1 and 2, it is revealed that the HERBAL method obtains rational solution with less number of exceptional elements than the one obtained using Gupta and Seffoddini weighted similarity coefficient.

References

[1]   Gupta, T. and Seifoddini, H. (1990) Production Data Based Similarity Coefficient for Machine-Component Grouping Decisions in the Design of a Cellular Manufacturing Systems. *International Journal of Production Research*. Vol. 28, no. 7, pp. 1247-69.

[2]   Nair, G. J. and Narendran, T. T. (1999) ACCORD: A Bicriterion Algorithm for Cell Formation Using Ordinal and Ratio-Level Data. *International Journal of Production Research*. Vol. 37, no. 3, pp. 539-56.

[3]   De Guio, R. and Barth, M. (1999) Cell Formation Using Production Flow Analysis. Irani S A, Editor. *Handbook of Cellular Manufacturing Systems*. New York: John Wiley & Sons, INC., pp. 69-109.

[4]   Marghalany, I. and Vitanov, V. (2002) Cell formation algorithm using heuristic rules-based logic to automate the production flow analysis. *Proceeding of the 19th International Manufacturing Conference*. Queen's University Belfast, 28th – 30th Aug. 2002.

[5]   Seifoddini, H. and Djassemi, M. (1996) A new grouping measure for evaluation of machine-component matrices. *International Journal of Production Research*. Vol. 34, no. 5, pp. 1179-93.

[6]   King, J. R. (1980) Machine-component grouping in Production Flow Analysis: An approach using a rank order clustering algorithm. *International Journal of Production Research*. Vol. 18, no. 2, pp. 213-232.

[7]   McAuley, J. (1972) Machine grouping for efficient production. *The Production Engineer*. pp. 53-57.

eMROM – A Web-based e-Manufacturing Modelling System

Dr. Zhijie Xu

School of Computing and Engineering, University of Huddersfield, Queensgate, Huddersfield HD1 3DH, z.xu@hud.ac.uk

Abstract: This paper reports an e-Manufacturing Reference Object Model (eMROM) system for modelling and simulating Web-based virtual e-Manufacturing enterprise. It focuses on the methodology in enabling e-Manufacturing asset modelling, system integration, and data management. Based on the approach, a prototype virtual e-Manufacturing authoriser is being constructed using the Java language and J2EE infrastructure to provide easy-to-use and flexible tools that enable users to construct and test Web-centric e-Enterprise rapidly and with minimum modelling effort.

1. Introduction

Broader product ranges, shorter model lifetimes, and the ability to process orders in arbitrary lot sizes are becoming the norm in today's manufacturing industry. The information processing capability to treat masses of customers as individuals is permitting more and more companies to offer individualized products while maintain high volumes of production [1,2]. The convergence of Intranet and Internet technologies is making it possible for groups of companies to coordinate geographically and institutionally distributed capabilities into a single "Virtual Enterprise", and in the process, achieve powerful competitive advantages. The key for ensuring the success of a "Virtual Enterprise" is to maintain agilities at all level in the organization and cross the whole spectrum of a complete product lifecycle [3,4]. The main difficulty for achieving the promised agility lies on the effort to not just loosely bind activities at the business and production levels but to integrate them into a unified and efficient infrastructure through a collaborative middleware where resource and information can flow freely and orderly to all sections in the virtual enterprise in real-time. This middleware is sometimes referred as an e-Manufacturing system.

e-Manufacturing integrates Internet technologies throughout production, from plant floor devices to corporate business systems. Deployed within an organization, the Internet can provide plant-floor personnel with real-time views of remote data. Embedded in factory equipment, it can produce smart sensors and intelligent machine tools. Connecting one company with another, it can foster collaborative designs and enhance just-in-time deliveries. When combined with supply-chain and customer-relationship management systems, manufacturing e-business enables employees and partners to move through the corporate intranet, checking product availability, placing orders, and tracking shipments anytime and anywhere [5].

2. Literature Review

2.1 Piloting projects

An industrial automation solution company – Emation – has been advertising a new Internet-based automation tool - Web@aGlance/DRM. It uses embedded Web servers to provide connectivity at the lowest level of plant floor control, and move real-time data from the device level to the controls level, and up to the enterprise tier. Without installing extra software on each and every client machines, Web@aGlance serves as the data integrator, or portal. It promises a rapid development environment that enables fast design and deployment of web applications.

Ravindranadh Vetsa [6] from Wipro Technologies described a Product Lifecycle Management (PLM) system for designing digital product models in a virtual enterprise by a collaborative virtual design team.

Rockwell Automation is researching on linking mechanisms between e-Business and e-Manufacturing, specifically on the implications of enterprise technologies and standards [7].

Harvey Wohlwend [8] from SEMATECH – a major semiconductor manufacturer - has developed an e-Diagnostic system which aims to integrate the processes of remote access and collaboration, data collection and control, problem analysis and predication within a global computer network system as well as the internal enterprise networks.

2.2 Standards and specifications

Among all those surveyed e-Manufacturing implementation cases, hardly a set of commonly recognised and unified e-Manufacturing standards and implementation specifications had been followed, except all seemed appreciating the Internet as a common data handling platform.

Efforts have been made by some leading manufactures and independent Standard Institutions in developing e-Standards for manufacturing applications. International SEMATECH - a global consortium of leading semiconductor manufacturers - has been organising an initiative in forming many e-Standards to improve semiconductor manufacturing technology, for example, equipment data acquisition and e-Diagnostic in a so-called "e-Manufacturing Standards Road Map". Some progress has been made since 2002. For example, interim DDA solution realized in December 2002; supporting standards (XML & CEM) approved in October 2002; and data interface definition by Spring 2003. So far, the work has mainly been focusing on the production level covering factory production management, plant control, and device and instrumentation.

On the customer, supplier and manufacturer collaborative level – where design meets production - the STEP (STandard for Exchange of Product model data) is a comprehensive ISO standard for CAD design data. STEP-NC, the manufacturing extension of this standard, annotates the design information with manufacturing data. STEP became a comprehensive ISO standard in 1994, and since then, all of the leading CAD software vendors have implemented STEP data translation. It is estimated that more than one and a half million CAD stations now contain STEP data translators. In the mean time, the STEP-NC is on its way to becoming a Draft International Standard in 2004.

It is the author's view that at the top of the e-Standard hierarchy - the level of business functions where sales, marketing, and business transaction are conducted - only third-party vendor-specialized software systems are seen rather than any standardized specifications for application implementation that benefits the aforementioned other two function levels.

2.3 New development tools

Although some of the Internet/Web-based e-Manufacturing functions may be slow off the mark, Web tools are expanding their capacity in an accelerated rate. For instance, Sun Microsystems created Java to provide a fast, simple way to program for the Web. It is claimed to be a portable, architecture-independent programming language that could be universally deployed—program once, read everywhere. The trade-off for those advantages existing in Java-based applications is its slow execution speed due to an interpreting process through a Virtual Machine layer. The justification for adopting Java in e-Manufacturing applications is of its ability in integrating different hardware and software systems.

For handling information and data in an enormous amount of varied formats thorough out an e-Manufacturing enterprise, an effective and universal toolset is required for the job. Among various approaches, the XML protocol is emerging and making headway to the Web-centric e-Manufacturing applications. Vendors such as Opto-22 are using XML to tag and transport data for interconnecting diverse software packages and operating systems from multiple manufacturers without the brittleness of point-to-point integration. Demands like these are driving the XML standardization process, including the ISA-95 standard, the new body of work that defines the interface between manufacturing control systems and the enterprise.

Among those progresses and problems described in this section, one of the biggest hurdle remained as how to upgrade the manufacturing organization's internal information architecture to adapt the "e-Volution". The goal of this project is to make it possible for the disparate pieces of the enterprise to be integrated so employees, customers, and suppliers have a unified view of the entire business, from sales to customer service to shipping and receiving.

3. eMROM - An Asset Modelling System

People, products, equipment, and processes make up the manufacturing assets of an enterprise. The e-Manufacturing Reference Object Model (eMROM) being developed in this project is a reference model that defines the interrelationship of e-Manufacturing based virtual enterprise asset components, data models and protocols so that e-Manufacturing objects are accessible and sharable across systems. The eMROM contains specifications to provide a comprehensive suite of e-Manufacturing capabilities enabling interoperability, accessibility and reusability of Web-based e-Manufacturing capacity. The eMROM system is consisted of two parts; feature aggregation model (FAM) and run-time environment.

3.1 Feature aggregation model

FAM is designed for modelling and integrating manufacturing asset models so that data "node" can be formed and data flow coordinated. The collated information will then be used for simulation and analysis.

The key specification of the FAM is the "Manufacturing Asset Object Metadata", which is a dictionary of XML tags used to describe the asset object models in a variety of categories. Those categories reflect features such as object ownership, function definition, process capacity, I/O interface, operational cost, and real-time features.

Another important specification is designed for XML "binding" so the metadata tags can be interpreted into required formats for machines or human users. In the project, Java and Extensible Style-sheet Language Transformation (XSLT) - an important part of the XSL Standards for transforming an XML document into another XML document, or other type of documents - have been used for realising this functionality.

The third specification in FAM is the Virtual Enterprise Packaging specification, which defines how to package together a collection of asset objects, their metadata, and information about how the system is to be delivered to the user. The packaging process consists of "zipping" all relevant files together with an XML "manifest" file that defines all of the contents and their relationship to one another.

3.2 Run-time environment

The e-Manufacturing implementation practices can be treated as the process of enabling any large-scale and complex information systems. It starts from defining data input (or information acquisition); and then data process and control; finally, data output (information product). Each stage will also interact with the surrounding environment and other sub-systems. A virtual enterprise implemented using the eMROM system and FAM specifications is displayed on the run-time interface as a series of "nodes" connected by a web of threads. Each node is a data island offering some kind of data processing and storage functionality as well as the data input/output abilities. The threads in this picture stand for the network facilities in the real world. A group of nodes could have been connected by a local area network (LAN) or Intranet. Others could have been linked using Internet and the World-Wide-Web (WWW). Instead of adopting the traditional pyramid-like hierarchical structure, this distributed and inter-connected structure enables the flexible formation of functional asset by grouping a number of nodes for a demanding task. For example, a virtual design team could be formed by connecting product designers, engineers, market experts or and/or even customer representatives. In the mean time, a virtual enterprise can be established by linking digital products, suppliers, and manufacturing workshops. In this research, a J2EE-based (Sun Microsystems' Java 2 Enterprise Edition) virtual enterprise information infrastructure has been adopted in realising the run-time environment for the eMROM applications.

4. eMROM Implementation

Figure 1 shows a virtual e-manufacturing enterprise structure where each module is modelled using the eMROM system and FAM specifications. At the top level, the client browsers and application programs access a virtual enterprise through the business functions such as product inventory, and on-line ordering services. The collected data is managed by an enterprise database system through various of functions in the format of enterprise beans.

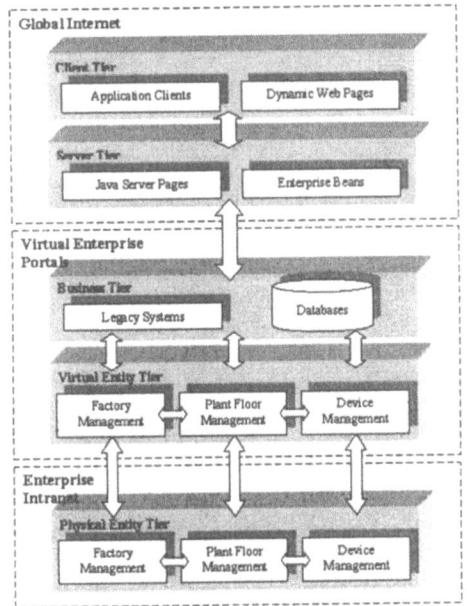

Figure 1: A Virtual Enterprise Structure

List 1: A machining planning class

```
Class Operation_Plan {
/*Resouces are declared as fields*/
private static final String factory_name = "VMC";
private String workpiece_file, nc_program_file,
tool_file, fixture_file;
private float x_offset, y_offset, z_offset;
private int machine_id, tool_id, fixture_id,
workpiece_id, nc_program_id;
/*Header info is essential for constructing a plan*/
public Operation_Plan(String plan_name,int part_id,
int machine_id, String date, int revision_no)
{
//Based on the plan information
this.machine_file= //i.e."WORKCELLS/VMC_1";
this.tool_file = //i.e."TOOLS/TWIST_DRILLS_1";
this.fixture_file = //i.e."FIXTURES/FIXTURE_1";
this.workpiece_file = //i.e."WORK/WORKPIECE_1";
this.nc_program_file = //i.e."NC/test_part1.cnc";
/*Procedures are defined as methods*/
public void Load_Tool(int tool_id)            {
this.tool_id = tool_id;
//other settings, i.e. slot_no
}
public void Load_Fixture(int fixture_id)            {
this.fixture_id = fixture_id//i.e. 1 for a clamp
//other settings, such as offset
}
public void Load_Workpiece(int workpiece_id)    {
this.workpiece_id = workpiece_id;
//other settings
}
public void Load_NC_Program(int program_id,
int machine_id) {
this.nc_program_id = program_id;
this.machine_id = machine_id;
}
public void Run_NC_Program(int program_id)    {
/*Corresponding set and get methods are defined
for setting and retrieving information*/ }
```

The middle level of the structure is the so-called Virtual Enterprise Portals, where eMROM FAM models are assembled to forming task-specific teams while exchanging data. The Internet/WWW and other enterprise Intranet network and protocols form the backbone of this level. The simulation tasks that can be performed here include organising the enterprise supply chain, managing manufacturing resources, performing process planning and scheduling, managing plant information, and control the manufacturing processes.

The bottom level of the proposed eMROM-based virtual enterprise infrastructure is Local Area Networks (LANs) or device controllers adopted by various manufacturing companies. It is consisted of factory management components, plant floor control components, and device control components. The simulation program designed and tested in a virtual environment can drive the virtual machines as well as being sent to the physical entity tier for plant floor and device control, diagnostic and maintenance.

A mock up e-manufacturing system has been developed based upon the proposed eMROM infrastructure. The application is made up of manufacturing asset components. Each component is a self-contained functional software unit that is assembled into a J2EE application with its related classes and files. As shown in List 1, a shop floor operation plan in the Java class format is interpreted from a XML file – FAM objects manifestos - where the tag provide information about operation sequence, and the attribute fields provide information such as "tool_id", "fixture_offset", and robot trajectory data.

5. Conclusions

In this research, a prototype e-Manufacturing modelling infrastructure and toolset has been established. A virtual enterprise can be quickly modelled using the eMROM system and its related specifications. The J2EE-based run-time environment provides open-architecture and operating system independent platforms to control, simulating, and analyse enterprise-level operational activities.

References

[1] Goldman, S. L., Nagel, R.N., Preiss, K. *Agile Competitors and Virtual Organisations.* Book Excert, *Manufacturing Review,* Vol.8, No.1, pp59-67, 1995.

[2] Xu, Z. J., Zhao, Z. X., and Baines, R. W. *Constructing Virtual Environments for Manufacturing Simulation. The International Journal of Production Research,* ISSN 0020-7543, Vol.38, No. 17, pp4171-4191, 2000.

[3] Koç, M. and Lee, J., 2001, *A System Framework for Next-Generation E-Maintenance System,* EcoDesign 2001, Second International Symposium on Environmentally Conscious Design and Inverse

[4] Waurzyniak, P., 2001, *Moving towards e-factory,* SME Manufacturing Magazine, v. 127, n. 5, http://www.sme.org/gmn/mag/2001/01nom042/01nom042.pdf

[5] Slansky, D. *Java Technology Powers E-Manufacturing. Jini Network Technology White Paper,* http://wwws.sun.com/software/jini/whitepapers, 2002.

[6] Vesta, R., 2003, *PLM - Rolling Stone for eManufacturing,* Wipro Technologies, http://www.wipro.com/itservices/industries/manufacturing/plm/pdm.htm

[7] Rockwell Automation *e-Manufacturing Industry Road Map,* http://www.rockwellautomation.com

[8] Wohlwend, H., 2002, *International SEMATECH e-Diagnostics,* http://www.sematech.org/public/resources/ediag/index.htm

Technology Oriented Costs Configuration in the Offering Process for Complex Products

F.-L. Krause[1]; Chr. Kind[2]; C. Müller[3]

[1] Fraunhofer Institute Production Systems and Design Technology, Pascalstr.8-9; 10587 Berlin, Germany; Frank-L.Krause@ipk.fhg.de
[2] Fraunhofer Institute Production Systems and Design Technology, Pascalstr.8-9; 10587 Berlin, Germany, Christian.Kind@ipk.fhg.de
[3] Fraunhofer Institute Production Systems and Design Technology, Pascalstr.8-9; 10587 Berlin, Germany; Cornel.Mueller@ipk.fhg.de

Abstract Due to aggravated competition, suppliers of complex single and small series products find themselves in a situation where more detailed offers in an even shorter period of time have to be prepared while the customers' requirements change constantly.

The article presents a methodology that supports the offering process by creating and comparing alternative technological solutions. Therefore all reasonable options have to be integrated in a generic machine base. By taking the customers requirements into consideration the user can transfer generic data into different specific solutions. The result is a relative cost information before the design process itself has begun to identify the cost optimal machine.

The methodology has been implemented in an industrial software prototype.

1. Introduction

For the machine and plant building industry, the globalisation of markets did not only lead to an increasing number of potential customers but also to a massively aggravated competition. Suppliers of single and small series products find themselves confronted with a situation where they have to face more competitors and have to prepare more detailed offers in an even shorter period of time. The continuously weak economic situation and the increasing number of crisis regions with warlike conflicts complicates the situation for companies additionally.

But not only the diffuse and constantly changing requirements, given by the customer, complicate the offer preparation process (OPP) for the supplier. He has to cope with a wide range of influence factors that have to be taken into consideration. The result of an offer preparation process is a technically configured solution developed according to the customers requirements and the add on information of the production cost of the offered product. The calculated price is subject to calculation risks as well as customer specific, external and technological influences that are difficult to manipulate and calculable for the supplier.

As a result, the preparation time for each offer is too short and is combined with a lack of cost transparency for the designer and a higher contingency risk. That, in turn, means a low ratio between submitted offers and received contracts for the offering company, whereas the cost pressure increases constantly.

Based on this, the following contribution describes requirements on a system to support an offer preparation processes and introduces a method for cost oriented configuration that meets these requirements.

2. Requirements on an Offer Preparation Process

Offer preparation processes normally aim at an optimal solution to a customer problem with respect to technological, economical and quality aspects. At the same time the manufacturers resources have to be taken into consideration. The offer itself should be prepared with the lowest effort whilst keeping up the highest possible quality standard. From this aim, one can derive a variety of requirements, which have to be met to increase the efficiency of the offer preparation process for single and small series products [1].

- Acceleration of the offer preparation process,
- Generation of alternative technical solutions,
- Appraisal of alternatives in the early phase of the design process,
- Comprehensive consideration of customers requirements,
- Increase of cost transparency,
- Progression of the offer process quality, and

➤ **Cost reduction within the OPP**

The method of resolution comprises the configuration of the optimal technological solution as well as the calculation of the resulting production cost and the fixing of the product price.

A variety of projects concerned with the topic of the offering process led to the conclusion that the crucial factor for offer preparation in many companies is time. In the post-processing and the search for causes for withdrawn offers, insufficient consideration of customers requirements was in second position after the time factor. And there is a correlation between quality and time as well. This is the reason why the designer just changes an already submitted offer or even an already realised order according to his understanding of the customers requirements. Because of time restrictions he is not able to develop a complete new solution. As a result, the quality of the given offers does not achieve its potential and the companies aim to get a contract will not be achieved.

If a customer declines an offer because the price of the product is to high, usually the calculation method is not the reason for this. The offer has not been accepted because alternative technical solutions have not been developed and evaluated. That is due to the above described time dilemma. If the calculation of the production cost begins after the design has been finished and it becomes clear that the cost will exceed the customers target it is usually to late to change the technical solution significantly because of a lack of time. A non-competitive offer will be submitted. Valuable resources have been used that will not contribute to the companies profit. Conceptual design and appraisal of alternative machines or plants before the elaboration of the technical solution could prevent a situation like this.

With such an approach, the demand for higher cost transparency and increased cost awareness of the designer would implicitly be fulfilled. Idle cost reduction potentials could be revealed and activated.

Furthermore, it is important for an unobstructed offer preparation process that besides design and calculation, all of the departments involved have access to a wide and integrated range of information. This information should not only be used by the people involved, but also be systematically cultivated to ensure the above described transparency and prevent the loss of information.

3. Cost Oriented Configuration Systems

The relations between companies get more and more individualised. To maintain competitiveness, it is important to comply with the steadily increasing importance of customer orientation and to incorporate this development into business processes [2]. The preparation of an offer and especially the calculation of the offer price with support of a cost oriented configuration system is a possibility to combine and ensure customer individualisation, quality and competitiveness at comparably low cost. A system like that, once implemented, is easy to use and can reliably generate and validate alternative technical solutions.

The development of configuration systems requires a systematic modularisation of the product and its sub-assemblies. But a simple modular product structure is only the first step. A simple modularisation only allows the designer to make variant designs, but do not go far enough to meet the needs of highly individualised and complex single and small series products [3]. What is even more important, also in terms of the later calculation of the production cost, is to take all customer requirements into account, because these are the cost drivers. To ensure this, combination and extension rules have to be developed, that are driven by the customers requirements and have a massive influence on the design and therefore on the cost of the offered product. This enables products to be configured virtually out of a given pool of sub-assemblies, modules and parts on a concept level [4][5].

In the context of effective offer preparation a configuration system should not only be able to develop a practicable technical solution to a customers problem, but should also include cost calculation procedures into the above mentioned configuration rules.

4. Optimisation of the Offer Preparation Process

The approach implemented for the offer preparation of complex single and small series products combines the above defined requirements with the characteristics of configuration systems. Before the design process itself starts, the project team gets information about which technological machine alternative is optimal under cost consideration and should be further elaborated.

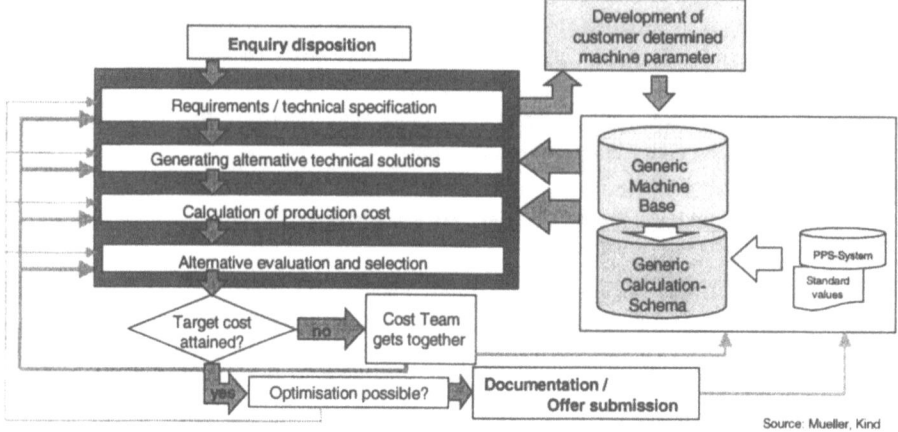

Figure 1: Optimised offer preparation process

The approach of the optimised offer preparation process on basis of a cost oriented configuration system is described in figure 1. The enquiry communicated by the customer represents his interest in a solution of a problem described by functional specifications. Suppliers of complex single and small series products in particular know the range of possible technological solutions to their customers' problems very well. The kernel of the cost oriented configuration is the generic machine base in which all possible machines or plants are included in a kind of basic configuration.

The functional specification describes the requirements from which customer determined machine parameters have to be derived. By using these parameters and executing the configuration rules mentioned above the relevant machines form the generic machine base that will get incorporated into specific machine solutions.

To make these generated machine alternatives comparable they have to be evaluated monetarily by using a calculation schema. For that purpose data such as cost information about manufacturing and material as well as prices from suppliers and other relevant cost information, have to be taken from the PPC system through an interface and have to be combined to calculate the production cost for the machine or plant. The decision for or against a technical solution will not be made on absolute cost information only. In this context it is much better to have relational cost information, for instance machine cost in relation to the number of parts that are supposed to be manufactured on or with the machine (figure 2). The designer has to remember at all times that costs are in most decision cases the crucial factor for or against a supplier. The monetarily evaluated machine solutions must under no circumstances exceed the given cost target. Generating and evaluating different technological machine alternatives opens a variety of possibilities for the supplier to enforce the relationship between him and his customer and to involve the customer into the decision making process of the technical solution that should be chosen before the design process itself has begun. The implementation of new and innovative technologies usually causes higher costs, but goes along with quite a few advantages such as higher speed of operation, smaller tolerances and a longer life cycle. The sales department has to communicate this to the customer to justify higher costs.

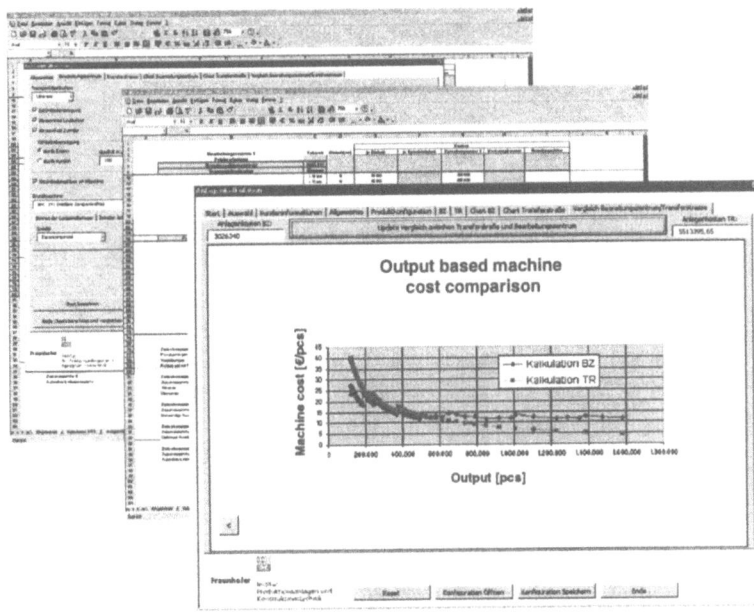

Figure 2: Output based comparison of alternative technical solutions

The application of the method by a designer will increase the cost transparency as well as his cost awareness. If the given cost target will be exceeded, the designer can react at this early phase appropriately and change the technical solution before he went through the complete design process. But it can also happen, that the designer realises that the customer was driven by a cheap supplier when fixing the cost target to such a low level. In this case the company has to withdraw from the bidding process, because it is not very promising. Here, it is only justified to go through the effort of preparing an offer if this is a 'political' necessity.

Furthermore, by applying the introduced method the designer gets a fairly detailed description of the machine. He just has to elaborate the given solution to assure the completeness and quality standard of the prepared offers on a constant high level.

Because actual numbers from production are used for the calculation the implementation of the method supports the possibility to optimise the utilisation of machines and specific production departments within the company. Well-utilised departments in the factory are rated with higher costs whereas less utilised departments can be rated with lower costs and therefore their capacity will be available for a lower price.

The accuracy of the cost calculation in the early phase of the offer preparation process is not that important because at this time of the OPP only relational cost information is required to compare alternative technical solutions and make a decision. Nevertheless when the final offer is about to be submitted to the customer, the calculation has to be as accurate as possible. This detailed level of calculation can only be reached after the design has been elaborated .

5.   Implementation Effort and Outlook

The method presented has been realised in a prototype version and has been tested in an industrial environment. A high level of acceptance within the group of users has been achieved because special effort was given to an intuitive scheme and an easy application for the user.

According to information from industrial project partners and the German Engineering Federation VDMA, the rate of return between submitted offers and received assignments in the German Machine and plant building industry had an average between 5% and 10%, sometimes even lower. The average time to prepare an offer with project partners was specified with about three month for complex single and small series products.

The effort to prepare conventional offers has to be compared with the implementation and utilisation effort for the introduced method within a revenue and expense analysis. Actually, the cost to elaborate the generic machine base is quit high. Within the realised projects, the average time to design one machine type for the generic machine basis was about half a day, excluding the intensive preparation from a research consultant such as the Fraunhofer IPK. The total effort for a company varies with the number of machine types and their variants. From the experience gained on this project, it can be said that the average effort to develop the generic machine base is as high as the preparation of one offer: about three months. If one takes this under consideration the cost for the implementation of the method is justified, because an increase of the return ratio of up to 20% and more seems to be realistic. This was emphasised by the consistently positive feedback from industrial partners to the method of technology oriented cost configuration.

References

[1] F.-L. Krause, Chr. Kind, C. Müller, 2003, *Produktkosten fest im Griff*, ZWF 6/2003, Carl Hanser Verlag München

[2] N. Gronau, E. Weber, 2001, *Marktüberblick Konfiguratoren in PPS-Systemen*, PPS Management 6/2001; GITO Verlag Berlin

[3] J. Schlingheider, 1994, *Methodik zur Entwicklung rechnerunterstützter Konfigurationssysteme*, Carl Hanser Verlag München as well as Dissertation TU Berlin

[4] A. Bronner, 1998, *Angebots und Projektkalkulation*, VDI Springer, Berlin

[5] H. Wang, X. H. Zhou; X.-Y. Ruan, 2003, *Research on injection mould intelligent cost estimation system and key technologies*, International journal of advanced manufacturing Technology, Springer Verlag London

Studies on the Benefits of Returns Policy

Hexin Wang[1], Khairy A. H. Kobbacy[2], Wenbin Wang[3]

Center for Operational Research and Applied Statistics (CORAS),
University of Salford, M5 4WT, UK
[1] H.Wang@pgr.salford.ac.uk [2] K.A.H.Kobbacy@salford.ac.uk [3] W.Wang@salford.ac.uk

Abstract: A returns policy is employed when suppliers allow retailers to return some of unsold products back, possibly for a partial credit. Returns policies are common in the distribution of perishable commodities. It has been shown that returns policies can lead the supply chains to maximize their expected profits. However, a returns policy does not necessarily benefit every partner of a supply chain, when the wholesale price is determined by the market. A one-supplier-one-retailer supply chain has been investigated using a simple newsvendor model. We prove that the retailer can always benefit from returns policies, but it is not necessarily true for the supplier. We further determine the condition under which suppliers can benefit from such a policy. The relationship between the benefit of a returns policy and the variability of demand is then investigated, using mean-preserve transformation. Finally the paper demonstrates numerically the relationship between the benefit of returns policies and the cost structure of the supply chain.

1. Introduction

A returns policy specifies that a retailer can return some of unsold product to the supplier, possibly for some partial credit. Returns policies are common in the distribution of perishable commodities, such as books, magazines, newspapers, recorded music, computer hardware and software, greetings cards, and pharmaceuticals [1]. Returns policies make suppliers share risks with retailers, and therefore, encourage retailers to stock more and probably sell more.

There exist five kinds of returns policies: full returns for full credit, full returns for partial credit, partial returns for full credit, partial returns for partial credit, and no returns. Pasternack [2] shows that both no returns and full returns for full credit are suboptimal for supply chain coordination, while full returns for partial credit can achieve coordination. Emmons and Gilbert [3] investigate the role of returns policies in the setting that the retailer has control over both the retail price and inventory decisions. They confirm that both the supplier and retailer can benefit from returns polices under certain conditions. Padmanabhan and Png [4] study returns policies in the situation of retail competition, and conclude that a returns policy intensifies retail competition and reduces retailer margins. They consider two policies: no returns, and full returns for full credit. Tsay [5] follows the framework of Padmanabhan and Png [4] and analyzes how risk sensitivity affects the use of return policies within a supply chain. He finds that a risk sensitive retailer does not always prefer the right to return excess product for full credit, and a risk sensitive manufacturer does not always oppose this, when the wholesale price is fixed. Lau and Lau [6] study a similar problem and conclude that a returns policy can often be manipulated by a shrewd manufacturer to increase his profit.

In this paper, we study the returns policy of full return for partial credit in a supply chain where retail price and wholesale price is exogenous, i.e. determined by the markets. We show the condition under which both the supplier and retailer can benefit from returns policies, and how the demand uncertainty and cost structure affect the benefits.

2. Model Development

The investigated supply chain, in this study, comprises a retailer and a supplier. The supplier fills the order from the retailer, who serves the final market. The product has a short life cycle, which means that the retailer has only one opportunity to replenish her inventory. Therefore, this model is developed based on a newsvendor model.

Notation

p : Retail price
w : Wholesale price
c : Variable production cost
a : Fixed production cost
b : Buyback price
x : Demand
$f(x), F(x)$: pdf. and cdf. of the demand x
Q : the retailer's order quantity
Q_n^*, Q_r^*, Q_j^* : Retailer's optimal order quantity without returns policies, with returns policies, and with the joint supply chain, respectively.
$\pi R_r(Q \mid x)$, $\pi R_n(Q \mid x)$: Retailer's profit with returns policies and without returns policies, given demand x
$\pi S_r(b \mid Q, x)$, $\pi S_n(\cdot \mid Q)$: supplier's profit with returns policies and without returns policies, given retailer's order quantity and demand x
$\pi J(Q \mid x)$: the profit of the whole supply chain, given demand x

Without returns policy

Given demand x, the retailer's profit is:

$$\pi R_n(Q \mid x) = \begin{cases} px - wQ & \text{if } x \le Q \\ (p-w)Q & \text{if } x > Q \end{cases}$$

The expected profit of the retailer:

$$E(\pi R_n(Q)) = \int_0^Q (px - wQ)f(x)dx + (p-w)Q\int_Q^\infty f(x)dx$$

$$= (p-w)Q - p\int_0^Q F(x)dx$$

(1)

Optimal order Q_n^* satisfies $F(Q_n^*) = \dfrac{p-w}{p}$

Since we assume that the wholesale price is fixed by the market, the supplier's profit is determined by the amount that the retailer orders:

$$\pi S_n(Q_n^*) = (w-c)Q_n^* - a$$

(2)

The expected profit of the whole supply chain is the sum of equations (1) and (2), that is: $E(\pi J(Q)) = (p-c)Q - a - p\int_0^Q F(x)dx$ (3)

To provide an efficiency benchmark, it is assumed that the supplier and retailer make decisions jointly as a single entity, which delivers the maximum expected chain profit. For the integrated firm the optimal order Q_J^* satisfies $F(Q_J^*) = \frac{p-c}{p}$.

Since $w > c$, and $F(\cdot)$ is an increasing function, $Q_n^* < Q_J^*$. Obviously, without returns policy the supply chain fails to achieve the optimal performance.

With returns policy

The expected profit of the retailer is:

$$E(\pi R_r(Q)) = \int_0^Q (px - wQ + b(Q-x))f(x)dx + (p-w)Q\int_Q^\infty f(x)dx \qquad (4)$$
$$= (p-w)Q - (p-b)\int_0^Q F(x)dx$$

The optimal order Q_r^* satisfies $F(Q_r^*) = \frac{p-w}{p-b}$.

For a given w, $F(Q_r^*) = \frac{p-w}{p-b} > \frac{p-w}{p} = F(Q_n^*)$, so $Q_r^* > Q_n^*$. That is, in the case with a returns policy, the retailer will order more than that in the case when no returns policy is offered.

The expected profit of the supplier is:
$$E(\pi S_r(b \mid Q_r^*)) = (w-c)Q_r^* - a - b\int_0^{Q_r} F(x)dx \qquad (5)$$

It is shown [7] that there is no such a unique solution for equation (5). Here we assume the wholesale price is determined by the market or the supplier has no control on the wholesale price. The supplier wants to sell as many as possible for various reasons, for instance, maximizing his profit or grabbing a larger share of the market demand. To induce the retailer to order as much as the integrated supply chain, i.e., $Q_r^* = Q_J^*$, the supplier should set b so that $\frac{p-w}{p-b} = \frac{p-c}{p}$, or

$$b = p \cdot \frac{w-c}{p-c}.$$

3. The Benefits of Returns Policies

In this section, we are going to investigate whether a returns policy can increase the expected profit of the retailer and the supplier.

Proposition 1: In the case *with a returns policy, the retailer's expected profit is more than that in the case without a returns policy.*

Proof: See Appendix ∎

Proposition 2: *the condition that the supplier will offer a return policy is*

$G(Q_J^*) > Q_n^*$, where $G(Q) = Q - \dfrac{\int_0^Q F(x)dx}{F(Q)}$.

Proof: See Appendix ∎

4. The Relationship Between the Benefits of Returns Policies and the Variability of Demand

The mean-preserving transformation is used to investigate the relationship between the benefits of returns policies and the variability of demand. This transformation is a common approach used for exploring the effects of changing variability (e.g. [8]).

Let X be a non-negative random demand with distribution function $F(x)$; $X_\alpha = \alpha X + (1 - \alpha)\mu_X$, $0 \le \alpha \le 1$, where $\mu_X = E(X)$; and $Var(X)$ be the variance of X. Then, $E(X_\alpha) = E(X)$ and $Var(X_\alpha) = \alpha^2 Var(X)$, which is non-decreasing in α. Thus, the random demand X_α becomes more variable when α increases while the mean is reserved. While X_α usually does not belong to the same parametric family as X, X can take any distribution.

Before we investigate the relationship between the benefits of returns policies and the variability of demand, we present some results similar to that of Gerchak and He [9], concerning the newsvendor problem when the demand undergoes a mean-preserving transformation. Let the retailer's profit $\pi R(Q^* \mid x)$ with optimal order quantity Q^*, given random demand X, be $\pi R^*(X)$.

Proposition 3: *For any non-negative random demand X and a non-negative constant* y, *we have* $E(\pi R^*(X + y)) = E(\pi R^*(X)) + y(p - w)$. *The corresponding optimal order size is given by* $Q^* + y$. *Also,* $E(\pi R^*(\alpha X)) = \alpha E(\pi R^*(X))$. *The optimal order size is* αQ^*. *Consequently,* $E(\pi R^*(X_\alpha)) = \alpha E(\pi R^*(X)) + (1 - \alpha)(p - w)\mu_X$. *The optimal order size is given by* $\alpha Q^* + (1 - \alpha)\mu_X$.

Proof: See Appendix ∎

The results in Proposition 3 are intuitive. When the random demand increases by a constant, the optimal decision is to add that constant to the original optimal solution, and the extra profit is then $(p - w)y$. When the random demand increases, the optimal order size and costs increase by the same multiplier.

Now, we consider our models with random demand X_α, and explore the benefits of returns policies. The retailer's expected increased profit is

$$\begin{aligned}
E(\Delta \pi R) &= E(\pi R_r^*(X_\alpha)) - E(\pi R_n^*(X_\alpha)) \\
&= \alpha E(\pi R_r^*(X)) + (1 - \alpha)(p - w)\mu_X - \alpha E(\pi R_n^*(X)) - (1 - \alpha)(p - w)\mu_X \quad (6) \\
&= \alpha (E(\pi R_r^*(X)) - E(\pi R_n^*(X))) \\
&= \alpha(p - w)(G(Q_r^*) - G(Q_n^*))
\end{aligned}$$

From equation (6), we can see that the benefit of returns policies for the retailer increases proportionally with α. Recalling that random demand X_α becomes more variable when α increases, therefore, the benefit of returns policies for the retailer increases when the demand becomes more variable.

5. Numerical Study

In this section we investigate the effect of various parameters on the percentage of the expected increased profit. For each parameter we find the optimal values of Q and b, and compute the expected profit numerically which is then compared with the case when no returns are allowed. The demand is assumed to follow a normal distribution with mean=1000, and standard deviation values of 100, 200, and 300. Other values used are retail price=40, variable production cost=10, fixed production cost=0, and wholesale price varies from 15 to 35.

First, the impact of the wholesale price on the performance of returns policies is investigated.

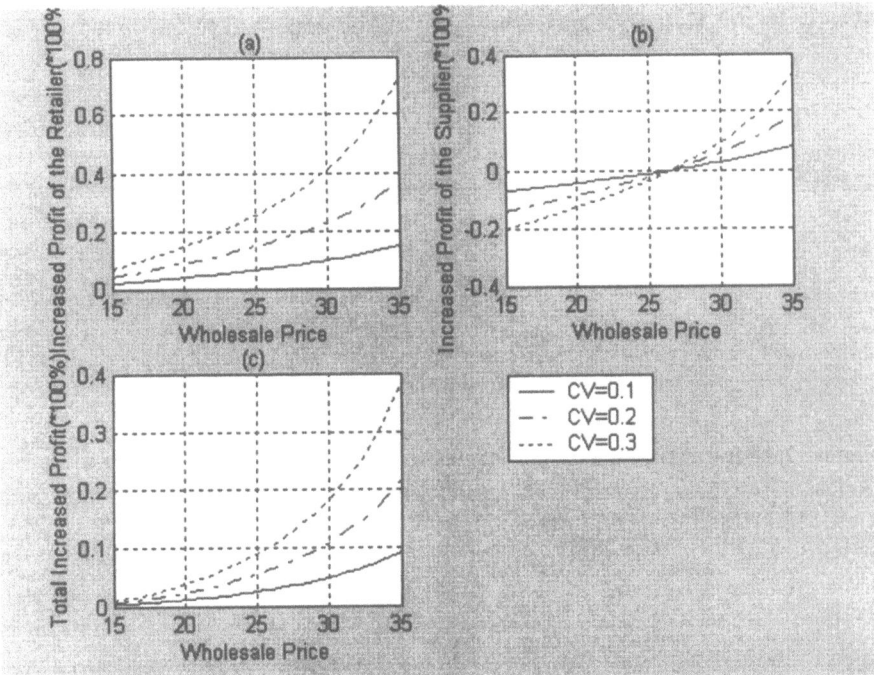

Figure 1 The impact of the wholesale price and demand uncertainty on the increased profit of the retailer, supplier, and the whole supply chain

Figures 1(a) and 1(c) show that a returns policy can always benefit the retailers and the whole supply chain. The higher wholesale price, the higher the retailer and the whole supply chain benefit. If the retail price and production cost are fixed, the wholesale price divides the margin profit between the retailers and the suppliers. The suppliers obtain large part of the margin profit with a higher wholesale price. Therefore, a higher wholesale price means that the supplier has a larger bargaining power. In this case, the retailer and the whole supply chain will gain more benefit if the supplier can offer a returns policy.

However, from Figure 1(b), it is clear that the supplier is not always better off with return policies. There exists a certain wholesale price threshold, below which the supplier is worse off. If demand uncertainty is taken into account, one can find that the demand uncertainty enlarges the impact of the wholesale price on the expected increased profits. That is, below that point, the more uncertain the demand is, the more the supplier loses. Above that point, the more uncertain the demand is, the more the supplier gains.

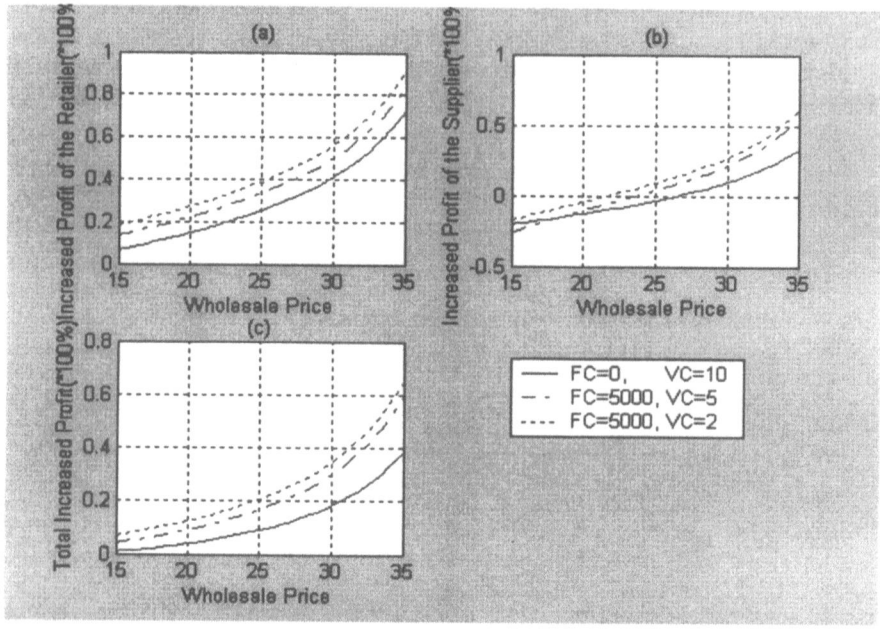

Figure 2 The impact of cost structures on the performance of returns policies

Figure 2 shows the effect of various cost structures on the benefits of returns policies. Here it is assumed that the demand follows a normal distribution with mean=1000, and standard deviation =200. Three sets of fixed and variable production costs (0, 10), (5000, 5), (5000, 2) are tested. The first two sets have the same production costs when the production quantity equals the expected demand, while the last two sets have the same fixed production costs. From Figure 2, it is clear that variable production cost can only move the curves in parallel. Fixed production costs have no effect on the retailer's curves but change the slope of the curves of the supplier and the whole supply chain.

6. Conclusions

Returns policies can always benefit retailers, but this is not always the case for suppliers. Only when the increased profit by a larger order from the retailer can cover the cost of buying back the unsold products, shall a returns policy benefit the supplier. It has been shown theoretically that the benefit of a returns policy for retailers increases as the demand becomes more uncertain. It has been further shown numerically that a returns policy performs better with the products of a lower variable production cost.

References

[1] Padmanabhan, V. and Png, I. P. L., 1995, Returns policies: make money by making good, *Sloan Management Review*, Fall, 65-72.

[2] Pasternack, B. A., 1985, Optimal pricing and return policies for perishable commodities, *Marketing Science*, Vol.4, No.2, 166-176.

[3] Emmons, H. and Gilbert, S. M., 1998, Note: The role of returns policies in pricing and inventory decisions for catalogue goods, *Management Science*, Vol.44, No.2, 276-283

[4] Padmanabhan, V. and Png, I. P. L., 1997, Manufacturer's return policies and retail competition, *Marketing Science*, Vol.16, No.1, 81-94.

[5] Tsay, A. A., 2002, Risk sensitivity in distribution channel partnerships: implications for manufacturer return policies, *Journal of Retailing*, Vol.78, 147-160.

[6] Lau, H.-S. and Lau, A. H. L., 1999, Manufacturer's pricing strategy and return policy for a single-period commodity, *European Journal of Operational Research*, Vol.116, No.2, 291-304.

[7] Lariviere, M. A., 1998, Supply chain contracting and coordination with stochastic demand, Tayur, S., Magazine, M. and Ganeshan, R.(eds.), *Quantitative models of supply chain management*, Kluwer Academic Publishers, Nowell, MA 233-268.

[8] Gerchak, Y. and Mossman, D., 1992, On the effect of demand randomness on inventories and costs, *Operations Research*, Vol.40, 804-807.

[9] Gerchak, Y. and He, Q. M., 2003, On the relation between the benefits of risk pooling and the variability of demand, *IIE Transactions*, Vol.35, 1027-1031.

Appendix

Proof for Proposition 1:

The retailer's expected increased profit is

$$
\begin{aligned}
E(\Delta \pi R) &= E(\pi R(Q_r^*)) - E(\pi R(Q_n^*)) \\
&= (p-w)(Q_r^* - Q_n^*) - (p-b)\int_0^{Q_r^*} F(x)dx + p\int_0^{Q_n^*} F(x)dx \\
&= (p-w)(Q_r^* - Q_n^* - \frac{p-b}{p-w}\int_0^{Q_r^*} F(x)dx + \frac{p}{p-w}\int_0^{Q_n^*} F(x)dx) \\
&= (p-w)(Q_r^* - \frac{\int_0^{Q_r^*} F(x)dx}{F(Q_r^*)} - Q_n^* + \frac{\int_0^{Q_n^*} F(x)dx}{F(Q_n^*)})
\end{aligned}
$$

Let $G(Q) = Q - \dfrac{\int_0^Q F(x)dx}{F(Q)}$. It can be shown that $G(Q)$ is an increasing function

since $\dfrac{dG(Q)}{dQ} = \dfrac{f(Q)\int_0^Q F(x)dx}{F^2(Q)} > 0$.

The retailer's expected increased profit can be rewrite as

$E(\Delta \pi R) = (p-w)(G(Q_r^*) - G(Q_n^*))$

Since $Q_r^* > Q_n^*$ and $G(Q)$ is an increasing function, $E(\Delta \pi R) > 0$. ∎

Proof for Proposition 2:

The supplier's increased profit

$$E(\Delta\pi S) = E(\pi S_r(b\,|\,Q_r^*)) - \pi S_n(\cdot\,|\,Q_n^*)$$

$$= (w-c)(Q_r^* - Q_n^*) - b\int_0^{Q_r^*} F(x)dx$$

$$= (w-c)(Q_r^* - \frac{b}{w-c}\int_0^{Q_r^*} F(x)dx - Q_n^*)$$

$$= (w-c)(Q_r^* - \frac{\int_0^{Q_r^*} F(x)dx}{F(Q_r^*)} - Q_n^*)$$

$$= (w-c)(G(Q_r^*) - Q_n^*)$$

Since $w > c$, if $G(Q_r^*) > Q_n^*$, then $E(\Delta\pi S) > 0$. ∎

Proof for Proposition 3:

This proposition can be applied to the profit functions of retailers with or without a returns policy. We prove the cases without a returns policy. The proof of the case with a returns policy is similar.

In the case without a returns policy, $\pi R^*(X) = \pi R_n^*(X)$ and $Q^* = Q_n^*$.

Let $Z = X + y$, Q_Z^* the optimal order size with random demand of Z, and $F_Z(z)$

the CDF of Z. Then, $F_Z(x+y) = F(x)$. Since $F_Z(Q_Z^*) = \dfrac{p-w}{p} = F(Q_n^*)$ and

$F_Z(Q_n^* + y) = F(Q_n^*)$, then $F_Z(Q_Z^*) = F_Z(Q_n^* + y)$ therefore, $Q_Z^* = Q_n^* + y$

$$E(\pi R^*(X+y)) = (p-w)Q_Z^* - p\int_0^{Q_Z^*} F_Z(z)dz$$

$$= (p-w)Q_Z^* - p\int_0^{Q_Z^*} F(z-y)dz$$

$$= (p-w)Q_Z^* - p\int_0^{Q_n^*} F(x)dx$$

$$= (p-w)(Q_n^* + y) - p\int_0^{Q_n^*} F(x)dx$$

$$= E(\pi R^*(X)) + y(p-w)$$

Let $U = \alpha X$, Q_U^* the optimal order size with random demand of U, and $F_U(u)$ the CDF of U. Then, $F_U(\alpha x) = F(x)$. Since $F_U(Q_U^*) = (p-w)/p = F(Q_n^*)$ and $F_U(\alpha Q_n^*) = F(Q_n^*)$, then $F_U(Q_U^*) = F_U(\alpha Q_n^*)$ therefore, $Q_U^* = \alpha Q_n^*$

$$E(\pi R^*(\alpha X)) = (p-w)Q_U^* - p\int_0^{Q_U^*} F_U(u)du$$

$$= (p-w)Q_U^* - p\int_0^{Q_U^*} F(u/\alpha)du$$

$$= (p-w)Q_U^* - p\int_0^{Q_n^*} F(x)d\alpha x$$

$$= (p-w)\alpha Q_n^* - \alpha p\int_0^{Q_n^*} F(x)dx$$

$$= \alpha E(\pi R^*(X))$$

∎

Performance and Cost Effective MRO Inventory

D. Tsakatikas[1], M. Sfantsikopoulos[2]

National Technical University of Athens (NTUA), School of Mechanical Engineering, Mechanical Design & Control Systems Department, Heroon Politechniou 9, 157 73, Zografou, Athens – Greece. dtsakati@central.ntua.gr

Abstract: Establishing maximum availability of use for industrial plant facilities while achieving, at the same time, minimum MRO inventory stock is a matter that addresses several challenges.

It is altogether difficult to optimise requirements of spare parts and relevant supplies through appropriate maintenance procedures and avoid the threats of a stock out or excess MRO material stock. On the other hand, the potential benefits of a cost optimum MRO material availability can certainly boost a company's competitiveness.

In the paper, a novel approach is presented for a performance and cost effective MRO inventory control. The system involves development of appropriate concepts for the definition and estimation of the criticalities that preside over an MRO inventory. Maintenance needs, equipment breakdowns, production schedules and actual output, MRO procurement and supply chain practices, material lead time demand are also considered. These entries lead to an MRO item "comprehensive identity tag" carrying vital information that can then be processed by the Decision Support System for performing an agile inventory control and also for benchmarking the MRO activities.

1. Introduction

Maintenance Repair and Operation / Overhaul (MRO) materials are those supplies responsible for sustaining an industry's working schedule, [1]. They include almost everything from bearings, fasteners, conveyor chains and truck tires up to office desks and personal computers. As MRO materials denote a very wide concept inside an industrial environment, a division between direct and indirect MRO exists separating those industrial supplies directly involved or assisting in the production process cycle from the rest, the non-technical ones, which nevertheless also constitute a very important part of the whole MRO context.

In the frame of this paper a new approach towards a performance and cost effective MRO handling system, is presented. The system (PEMROT-Performance and Cost Effective MRO Inventory), currently under development, focuses on direct MRO spare parts because of their particular weight on the production run. Within this frame, the establishment and evaluation of criticalities for equipment availability and MRO management and handling issues, are presented and discussed. The thus produced criticalities serve as a major input to the new tool that acts complementarily to existing MRO procurement and supply systems.

2. Current Status

MRO materials procurement and acquisition issues are currently being carried out in accordance to the maintenance working schedules executed and tracked by integrated IT systems such as ERP, CMMS and EAM. In certain applications, forecoming MRO need is predicted by the use of probabilistic models following past time consumption and isolating a requirements pattern, [2].

Although maintenance activities and the mentioned transactions are being recorded in the corporal IT infrastructure, this is rather done in a reporting way without the ability of proper intervention in terms of actually translating the various trend patterns. It is not at all certain that all spares held in stock are actually needed, as rarely there is any sound evidence linking the operational characteristics of the equipment (reliability, actual time of use, load conditions) with MRO consumption. Moreover, the lack of a proper tool to classify spares according to their criticality, in a systematic and representative from an engineering and managerial point of view, method, frequently results in having unimportant SKUs (Stock Keeping Units) being held with no apparent reason in the warehouses. At the same time items for carrying out critical maintenance operations may not be properly managed. This results in having an idle investment threatened by obsolescence while still not having resolved the problem of accessibility of the right part at the right time and at the right cost, [3].

Certain industrial companies, in their effort to minimize the considerable capital tied down in MRO supplies, have attempted the reverse strategy of trying to eliminate "unnecessary" purchases. Such an approach has been proven, though, very problematic without an appropriate knowledge methodology behind, [4]. This is profoundly illustrated with the vital spares, whose unavailability in case of an unexpected stock out may cause a whole plant unit to cease operation.

The inefficiencies mentioned above demonstrate the necessity of performing a validated, concise and cost effective provisioning of MRO spare parts. A method for dealing with this problem is outlined in the following sections.

3. System Overview

3.1 Concept

Currently, attributes assigned to an MRO spare part belong to two main categories. "Technical specifications" include information on engineering drawings, standards, physical and chemical properties, handling, eventual life expectancy etc. "Commercial characteristics" include information on cost, delivery lead-time, minimum order quantity, certification, etc. Attributes of both categories can be either rigid (e.g. material properties, packaging instructions) or flexible (e.g. cost, lead time). Within this frame, PEMROT mainly acts as an intelligent agent that establishes a pathway for spare parts classification, copes with logistics and procurement procedures and at the same time takes into account actual plant operation data (production schedules, product mix, time of use, loads, breakdowns, maintenance activities) as well as mechanical reliability considerations. The MRO supply and procurement system becomes thus pragmatic in its response and contributes to a continuous maintenance improvement. It does not only monitor spare parts requirements, but is also able to optimise MRO stocks time and cost wisely.

PEMROT, in order to meet the above objectives, introduces as a basic concept an additional attribute category that of the "MRO item criticality attributes". The attributes of this category address either the criticality for equipment availability or the criticality for MRO management issues.

3.2 Criticality for equipment availability

For a typical industrial production line, Asset Configuration Management provides both a functional and an architectural breakdown of the equipment. Five Indenture Levels are used in PEMROT as shown in Table 1.

Table 1: Levels of functional/architectural breakdown

Level 1	SYSTEM	Principal Production Facility (Machine Tool, Rolling Mill, Mine Excavator, ...)
Level 2	SYSTEM	Secondary Production Facility (Air Compressor, Boiler, Crane, ...)
Level 3	SUBSYSTEMS	Mechanical Main/Secondary, Power Supply, Automation/Controls, Hydraulics/Pneumatics, Auxiliaries, ...
Level 4	ASSEMBLIES	Hydraulic Power Packs, Gear Units, Switching Boards, Conveyors, ...
Level 5	SPARES	Hydraulics, Bearings, Gears, O-rings, Gaskets, Switches, Sensors, Couplings, PCB's, ...

A Level 1 Principal Production Facility and its Level 3 subsystems attain maximum availability within their nominal performance specifications at all times. A Level 2 system, on the other hand, depending on the particular application, may not require maximum availability at all times, [5]. The same obviously applies for its related subsystems. In this way, analysis of an MRO item criticality in terms of equipment availability is performed at Level 4 assemblies going down to Level 5 spares (if needed).

There are two alternative approaches in carrying out the analysis, Fig.1. The first path is to proceed with FMECA, identifying Failure Modes (i.e. different ways the assembly may fail to provide its intended function) and corresponding Failure Effects (i.e. the consequences of a Failure Mode on the availability of the subsystem), [6]. A criticality assessment (C) then results as a function of the effect's severity (S) (i.e. evaluation of consequences over the lower analysis level) and its relevant probability of occurrence (O), [7]. The probability of occurrence is here taken as the number of replacements - due to either regular maintenance specifications and/or failures - over the actual operation hours (h) during a given time period (e.g. one calendar year).

430

The Criticality for Equipment Availability of an MRO item is obtained as the product $C = S \cdot O$, where S-values range from 1(minor consequences effect) up to 5 (catastrophic effect). In case that an MRO item's actual working time is less than (h), the value of $O = 1$, taking into account the lack of sufficient operational statistics.

The second, yet less systematic, path, involves a direct assignment of severity values to MRO items by the maintenance engineer himself using his judgment, expertise and past experience.

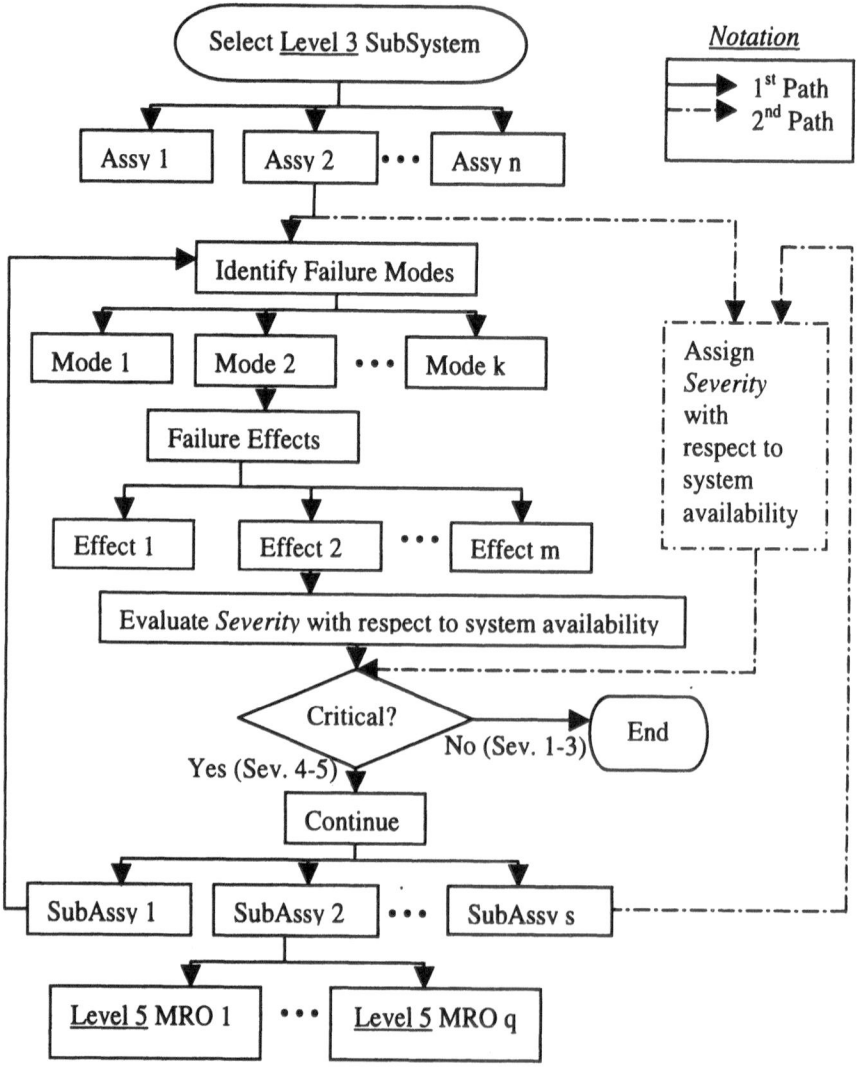

Fig.1 MRO criticality for equipment availability

With the end of the Fig.1 analysis, MRO items inherit the severity value of the last critical higher level, that is where the analysis ends, Fig.2.

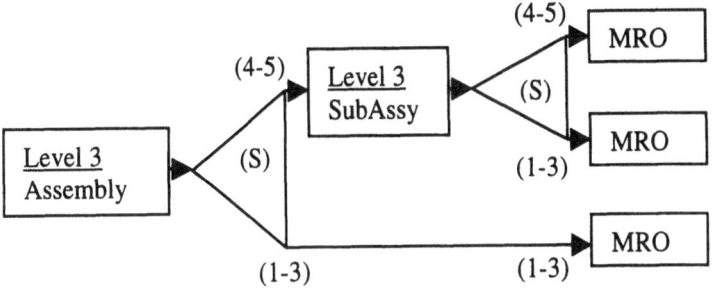

Fig.2 MRO item severity value assignment

It is pointed out that not only the importance of the next higher level of an MRO spare part to a Level 1 or Level 2 Facility should be considered but also the relative importance of the item to its parent assembly, [8]. The issue is included in the system algorithm at a subsequent stage.

3.3 Criticality for MRO management issues

In order to obtain the criticality of an MRO item in respect to material management and handling related issues, a blend of attributes from both technical specifications and commercial characteristics are taken into account. These are classified into three categories named as Uncertainty (U), Application (A) and Sensitivity (SN), Table 2. The contributors of each category are assigned a value from 1-5 that represents the degree that each of the attributes fulfils its description, 1 being the least and 5 the most.

Table 2: MRO criticality for management issues

MRO Attributes	Category	Value	$n_{(U,A,SN)}$
Equipment obsolescence	UNCERTAINTY (U)	U_1	4
Supplier unreliability		U_2	
Unique/Dedicated supplier		U_3	
Item out of production		U_4	
Diversity of application	APPLICATION (A)	A_1	5
Cost		A_2	
Alternatives sharing same properties		A_3	
Lead time dependency		A_4	
Product loss impact if not available		A_5	
Cost as a % of asset value	SENSITIVITY (SN)	S_1	4
Age sensitive		S_2	
Special holding/storing needs		S_3	
Irreparable		S_4	

To obtain the criticality for an MRO spare part, the mean values from the three categories, after a weight adjustment, are added together,

MRO item Management and Handling Criticality =

$$= \frac{1}{4} * \left(\sum_{i=1}^{4} U_i \right) * W_U + \frac{1}{5} * \left(\sum_{i=1}^{5} A_i \right) * W_A + \frac{1}{4} * \left(\sum_{i=1}^{4} SN_i \right) * W_{SN}$$

W_U, W_A and W_{SN} are the weights of the three categories, adjustable according to company goals, with $W_U + W_A + W_{SN} = 1$. The output is distributed on a 1-5 scale.

4. Discussion and Conclusions

In this paper, an approach for establishing criticalities for the procurement and supply of MRO spare parts is presented. They constitute the introductory part of a Performance and Cost Effective MRO Inventory Control Tool (PEMROT), currently under development. The introduced criticalities are classified in two types according to their role in the plant facilities availability and the plant maintenance management and economics. Creation and updating of a criticality database for MRO spare parts in conjunction with appropriate processing of performance IT data of the production facilities, provides a robust and validated procurement methodology for MRO spares. Such a systematic approach has also apparent referral and activity advantages for the maintenance of an industrial plant. Benchmarking, finally, of the expected results against those actually achieved acts as a motivator for improvement in both technical and financial aspects and also for constructive comparisons against competition.

References

[1] M. Le Sueur, B.G. Dale, The procurement of maintenance, repair and operating supplies: a study of the key problems, European Journal of Purchasing & Supply Management 4 (1998) pp. 247-255.

[2] Muhammad A. Razi, J. Michael Tarn, An applied model for improving inventory management in ERP systems, Logistics Information Management Volume 16 Number 2 (2003) pp. 114-124.

[3] Angel Diaz, Slow flow logistics: Optimizing MROs, White paper, Bordeaux School of Business, July 2000 V2.

[4] Robert Matusheski, The Role of Information Technology in Plant Reliability, P/PM Technology, June 1999.

[5] Richard G. Lamb, Availability Engineering & Management for Manufacturing Plant Performance, Prentice Hall, 1995.

[6] MIL-STD-1629A: Procedures for Performing a Failure Mode, Effects and Criticality Analysis, U.S. Department of Defence, Washington D.C., Nov.1984.

[7] SAE Surface Vehicle Recommended Practice J1739: Potential Failure Mode and Effects Analysis in Design, Manufacturing and Assembly Processes, SAE International, Warrendale, PA, June 2000.

[8] Tim Exton, Ashraf W. Labib, Spare parts decision analysis – The missing link in CMMS's (Part II), Maintenance & Asset Management Vol. 17 No 1 (2002) pp.14-21.

PRECISION MACHINE TOOL DESIGN

A Single-point Diamond Turning Machine with Micro Cutting Mechanism using Active Aerostatic Guideway

Hiroshi Mizumoto[1], Shiro Arii[2], Yoshihoro Kami[3] and Makoto Yabuya[4]

[1] Tottori University, Koyama-Minami Tottori Japan, mizu@ike.tottori-u.ac.jp
[2] Tottori University, Koyama-Minami Tottori Japan, arii@ike.tottori-u.ac.jp
[3] Nachi-Fujikoshi, Okake Namerikawa Toyama Japan, ykami@po.hitwave.or.jp
[4] Nachi-Fujikoshi, Okake Namerikawa Toyama Japan, myabuya@po.hitwave.or.jp

Abstract: A unique three-axis ultraprecision CNC machine tool is proposed, where two innovative technologies invented by the authors are employed. The first technology is an Angstrom positioning system using the Twist-roller Friction Drive (TFD), and the second technology is an active aerostatic guideway incorporating the Active Inherent Restrictor (AIR). This active aerostatic guideway was invented to improve the straightness of the table movement. In the present paper, the guideway is used as a micro cutting mechanism. During single-point diamond turning of an aluminium alloy, micro steps of 10nm to 100nm are cut on a mirror finished surface. It is shown that the proposed ultraprecision CNC turning machine with the micro cutting mechanism can be used for machining optical elements.

1. Introduction

Ultraprecision CNC (computer numerical control) machine tools are the essential industrial equipments in the field of nanotechnology. For example, two-focused pick-up lenses for CD and DVD are mass-produced by using a kind of ultraprecision machine tool called an aspheric generator. Both the aspheric profile and the grooves for Fresnel lens are machined under numerical control. Such CNC machine tools should have at least three control axes for generating 3-D object; the straightness and the positioning accuracy of each axis are of importance. Several types of ultraprecision CNC machine tools have been designed for machining mirror finished surfaces [1-3]. These ultraprecision machine tools are effective for mass-production of optical elements such as aspheric lenses, because surfaces of high quality can be obtained without time consuming lapping or polishing process.

In the present paper, a unique three-axis ultraprecision CNC turning machine is proposed, where an Angstrom positioning system using the Twist-roller Friction Drive (TFD) and an active aerostatic guideway incorporating the Active Inherent Restrictor (AIR) are employed. In the conventional capstan friction drive, the driven roller is pressed against the drive shaft with a right angle. On the other hand, the driven roller of the TFD is pressed against the drive shaft with a minute crossing

436

angle. Consequently, the lead of the TFD can be less than 0.1mm, and the positioning resolution is improved to be 0.5nm [4]. The Active Inherent Restrictor, the AIR, is driven by a piezoelectric actuator for improving the straightness of the table movement [5]. The active aerostatic guideway can also be used as an ultraprecision positioning mechanism. The positioning resolution of this mechanism has been reported to be 10pm [6]. In the present paper, this active aerostatic guideway is utilized as a micro cutting mechanism. During single-point diamond turning of an aluminium alloy, the micro cutting mechanism reduces the influence of the geometrical error in the guideway and the fluctuation caused by the feed drive device (TFD) on the machined surface. At the same time, micro steps of 10nm to 100nm are formed on the mirror finished surface by using the mechanism.

2. Turning Machine with Micro Cutting Mechanism

A three-axis ultraprecision CNC turning machine with active aerostatic guideway is shown in Fig. 1. A work spindle is mounted on the horizontal X-table, and a tool holder is mounted on the vertical Y-table; the Y-table is mounted on the horizontal Z-table. The positioning system using the TFD is employed as the feed drive device

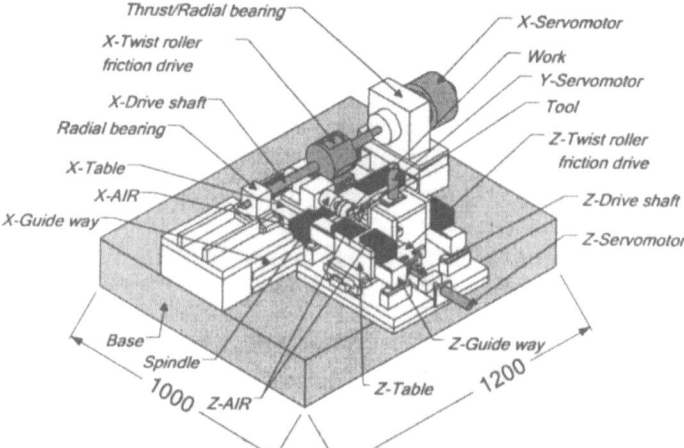

Fig. 1 Three-axis ultraprecision CNC machine tool

of the X and Z axes. The Y-axis is driven by a ball screw. Each table is guided by the active aerostatic guideway with the AIR. The displacement of the table is detected by a laser scale for full stroke. The turning machine is mounted on a granite base with vibration isolators in an air-conditioned room. Detailed view around the cutting point is shown in Fig. 2. A cylindrical work

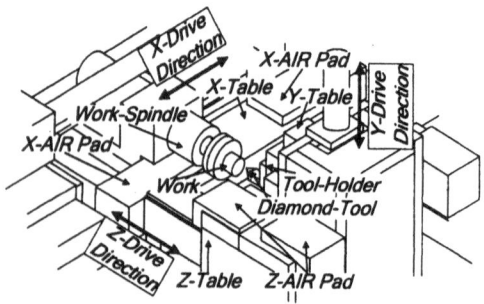

Fig. 2 Detailed view of cutting point

of aluminium alloy is held in the rotating air-spindle on the X-table. For face turning, a single-point diamond tool is held in the tool holder on the Y-table. Coarse cutting is controlled by the Z-axis and fine cutting is controlled by the X-AIR pads attached to the aerostatic guideway. These AIR pads construct the micro cutting mechanism.

Figure 3 shows the TFD employed in the turning machine. Three rollers are pressed against the drive shaft with a minute crossing angle. The TFD is mechanically a kind of lead screw, and it converts the rotation of the drive shaft into linear motion. The advantage of the TFD is its small lead. The lead of the X-axis is 60 μm and that of the Z-axis is 0.2 mm. Small lead is advantageous for improving positioning resolution. An example of Angstrom step positioning is shown in Fig. 4, where a fiberoptic sensor is used for the measurement.

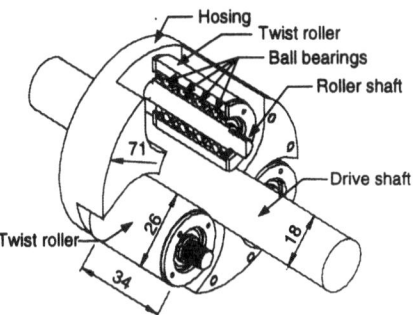

Fig. 3 Mechanism of TFD

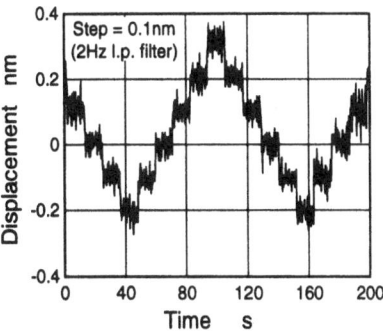

Fig. 4 Step positioning using TFD

3. Active Inherent Restrictor

A cross sectional view of the AIR unit in the AIR-pad is shown in Fig. 5. The AIR unit consists of a piezoelectric transducer (PZT) and a diaphragm case that has an air outlet. The outlet is small enough to function as an orifice when the unit is embedded in the guideway as shown in Fig. 6. The orifice area of the inherent restrictor formed on the bearing surface is $\pi d\, h_d$, where d is the diameter of the air outlet and h_d is the restriction gap. In the conventional inherent restrictor, the restriction gap h_d is always equal to the air film thickness h. In the AIR unit, h_d can be changed independently by using the PZT. When the

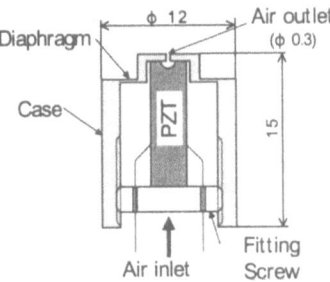

Fig. 5 Structure of AIR unit

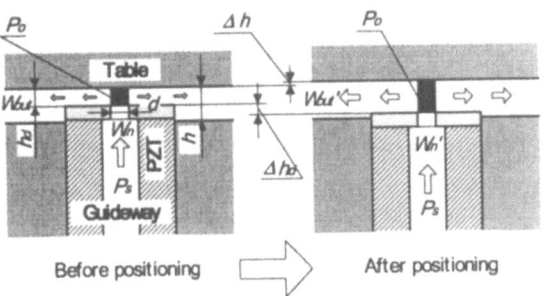

Fig. 6 Micro positioning driven by AIR

length of the PZT decreases by Δhd, the orifice area increases and the pressure on the bearing surface increases. Then, the table moves upward by Δh. This table movement is utilized as the positioning and micro cutting motion for the turning machine.

Figure 7 illustrates the positioning and micro cutting system using the

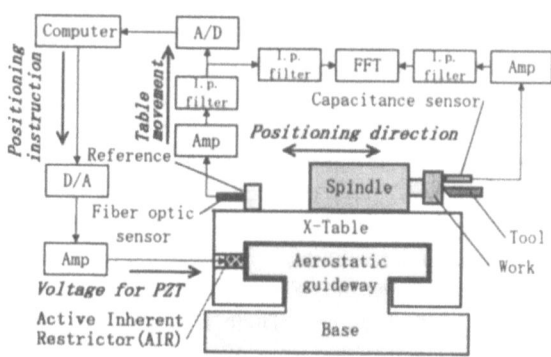

Fig. 7 Positioning and micro cutting system

AIR. A ceramic X-table (size: 300mm × 350mm × 100mm, mass: c.50kg) of the turning machine is guided by the aerostatic guideway in the X direction (normal to the paper). This X-table is also positioned in the Z-direction (horizontal, cutting direction) by the AIR unit incorporated into the aerostatic guideway. A positioning instruction output from a microcomputer is converted into the supply voltage to the PZT in the AIR. The X-table moves according to the deformation of the PZT. The table movement in the Z-direction is detected by the fiberoptic sensor and fed back to the computer. Thus a closed-loop control system is completed. Then the computer outputs the instruction for correcting the table position. The movement of the table is monitored by an FFT analyzer.

An experimental relationship between the deformation of PZT Δhd and displacement of the table Δh is

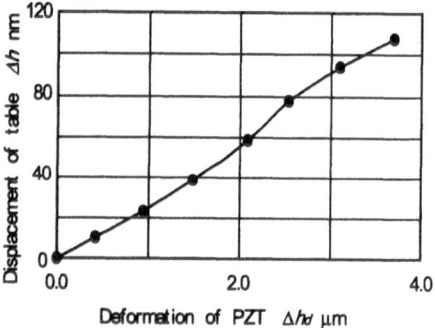

Fig. 8 Motion reduction by using AIR

shown in Fig. 8. Full stroke of PZT is 3.5 μm, while the maximum movement of the table is 110nm. Owing to the aerostatic mechanism, the table displacement can be much less than the deformation of the PZT. This means that the aerostatic guideway with the AIR acts as a motion-reduction mechanism. The reduction rate of the mechanism calculated from Fig. 8 is about thirty.

4. Ultraprecision Positioning Using AIR

The performance of the positioning and micro cutting system shown in Fig. 7 is examined by step response. The result of dynamic response shows that the slew rate is about 4.5nm/ms, and settling time is about 20ms. The result of static step positioning with ultra small step widths is shown in Fig. 9. Steps of the same width are repeated five times in one direction then in the reverse direction. Low pass filter is used to decrease the noise level of the fiberoptic sensor. Steps of 20pm are clearly resolved. Steps of 10pm are degraded by noise, however, every step can be recognized. Therefore, the

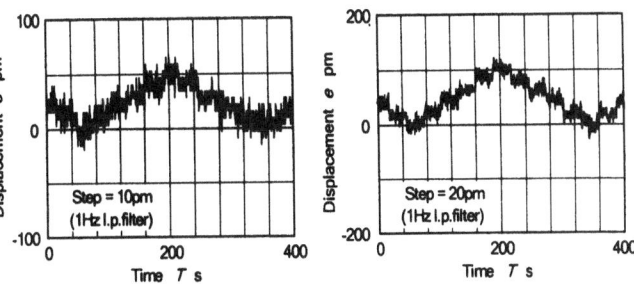

Fig. 9 Ultraprecision step positioning using AIR

positioning resolution of the positioning system is considered to be 10pm.

5. Single Point Diamond Turning

Unbalanced force caused by the TFD vibrates the machine table considerably. Horizontal vibration of the table directly influences the depth of cut. The upper diagram of Fig. 10 shows the profile of the surface turned without the active control of the guideway. Significant undulation of the turned surface can be seen. The

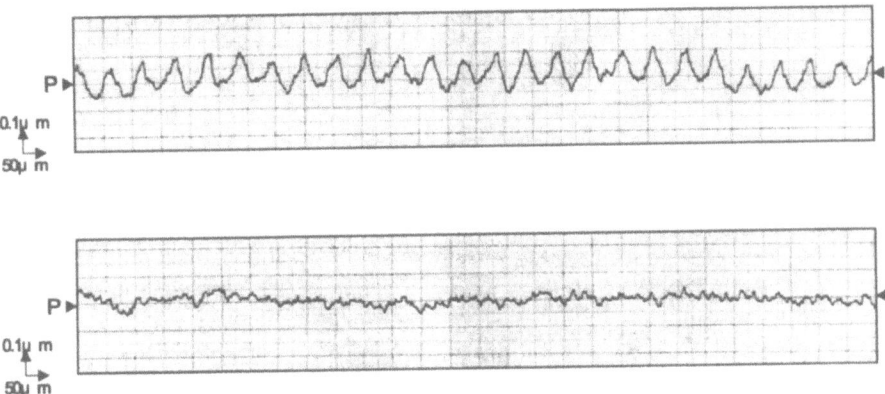

Fig. 10 Effect of active aerostatic guideway on profile of turned surface
(upper ; without active control, lower ; with active control)

wavelength of this undulation coincides with the lead of the TFD used in the X axis. On the other hand, the lower diagram of Fig. 10 shows the surface turned with the active control of the guideway. Owing to the facility of the positioning and micro cutting mechanism, the amplitude of the undulation decreases. Thus, the system can diminish the influence of the TFD on machining, and improve the quality of the turned surface. Surface roughness of the turned surface is less than 10nm Ra.

During the single-point diamond turning, micro grooves and steps can be made on the turned surface by using the micro cutting mechanism. Figure 11 shows a series of micro grooves on a mirror finished surface; the depth of the groove is

100nm. In spite of the dispersion in the depth, the grooves are arranged periodically according to the instruction of the micro cutting system. Pitch of the groove is about 0.3mm

The profile of a turned surface with micro steps is shown in Fig. 12, where steps of the same width are repeated five times in one direction then in the reverse direction. The width of the step is about 18nm. The resolution of the proposed micro cutting mechanism is at least 10nm.

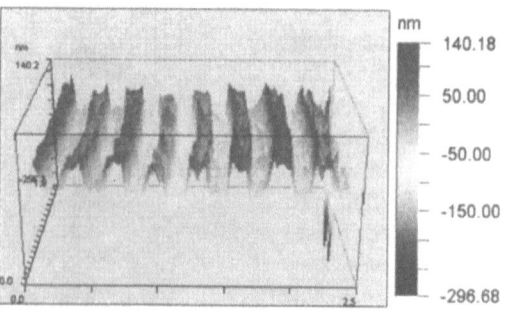

Fig. 11 Micro grooves on turned surface (WYKO)

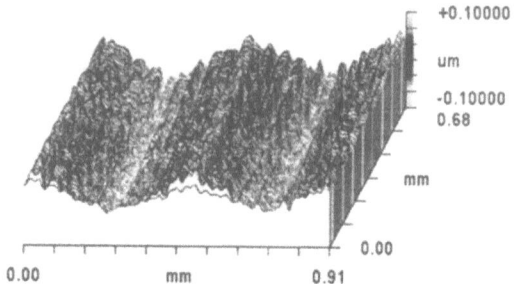

Fig. 12 Micro steps on turned surface (ZYGO)

6. Conclusions

The effect of the ultraprecision positioning and micro cutting mechanism on the positioning resolution and cutting performance of the ultraprecision turning machine is analyzed and following results are obtained:

(1) The motion-reduction action of the ultraprecision positioning and micro cutting mechanism effectively reduces the deformation of the piezoelectric actuator. Measured positioning resolution of the mechanism is 10pm.

(2) By using the micro cutting facility of the mechanism, the influence of the feed drive device on the turned surface can be reduced, and 10 – 100 nm depth of micro grooves and steps can be machined on a mirror finished surface made by single-point diamond turning.

References

[1] Home page of Cranfield Precision, http://www.cranfieldprecision.com/
[2] Home page of Nachi-Fujikoshi, http://www.nachi-fujikoshi.co.jp/eng/
[3] Home page of Precitech, http://www.precitech.com/
[4] H. Mizumoto, et al., Ann. of the CIRP, Vol. 45/1 (1996) pp. 501-504.
[5] H. Mizumoto, et al., Proc. of 1st euspen, Vol.1 (1999) pp.28-31.
[6] H. Mizumoto, et al., Proc. of 16th ASPE, Vol.1 (2001) pp. 119-122.

Feed Drive System Control using Complex Servo Mechanism

Takanori Yamazaki[1], Yuuichi Okazaki[2]

[1]Oyama National College of Technology, 771 Nakakuki, Oyama, Tochigi, JAPAN,
yama@oyama-ct.ac.jp
[2]The National Institute of Advanced Industrial Science and Technology, 1-2-1 Namiki,
Tsukuba, Ibaraki, JAPAN

Abstract: Many complex servo systems for precise motion control have been proposed so far. Generally, the coarse motion and the fine motion are designed separately. In this paper, we use the only one displacement for precise motion control and design of compensators. The structure of our servo system is as simple as a traditional control system. The simulation results show that the precise motion control works well for two types of reference trajectories.

1. Introduction

In modern precision positioning systems, a highly accurate control is required so as to attain the precision at nanometer level. To do this, the complex servo system combined with a coarse stage and a fine one has been proposed. In our experimental system, the coarse stage consists of a AC servo motor and a ball- screw, and the fine stage using a piezo-actuator mounted on the coarse stage.

The key idea in this study is that the only one displacement combined with the coarse displacement and fine one would be needed to achieve the precise position control. In other words, two mechanisms can be controlled by using of the only one sensor, and this control system exhibits an interaction between the coarse stage and fine one. The output of the fine stage will get to saturate since its stroke is limited within small ranges. The overall performance of this control system will be heavily affected by the saturation. Thus, an entirely new design method to compensate the saturation for the fine stage is proposed in this paper.

Fig. 1 Complex serve mechanism

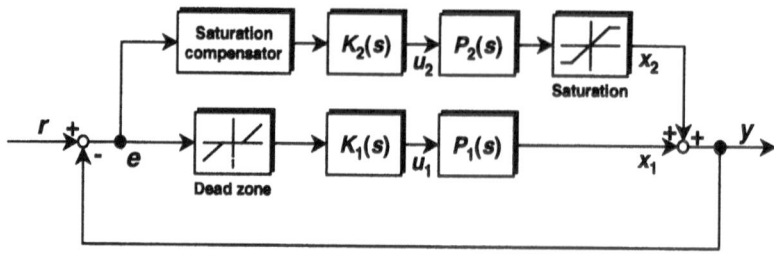

Fig. 2 Block diagram of complex serve system

2. Plant and Control System

The controlled plant, shown in Fig.1, is a complex servo mechanism. The plant is of the structure that the plant of fine motion is mounted on the plant of coarse motion. Block diagram of complex servo system is described as shown in Fig. 2, where, r: reference trajectory, e: error, y: displacement, u_1: control input of coarse motion, u_2: control input of fine motion, x_1: displacement of coarse motion, x_2: displacement of fine motion, $K_1(s)$: transfer function of coarse motion compen- sator, $K_2(s)$: transfer function of fine motion compensator, $P_1(s)$: transfer function of plant of coarse motion and $P_2(s)$: transfer function of plant of fine motion compensator. The plant dynamics in experimental results are given by, respectively,

$$P_1(s) = \frac{0.0987}{s(s^2 + 0.3142s + 0.0987)} \quad f_1 = 0.05\text{kHz}, \ \zeta = 0.5, \tag{1}$$

$$P_2(s) = \frac{114.1}{s^2 + 15.10s + 114.1} \quad f_2 = 1.7\text{kHz}, \ \zeta = 0.707. \tag{2}$$

In addition, we consider the dead zone in coarse motion and the saturation in fine motion, and install saturation compensator before the fine motion compensator.

3. Design of Compensators

The design of two compensators for complex servo system will be described. The compensators should be designed on the basis of a simple structure without considering saturation.

3.1 Design of coarse compensator

In the design of a coarse motion compensator, it is desirable to have a fast response with no overshoot, and its transfer function is given in the following

$$K_1(s) = \frac{0.02}{s + 0.3}. \tag{3}$$

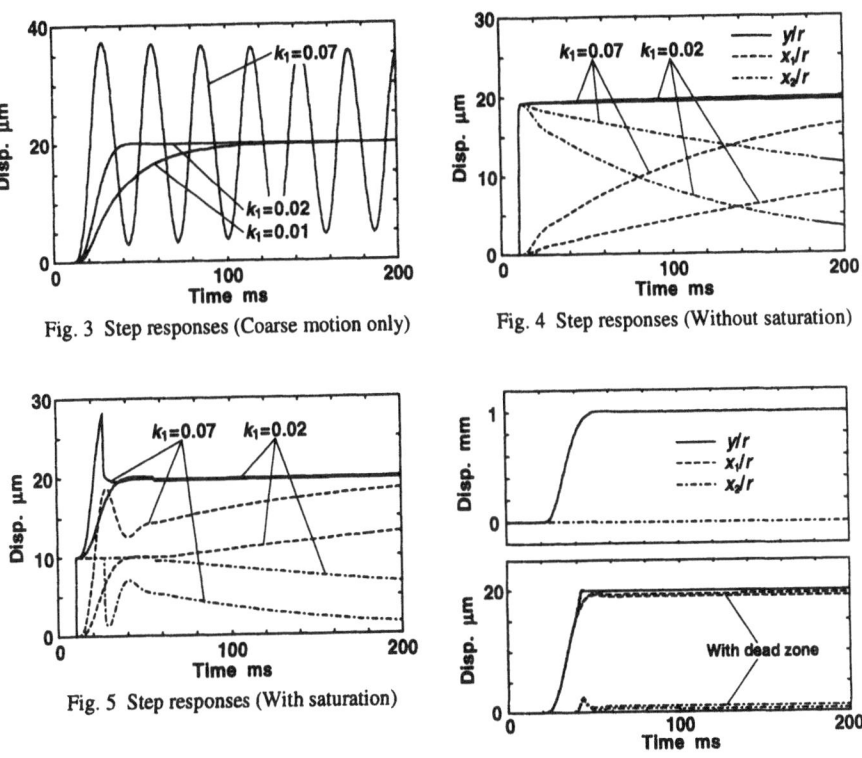

Fig. 3 Step responses (Coarse motion only)

Fig. 4 Step responses (Without saturation)

Fig. 5 Step responses (With saturation)

Fig. 6 Step responses with saturation compensator

Fig. 3 shows the displacement x_1 to the step-wise change in the reference that is set to 20μm for the only coarse motion. In this figure, the responses for various values of gain k_1 in the coarse motion compensator are compared and the responses become oscillatory as k_1 increases.

3.2 Design of fine motion compensator

The fine motion compensator is designed so that the coarse motion compensator should be included in the closed loop. It should have the smooth frequency response and its transfer function is given by

$$K_2(s) = \frac{2.4}{s+0.1} . \tag{4}$$

Fig. 4 shows three displacements, y, x_1, and x_2, to the step-wise change in the reference without the effect of the limiting saturation in fine motion. In this case, the displacement y is not almost affected by the adjustable gain, and as a result the fine motion suppresses the vibration of coarse motion.

Considering the effect of the saturation, the step responses of the complex servo system are shown in Fig. 5. In this case, the stroke of fine motion ranges from

444

-10μm to +10μm. In case that $k_1=0.07$, the displacement of coarse motion is increased by saturating displacement of fine motion. On the other hand, in $k_1=0.02$ the displacement has an offset after 200ms. These results suggest that the saturation compensator for fine motion reduces the offset.

4. Saturation Compensation for a Step Response

When an error e is longer than the stroke of fine motion in order to prevent the saturation of fine motion, we propose saturation compensator of fine motion as follows:

$$f(e) = 2e \cdot \exp(-|1000e|),\qquad(5)$$

where the coefficients of exponential function in Eq. (5) is required for tuning the stroke of fine motion.

Fig. 6 shows three displacements to step-wise change in the reference using saturation compensator of fine motion, and in upper figure the reference is set to be 1mm and in lower figure the reference is set to be 20μm. In these figures, as the result applied the saturation compensator, the responses of fine motion are not saturated for various reference inputs and the still worse response of coarse motion by the dead zone are compensated.

The offset of displacement y is decreased by applying the saturation compensator. However, it is possible by moving the coarse motion to the inside of the stroke of fine motion without overshoot. Moreover, the offset at the time 100ms is 50nm in case that the reference is set to be 1mm and the offset is 10nm in case that the reference is set to be 20μm, and the values decay with time.

5. Saturation Compensation for a Circular Motion

Compensating the saturation of fine motion depending on the nonlinear equation by using Eq. (5), the displacement x_2 to sine wave input is distorted.

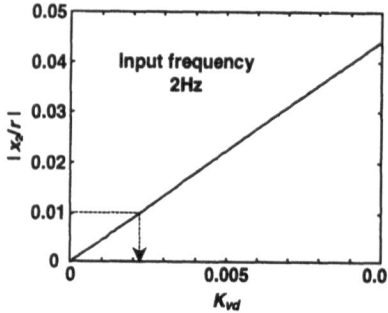

Fig. 7 Relation between K_{vp} and $|x_2/r|$

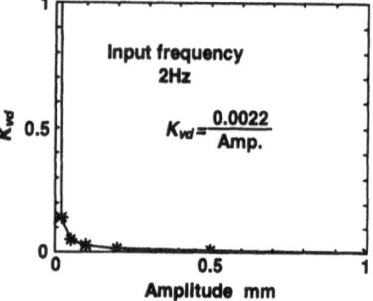

Fig. 8 Relation between amplitude and K_{vp}

Therefore, the saturation compensation of fine motion in circular motion is applied to the variable gain.

Assuming that the circular motion (1.59Hz) where the radius is equal to 1mm and the feed speed is 10mm/s, the relation between the variable gain K_{vd} and the gain $|x_2/r|$ from reference r to displacement x_2 of fine motion in the input frequency of 2Hz is shown in Fig. 7. As a result, if the amplitude of input frequency is 1mm, the displacement x_2 of the fine motion is almost equal to 10μm in K_{vd}=0.0022. Fig. 8 shows the value of K_{vd} tuned up to the displacement x_2 of fine motion is less than 10μm for the amplitude of input frequency (2Hz) range from 0.01mm to 1mm. In this figure, the amplitude is in inverse proportion to variable gain K_{vd}.

Fig. 9 and Fig. 10 show the gain diagrams in the upper figure and three displacement to the input frequency of 2Hz in the lower figure, where using result of Fig. 8, the value of K_{vd} in Fig.10 is fifty times the value of K_{vd} in Fig.9. These results of the displacement to the sine wave input shows that the saturation of fine motion is vanished and the responses of coarse motion affected by the dead zone are compensated.

The upper figure in Fig.11 shows the gain diagram that K_{vd}=1, represent that the frequency band is wide compared with the frequency band of K_{vd}=0.002 or K_{vd}=0.1 in Fig.9 or Fig.10, respectively. For the lower figure in Fig.11, assuming that the circular motion (159Hz) in case that radius is equal to 10μm and the feed speed is 10mm/s, three displacements to the input frequency of 200Hz are shown. This result means that it is possible to be circular motion in with 159Hz.

As a result, using the variable gain for fine motion, the motion with large amplitude in low frequency and the motion with small amplirude in high frequency are possible.

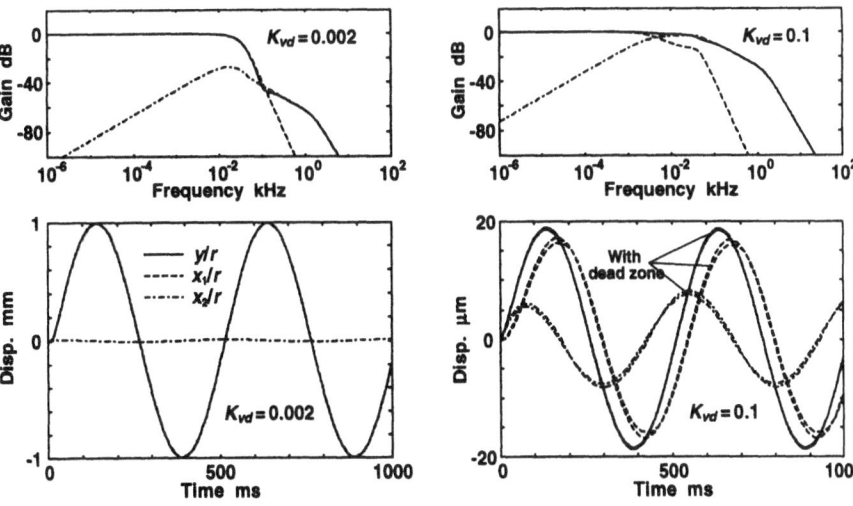

Fig. 9 Gain diagram and frequency responses
(K_{vd}=0.002, Amp.=1mm, 2Hz)

Fig. 10 Gain diagram and frequency responses
(K_{vd}=0.1, Amp.=20μm, 2Hz)

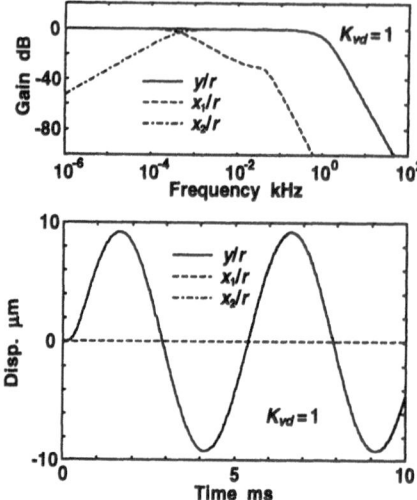

Fig. 11 Gain diagram and frequency responses
$(K_{vd}=1$, Amp.$=10\mu m$, 200Hz$)$

6. Conclusion

We examined the saturation compensators for two types of reference trajectories, namely a step input and a circular curve. Consequently, the following conclusions are summarized:

1) For a step input, the good response can be achieved by the saturation compensator provided by the function of error signal regardless of the values of step-width.

2) For the circular curve, we show a tuning method of variable gains for the fine stage. The control performance can be improved to avoid the saturation for various amplitudes of circular trajectories.

References

[1] Fan L.-S., Ottesen, H. H., Reiley, T. C. and Wood, R. W., 1995, *Magnetic Recording Head Positioning at Very High Track Densities Using a Microactuator-Based, Two-Stage Servo System*, IEEE Transactions on Industrial Electronics, Vol. 42, No. 3, 222-233.

[2] Okazaki, Y., 1998, *Fast tool servo system and its application to three-dimensional fine surface figures*, Proceedings of ASPE 13th Annual Conference.

[3] Ding, J., Tomizuka, M. and Numasato, H., 2000, *Design and robustness analysis of dual stage servo system*, Proceedings of the 2000 American Control Conference, 2605-2609.

[4] Kobayashi, M. and Horowitz, R., 2001, *Track Seek Control for Hard Disk Dual-Stage Servo Systems*, IEEE Transactions on Magnetics, Vol. 37, No. 2, 949-954.

[5] Otsuka, J., 2002, *Ultraprecision Positioning Technology in Japan*, The 1st Korea-Japan Conference on Positioning Technology, 3-7.

Seismically Balanced High Precision Machine Tools

Jaromir Zeleny[1], Lukas Novotny[1]
[1] RCMT- Research Center of Manufacturing Technology, Czech Technical University in Prague, j.zeleny@rcmt.cvut.cz, l.novotny@rcmt.cvut.cz

Abstract: The paper contains description and analysis of a new, fully original "Floating Principle" for design of "Seismically Balanced Machine Tools" as a new class of future HSC machine tools. Along this principle, the machine has two parts, moving in opposite directions and supporting respectively the workpiece and the tool. The feed driving force acts only between these moving parts and the transfer of its reaction force to the foundations is in the respective direction completely eliminated.

1. Introduction

Recently applied high-speed cutting (HSC) technologies bring about new requirements on design of drives and machine tool structures. To keep pace with fast rotating spindles, feed drives for HSC machines have to be designed for generation of highly dynamic forces for fast acceleration and deceleration of moving masses in all NC axes. In machine tools for HSC technologies, acceleration forces may considerably surpass cutting forces as well as weight and friction forces and have to be considered as a main parameter for design of corresponding machine frames and drives.

Other critical problems in performance of HSC high precision machine tools arise from mechanical shocks, jerks and vibrations acting within the machine structure and foundations caused by reaction forces to the acceleration and deceleration forces of feed drives. The paper describes a new, fully original "floating principle" for design of feed drives and machine tool structures for highly dynamic and highly precise machine tools. The principle brings about elimination of dynamic shocks within the machine tool structures and their foundations. Other positive effects are reduction of energy for acceleration of moving masses, reduction of installed power, reduction of noise, better quality of feedback control a.s.o.

The principle has been invented by Prof. Zeleny, main author of this paper and experimentally tested in the RCMT – Research Center of Manufacturing Technology in Prague, Czech Republic

2. Problems of Highly Dynamic Machine Tools

Highly dynamic machine tools have fast rotating spindles, fast indexing tool turrets and fast indexing or rotating tables. High acceleration and jerks of these rotary movements generate torque shocks into the machine frame. Fortunately, there disturbing torques are usually tolerable because the maximum angular accelerations rarely coincide with maximum moments of inertia. As to the translatory movements, we have often to face much worse combinations with heavy masses moving at highest velocities, accelerations and jerks. For bigger machines, the moving masses and accelerating forces reach at least in one NC axis untolerable values. Corresponding reaction forces generate shocks and vibrations, which act on the machine frame, bed and foundation and strongly reduce the working accuracy of the machine. The situation for typical conventional case of a horizontal milling machine is depicted in the Fig 1 a, b. Biggest masses are the bed, fixed to the foundation, the tool supporting structure and the workpiece supporting structure with the mass MT. The workpiece supporting structure is connected to the stationary bed with the mass MS. Tool structure in the shown arrangement moves in the "z" direction. The accelerating, force F_A is generated by a feed drive of any arbitrary type acting between the combined stationary mass (MW + MS) and the moving mass MT. Naturally, the drive generates as well the reaction force F_R, which equals in its value to the accelerating force F_A and acts back to the bed and foundations. The moving, tool supporting structure with the mass MT generates in any moment an acceleration of: $a = F_A/MT$.

3. Function of the "Floating Principle" for Linear Movements

The new term of "Seismically Balanced Machine Tools" means in general machines, which would be able to perform its machining tasks without disturbing its environment by shocks and vibrations acting into foundations. The question is whether there is a chance to eliminate or substantially reduce harmful effects of reaction forces or even reaction torques arising at high accelerations and velocities of moving masses. The "floating principle" shows that, at least in machine tools such possibility for linear movements exists.

Fortunately, in machine tools we don't need to get absolute velocities or absolute accelerations of moving masses. It is important to realize, that what is for machining processes actually needed is only generation of relative movements, relative velocities and relative accelerations between the workpiece and the tool.

Figure 2 explains the function of the floating principle in a simplified case in which the mass MW equals the mass MT. Along the "floating principle", the machine tool consists of a stationary part and two moving parts with comparable masses, one of them supporting the workpiece and the other supporting the tool. The idea is to let these parts mutually move in opposite directions at least along one NC axis. If we then put the measuring system and the feed drive between both moving parts, we would be able to control their relative position, velocity and acceleration regardless of their absolute movements. This is the key for complete elimination of shocks and vibrations into foundations.

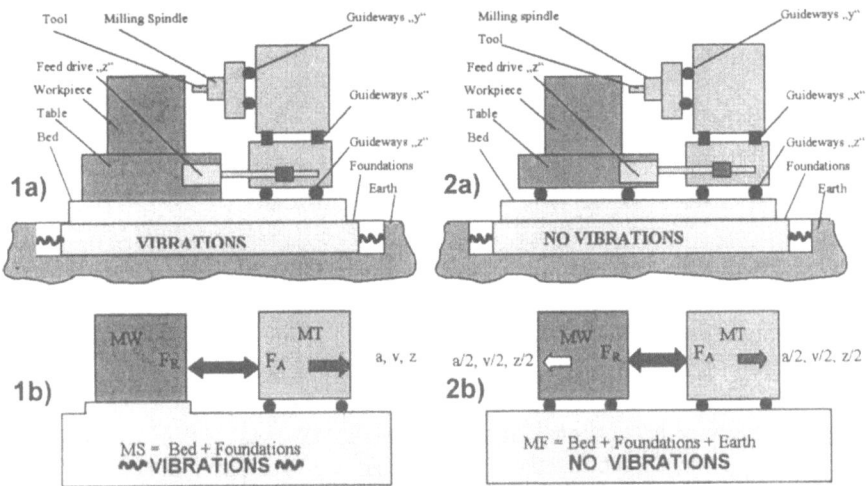

Figure 1: Structure of a conventional machine tool (1a) and its dynamic model (1b) for movement in the "z" axis

Figure 2: Structure of the machine from the Figure 1, modified along the "floating principle" (2a) and its dynamic model (2b)

Accelerating forces and cutting forces act here only as internal forces between two moving masses and no recognizable shocks can be transferred to the foundations, regardless on how big and how different the individual moving masses may be. The machine tool is now in the chosen "z" coordinate direction completely "seismically balanced". The accelerating force F_A accelerates the first moving part in one direction, whereas the corresponding reaction force F_R, until now useless and harmful, accelerates the other part in the opposite direction. Another positive contribution of this arrangement is that the relative acceleration, generated by certain accelerating force will always be higher by application of the "floating principle" when compared to the conventional arrangement. For the case where MW = MT and for negligible inertia of the drive, relative acceleration will double because the equally high reaction force could be fully utilized for acceleration of the mass MW in the opposite direction.

If we on the contrary evaluate kinetic energy for the inicially required relative velocity "v", we get for two equal masses moving in opposite directions with the reduced velocity "v/2" the relation:

$$E_k = \frac{1}{2} \cdot MW \cdot \frac{v^2}{4} + \frac{1}{2} \cdot MT \cdot \frac{v^2}{4} = 0{,}25 \cdot MT \cdot v^2 \tag{1}$$

Evidently, masses MW and MT moving in opposite directions with reduced absolute velocities "v/2"have together only one half of kinetic energy with respect to the conventional case with the mass MT, moving with the full velocity of v. From this follows, that generation of relative velocity between the tool and the workpiece can be reached by application of the "floating principle" with only one half of externally applied energy. For different values of MW and MT, the savings of energy will be variable but always positive. Shocks and vibrations into the foundations will be at any mass ratios eliminated. For the extreme, conventional case, when one of the masses would approache infinity, energy savings would disappear and the machine would start again to vibrate.

4. Absolutization of the Controlled Position

As the dynamic forces and cutting forces acting between the tool and workpiece are only internal forces of the dynamic system, the common center of gravity of both moving masses MW and MT will keep its original position independently on values of these forces. Forces F_A and F_R generated by the feed drive are always equal and absolute accelerations, velocities and distances from the common center of gravity will be inversely proportional to the values of MW and MT. Nevertheless, the moving masses can, both together, slightly float due to differences between friction forces in opposite movement directions, because these friction forces are external forces acting between the moving masses MW, MT and the stationary bed. This slow floating of the common center of gravity is acceptable with respect to machining requirements, but it should be reduced to minimum or fully eliminated for the purpose of tool or workpiece change. This can be secured by adding another measuting system, evaluating the absolute position of one of the moving masses MW, MT. A low power, auxiliary actuator or servosystem then uses these data for reduction of floating effects to acceptable limits.

5. Experimental verification

The floating principle has been tested on a stand with an AC servomotor and a steady ball bearing screw and rotating nut. Two mechanical tables 1 and 2, represent the masses MW and MT with respective values of 90 and 60 kg. These tables are moving on a stationary bed and interconnected by the feed drive and its positon feedback control loop. Auxilliary DC motor is used for reduction of floating phenomena. One of the tables is equipped with a brake, which makes both conventional and floating operations on the same experimental stand possible.

Comparative measurements have been performed at periodically repeated working cycles, shown in the Fig. 3. In the left hand part, the programmed path, velocity and accelerations are shown. The lower right hand part depicts the position response in conventional arrangement with engaged brake of the table 2. In the upper right hand part, the position respose of the floating system to identical programmed acceleration pulses is shown. The excited vibrations of the bed in floating arrangement has been far more quiet, smooth and very well damped. Experiments have proved substantial elimination of shocks into foundations. Maximum accelerating force of 6500 N has been used during the experiment. The

force of auxilliary drive for absolutisation of position and elimination of floating
phenomena has reached only 0.3% of the maximum accelerating force in this case.

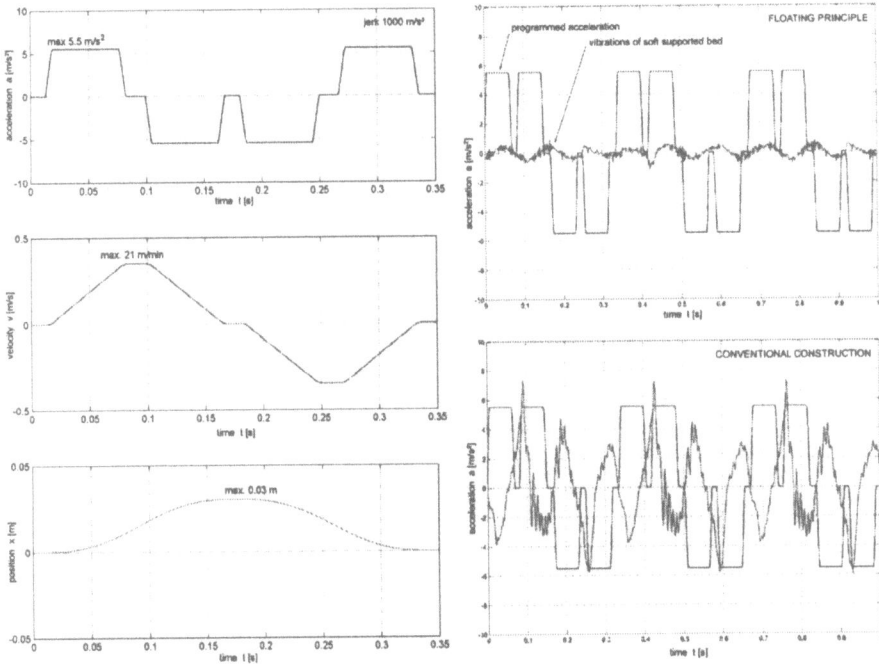

Figure 3: Dynamic behavior of "floating" and conventional systems

6. Design of "Seismically Balanced Machine Tools"

It is technically possible to design machine tools, completely seismically balanced
in all linear and rotary NC axes. Nevertheless, it is in most cases not necessary. The
HSC technology needs comparable dynamic responses of all movements and the
simplest way to fulfill this requirement is improving the machine behavior in the
most critical NC direction. Other NC axes can be then tuned to a comparable
dynamic quality. Most of reaction torques acting to the foundations can be reduced
by symmetric design of the machine, by application of gantry frames and by double
arrangements of feed drives. Masses MW and MT can be influenced by suitable
feed drive arrangements. Rotating masses of feed drives have to be considered as
parts of linearly moving masses to which they are kinematically connected. Best
results can be reached with directly operating servomotors as linear motors or high-
torque multipole motors. Installation of auxiliary subassemblies as tool magazines
or pallet manipulators can as well positively influence the balance of moving
masses. Figures 4 shows example of a three axis horizontal milling machine using a
couple of feeddrives with rotating ball-bearing nuts in the "floating z" axis.
Research and development of seismically balanced machines continues in the

452

RCMT Praha within the LN00B128 project supported by the Czech Ministry of Education, Youth and Sports.

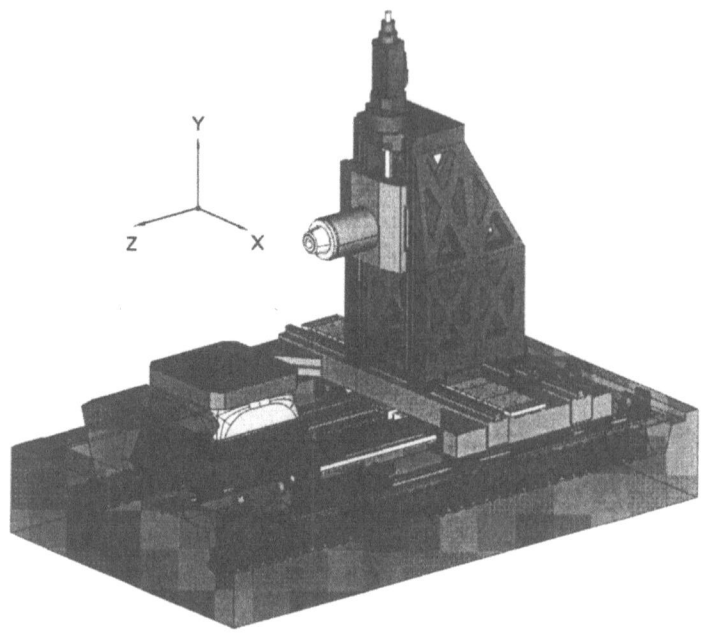

Fig. 4: Design of a three-axis horizontal milling machine with the floating axis "z"

7. Conclusion

Higly dynamic machine tools impose new requirements on design of feed drives and machine tool structures and bring about new kind of problems. High accelerations and jerks generate strong reaction forces and torques, which excite shocks and vibrations within the machine structures and foundations. The paper defines the new term of "Seismically Balanced Machine Tools" for machines in which these phenomena are suppressed or eliminated and describes the original "Floating Principle" as a new approach to design machines with this new quality.

References

[1] Zeleny J.,2002, *Patent Pending and Industrial Proprietary Formula*
[2] Bubak A., Soucek P., Zeleny J., 2003, *New Principles for Design of Highly Dynamic Machine Tools,* International Conference ICPR-17, Blacksburg, USA
[3] Zeleny J., 2003, *Influence of HSC Technologies on the Design of New Generation Machine Tools,* High-Speed Machining, 11. Oesterreichische Tagung, Steyr, Austria
[4] Novotny L., 2003, *Horizontal Machining Centre with Floating Axis (in Czech),* Research Report 2-04-03, RCMT, CTU in Prague, Czech Republic

Modelling of Milling Spindles for Optimizing the Spindle Cutting Performance

Pavel Bach

Research Centre for Manufacturing Technologies,
Czech Technical University in Prague,
Horska 3, 128 00 Praha, Czech Republic
P.Bach@rcmt.cvut.cz

Abstract: Optimization of spindle units to meet required productivity is one of most important tasks for design and testing laboratories. The paper presents selected research results on this field recently obtained. Dynamic modal models extracted from measured data are used to study dependence of the spindle performance on modal parameters. FEA dynamic models are utilised to show how the basic design elements of a spindle unit involve in performance.

1. Introduction

Metal removal rate (MRR) of milling depends considerably on the axial depth of cut. Usually, the axial depth of cut is considered as a static parameter in the calculation. But if is considered as a dynamic parameter, then

$$MRR_{\lim} = b_{\lim}.a_e.z.f_z.n \,, \tag{1}$$

where a_e is width of cut, f_z is chip load, n is spindle speed and b_{lim} is the stable depth of cut for the limit of stability. It is given by

$$b_{\lim} = -\frac{1}{2.K_s.z.G_o(p)} \,, \tag{2}$$

where z is the averaged number of the effective teeth in cut. The function $G_o(p)$ is the real part of the complex frequency response function (FRF) of the spindle-holder-tool structure (SHT), and K_s is the specific cutting force, while $G_o(p)$ is calculated by means of the formula

$$G_o(p) = \sum_{i=1}^{m} u_i.G_i(p) \,, \tag{3}$$

where u_i is directional orientation factor and $G_i(p)$ is the FRF of a particular mode. Equations (1) and (2) express the limitation of the milling process performance when dynamic properties of the SHT structure are considered [1-3].

2. Models Based on Measured Data

Many examples have been measured in which the $G_o(p)$ of the spindle-holder-tool structure summarises only two or three dominant modes. FRF dependence on the modal static stiffness, damping ratio and natural frequency of each mode allows one to optimize $G_o(p)$ to achieve higher value of MRR_{lim}. The modal models derived from the measured data are used for optimising the structure. One example is shown in Figure 1 and Figure 2. The results shown in Figure 1 have been obtained for the SHT structure with modal parameters for 9 modes. is shown on the upper chart. In Figure 2, the MLInc's package Metalmax has been used for FRF data measurements, as well as for the stability lobe calculation and for the optimum spindle speeds evaluation [2]. The structure has two decisive modes in the X direction and one in the Y direction. These very flexible X-modes cause the parameters b_{lim} and MRR_{lim} to be quite low. An example of the optimized case is depicted in Figure 3.

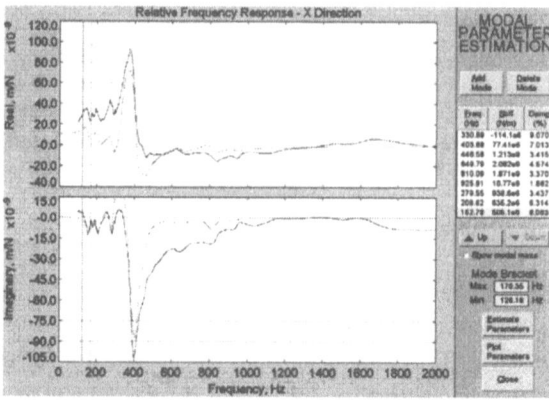

Figure 1: Real and imaginary FRFs measured in the X and Y directions at the tool tip for the SHT structure modal parameters for 9 modes.

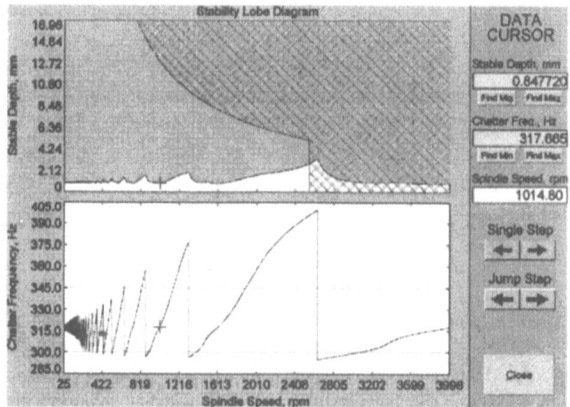

Figure 2: Stable region for grey cast iron machining and optimum spindle speeds.

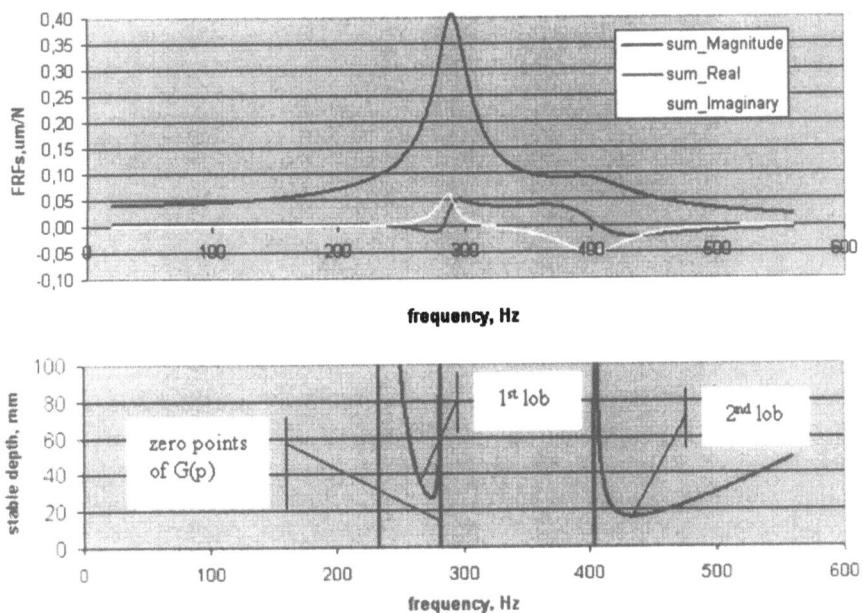

Figure 3: The optimized model consisting of 3 modes and stable depths.

Minimum b_{lim} is equal to 16 mm in this case. This value has been achieved by modification of modal parameters. The directional orientation factors have not been changed. A single point cutter has been considered. The transfer of these results to a unit structure requires knowledge of the modal shapes.

3. Design and Dynamic Modelling of Milling Spindle

In order to define the influence of the spindle shaft, bearings, motor and SHT system configuration on the cutting performance of the spindle, FEA 2D models have been studied by using the Spindle Analyser Program of MLI [2]. Concurrently, the relationship of structure parts and FRFs was investigated.

A motorised 25 000 rpm spindle with a power limit of 22/15kW (30 minutes/continuous) has been assumed. A spindle base speed is supposed to be 1000/700 rpm. A DN number should achieve a value of about $1,6.10^6$. A couple of front angular bearings of inner-race diameter 70 mm will satisfy this specification. The rear angular bearing will have an inner-race diameter of 60 mm. Based on these proposed parameters, a standard clamping mechanism and a motor with the above characteristics, an initial model of the spindle system was developed (see Figure 4). The model is considered to be axi-symmetric. In the model, the interface between the tool and holder interface is ISO40, with the tool diameter =20mm and length=30mm.

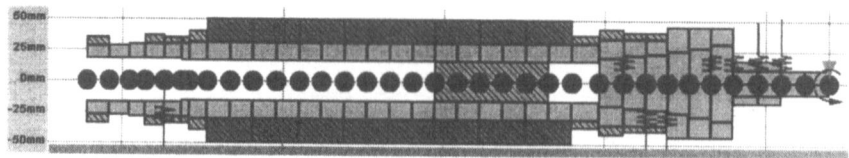

Figure 4: The initial FE 2Dmodel

As the structure is axi-symmetric, there are always two identical FRF calculated in the direction of the X and Y axes. It means that the modes have a stable direction of either 45° or 135° from the X-axis. In this case, the directional orientation factors and also other results depend only on cutting conditions, which have to be chosen in the programme setup. The slotting of steel (K_s=2100 N/mm^2) with an end-mill was considered as the milling mode for all our stability calculations.

In order to differentiate the modes of the shaft from those of the tool, the shaft, with its bearings, has been calculated first and then the full spindle. Figure 5 shows the FRF of the full model. The first four modes belong to the shaft. The other modes are strongly influenced by the tool. Figure 6 shows three modal shapes; the top modal shape is that of the shaft, which is the third dominant mode in Figure 5; the other two modal shapes in Figure 6 are that of the tool.

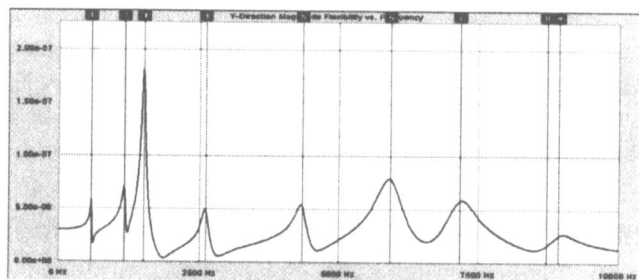

Figure 5: The FRF-magnitudes of the complete model.

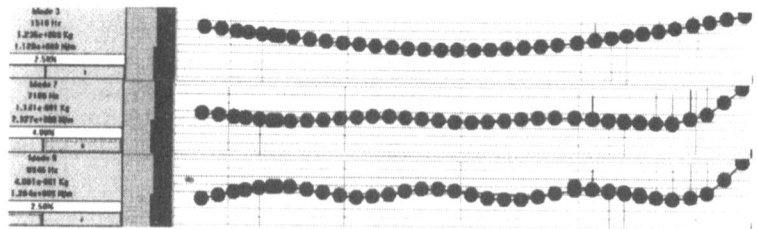

Figure 6: Some modal shapes of the shaft and tool.

A calculated stability diagram of the initial FE 2D model is shown in Figure 7. The achieved critical axial depth of cut of about 2,4 mm ensures a MRR of about 180 cm^3/min with a conventional cutting speed of steel 190 m/min, spindle speed 3000 rpm and a chip load f_z=0.3 mm/z.

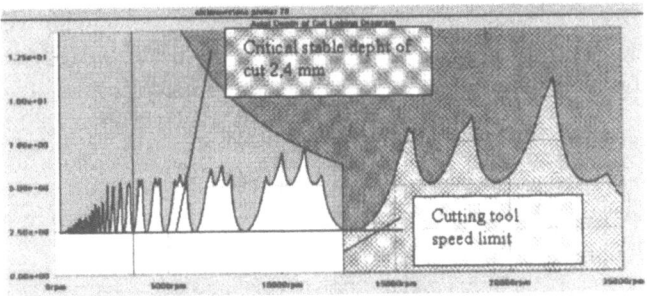

Figure 7: The calculated stability diagram

4. Results and Discussion

A significant improvement was obtained by optimising the measured model. The stable depth of cut was increased from 0.85 to 16 mm. The changes, suggested in the model might be implemented in the spindle design, if one could measure all the dominant modal shapes of the SHT system. But, this was not feasible as there is no access to the internal parts of the spindle body. One cannot assign modes to these parts of the SHT structure. Of course, it can be estimated that the most dominant mode belongs to spindle, which vibrates together with the holder and the tool. But, assignment of modes to other parts would be much more difficult.

The problem was clarified by calculations. Comparing the calculated modal shapes of the shaft and those obtained for the full model, it is clear that the first four "low" frequency modes belong to the shaft. The others are strongly influenced by tool. So, in the FRF, the first four peaks are "shaft modes" while the others are "tool modes". Generally speaking, if the FEA modes are calculated at individual stages of the design procedure, the peaks of FRF could be identified with the parts of SHT system more clearly. Identification is the key condition for an optimization.

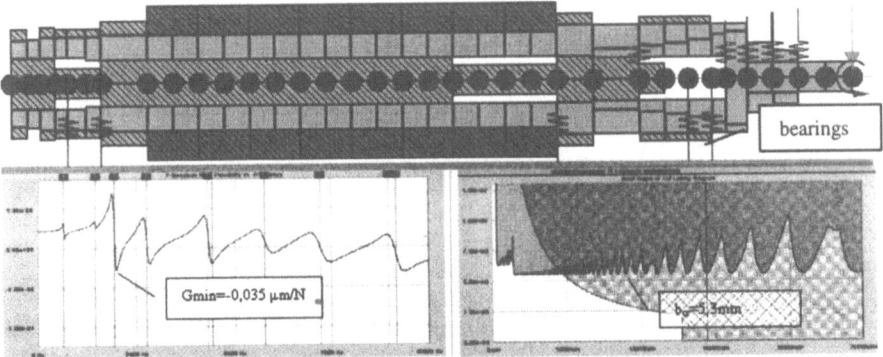

Figure 8: The final spindle design, its frequency response and stability diagram.

Calculating the spindle power of the first version of the spindle design, one obtains 6.3 kW which is only 42% of the available continuous spindle power. In order to improve this unacceptable power utilization, the spindle unit has been optimized changing its bearing and shaft stiffness, bearing configuration, tool-holder interface stiffness, tool clamping stiffness, tool-holder mass, tool mass and spindle nose mass. The resulting spindle model is shown in Figure 8 including the $G_o(p)$ and the stability diagram.

Additional dampers at the bearings can be used to gain better performance. In our case, increasing the damping ratio from 2.5% to 3% is enough to improve the critical depth of cut as shown in Figure 9.

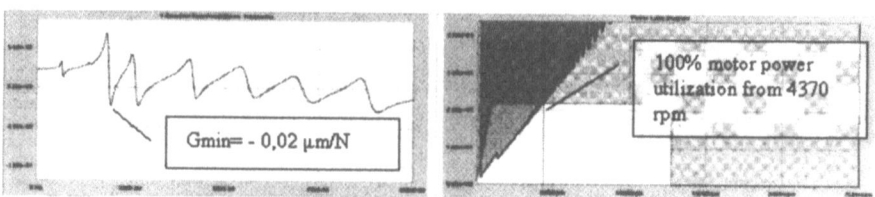

Figure 9: Improved spindle performance and machine power utilisation due to bearing dampers.

5. Conclusion

Milling spindles can be optimized by means of a model based on measured data. The disadvantage of this procedure lies in the fact, that the modal shapes can be measured for only the externally accessible parts of the SHT system, namely the tool, tool-holder and spindle nose. This limits the knowledge of individual modal shapes and their identification.

A more precise optimization of the spindle structure can be performed if identification of the main modal shapes is made possible during the conceptual phase of design. Computer-aided FE programmes available today make this procedure feasible.

References

[1]   Tlusty J., 1999, *Manufacturing Processes and Equipment*. Prentice Hall, New Jersey.
[2]   Delio Th., 2002, *Metalmax Operational and Tutorial Manual*. Manufacturing Laboratories Inc., Gaineswille, USA.
[3]   Smith S., 2003, *Spindles for Air Frame Machining*. Proceedings of the conference The Dominance of Spindle Performance, Dearborn, Michigan.
[4]   Badrawy S., *Dynamic Modeling and Analysis of Motorized milling Spindles for Optimizing the Spindle Cutting Performance*. Proceedings of conference The Dominance of Spindle Performance, Dearborn, Michigan.
[5]   Bach P., 2003, *Milling Machine Tool Stability and Performance*. Proceedings of LAMDAMAP Conference, Huddersfield, UK.

This research has been supported by the Czech Ministry of Education under grant LN00B128.

Modelling and Design of Main Spindle Box Made of Polymer Concrete

Gligorce Vrtanoski[1] and Vladimir Dukovski[2]

[1] University of Skopje, Faculty of Mechanical Engineering, Karpos II b.b. or P. O. Box 464, MK-1000, Skopje, Republic of Macedonia, gliso@mf.ukim.edu.mk
[2] University of Skopje, Faculty of Mechanical Engineering, Karpos II b.b. or P. O. Box 464, MK-1000, Skopje, Republic of Macedonia, dukovski@mf.ukim.edu.mk

Abstract: Advanced composite materials have been widely used to improve various properties such as impact resistance, strength, durability and vibration damping. Polymer concrete is one of the new materials, which has been developed for potential application in machine tool industry.
This paper describes modeling and analysis of machine tool main spindle box made of different materials. The casting metal structure of the original structure is substituted by the structure made of polymer concrete. The analysis was carried out using FEA commercial packages ALGOR and I-DEAS Simulation system.

1. Introduction

International competition in machine tool manufacture demands flexibility, high quality, reaction to market demands and low costs. Therefore, the machines are designed with modular structure. In addition, recent developments in the fields of electronics, micromachining, material science, sensors and also production engineering make these aims achievable. A main spindle box could be such a module of a machine tool. The most important factor affecting machining accuracy is the accuracy of the machine tool itself resulting from low stiffness and damping ratio of machine tool structure. In this paper the feasibility of applying new material is studied in order to raise the static and dynamic performances of the structure to a desired level. The new design of a main spindle box structure is made of polymer concrete instead of cast iron structure.

Polymer concrete consists of a mineral filler and a polymer binder, which may be a thermoplastic, but more frequently, it is a thermosetting polymer. When sand is used as filler, the composite is referred to as a polymer concrete. Other fillers include crushed stone, gravel, limestone, chalk, condensed silica fume (silica flour, silica dust), granite, quartz, clay, expanded glass, and metallic fillers. Generally, any dry, non-absorbent, solid material can be used as a filler [1].

A comparison of different construction materials is shown on Table 1. It can be seen that the major advantage of polymer concrete in comparison with cast iron is a much better damping ratio, which may lead to improved dynamic performances of the machine. The other characteristics of polymer concrete are substantially worse in comparison to cast iron, which requires appropriate redesigning of the structure to be able to achieve desirable performances. As a result of the implementation of polymer concrete in the structure design many advantages may occur.

Table 1: General Characteristics of Construction Materials

Characteristics	Unit	Steel	Cast Iron	Granite	Polymer concrete
Compression strength	MPa	250-1200	600-1000	70-300	140-170
Flexural strength	MPa	400-1600	150-400	up to 35	25-40
Young's modulus	GPa	210	80-120	35- 85	30-40
Heat transfer coefficient	W/mK	45-50	45-50	2.4	1.3-2.0
Coeff. of heat expansion	10^{-6}/K	12	10	6.5- 8.5	12-20
Density	kg/m^3	7850	7150	3000	2100-2400
Damping ratio	-	0.002	0.003	0.015	0.02-0.03

The sand and the fillers are cheap and natural materials available for easy use. The overall process of production of polymer concrete is performed at room temperature and recycling of the materials is much easier than cast iron. Also, the control of the properties of polymer concrete may be much easier and less costly in comparison with cast iron. As a result of the above and some other advantages (flexibility of the design, short development time, simple and less costly production process) the overall competitive advantages of the company may significantly improve [3].

2. Redesign of Existing Spindle Box for Substitution of Cast Iron with Polymer Concrete

The aim of this research is the investigation of the possibility of using polymer concrete in building machine tools structures. For that purpose, the NC lathe Mazak QT 10 headstock's housing is completely redesigned and polymer concrete constructive material has been chosen and applied.

The available references [1,2,3] have shown similar examples of the use of polymer concrete as a substitute material for cast iron in the design of machine tool beds. We have decided to investigate the possibilities of the use of polymer concrete in main spindle housing design due to the more demanding requirements connected with the dissipation of temperature, damping and high accuracy of the structure.

To be able to redesign the original housing for the purpose of material substitution, we have performed static, dynamic and thermal analysis of the structure. The differences in the material properties (Table 1) have initiated an iterative process of redesign with the aim to achieve properties of the original design. Both models, of original and redesigned housing are shown on Figure 1.

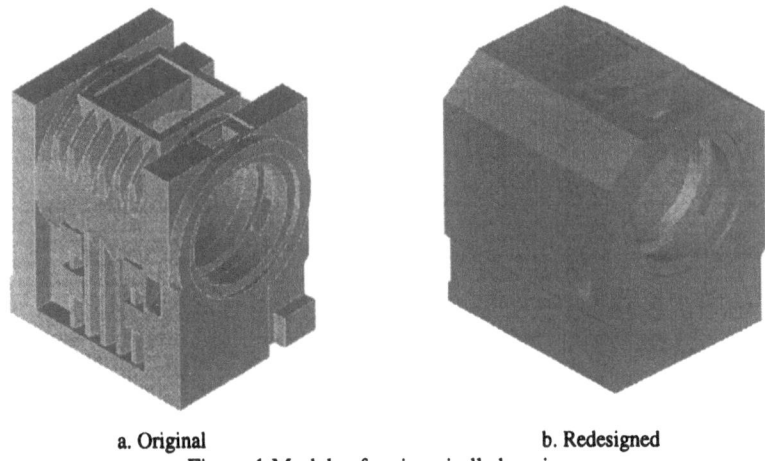

a. Original   b. Redesigned

Figure 1 Models of main spindle housing

3. Theoretical and Experimental Study of New Design

We have performed wide numerical and experimental investigations of static, dynamic and thermal behaviour of original and redesigned housing, part of which are presented in another publication [4].

3.1 Statical modelling and analysis

Modelling and processing of the housing were performed with the use of ALGOR and I-DEAS commercial packages. Figure 2 shows displacement of both models, and Figure 3 maximal principal stresses.

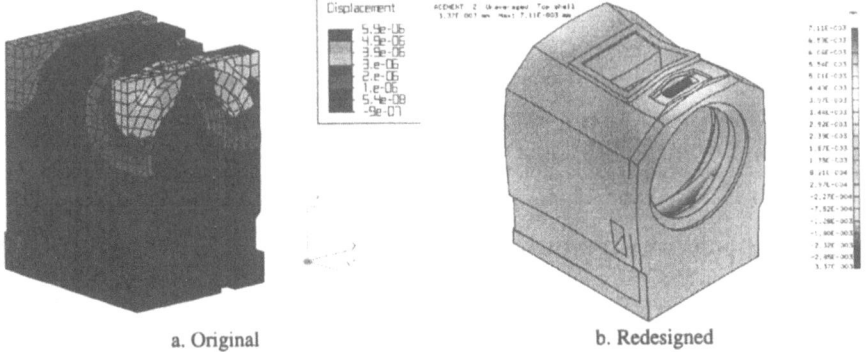

a. Original   b. Redesigned

Figure 2. Displacements of models

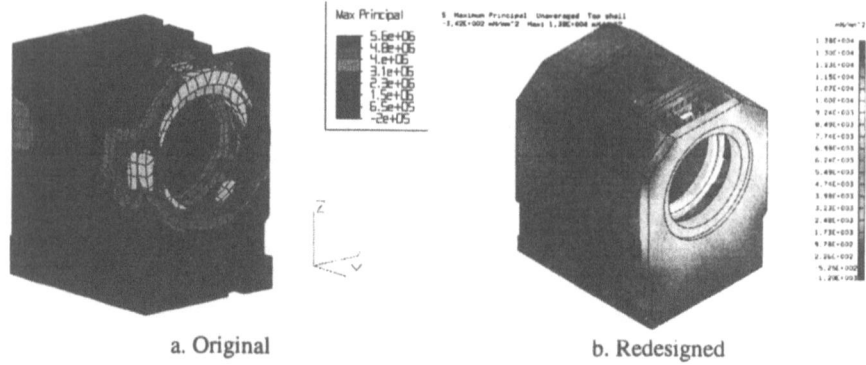

a. Original   b. Redesigned

Figure 3. Maximal Principal Stress of Models

Comparative analyses of the statical characteristics (displacements, Von Mises, Maximal Principal Stress, Mass and Young's module) of both housings (original and redesigned) are shown on Figure 4. Despite worsening of Von Mises and maximal Principal Stress, the new design satisfied the limits. We have a reduction of the mass of the redesigned housing of approximately 50% in comparison with the original design which strongly recommend the use of polymer concrete.

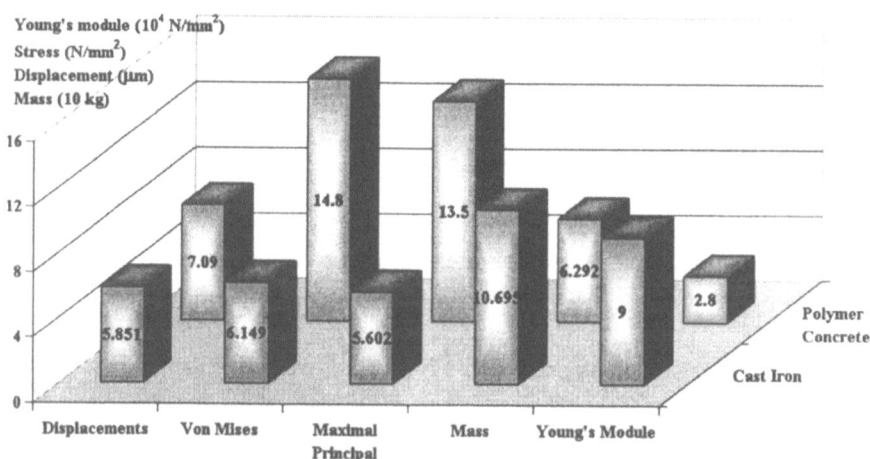

Figure 4. Comparative Analysis of Statical Characteristics

3.2 Dynamical modelling and analysis

We have used the same models, developed with ALGOR and I-DEAS, for dynamic analysis of both housings [4]. Modes of vibration are shown in Figure 5. The comparative analysis of Eugene frequencies of both models is shown in Figure 6. As we can see the Eugene frequencies of polymer concrete design are twice higher then those of cast iron design which represents certain improvement of the design.

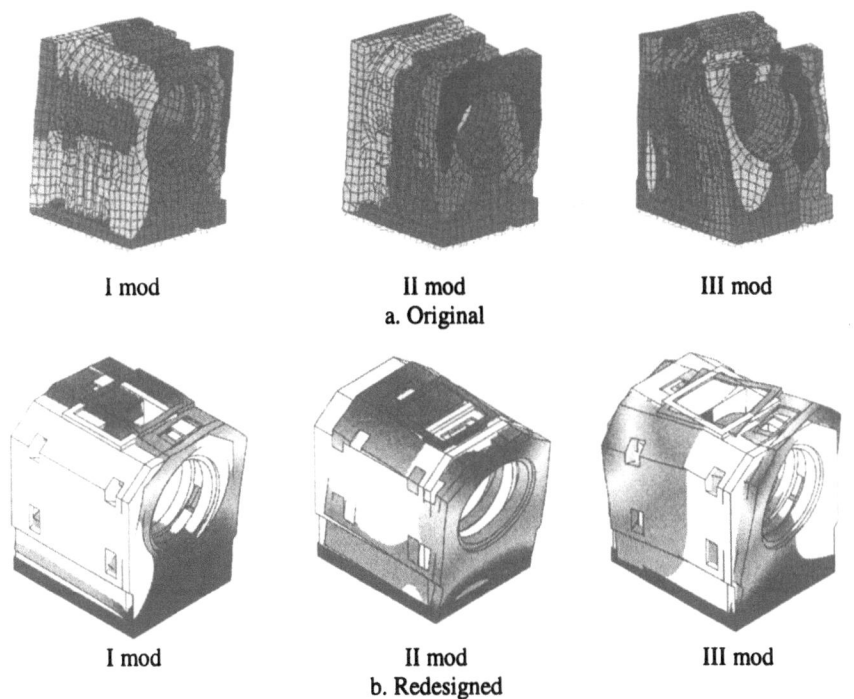

I mod   II mod   III mod
                    a. Original

I mod   II mod   III mod
                    b. Redesigned

Figure 5. Modes of Vibration

Figure 7 shows comparative analysis of the experimental data of the damping ratio of both designs. The results show superior performances of polymer concrete housing in comparison with cast iron structure.

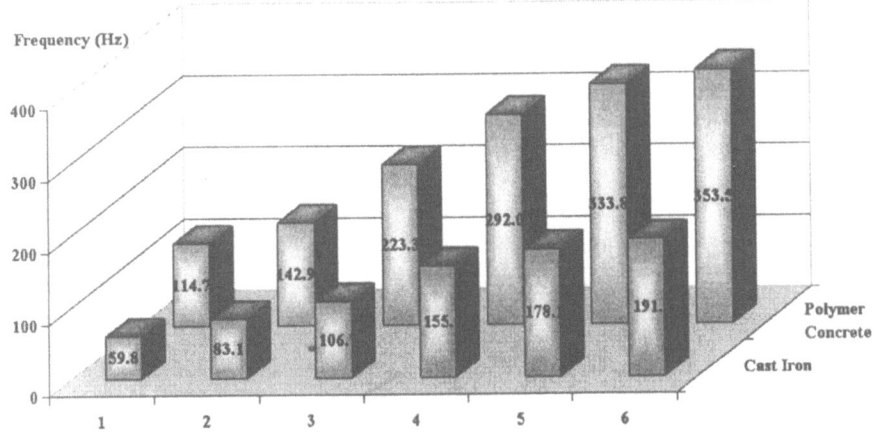

Modes of Vibration
Figure 6. Comparative Analysis of Dynamical Characteristics

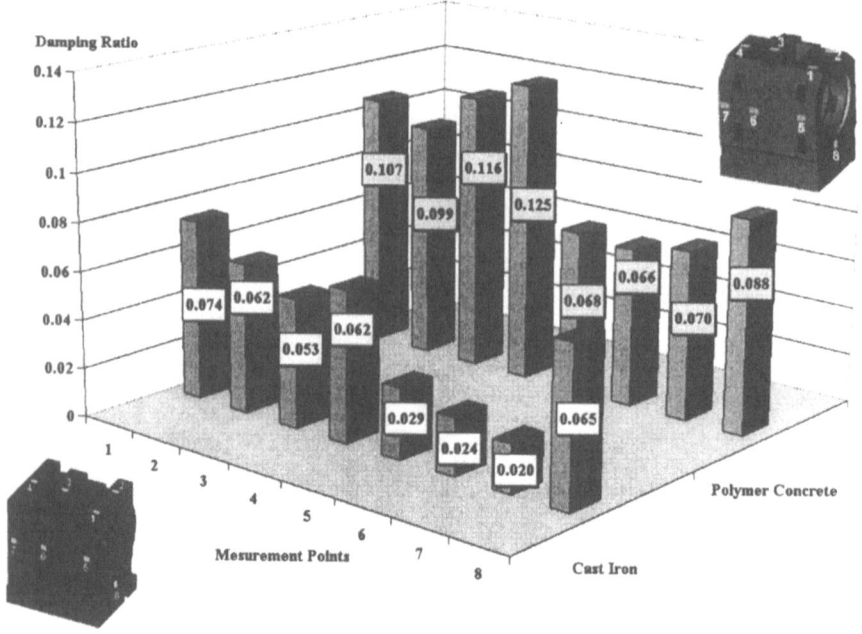

Figure 7. Comparative Analysis of Experimental Damping Ratio

6. Conclusions

The substitution of cast iron with polymer concrete in machine tool main spindle housing has been investigated. Some results of static and dynamic analysis are shown in this article. Without considerable reduction in static performances a significant improvement in damping and Eugene frequencies has been achieved. The superiority of polymer concrete has been demonstrated in the 50% reduction of the weight. Also, the new product development time and the manufacturing cost have been dramatically reduced due to the simplification of the production process.

The ecological aspect of polymer concrete gives another key advantage in comparison with cast iron.

References

[1] Sugishita H., Nishiyama H., Nagayasu O., Shin-nou T., Sato H., O-hori M.: Development of Concrete Machining Center and Identification of the Dynamic and the Thermal Structural Behavior, *Annals of the CIRP*, Vol. 37/1/1988, 377 – 380.

[2] Saljé E., Gerloff H., Mayer J.: Comparison of Machine Tool Elements Made of Polymer Concrete and Cast Iron, *Annals of the CIRP*, Vol. 37/1/1988, 381 – 384.

[3] Mason F.: Cast Polymer Machine Bases, *Machine Shop Guide Magazine*, Vol. 5, Issue 6, June 2000, 1 – 6.

[4] Vrtanoski G.: Investigation and Development of Machine Tool's Polymer Concrete Structure, doctoral dissertation, Faculty of Mechanical Engineering-Skopje, October 2003.

Challenges and Rationale in the Design of a Miniaturised Machine Tool

S. Mekid, A. Gordon, P. Nicholson

Dept. of Mechanical, Aerospace and Manufacturing Engineering, UMIST.
PO Box 88, M60 1QD, Manchester, UK. s.mekid@umist.ac.uk

Abstract: This paper discusses design challenges one has to overcome to build a machine with acceptable ranges of forces and accuracy expected in manufacturing. Relevant reasons leading to miniaturisation and improving robustness of machine tools are presented. The actuation forces scale well into the micro-domain especially the electrostatic force, which remain constant, hence its importance.

1. Introduction

Extreme miniaturisation has taken place in the last decade in several domains, requiring mesoscopic parts with complex microscopic features. Therefore, micro machining becomes an important requirement in the new concept of the micro-factory. The growing markets and challenges of the future are in biotechnology, life science, telecommunication and mobility. The arising products are characterized by miniaturization and integration of mechanics, electronics and information processing. Europe can gain leadership in these areas with the possibility of implementing such fields as an added value for the development and production of micro parts. Such sustainable production structures and standards do not exist yet worldwide.

The required small parts have delicate features and tight tolerances. It is inefficient to produce parts such as moulds and dies with electrical discharge machining (EDM) or investment casting techniques, as is the current practice. The actual production of micro-parts is performed by standard machine tools consuming high energy, occupying large space and requiring

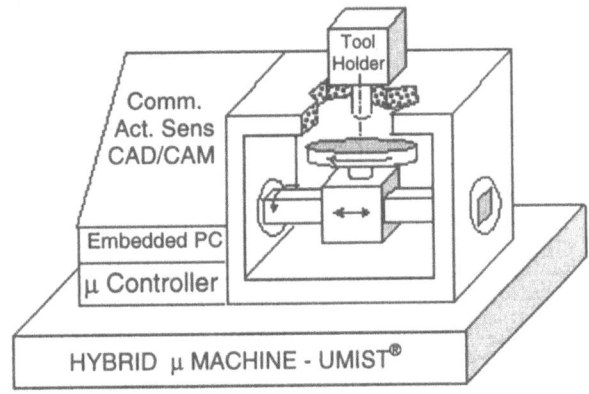

Figure 1: Hybrid concept of micro-machine.

large air-conditioned environments; therefore it is

important to save driving energy, space and resources in controlling the surroundings (e.g. temperature, humidity) by applying the concept of the micro-factory and robust design of micro-machines with a suitable volume ratio of machine/workpiece.

The last decade has seen the emergence of a few attempts in designing micro machine tools in Japan [1][2] and Mexico [3]. The next way forward is the design of flexible reconfigurable and cost-effective miniaturised production units. This will lead to the development of new micro-technologies based on precision machine tool mechanics aided with various attachment units to accomplish the micro-machining on demand for applications in electronics, MEMS, bio-medical, automotive and micro-machines industry. The developed technology will be further evaluated to test its application capabilities as well as its limitations. Hence, the main goals would be as the following.

- Development of a fabrication hybrid facility of meso-scale components and to widen the horizon of Micro Electro Mechanical Systems (MEMS).
- To investigate the design challenges and scalability in miniaturization of CNC machine tools.
- Establish general guidelines for robust miniaturisation of CNC machines and their implementation in a flexible micro-factory.

As a consequence, this analysis will result in a low cost production by high speed machining due to reduction in distance and mass with an increase of precision by smaller forces and high natural frequencies of the structure.

2. Reasons for Miniaturisation

Many companies offer services to produce micro-components to precision accuracies from processes using contemporary CNC machines, EDM and Laser Machining, to industries such as biomedical applications, components of micro-devices and part of the MEMS production. Logically, if one requires a small component to be manufactured, a suitable solution would be to have a proportionally sized machine to perform the manufacturing tasks. The relevant reasons are given hereafter.

a) *Negligible thermal drift*: Thermal drifts that are generated by the machining process causing deformations directly affect the accuracy of standard machines. These effects are reduced in micro-machines due to the miniature nature of the components, and can often be regarded as negligible.

b) *Reduced vibration amplitude with high natural frequencies.* Reasonable decrease in vibration amplitude is due to reductions in mass of moving parts and therefore larger natural frequencies of the machine. As an estimation, the stiffness $K \approx L$ and the mass $m \approx L^3$ where L is the size of the component, therefore if the machine is miniaturised with a scaling factor of s, the natural frequencies will be s time larger (eq.1). This will depend, of course, on the mode of vibration.

$$f_\mu = s f_o \qquad (1)$$

c) *Reduction of material consumption*: As micro machine tools are dedicated to manufacturing components of micro dimensions, they are capable of using billets of comparable size to the end product. This is not the case in standard size machines where limitations on work-holding devices may impose a restricting minimum diameter of billet.

d) *Magnitude of induced forces*: The eccentric loading of a shaft induces a deflection δ, due to the centrifugal force F_i [3]. Analysis of a miniaturised shaft shows that the deflection is reduced by a power of the scaling factor s depending on the scale of speed. Hence, the magnitude of the deflection δ has been analysed for three different rotating speeds scaling as shown in table 1.

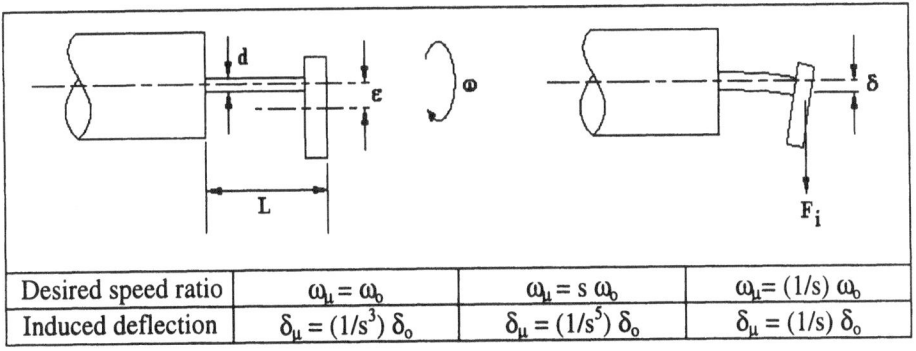

Desired speed ratio	$\omega_\mu = \omega_o$	$\omega_\mu = s\,\omega_o$	$\omega_\mu = (1/s)\,\omega_o$
Induced deflection	$\delta_\mu = (1/s^3)\,\delta_o$	$\delta_\mu = (1/s^5)\,\delta_o$	$\delta_\mu = (1/s)\,\delta_o$

Table 1: Magnitude of eccentric deflection.

If the rotating speed of the shaft is increased or decreased, δ is reduced by a factor of s^5 or s, respectively. The eccentric force is given by $F_i = m\omega^2\varepsilon$ and the deflection $\delta = c\dfrac{L^3 F_i}{Ed^4}$ the standard axis is down scaled as: $L_o = sL_\mu$, $d_o = sd_\mu$, $\varepsilon_o = s\varepsilon_\mu$ and $m_o = s^3 m_\mu$ where the indices are: o for standard scale and μ for micro-machine.

e) *Enhanced machining accuracy*: This is improved by the inherent reductions of machine component inertia, negligible thermal drift and larger eigen frequencies. As an example, any eccentricity in a lathe machine will be reduced by s, s^3 or s^5, where s is the scaling factor. The effect of miniaturisation upon inertial, elastic, electromagnetic and electro-static forces can be quantified as represented by the equations in table 1. As can be seen, miniaturisation can play a beneficial role by reducing these forces in varying magnitudes.

f) *Agility*: As the inertia of moving parts decreases it induces larger accelerations to enhance the agility of such small machines

g) *Space and energy consumption*: Conventional machine tools can occupy up to approximately 5m² (including service area), whereas a single micro machine from the Japanese micro factory uses only 0.09m² [1], thus providing considerable spatial savings. This translates into economical advantages in terms of the costs of rent of premises. This provides a driving force for the replacement of conventional machine tools with miniature machines. Issues of portability of such a machine can also be addressed, in that the costs of relocation and installation of large machines are considerable.

On the other side, the saving in energy is of 10% in a micro-lathe [4], whereas others have claimed 60% [5]. However, precision manufacturing often requires special working environments (e.g. controlled temperature). The reduction of the working volume of the manufacturing system would diminish the complexity of the maintenance of such operational conditions.

3. Design Approach

Optimal parts manufacturing at meso-scale will not be simply achieved by scaling-down versions of conventional manufacturing devices. As the micro machine has larger natural frequencies and the thermal effect is insignificant, a direct robust design method, implementing, for example, Taguchi's approach will be setup to directly identify the design parameters, which significantly affect the machining tolerances. After the specific concept with all functionalities is defined, it is necessary to ensure that the working volume of the machine is located within the region of the highest stiffness of the mechanical system. The machine is then divided into sub-systems to be designed accurately by defining design criteria derived from the main specifications. Each sub-system will be evaluated and optimised to meet these criteria. Needless to mention issues related to friction and induced micro-dynamics are addressed at this scale. The overall steps of the design methodology are described as follows.

- Problem identification and concept development.
- The functional requirements of the machine will be defined based on the main specifications, followed by an outline viable concept, which includes combined axes, motion generation and minimisation of errors.
- The system is split into critical sub-systems with detailed sub-specifications.

Fundamentally, the two strategies that are most likely to succeed are: one design that is most mechanically simple, but would require more controls effort, and one design that is simple to control, but may require more mechanical complexity [11]. The selection will be based on risk assessment of preliminary calculations of accuracy.

4. Design Challenges

The stringent specifications in quality, precision and time are directing efforts towards more integrated machines as a general concept even for meso-scale workpieces. A machine that can fabricate a component with a variety of operations without the need to transfer the artefact in any way would provide advantages in terms of positional accuracies and setup times. The design challenges are revealed by the sub-systems of the micro-machine and are succinctly described hereafter:

a) Kinematics: Machines will parallel kinematics have difficulties in achieving stiffness [10] and avoiding singularities in the workspace where some of the degrees of freedom are lost. Serial machines accumulate errors from each axis. Stacked axes lead to larger machines generally with small working volumes; hence a compromise in kinematic concept is acceptable for hybrid configurations (fig.1). The structure of the machine must be robust in order to obtain high machining accuracy and repeatability especially in line productions.

b) Interactive forces: With micro-scale components, interactive forces (e.g. Van der Waals, surface tension, electrostatic) will exist between components, which instigate difficulties in manipulation and control. To overcome these problems, contact type manipulators such as ultrasonic travelling waves, or mechanical grippers; or non-contact type manipulators for example magnetic fields, aerostatic levitation, or optical trapping could be used in place of conventional solutions [6].

c) Actuators: When scaling down actuators, it is important to keep an acceptable range of forces required to cut materials or to move and position devices (e.g. the tool). A rapid comparison between different types of forces shows that the electrostatic force remains exactly the same if it is scaled down.

1. *Inertial force:* the miniaturisation of an actuator delivering this type of force will result in a reduction of the force by s^4 where s is the scaling factor.

$$F = m \cdot a = m \cdot \frac{dx^2}{dt^2} \Rightarrow [L]^3 \cdot [L] = [L]^4 \tag{2}$$

2. *Elastic force:* the reduction of the force is about s^2.

$$F = \frac{E \cdot A}{L} \delta x \Rightarrow [L] \cdot [L] = [L]^2 \tag{3}$$

3. *Electromagnetic force:* the reduction of the force is about s^2.

$$F = \frac{\beta^2}{2\mu} \cdot A \Rightarrow [L]^2 \tag{4}$$

4. *Electro static force:* the force remains constant.

$$F = \frac{\varepsilon V^2}{2} \cdot \frac{A}{d^2} \Rightarrow [L]^2 [L]^{-2} = [L]^0 \tag{5}$$

The forces incurred in the cutting process must be evaluated so that suitable systems can be designed. The forces in a metal turning operation can be evaluated as described in fig.3. The forces are: F_s; axial force, F_a; radial force, and F_v; tangential force. The forces are dependent on cutting speed (V), feed rate (S), and depth of cut (a).

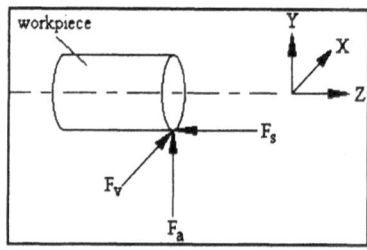

Figure 2.

Governing equations for these forces on steel work are: $F_v = 1750S^{0.75}a$, $F_s = 650S^{0.35}a$ and $F_a = 250S^{0.25}a$.

Figure 4 shows the effect that varying the feed rate has upon the cutting forces incurred in the cutting process. Cutting speeds for turning operations may also cause problems at the miniature scale; this is due to the reduction of the size of products to be machined.

$$N = 1000V/\pi D \tag{6}$$

If spindle speed N is obtained via equation 6, where V is cutting speed, and D the work diameter, then it can be seen that the required speed is about 300,000 rpm for a diameter of 1 mm using Al-alloys with TiN coated carbide tool. The speed decreases hyperbolically with the diameter. Selection of such motors for spindles is obviously problematic but one has to design direct drive motors for better torques and acceptable speeds.

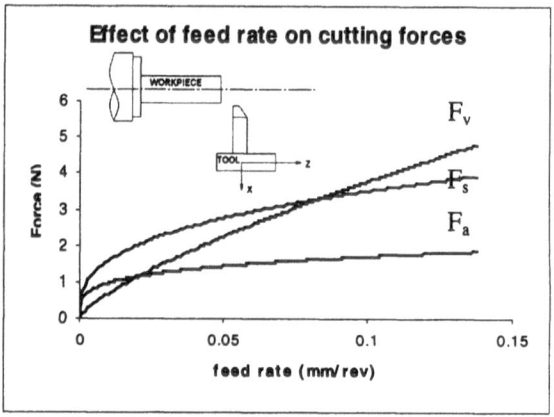

Figure 3: Effect of federate on cutting forces.

5. Miniaturised Controller

An important aspect to be considered is the system of control, which is increasingly being required to perform a wide variety of complicated tasks under varying operating conditions and in different environments, while at the same time achieving higher levels of precision, accuracy, repeatability and reliability [7].

Modern CNC approaches employ PC-based solutions to incorporate extensive functionality in order to combine high quality and flexibility, with reduced processing time. The controller is required to satisfy several key issues set out in the "Technical Committee of Open Systems" of the IEEE 1003.0. These criteria address portability, extendibility, interoperability and scalability in order to specify flexibility of the controller to varying applications [8]. To achieve interoperability, one must ensure that the controller is vendor-neutral, achievable by employing a modular approach. However there is a compromise between the degree of openness and the cost of integration [9]. One must also consider the processing power of the controller hardware, as too great a modularity can result in deterioration in the real-time performance of the system.

Integration of the PC-based control system into a CAD/CAM manufacturing system is a fascinating area of research (fig.1). The CAD/CAM system incorporates a feature recognition program which links directly to a computer-aided process planning (CAPP) software tool; hence the manufacturing process can become totally automatic thus improving efficiency.

Concluding Remarks

Miniaturisation of machine tools for meso-scale components is very tempting due to the reasons presented earlier. For high efficiency and robustness, the challenges discussed earlier have to be addressed for hybrid configurations. Exploitation of existing development in micro-systems will tremendously help the implementation of specified devices to ensure superior performance in their functionalities. Performance assessment of micro-machines is required today to evaluate the real performance and limitations as no results have been published so far.

References

[1] Mishima N., Ashida K., Tanikawa T., Maekawa H., Development of desktop machining microfactory, 2000, Japan-USA Flexible Automation Conference, Michigan.

[2] Okazaki Y., Mishima N., Ashida K., 2002, Micro Factory and Micro Machine Tools, The 1st Korea-Japan Conference on Positioning Technology, Daejeon, Korea.

[3] Kussul E., Baidyq T., Ruiz-Huerta L., Caballero-Ruiz A., Valesco G., Kasatkina L., 2002, Development of micromachine tool prototype for microfactories, Journal of micromechanics and micro engineering, p.797.

[4] Ooyama Naotake et al., 2000, Desktop machine microfactory, Proc. 2nd Int. Workshop on microfactories (Switzerland 9th-10th October), pp14-7.

[5] T. Kitahara, Y. Ishikawa, K. Terada, N. Nakajima and K. Furuta, 1996, Development of Micro-lathe, Journal of Mechanical Engineering Laboratory, Vol. 50, No. 5, pp. 117-123.

[6] Alting L., Kimura F., Hansen H. N., Bissacco G., 2003, Micro Engineering, Keynote papers, p. 17.

[7] Smith M. H., Annaswamy A. M., Slocum A. H., 1995, Adaptive control strategies for a precision machine tool axis, Precision engineering Vol 17 No. 3.

[8] Pritschow G., Altintas Y., Jovane F., Koren Y., Mitsuishi M., Takata S., van Brussel H., Weck M., Yamazaki K., 2001, Open control architecture – past, present and future, Keynote papers, p3.

[9] Koren Y., 1998, Open architecture controllers for manufacturing systems, Open architecture controllers, ITIA series.

[10] Zhang D., Xi F., Mechefske C. M., Lang S. Y. T., 2004, Analysis of parallel kinematic machine with kinetostatic modelling method, Robotics and Computer-Integrated Manufacturing, Volume 20, Issue 2,

[11] S.Mekid, Design Strategy for Precision Engineering: Second Order Phenomena., under press in J. Eng. Design.

Rapid Conceptual Design: Towards a Viable Informed, Automated and Interactive Process

Nick Kalargeros[1], George Haritos[2], David K Harrison[3]

[1]Cranfield University, empnefsi@clara.co.uk
[2]Softnet Consultancy Ltd, softnet@lineone.net
[3]Glasgow Caledonian University, D.Harisson@gcal.co.uk

Abstract: The present automotive product development environment is characterised by the customer's thirst for new, innovative and affordable products. This imposes a shorter Product Development Process (PDP). A healthy concept definition, and control, at the early PDP stages results in successful end products. A key factor of which is an upfront knowledge assessment and integration between the 'true' customer's needs and marketing / finance, product design / engineering and manufacturing functions. The Common workspace in all these functions is considered to be the product model⌑.

Numerous software tools have been developed for an efficient PDP, though these are more devoted to reliable product model data transfer and representation. Consequently less emphasis has been placed on the tools in the early PDP stages. Yet it is at these stages where the knowledge can be complex and disjoined, hence demanding greater effort in their utilisation within the product model. This paper addresses this issue by introducing results from ongoing research aimed at a system, which enhances these early PDP stages by acquiring and refining the customer's needs and expert knowledge iteratively, hence refining and enhancing the product model.

1. Introduction

Today's automotive product market place is flooded with conforming products obeying fashion and marketing statements rather than the 'true' customer's needs. This very fact has resulted in an increased customer thirst for new, innovative and affordable products. The Institute for the Future (IFTF) confirms this by identifying a new and increasing group of consumers who have less trust in brands [1] and, as a result, are willing to experiment with new products.

To satisfy these customer needs, products have to be fresh, innovative and affordable. However, customer's lifestyles and needs are evolving rapidly and one way to address them is by capturing these needs by introducing products into the market place as quickly as possible, to satisfy these demands. To achieve this, a more direct and efficient method would be to attain transparent translation of the customers needs into quantifiable product attributes. This will enhance the refinement and recycling of the existing product attributes and knowledge, which in return will help to focus the PDP on those product attributes, which vastly differentiate a product from the competition,

---

⌑ The definition of a product model can vary according to the application; generally it should contain data, algorithms and a data structure suitable for the representation of the end product.

yet most of the fundamental product attributes and components remain the same. The process to attain transparent translation of the customer's needs is by quantifying and mapping each need into a realisable product attribute or feature. Considering the multitude of product attributes and features at the conceptual phase of the product development process, one can appreciate the level of this challenge.

The proposed system attempts to handle these issues, utilising knowledge intensive processes and soft computing techniques. This is accomplished iteratively by capturing the customer's needs and the product related knowledge. Then it processes the data and provides an immediate response in a multitude of ways (visually, textually, functionally, hierarchically etc.) using a number of soft computing reasoners and assessing the probable satisfying products using genetic algorithms.

2. Systems Methodology

The overall system has been developed as a requirement[1] driven process, in which the voice of the customer[2] and the expert[3] is paramount. This approach relies heavily on knowledge intensive processes, which reflect on existing or future product development capability and its corresponding potential to satisfy the customer and PDP needs.

The conceptual product is characterised as a potential result based on the information provided (value assessments / judgements, product knowledge, benchmark results etc.). Pugh [2], a pioneer in the field, proposes that lack of breadth & depth to detailed requirements will lead to conceptual weakness and a corresponding fruitless PDP. However, the 'one best conceptual result' can be difficult to obtain when this result is based on several conflicting, disjoined and in some instances not adequately defined requirements. An attempt to tackle this is by deriving a quantifiable property based model based on attributes and features. Consequently the overall conceptual product can be based on the selection and local adjustment of these attributes and features such that their balance will fulfil the global product requirements. In the realisation of this structure, the expected / known performance of each attribute and feature is used as a requirement in the next PDP detailed phase.

According to Seifert & Drisis [3] a fully Object Oriented (OO) description of product elements i.e. (product functional properties, function specific features), assures the formal exactness needed to create and represent a property-based model which reflects the customer's as well as the PDPs idiosyncrasies. The developed system uses, as a core, an iterative OO soft computing process between the customer, the expert and the Explanatory & Representation Module.

The fundamental system assumption is that a product can be represented solely on a predefined and customisable set of functional properties. This will be the case where

---

[1] "A statement identifying a capability, physical characteristic, or quality factor that bounds a product or process need."
[2] A customer can be defined as anybody downstream of the process that has equity in its outcome.
[3] An expert can be defined as anybody in the PDP who has ownership in a product attribute.

these properties do not exist or they do not cover faithfully certain product aspects. In such a case the system will synthesise new ones from the existing, failure to do so, will prompt the user to define new ones based on first principles. Consequently the transition from these functional properties to more precise geometrical, textual, visual etc. would be feasible. A potential limitation of the system is that large product specific libraries are needed.

2.1 System implementation

The overall implementation starts with the data capturing process, involving data captured from anyone that has equity in the PDP commencing with the consumer and concluding with the raw material supplier. Each product need or related knowledge is captured and organised in a hierarchical data abstraction. Each data abstraction is organised in a predefined object, incorporating data structure and behaviour. Each object is defined by a class which means that by this stage all OO behaviour and data manipulation will be inherited.

Each product need can be expressed in various forms, judgments, evaluations, comparisons, generic descriptions all of which are subject to vagueness and comprehension. This vagueness and comprehension is modelled using Fuzzy Logic (FL). FL linguistic variables have the inherent OO ability and flexibility to define words or sentences in a natural or synthetic language. At this stage Genetic Algorithms (GA) are utilised in order to create, evaluate and synthesise potential satisfying concept designs. An endowed ability from this approach is the utilisation of hidden or implicit information based on inference principles i.e. if a customer asks for a diesel engine, the system then implicitly considers that vehicle running costs are important.

2.2 System description

Overall, the system's architecture can be viewed by the following Figure 1.

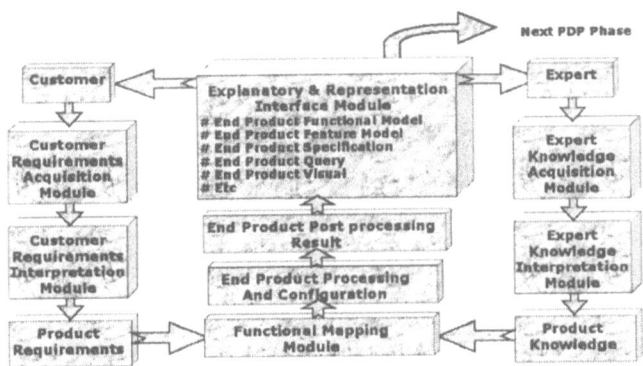

Figure 1 – Proposed System Architecture

2.2.1 Customer requirements acquisition module

This is the Marketing & Product Orientated module, it acquires and updates the customer's profile and the detailed requirements from the product. This is achieved by a series of rule based questions related to: product performance, product preferences, current satisfaction levels, future expectations for competitive advantage etc. These requirements are captured using the customer's language. Each requirement has an associated quantitative measure of desired performance. Early in the design process, these measures may be subjective, or relative, (best-in-class, stronger than current etc.). The processed informational outcome from this module is: *Detailed Customer Profile* and *Detailed Customer Requirements*

2.2.2 Customer requirements interpretation module

This module facilitates calculations and interpretations for the qualitative and quantitative prioritisation of the *Product Requirements*. Here is where the first production rules are established or improved. These rules process precedence or affiliation for a customer segment or a customer need, i.e. customers of this segment normally go for the colour red.

2.2.3 Expert knowledge acquisition module

This module acquires and updates the expert's profile and expertise related to the product requirements as defined by the customer. This is achieved by a series of rule based questions established at the Functional Mapping Module and fed back by the Explanatory & Interface module. The overall benefit of which is that everyone that has equity in the PDP influences the outcome according to their function and expertise.

2.2.4 Expert knowledge interpretation module

This module facilitates calculations and interpretations for the qualitative and quantitative prioritisation of the *Product Knowledge*. Here is where more production rules are established or improved and where existing ones can be answered. These rules process precedence or affiliation for expertise or correspondence to a product requirement, i.e. a need for a diesel engine is allocated to the power train department, diesel engines are cheaper to run, but they have trade offs such as being heavier, harsh NVH, etc.

2.2.5 Functional mapping module

Product Knowledge and Product Requirements are the inputs for this module. The idea of this module is that for every Product Requirement a Functional Element is created based on the corresponding Product Knowledge. In aid of this, the module processes and retrieves the closest analogous case (source case) and then applies a set of transformation operators to reduce the difference between the source case solution and the product requirement specification. If the transformation is successful, the target case is solved and the Functional Element is created and further in the system a feedback in the form of product representation (visual and textual) will be provided to the customer and the expert. If not then the second level will acquire and improve its own knowledge base with a series of rule based questions to the expert (application specific cases) in the following order:

1. For every sub-function define an appropriate functional element (a component or a set of components that perform a function)
2. Compose a design solution from these defined elements.

If a solution cannot be reached then a third level of knowledge acquisition process is engaged (first principal case) and basic design knowledge rules have to be imported so they can be used as building blocks for the Functional Element. The overall operation is controlled by a soft computing process, which maintains the flow of data and knowledge and ensures that the knowledge is compatible before it is transferred between modules. All transformation levels eventually use knowledge acquired at the first principal case. This process is iterative and problem solving will take place by a series of successive transformations from an initial state to a solution state at the Explanatory and Representation Interface Module.

2.2.6 End product processing and configuration module

This is an OO based module. Functional Elements are organised in predefined objects, which incorporate data structure and behaviour. Each object is defined by a class. Each class is a blueprint of the objects that a system requires to work with i.e. a product function, a geometric feature etc. Each class encapsulates the necessary data items and functions, which provide control in their access from other parts, functions or classes. However, different parts of the same product should understand differently their existence. Polymorphism gives such an ability. Using this approach the module iteratively produces an *End Product Result*. However, the *Explanatory and Representation Interface Module* can only view this result, which can be at an abstract or detailed level. The loop must be repeated according to the specific needs and the available time.

2.2.7 Explanatory and representation interface module

In this module communication between the customer and expert is facilitated in a number of ways (i.e. visual, textual, functional, hierarchical etc.). This then enables both the customer as well as the expert to determine how the conceptual product model [4] has been developed and how the system obtained such conclusions or why specific information has been requested. This module is crucial because it improves the quality of the collected data and eliminates misunderstandings or omissions by either the expert or the customer. Additionally any system errors or inconsistencies can be checked.

3. Conclusions

This paper explains the principles of the proposed system The immediate benefits of which are:
- Up-front product assessment for potential product viability in the market.
- Traceability of the decision-making process
- End product is immune from idiosyncratic and unsubstantiated responses.
- Emphasis of the importance of interpretation and substantiated knowledge.

References

[1] Story,V., Hurdley, L., Smith, G. & Saker, J., (2001), Forecasting the Future of European Automotive Retailing. Academy of Marketing Conference, Cardiff, July.
[2] Pugh, S., (1991), *Total Design*, Addison-Wesley Publishing Limited.
[3] Seifert, H. & Drisis, L., (1995), "Object-Oriented Product design", Proceedings of ICED 95, pp 1390-1396
[4] Elshennawy, A.K., Krishnaswamy, G.M. & Mollaghasemi, M., "Concurrent Engineering Deployment: a virtual reality approach", Integrated Manufacturing Systems Volume 4 No 4, pp 24-28

Declaration
Any ideas, opinions, findings and conclusions or suggestions are those of the authors and do not necessarily reflect Cranfield University, Softnet Ltd. or Glasgow Caledonian University views on this subject.

Quantifying Experiential Design and Knowledge in the Tool, Die and Mould Industry

Diane J Mynors[1], Julia Moore[2], Brian Griffiths[1], Colin Piddington[3], Dean Etheridge[4], and Tom Griffiths[2]

[1]Department of Design and Systems Engineering, Brunel University, Middlesex, UB8 3PH, United Kingdom, D.J.Mynors@brunel.ac.uk
[2]Gauge and Tool Makers Association, 3 Forges House, Summerleys Road, Princes Risborough, Buckinghamshire, HP27 9DT, United Kingdom, Julia@gtma.co.uk
[3]Technology Application Network, TANet, 2 Station Road, Hoghton, Preston, PR5 0DD, United Kingdom, Colin.Piddington@cimmedia.com
[4]Dzus Fasteners, Farnham Trading Estate, Farnham, Surrey, GU9 9PL,United Kingdom, Dean.Etheridge@dzus.co.uk

Abstract: The design of components, manufacturing processes, associated equipment and tooling is frequently dependent on the experience and skills of individuals. Researchers and industrialists rarely, if ever, dwell upon or try to quantify the experience or knowledge state of a company. This paper considers the type of knowledge and experience that can exist in companies that design, make, and/or use tools, dies, or moulds; this paper also proposes a knowledge coding system as the basis for future benchmarking development.

1. Introduction and Background

The UK tool, die, and mould (TDM) industry exists partly as standalone and partly incorporated, almost hidden, within companies whose products may be diverse. The GTMA through the United Kingdom's Department of Trade and Industry (DTI) supported Com-Met 2005 project [1] is presently working to assess the true enormity of the UK TDM industry. The TDM industry produces tools for composite layups, jigs, and fixtures in addition to the tools, dies, and moulds (TDMs) that are used by industries including the automotive, aerospace, packaging, pharmaceuticals, medical, and electronics to give the final form to *inter alia* plastic, metal, rubber, glass, food, medical, mineral, and building products. TDMs can be very sophisticated with complex kinematics, heating, cooling, pressure sensors, and retractable elements. TDMs are not just cavities or holding mechanisms but thermo-mechanical machines in their own right, which if designed, understood and used effectively, are a means of wealth generation. TDMs are the foundation of nearly every industry and it is reasonable to say that if the TDM industry in the UK was eliminated then the majority of UK manufacturing would suffer and possibly disappear.

2. The UK Tool, Die and Mould Industry

2.1 Present state of the industry

It has been estimated that the size of the UK TDM industry has halved in the last ten years and will halve again in the next ten. This is inline with economic indicators showing the UK manufacturing sector is shrinking as a proportion of the Gross Domestic Product (GDP) and workforce. The UK manufacturing workforce is now well below 3.5 million [2]. This is partially fuelled by the slowdown in the world automotive industry. However, there are clearly bright spots in the industry [3]. For instance, there is a growth in the demand for moulds for light metal casting and a consensus view that TDM companies supplying the pharmaceutical and medical industries are in a stronger position than those reliant on the automotive industry. There are also reports of growth in the electronics sector [4].

2.2 Training

Twenty years ago most companies had some form of in-house apprenticeship scheme ensuring the *future* of the industry. There are now significantly fewer apprentices and a distinct lack of young people both at engineer and technician levels, particularly in SMEs. In a survey of the state of UK manufacturing, the situation was summarised as: *"the life-blood of our national engineering expertise – our reservoirs of qualified and experienced engineers – has begun to leak away, particularly in the last two years. ... The DTI says that our future prosperity depends upon this expertise but seems unable to prevent its erosion."*[5].

2.3 Foreign Competition

The TDM industry is a global industry. The use of information and computer technology means that design information can be transmitted anywhere in the world such that a TDM designed in the morning in the UK can start to be manufactured in the afternoon on the other side of the world. An option, depending on TDM weight, when the wages in China are 5 to 10% of those in the UK and 50% in South Korea, yet the proportion of graduates in the working age population is identical [5]. Although, what is often overlooked is that the cost of sub-contracting abroad can often be more expensive when rework as well as other 'hidden' costs are taken into account [6].

The availability of enhanced computer-aided design (CAD) and knowledge-based engineering (KBE) software, its link to process simulation and computer-aided manufacture (CAM) as well as testing/inspection (CAT) means that design tools are available to all. Hence, design capabilities and the complexity of tools produced in low-cost based countries will increase.

2.4 The future of the UK Tool, Die, and Mould industry

If the UK TDM industry, or indeed any part of UK manufacturing industry, is to survive, it must be capable of doing more than the competition. One way a company is able to add value, as opposed to reducing waste, is to allocate time for innovation and the provision of full engineering services to customers, creating customer dependence and hence long-term stability. This must happen, since results from methodologies such as shop floor lean will eventually converge for all companies – with the only differentiator being the cost base, typically the geographic location of a company.

To innovate and provide full engineering services, a company must allocate/create time. Hence, eliminate or automate repeatable design activities and build upon the intellectual capital, experience and knowledge, within a company. This form of intangible capital must not be lost, or allowed to drift away as a result of retirements, etc. To assess the type of knowledge and experience within a company it might be appropriate to complete a knowledge audit. The following considers methods of categorising knowledge and considers how the definitions can be related to the environment of a TDM company.

3. Categorising Knowledge

If a company is to assess its knowledge state, there must be something against which to compare. The concept of knowledge can be quite complex and can include facts, calculations, accumulated non-quantifiable experience, etc.

3.2 The definitions of 'Knowledge'

The following definitions [3, 7] are used here.

- **Implicit Knowledge** is experiential know-how based on intuition, personal experience and insights. This is also referred to as Tacit Knowledge. Implicit knowledge often relates to either intuitive or creative processes and is often difficult to articulate in a form that allows ease of use and transfer.

- **Explicit Knowledge** is knowledge that is easily codified and conveyed to others. Explicit knowledge can be present as a set of engineering attributes, rules, relations or requirements. Inside knowledge deals with observations and may be available in papers, books, product design guides and standards

- **Derived Knowledge** is knowledge that is discovered only by running external programs, such as analyses, simulation, etc. Derived knowledge may be seen as interpolation or extrapolation of explicit knowledge.

- **Knowledge-Based Engineering** (KBE) is the process of combining engineering knowledge, methodologies, rules and best practices with process knowledge and best practice to create product models that describe how product designs are created or engineering analyses are undertaken. Closely linked to KBE is a Knowledge Based Engineering System which is a particular type of Knowledge Based System, based upon an object oriented programming language and tightly integrated with a geometric modelling tool. It is a software application that models engineering processes, enabling design and analysis automation to be carried out. A key feature of a Knowledge Based Engineering system is attribute dependency tracking which ensures that all elements of the model are consistent.

4. Forms of Knowledge in the TDM Industry

Implicit knowledge is in a designer's or toolmaker's head, gained through training and experience. The volume of knowledge is typically as large as the world a person can scan. This type of knowledge is at every level as it is simply people at work. A problem is that this is not codified and can be lost. There is a need to get this knowledge into a form that allows it to be built upon and reused. When implicit knowledge is written down and can be used by others, it becomes explicit knowledge. *Explicit* knowledge in the form of 'little black books', design guides, textbooks, and standards are used by the industry. Knowledge is also being processed, analysed, interpolated, or extrapolated, into *derived* knowledge via software packages.

Companies also manipulate knowledge to enhance understanding using machine intelligence. For example, companies have designed systems for integrating new knowledge into databases containing previous knowledge, previous lessons learnt. However, such examples do not coincide easily with the above definitions, although it is TDM knowledge.

TEDI knowledge code	People Equivalent	Software Equivalent
0	A technician	
1	A technician plus 'calculator'	Spreadsheet or Statistical Process Control
2	An analyst working for months	3D modelling capability, manufacturing process simulation, machine path verification
3	A team of professional engineers working for years	KBE type capability integrated with a 3D modelling system

Table 1: Proposed (TEDI) knowledge coding system.

A large range of knowledge approaches and techniques are used in the TDM industry, both formal and informal, from card type systems to advanced computation. Hence, to categorise this type of *knowledge* engineering, a coding system has been devised based on the number sequence 0, 1, 2, and 3. The lower the number, the more knowledge is in a person's head or documented. The higher the number, the more the knowledge can be related to KBE. The proposed conceptual definitions shown below, and the coding system in Table 1, enables comparisons to be made and hence potentially, benchmarking methods to be developed.

Conceptual knowledge definitions:

Implicit =	Within a person's head.	
Explicit =	Within books.	
Derived =	Extrapolated or interpolated.	
KBE =	Based on combining engineering knowledge, methodologies, process knowledge, and best practice.	

	Casting company		A	B	C	D
	Stage					
1	Component design		0 1	0 1 2 3	0 1 2	0 1 2
2	Mould or tool design		0 1	0 1 2 3	0 1 2	0 1 2
3	Mfrg process design		X	0 1 2	0 1 2	0 1 2
4	Mfrg process simulation		X	0 1 2	0 1 2	0 1 2
5	Rapid manufacture	TEDI knowledge code	X	0 1 2	0 1 2	0 1 2
6	Pattern making		0 1	-	0 1	0 1 2
7	Tool path generation		X	0 1 2	0 1 2	0 1 2
8	Tool path verification		X	0 1 2	0 1	0 1 2
9	Machining die/mould		X	0 1 2	0 1 2	0 1 2
10	Inspection		0 1	0 1	0 1	0 1 2

Table 2: TEDI knowledge coding system applied to casting companies.

Table 1 defines four levels of knowledge use within a company in terms of the skill level of the people undertaking the work and the software they may be using. Once a company has assessed its knowledge status it is possible for decisions to be made about the type of knowledge, the risk of losing the knowledge and the possible options relating to capturing and building upon the knowledge. Table 2 shows an example of a knowledge assessment, based on the coding system shown in Table 1, for four casting companies. With an assessment of knowledge complete it is possible to assess the need versus return to move up a knowledge code.

5. Conclusion

Consideration of *knowledge*, its capture and redeployment in some form needs to become a priority for the TDM industry bearing in mind the aging workforce at both technician and engineer level. A coding system has been presented that allows a company to assess its potential knowledge status and therefore identify if there is a need to consider tools to minimise potential knowledge loss and enable time to be allocated to innovation.

Acknowledgements

The work presented was undertaken as part of a UK Department of Trade and Industry ICT Scoping Study: Tooling and Experiential Design Initiative (TEDI) [3].

References

[1]   UK's Department of Trade and Industry supported Com-Met 2005 project. http://www.com-met2005.org.uk/

[2]   Anon, 2003, Employment figures confirm that manufacturing is withering away, Professional Engineering, Institution of Mechanical Engineers, November, 2003, p10.

[3]   Mynors D J, Griffiths B J, Twelves M, Etheridge D, Moore J, Griffiths T and Piddington C. *Final report of the Tooling and Experiential Design Initiative (TEDI)*. (ICT Scoping Study 0116). March 2004. Obtainable from either the UK Department of Trade and Industry or the Department of Design and Systems Engineering, Brunel University, Uxbridge, Middlesex, UB8 3PH, UK

[4]   IEE journal the Manufacturing Engineer 2003/2004, p6.

[5]   Benchmark Research, *Trends in UK manufacturing – 1997 to 2003*, Findlay Publications, January 2004.

[6]   GTMA, *The Real Cost of Tooling*, GTMA, 3 Forges House, Summerleys Road, Princes Risborough, Buckinghamshire, HP27 9DT, United Kingdom March, 2002

[7]   Javed Y, *Knowledge Based Engineering*, MSc Dissertation, Department of Design and Systems Engineering, Brunel University, Middlesex, UB8 3PH, United Kingdom, to be submitted in 2004.

Geometry of the Helicoidally Bevel Gear: Analysis of Geometrical Errors in Coefficients of Correction of the Adjustment Parameters of the Cutting Machine

M Bouaziz [1], R Ifrah [2]

[1] Département de Génie Mécanique, Ecole Nationale Polytechnique, 10 avenue Hacène Badi, 16200 El Harrach (Alger), Algeria. mbouazizdz@yahoo.fr
[2] Institut de Génie Mécanique, Université M'Hamed Bouguera, Boumerdès, Algeria. E-mail : ifrid@yahoo.fr

Abstract
This article treats, by analogy with the right conical denture, the theoretical teeth with spherical involute and the octoïde teeth of 1st species (practical teeth). For each of these teeth, one determines its profile using its co-ordinates and the associated normal vectors. Relationships will be established to calculate the basic thickness, the thickness on an unspecified cone, the apparent angle of pressure, the basic helix angle and the apparent module from the elements characteristic of teeth. A short description of a procedure to evaluate the corrections to be brought to the parameters of adjustment of the cutting machine from the cartography of teeth surface is also presented.

1. Introduction

Compared with the right bevel gear, the helicoid bevel gear has many advantages, for example, a silent working and a higher conduct report. Compared with the spiral bevel gears and hypoïdes, their geometry is relatively simple. This study could be of a significant practical interest. Thus, the knowledge of the helicoid bevel gear geometry would permit its development and inspection. Contrary to the other types of gears, there is only one elementary documentation on the helicoid bevel gear in the literature, which encouraged us to present this study related to its geometry and being given the possibility to inspect it using and coordinate measuring machine.

2. Geometry of the Spherical Involute Teeth Surface

Figure 1 illustrates the principle of generation of teeth. The plan (P_b), called action plan or basic plan, rolls without slipping on the basic cone of half angle at the top δ_b of the part to cut. The profile of teeth is described on the sphere of center S and from ray R by the point M. The generator of teeth in this point M passes by the point such as SM' and SM forms an angle β_b. The plan of generation (N) passes by S'. It is normal with (P_b) and is invariably dependent for him. The trace of (P_b) on the sphere, called line of action, is a large circle of radius R.

486

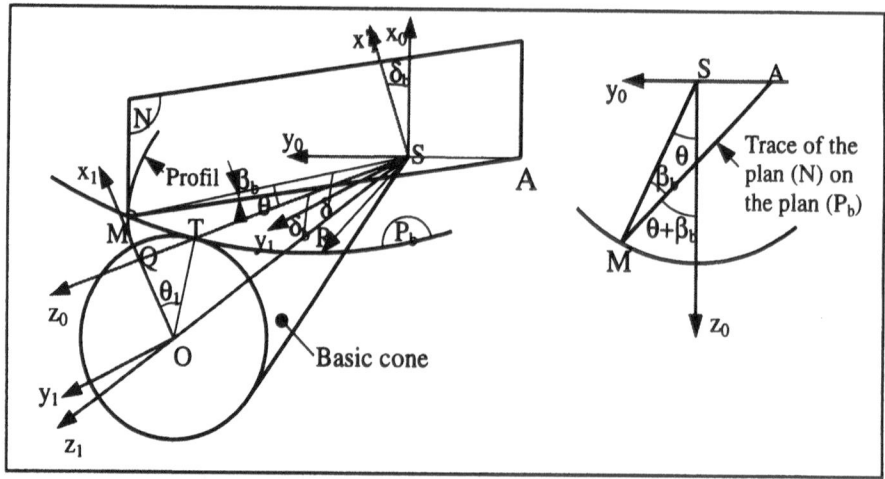

Fig. 1 : Generation of helicoid conical teeth with spherical involute

2.1. Co-ordinates of point M in reference (S, x_1, y_1, z_1)

The point M being in the plan (P_b), its co-ordinates expressed in the reference (S,x_0,y_0,z_0) (see : fig.1), are :

$$x_0 = 0 \; ; \; y_0 = R \sin\theta \; ; \; z_0 = R \cos\theta$$

One determine its co-ordinates in the reference (S,x_1,y_1,z_1) by changing of reference frame (rotation of δ_b around y_0 and θ_1 around z_0). One obtains:

$$\overrightarrow{SM} \begin{cases} x_1 = R(\sin\theta\sin\theta_1 + \cos\theta\sin\delta_b \cos\theta_1) \\ y_1 = R(\sin\theta\cos\theta_1 - \cos\theta\sin\delta_b \sin\theta_1) \\ z_1 = R \cos\theta\cos\delta_b \end{cases}$$

The angle θ_1 is obtained by considering the rolls without slipping of the plan (P_b) on the basic cone. That yields: $\theta = \theta_1 \sin\delta_b$

2.2. Normal on the teeth surface

This is obtained by a vector $\overrightarrow{V}$ carried by AM and one vector $\overrightarrow{U}$ contained in the plan (N) and parallel to axis x_0. We obtain: $\overrightarrow{N} = \overrightarrow{U} \wedge \overrightarrow{V}$. This relation permits to express in the reference (S,x_0,y_0,z_0) by carrying out the changes of reference marks indicated previously, that is written in the reference (S,x_1,y_1,z_1) as follow:

$$\overrightarrow{N} \begin{cases} N_{x1} = \cos(\theta+\beta_b)\sin\theta_1 - \sin(\theta+\beta_b)\sin\delta_b \cos\theta_1 \\ N_{y1} = \cos(\theta+\beta_b)\cos\theta_1 + \sin(\theta+\beta_b)\sin\delta_b \sin\theta_1 \\ N_{z1} = -\sin(\theta+\beta_b)\cos\delta_b \end{cases}$$

2.3. Thickness connects teeth on a cone of a half angle at the top δ

With the help of the fig.2 and without spreading the development of calculations, one finds:

$$S_t = R \sin \delta \left\{ \frac{S_{pt}}{r_p} - 2 \left[(\theta_1 - \theta_{1p}) + \arccos\left(\frac{tg\delta_b}{tg\delta_p}\right) - ar\cos\left(\frac{tg\delta_b}{tg\delta}\right) \right] \right\} \qquad (1)$$

δ_p is the half angle at the top of the primitive cone.

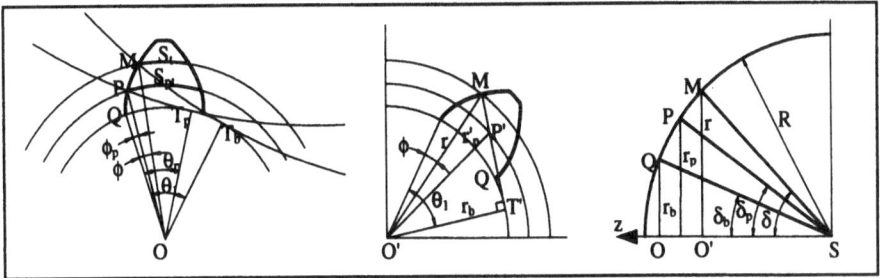

Fig.2 : Thickness connects spherical involute teeth

2.4 Real thickness of teeth

One determines his expression with from fig.3 where we have : $AS = BS = R$; $SS' = R\sin\beta_p$; $AB = S_t$; $BS' = CS'$; $BC = S_n$
After any made calculation, we obtain :

$$S_n = 2R \sin\left[\frac{1}{2}\arcsin\left(\frac{\cos\phi - \sin\beta_p}{\lambda}\right)\right] \cdot \lambda \qquad (2)$$

with : $\quad \phi = \dfrac{\pi}{2} - \beta_p - 2\arcsin\left(\dfrac{S_t}{2R}\right)$

and $\quad \lambda = \sqrt{1 + (\sin\beta_p - 2\cos\phi)\sin\beta_p}$

β_p is the slope of the generator of teeth defined on the primitive level

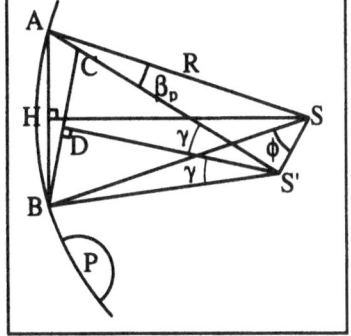

Fig.3 : Real thickness, Apparent thickness

2.5 Relation between the angles β_b and β_p

In practice, the angle β_p generator on the primitive cone is known. One determines the angle β_b by taking into account that the generating is convergent at the top S' such as $SS' = R\sin\beta_p$ and the tracing of plan (N) on the primitive plan (P) is tangent with the circle of radius SS'. For this, one introduce angle Δ (fig.4) such as $\phi_p r_b = R\Delta$ (the point P' on (P_b) corresponds to point P of profile). That gives:

$$\Delta = \arccos\left(\frac{\cos\delta_p}{\cos\delta_b}\right) - \arccos\left(\frac{tg\delta_b}{tg\delta_p}\right)\sin\delta_b \tag{3}$$

Knowing Δ et β_p, one determines angle β_b from the relation:

$$\beta_b = \frac{\pi}{2} + \beta_p - \left\{\Delta + \arccos\left[\frac{\sin\beta_p - \sin(\beta_p - \Delta)}{\sqrt{1+\sin\beta_p\left[\sin\beta_p - 2\sin(\beta_p - \Delta)\right]}}\right]\right\} \tag{4}$$

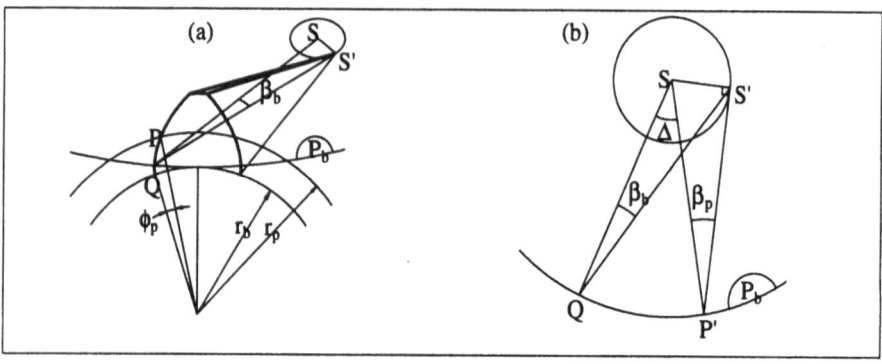

Fig.4 :Geometrical layout for the calculation of the angle β_b

3. Geometry of the Surface of Teeth in Octoïde of the First Species

The fig.5 illustrates the principle of this generation type of tooth. The plan (P) rolls without slipping on the primitive cone (C_p) of half angle at the top δ_p and the top S superimposed with the centre of the sphere of radius R. It forms an angle a' with the basic plan (P_b). Its trace on the sphere is a large circle of radius R. It is called line of action. Trace AB of the plan of generation (N) on (P) is the generating line. It forms an angle B with the tangent S_B of (p) on (C_p). The angle A_t is the apparent angle of pressure. The profile of teeth is generated by the point M during the bearing without slip of (P) on (C_P).

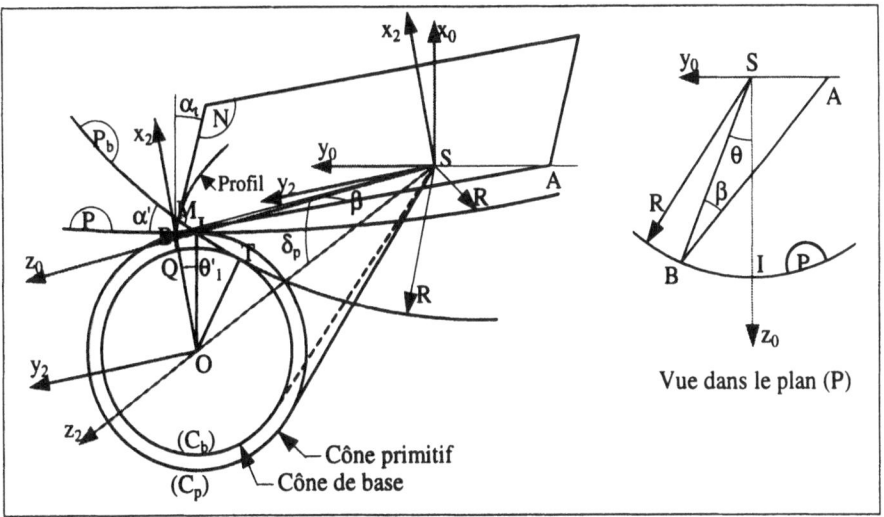

Fig. 5 : Generation of helicoid conical teeth in octoïde of first species

3.1 Normal on the surface of teeth

This is obtained by building two vectors, carried by, $\vec{U}$ is in AB and $\vec{V} \perp \vec{U}$ and contained in the plan (P). The normal forms an angle $\vec{N}$ carried an angle α_t $\vec{V}$. One has then $\vec{N} \wedge \vec{V} = \lambda \vec{U}$. The components of normal $\vec{N}$ are obtained by carrying out the products vector and respectively scalar $\vec{N} \wedge \vec{V}$, $\vec{N} \cdot \vec{V}$. All made calculations, one obtains:

$$N \begin{cases} N_{x2} = \cos\alpha_t \cos(\theta+\beta)\sin\theta_1' + \left[\sin\alpha t \cos\delta_p - \cos\alpha_t \sin(\theta+\beta)\sin\delta_p\right]\cos\theta_1' \\ N_{y2} = \cos\alpha_t \cos(\theta+\beta)\cos\theta_1' - \left[\sin\alpha t \cos\delta_p - \cos\alpha_t \sin(\theta+\beta)\sin\delta_p\right]\sin\theta_1' \\ N_{z2} = -\cos\alpha_t \sin(\theta+\beta)\cos\delta_p - \sin\alpha_t \sin\delta_p \end{cases}$$

The angles θ and θ_1' are bound by the condition of bearing without slip of (P) on (C_p), let: $\theta = \theta_1' \sin\delta_p$

3.2 Co-ordinates of the point M of the teeth profile

The point M belongs at the same time to the plan of generation (N) and with the sphere of radius R. this verify their equations. The trace of the plan (N) on The plan (P_b) is the line AM. The equation is obtained by: $\vec{N} \cdot \left(\vec{SI} \wedge \vec{SM}\right) = 0$. While indicating by M (x_0, y_0, z_0), one obtain : $x_0 \cos(\theta+\beta) - y_0 tg\alpha_t = 0$.

Knowing the components of the normal in the plan of generation (N), one easily can determine the equation of this plan, as follow:

$$x_0 tg\alpha_t + y_0 \cos(\theta+\beta) - z_0 \sin(\theta+\beta) + R\sin\beta = 0$$

The equation of the sphere with centre S and radius R supplements the equation system to be solved. Finally, the co-ordinates of point M can be written by:

$$x_0 = \frac{[z_0 \sin(\theta+\beta) - R\sin\beta]tg\alpha_t}{tg^2\alpha_t + \cos^2(\theta+\beta)} \quad ; \quad y_0 = \frac{[z_0 \sin(\theta+\beta) - R\sin\beta]\cos(\theta+\beta)}{tg^2\alpha_t + \cos^2(\theta+\beta)}$$

$$z_0 = R\cos^2\alpha_t\left\{\sin(\theta+\beta)\sin\beta + \sqrt{[tg^2\alpha_t + \cos^2(\theta+\beta)][tg^2\alpha_t + \cos^2\beta]}\right\}$$

The co-ordinates of point M expressed in the reference frame (S,x_2,y_2,z_2) can be obtained by changes of reference frame (rotations of δ_p around of y_0 and θ'_1 around of z_2).

3.3 Apparent thickness, real thickness

A similar reasoning to that developed for the calculation of thickness of spherical involute teeth gives an expression having the relation form (1) in which it is necessary to substitute the quantity $(\theta_1 - \theta_{1p})$ by θ'_1. The real thickness S_n, preserves the form by given the relation (2).

3.4 Apparent module m_t, apparent angle of pressure α_t

The real module m_n, the real angle of pressure α_n, the slope of the pitch helix β and the half angle to top δ_p of the primitive cone are supposed known.

One calculates the real thickness by the relation $S_n = \pi m_n /2$. The apparent module is given by: $M_t = 2S_t /\pi$. One establish one expression of apparent thickness from fig.3. The developments of the calculations are:

$$S_t^2 = \left(R\cos\beta - \frac{S_n}{2\sin\gamma}\right)\left(R\cos\beta - \frac{S_n}{2\sin\gamma} + 2S_n \sin\gamma\right) + S_n^2 \tag{5}$$

γ is determined from the following expression:

$$S_n^2 + 4RS_n \sin\beta\sin\gamma\sin(2\gamma) - 4R^2\cos^2\beta\sin^2\gamma = 0 \tag{6}$$

The angle of apparent pressure is such as: $tg\alpha_t = \dfrac{S_n}{S_t}tg\alpha_n = \dfrac{M_n}{M_t}tg\alpha_n$

4. Optimisation of the Correction of the Adjustment Parameters of the Cutting Machine

A statement of variations on surfaces of teeth is not easily interpretable, but by applying the method of optimisation of J.M. David, it is possible to translate these variations into coefficients of correction parameters of adjustment of the cutting machine. This method requires the construction of vectors of analysis of the adjustment defects. Two procedures can be employed, one, called by simulation, consists in considering the mathematical model in which one varies the parameter corresponding to the adjustment to study, the other, known as analytical, which uses the expression of the teeth thickness since any variation of the adjustment parameters appears there. In reference [1], detailed description of these two methods could be found. Numerical result are given hereafter and found by optimising the field of the imagined variations raised on 4 regularly spaced tooth spaces and by considering a variation of 0,001 rad of the angle β (the vector of analysis is built according to the analytical method).

Translation following X (mm)	–0,0273828
Translation following Y (mm)	0,0060999
Rotation around de X (mm/m)	0,0863832
Rotation around de Y (mm/m)	0,0376094
Rotation around de Z (mm/m)	0,5582464
Vector of analysis of the angle β	0,0079499

5. Conclusion

It was possible in this study to establish the geometry of the surface of helicoid conical teeth and the relations expressing the geometrical parameters (basic taper angle, modulates apparent, apparent angle of pressure, apparent thickness). It is thus possible to plan control compared to the mathematical model of teeth on coordinates measuring machine and to use mathematical methods worked out to analyse the variations raised in correction coefficient of the parameters to adjust the cutting in machines.

References

[1] M. Bouaziz : Contribution au contrôle d'engrenages sur machine à mesurer tridimensionnelle. Thèse de doctorat d'état, ENP, janvier 1996, Alger, Algérie
[2] R. Ifrah : Géométrie des engrenages coniques hélicoïdaux, Analyse des erreurs géométriques en coefficient de correction des paramètres de réglage de la machine de taillage. Mémoire de magister, ENP, octobre 2003, Alger, Algérie

Robotic Jointing Of Composite Materials

J D Tedford, G Cho and J Helmink

The University of Auckland, Department of Mechanical Engineering, Auckland, New Zealand. d.tedford@auckland.ac.nz

Abstract: With the increasing use of Composite materials in industry, the joining together of components to produce larger products, is becoming an increasingly time consuming, difficult and expensive task. The main objective of this paper is to demonstrate that an automated solution to joining components is feasible and produces consistently strong joints. A robotic cell has been developed to produce double lap joints for testing consistency and repeatability. Comparing these robotic produced joints with identical joints produced manually, it has been mechanically and statistically shown that the automated system is reliable, consistent and competitive with the alternative labour intensive manual jointing techniques.

1. Introduction

Composite material structures and components are growing in popularity due to their light weight and excellent material properties. They are strong, have good wear and environmental resistance and can be used in many applications. However, their use, in some cases, is limited by several factors related to their construction. Large structures can be too big to lay-up in one mould, or the necessary moulds can become too complicated and/or expensive to be feasible [1]. In some instances, dissimilar materials need to be connected and difficulties can arise due to differing thermal expansion properties, thus making it difficult to create consistent results [2]. To overcome these problems, components have to be manufactured in smaller parts, which need to be subsequently joined.

Conventional mechanical joining methods such as rivets, bolts or screws are not conducive to composite material bonding [3]. Drilling holes usually weakens the substrate and tend to cause damage, while fastener bonded joints pass loading to the substrate materials through the fasteners, which create stress concentration points in the structure. In most materials this problem is reduced by localised substrate yielding at these points, but brittle materials such as thermosetting composites cannot yield in this way and matrix-fibre debonding or interply splitting can occur [4,5,6]. Other connection methods such as welding (fusion bonding) are not applicable in thermosetting matrix-fibre composites since the matrix material cannot be re-melted once cured [7].

2. Adhesive Jointing

The solution to this jointing problem is to use an adhesive to bond pieces together. A primary advantage of adhesive jointing over other composite jointing methods is that it passes stresses to the substrate in a distributed manner [3,8,9].

The strength obtainable from an adhesive bond loaded in shear is limited by the interlamina shear strength of the substrate material and the area over which adhesion occurs. In practice, a maximum of only about 70% of this can be achieved [10].

The percentage of joint area that actually undergoes adhesive-substrate bonding is obviously directly proportional to the joint strength, this is known as the effective joint area as shown in Figure 1, and therefore, a reduction in the effective area, increases the maximum stresses encountered in the joint for a particular load. The bond gap thickness affects bond strength with a thicker bond gap reducing the strength.

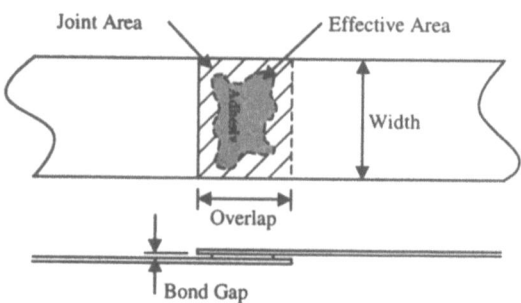

Figure 1: Joint area vs. effective area

The quality of adhesion achieved between the adhesive and the substrate is another strength factor influenced by the adhesive's ability to 'wet' the complete surface of the joint area, and is also influenced by the substrate surface roughness and the surface preparation prior to bonding. These factors affect the quality of the chemical bond achieved between substrate surfaces [9,11,12].

Finally, the cohesive strength of the adhesive itself is also an important factor, especially if the bond gap is large.

2.1 Automated adhesive jointing

Adhesive bonding can be an extremely labour intensive process. When bonding many joints together repeatedly, it is difficult to get consistent results with the high level of skill that is required to do the job effectively. There are also health and safety issues that arise from working with solvents in the adhesives [1,7,9].

These problems can be overcome if the human element is eliminated and some form of automated robotic system is used to do the bonding. Thus, the reason for undertaking this research study was to investigate whether there is a viable economical automated solution to the composite jointing problem. The key objectives of the study, therefore, were to prove whether an automated robotic jointing system could:

- Consistently produce joints of an equivalent or greater strength than the same joints produced manually.
- Perform the jointing operation in equal or less time than the manual operation.

2.2 Test specimen manufacture

There are several different joint type configurations available for Tensile Shear Testing. For the purposes of this study, however, the Double Lap joint was selected.

As this investigation was concerned with the jointing of composites, a suitable composite substrate was required. Since the actual tests themselves are not of the composite but of the adhesive which bonds the joint together, the composite used is not critical as long as it has greater strength, in tension, than the adhesive. On this basis, it was decided to manufacture the composite substrate from 6-ounce plain weave bi-directional glass-fibre and Hetron 922PA vinyl ester.

Since, when laying-up, there was always a perfectly smooth surface on the bottom of the substrate, two separate single layer 'skins' were laid-up on two smooth thick sheets of plastic. Once the skins cured, three 'core' layers were laid-up on one of the skins and the second skin, smooth side up, laid down on top. These sheets were then trimmed and cut into 100 x 40mm samples.

2.3 Adhesive selection

Selecting the correct adhesive is critical to the overall joint performance and is affected by varying aspects of the substrate material, environmental factors, temperature, humidity etc. Here, a number of different adhesive types could be used. Since a fast curing time is essential for an automated system, it was decided that a Cyanoacrylate type of adhesive should be used. This type of adhesive starts to cure when it comes in contact with the moisture in ambient air.

Loctite 401 is one such glue with a viscosity similar to water and low liquid surface tension for excellent wetting properties. It has an extremely short curing time in small bond gaps of between 5 and 30 seconds to full bond strength. It has an infinite pot life and does not require priming of the substrate surfaces [11]. The configuration and dimensions of the final test joint is shown in Figure 2.

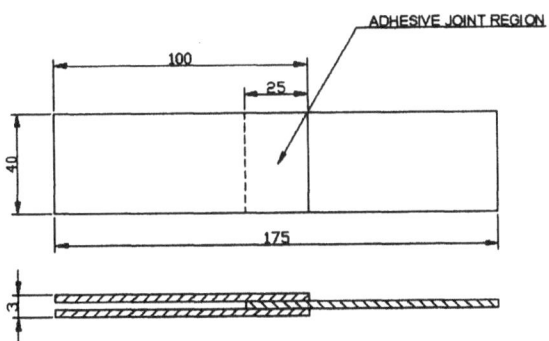

Figure 2: Substrate sample and joint dimensions

3. Experimental Equipment

A key component of the automatic jointing process is the Kuka-15 six-axis industrial robot arm. It is capable of handling payloads of 15kg at speeds of up to 2 m/s with an accuracy of better than one tenth of a millimetre.

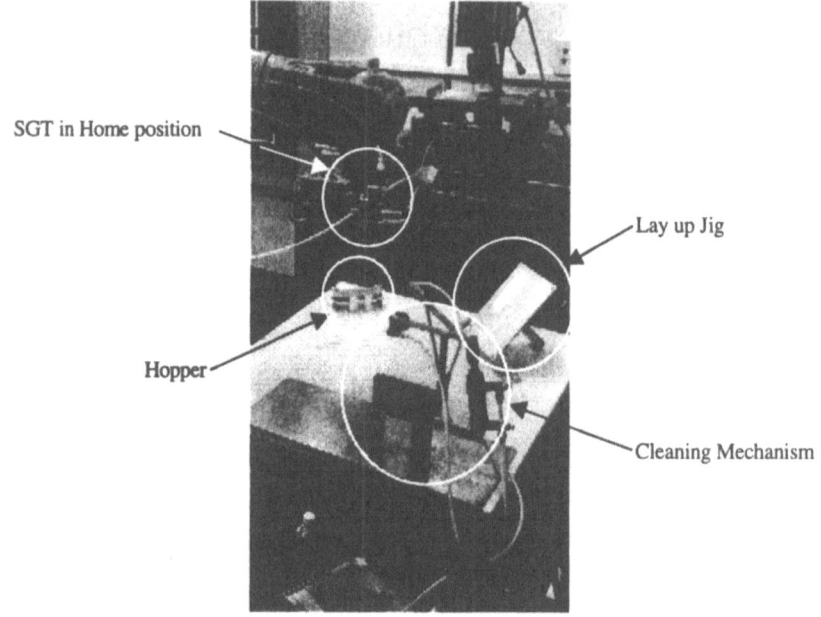

SGT in Home position

Lay up Jig

Hopper

Cleaning Mechanism

Figure 3: The joint manufacturing cell.

The robot's control system also has the facility to connect several other devices to it. These can be programmed to be activated by the robot's own controller.

A spring-loaded 'hopper' was designed to hold a stack of substrate samples for the robot to use along with a suction cup Specimen Gripping Tool (SGT), a solvent cleaning mechanism and bath, a lay-up jig, and adhesive applicator. The complete jointing system is shown in Figure 3.

Cyanoacrylate adhesives are inherently difficult to work with since they cure rapidly once exposed to moisture in the air. An adhesive feeding system was developed therefore using pressurised Nitrogen to provide an inert environment. The Nitrogen pressure forces the adhesive along a feed line through a pinch valve to a fixed position needle applicator.

Enough substrate samples were made to create 15 automated joints and 15 manual joints from each batch of substrate composite, giving a total of 60 double lap joints to test. During the construction of each joint temperature and relative humidity were controlled to minimise the effects that these factors could have on the joint strength. To help reduce these effects further, the manually manufactured joints were assembled simultaneously with the automated assembly.

4. Results

An Instron 5500R hydraulic tension/compression testing machine was used to test all joints. Because an adhesive tensile test was being conducted, the standard ASTM D 3528 Double Lap Joint Adhesive Shear by Tension Loading procedure was used [13].

The temperature and relative humidity were held constant during testing, thus the effects of these factors were ignored. The failure modes that occurred during the testing were primarily adhesive failures although in some instances a combination of adhesive and substrate failure did occur. These results were ignored in the subsequent analysis.

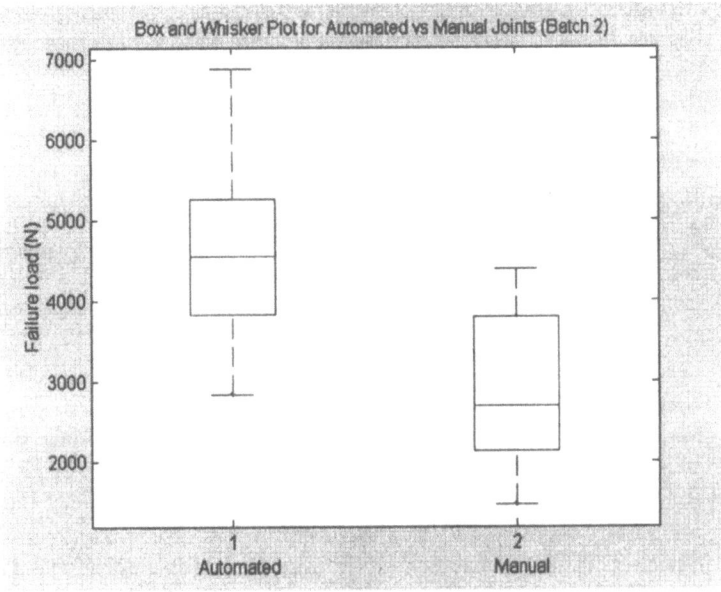

Figure 4: Box and whisker comparison of joints

A statistical analysis was conducted on the data using MATLAB for both automated and manual joints and presented in the form of a Box and Whisker plot shown in Figure 4. It is obvious from this plot, that joints produced by the robot are stronger than the manually produced joints. The shapes of the plots also show that the sample sets are independent and have been taken from a statistically 'normal' population, although the manually produced joints are slightly skewed. The results of a T-test proved that the average strength of the automated joints was significantly higher than the average strength of the manually produced joints.

5. Conclusions

By developing a lay-up procedure which ensured that bonding could take place on consistently smooth substrate surfaces, a reliable method for adhesive shear testing was successfully achieved.

The developed robotic joint manufacturing cell, has created a fully automated system that can repeatedly produce consistent, adhesively bonded, double lap joints.

The system produced joints which statistically showed that the automated system could produce a higher and slightly more consistent joint strength than that of a manual assembly process.

References

[1]   Mazumdar, S.K. (2002) Composites Manufacturing: Materials, Product and Process Engineering. CRC Press, New York. pp. 309-327.

[2]   Comprehensive Composite Materials (2000) Elsevier Science Ltd., New York. pp. 248-258.

[3]   Sampath, K. (1990) ASTM Handbook: Selection and Design, Design for Joining. Vol. 20. ASM International, U.S.A. pp. 762-773

[4]   Schwartz, M.M. (1983) Composite Materials Handbook. McGraw-Hill, U.S.A. pp. 6.38-6.49

[5]   Concise Encyclopaedia of Composite Materials (1994) revised edition, Elsevier Science Ltd., New York. pp. 155-158.

[6]   Lee, S.M. (Ed) (1991) Encyclopaedia of Composites. VCH Publishers Inc. U.S.A. 2, pp. 438-450, 509-520

[7]   Mathews, F.L. and Rawlings, R.D. (1994) Composite Materials: Engineering and Science. Chapmann & Hall, London. pp. 392-412

[8]   Brandon, D. and Kaplan, W.D. (1997) Joining Processes – an Introduction. John Wiley and Sons Ltd. England, pp. 292-300

[9]   Skeist, I. (1977) Handbook of Adhesives. Lifton Educational Publishing, New York. pp. 573-592.

[10]   The Loctite Design Guide for Bonding Plastic (1998), Vol. 2. Loctite North America, U.S.A. pp. 7, 70-71

[11]   Loctite Worldwide Design Handbook (1998) second edition, Loctite North America, U.S.A. pp. 21-50, 147-170.

[12]   Donaldson, S.L. and Roy, A.K. (1996) Moisture and Temperature Effects on Bonded Composite Double-Lap Shear Specimens. ASME Mater Division Publishing MD, 74 pp. 73-74

[13]   Standard Test Method for Strength properties of Double Lap Shear Adhesive Joints by Tension Loading, (1996). Designation D-3528-96. American Society for the Testing of Materials, Annual Book of ASTM Standards, 1999, 15.06 Adhesive, pp. 240-243